IFIP Advances in Information and Communication Technology 317

T0181199

IFIP – The International Federation for Information Processing

IFIP was founded in 1960 under the auspices of UNESCO, following the First World Computer Congress held in Paris the previous year. An umbrella organization for societies working in information processing, IFIP's aim is two-fold: to support information processing within its member countries and to encourage technology transfer to developing nations. As its mission statement clearly states,

> IFIP's mission is to be the leading, truly international, apolitical organization which encourages and assists in the development, exploitation and application of information technology for the benefit of all people.

IFIP is a non-profitmaking organization, run almost solely by 2500 volunteers. It operates through a number of technical committees, which organize events and publications. IFIP's events range from an international congress to local seminars, but the most important are:

- The IFIP World Computer Congress, held every second year;
- Open conferences;
- Working conferences.

The flagship event is the IFIP World Computer Congress, at which both invited and contributed papers are presented. Contributed papers are rigorously refereed and the rejection rate is high.

As with the Congress, participation in the open conferences is open to all and papers may be invited or submitted. Again, submitted papers are stringently refereed.

The working conferences are structured differently. They are usually run by a working group and attendance is small and by invitation only. Their purpose is to create an atmosphere conducive to innovation and development. Refereeing is less rigorous and papers are subjected to extensive group discussion.

Publications arising from IFIP events vary. The papers presented at the IFIP World Computer Congress and at open conferences are published as conference proceedings, while the results of the working conferences are often published as collections of selected and edited papers.

Any national society whose primary activity is in information may apply to become a full member of IFIP, although full membership is restricted to one society per country. Full members are entitled to vote at the annual General Assembly, National societies preferring a less committed involvement may apply for associate or corresponding membership. Associate members enjoy the same benefits as full members, but without voting rights. Corresponding members are not represented in IFIP bodies. Affiliated membership is open to non-national societies, and individual and honorary membership schemes are also offered.

Daoliang Li   Chunjiang Zhao (Eds.)

Computer and Computing Technologies in Agriculture III

Third IFIP TC 12 International Conference, CCTA 2009
Beijing, China, October 14-17, 2009
Revised Selected Papers

 Springer

Volume Editors

Daoliang Li
China Agricultural University
EU-China Center for Information & Communication Technologies (CICTA)
17 Tsinghua East Road, Beijing, 100083, P.R. China
E-mail: dliangl@cau.edu.cn

Chunjiang Zhao
China National Engineering Research Center
for Information Technology in Agriculture (NERCITA)
Shuguang Huayuan Middle Road 11, Beijing, 100097, P.R. China
E-mail: zhaocj@nercita.org.cn

CR Subject Classification (1998): H.2.8, H.2, H.4, H.3, I.4, I.6

ISSN   1868-4238
ISBN-10   3-642-42233-0 Springer Berlin Heidelberg New York
ISBN-13   978-3-642-42233-1 Springer Berlin Heidelberg New York

springer.com

© IFIP International Federation for Information Processing 2010
Softcover re-print of the Hardcover 1st edition 2010

Typesetting: Camera-ready by author, data conversion by Scientific Publishing Services, Chennai, India
Printed on acid-free paper   06/3180

Preface

I want to express my sincere thanks to all authors who submitted research papers to support the Third IFIP International Conference on Computer and Computing Technologies in Agriculture and the Third Symposium on Development of Rural Information (CCTA 2009) held in China, during October 14–17, 2009.

This conference was hosted by the CICTA (EU-China Centre for Information & Communication Technologies, China Agricultural University), China National Engineering Research Center for Information Technology in Agriculture, Asian Conference on Precision Agriculture, International Federation for Information Processing, Chinese Society of Agricultural Engineering, Beijing Society for Information Technology in Agriculture, and the Chinese Society for Agricultural Machinery. The platinum sponsor includes the Ministry of Science and Technology of China, Ministry of Agriculture of China, Ministry of Education of China, among others.

The CICTA (EU-China Centre for Information & Communication Technologies, China Agricultural University) focuses on research and development of advanced and practical technologies applied in agriculture and on promoting international communication and cooperation. It has successfully held three International Conferences on Computer and Computing Technologies in Agriculture, namely CCTA 2007, CCTA 2008 and CCTA 2009.

Sustainable agriculture is the focus of the whole world currently, and therefore the application of information technology in agriculture is becoming more and more important. 'Informatized agriculture has been sought by many countries recently in order to scientifically manage agriculture to achieve low costs and high incomes.

The topics of CCTA 2009 covered a wide range of interesting theories and applications of information technology in agriculture, including simulation models and decision-support systems for agricultural production, agricultural product quality testing, traceability and e-commerce technology, the application of information and communication technology in agriculture and universal information service technology, and service systems development in rural areas. We selected 80 papers among all those submitted to CCTA 2009 for this volume. In these proceedings, creative thoughts and inspirations can be discovered, discussed and disseminated. It is always exciting to have experts, professionals and scholars with creative contributions getting together to share inspiring ideas and hopefully accomplish great developments in these technologies of high demand.

Finally, I would also like to express my sincere thanks to all the authors, speakers, session chairs and attendees for their active participation and support of this conference.

January 2010

Daoliang Li
Zhao Chunjiang

Conference Organization

Sponsors

- China Agricultural University
- China National Engineering Research Center for Information Technology in Agriculture
- Asian Conference on Precision Agriculture
- International Federation for Information Processing, Laxenburg, Austria
- Chinese Society of Agricultural Engineering, China
- Beijing Society for Information Technology in Agriculture, China
- Chinese Society for Agricultural Machinery

Supporters

- Ministry of Science and Technology of PRC
- Ministry of Agriculture of PRC
- Ministry of Education of PRC
- National Natural Science Foundation of China
- Beijing Municipal Natural Science Foundation, China
- Beijing Association for Science and Technology, China
- Beijing Economic and Information Technology Commission, China
- Beijing Academy of Agriculture and Forestry Sciences, China

Conference Co-chairs

Maohua Wang
Sakae Hibusawa
K.C. Ting
Chunjiang Zhao

Organizing Committee Co-chairs

Minzan Li
Jihua Wang
Daoliang Li

Academic Committee Co-chairs

Maohua Wang
Noguchi, Noboru

Simon Blackmore
Chenghai Yang
Chunjiang Zhao

Conference Secretariat

Lingling Gao

Table of Contents

Target Detection in Agriculture Field by Eigenvector Reduction Method of Cem

Chun-hong Liu[*] and Ping Li

College of Information and Electrical Engineering, China Agricultural University,
Beijing 100086, P.R. China,
Tel.: +86-10-82395850
sophia_liu@cau.edu.cn

Abstract. Constrained Energy Minimization algorithm is used in hyperspectral remote sensing target detection, it only needs the spectrum of interest targets, knowledge of background is unnecessary, so it is well applied in hyperspectral remote sensing target detection. This paper analyzed the reason of better results in small target detections and worse ones in large target detections of CEM algorithm, an eigenvector reduction method to increase the ability of large target detection of CEM algorithm was proposed in this paper. Correlation matrix R was decomposed into eigenvalues and eigenvectors, then some eigenvectors corresponding to larger eigenvalues was choosed to reconstruct R. In order to test the effect of the new method, experiments are conducted on HYMAP hyperspectral remote sensing image. In conclusion, by using eigenvalue reduction method, the improved CEM method not only can detect large targets, but also can well detect large/small targets simultaneously.

Keywords: Hyperspectral remote sensing, Constrained Energy Minimization algorithm, Eigenvector reduction method.

1   Introduction

With the development of Hyperspectral image spectrometers, hyperspectral remote sensing image can be obtained. An example of such hyperspectral sensors is 224 bands Airborne Visible Infrared Imaging Spectrometer (AVIRIS) of U.S. NASA Jet Propel Lab, another example is 210 bands HYDICE sensor [2] of Naval Research Laboratory. The resolution of an image got by high resolution spectrometers is much better than that got by multispectral sensors, people can acquire information that can not be acquired in multispectral image. This development brought hope to remote sensing target detection.

Methods based on constraint came from linearly constrained minimized variance adoptive beam-forming in digital signal processing(Chang, C.I,1999). This method also called Frost beam-forming algorithm, it constrains weight vector according to

---

[*] Corresponding author.

D. Li and C. Zhao (Eds.): CCTA 2009, IFIP AICT 317, pp. 1–7, 2010.
© IFIP International Federation for Information Processing 2010

statistic characteristic of receiving array data, when output power of beam-forming reaches minimum, beam-forming remains unit response in signal direction. Reasons of good detection in small target detection by constrained energy minimization algorithm were analyzed in this paper, then proposed an method for large targets detection based on analysis of bad reasons in large target detection. Computer simulated are carried to test the proposed algorithm.

2   Materials and Methods

2.1   The Constrained Energy Minimization

Harsanyi proposed the Constrained Energy Minimization algorithm, he considers target detection as linearly adaptive beam-forming problem(Heinz, D.C. 2000), whose characteristic is increasing pixel by pixel and separating background interfere signal from the target signal and background signal, so that target signal can be extracted in per-pixel at maximum. The step of CEM algorithm is: by designing a finite response filter, minimizing the total output energy of the linear combining process subject to a linear equality constraint applied to desired target $d_\circ$ This problem can be converted to an unconstrained minimization using the method of Lagrange multipliers, using the filter to all pixel vectors to get the target detection.

Set $w = [w_1, w_2, \cdots, w_L]^T$ as the vector of weights, where L is the dimension of the data. When input is $r_i = (r_{i1}, r_{i2}, \cdots, r_{iL})^T$, then the output of filter is

$$y_i = w^T r_i \tag{1}$$

The total output energy is

$$E = \frac{1}{N}\sum_{i=1}^{N} y_i^2 = \frac{1}{N}\sum_{i=1}^{N} \mathbf{w}^T \mathbf{r}_i \mathbf{r}_i^T \mathbf{w} = \mathbf{w}^T (\frac{1}{N}\sum_{i=1}^{N} \mathbf{r}_i \mathbf{r}_i^T)\mathbf{w} = \mathbf{w}^T \mathbf{R}\mathbf{w} \tag{2}$$

Where N is all the pixel numbers in hyperspectral image data, R is sample autocorrelation matrix.

Filter w is constrained as formula (3)

$$w^T d = 1 \tag{3}$$

In order to get the minimum of E (condition minimum), using Lagrange multiplier to compose target function $J(w)$

$$J(\mathbf{w}) = \frac{1}{2}\mathbf{w}^T \mathbf{R}\mathbf{w} + \beta(\mathbf{w}^T \mathbf{d} - 1) \tag{4}$$

Where β is Lagrange multiplier, partial derivative for w and set formula (4) equal to 0, then

$$\frac{\partial J(\mathbf{w})}{\partial \mathbf{w}} = \frac{1}{2}(\mathbf{R} + \mathbf{R}^T)\mathbf{w} + \beta \mathbf{I}_L \mathbf{d} = 0 \tag{5}$$

Where I_L is $L{\times}L$ unit matrix, because R is a symmetrical matrix and non-singular, then

$$w = -\beta R^{-1}d \tag{6}$$

Put formula (6) into formula (3), β can be decided

$$\beta = -(d^T R^{-1}d)^{-1} \tag{7}$$

Put formula (7) into formula (6), the filter coefficient w_{CEM} is

$$\mathbf{w}_{CEM} = \frac{\mathbf{R}^{-1}\mathbf{d}}{\mathbf{d}^T\mathbf{R}^{-1}\mathbf{d}} \tag{8}$$

By using filter coefficient in formula (8), a CEM detector can detect desired target d by using $CEM(r) = w_{CEM}^T r$, at the same time, minimum output energy caused by interfere background and unknown signal.

2.2 Analysis of Good Detection for Small Targets by CEM

By applying CEM algorithm into hyperspectral remote sensing target detection, CEM algorithm can detect small target better, the reason is analyzed as following.

Essentially, CEM can be seem as the inverse of Principal Component Analysis. By using PCA, main direction of original data distribution can be got, that is several eigenvectors corresponding to large eigenvalues can be decided(Geng X.R.2005). That is, PCA compresses most meaningful information from original image feature space to space that several non-correlation principals composed. Obviously, targets with low probability(small targets) in image will not included in these principals, on contrary, they often appear at feature direction corresponding to small eigenvalues of correlation matrix (Joseph C. H. 1993).

CEM finds targets by calculating $CEM(r) = w_{CEM}^T r$, if there are small targets whose corresponding energy are small, and they are corresponding to small eigenvalues of autocorrelation matrix R. while smaller the eigenvalue, $CEM(r) = w_{CEM}^T r$ the bigger, that is the reason why CEM can detect small targets. The above discuss can be deduced as following.

Decomposing R into eigenvalues and eigenvectors

$$R = V\Lambda V^T \tag{9}$$

where $\Lambda = diag(\lambda_1, \cdots, \lambda_i, \cdots, \lambda_L)$ is $L{\times}L$ diagonal matrix which contains eigenvalues of R. $V = (v_1, \cdots, v_2, \cdots, v_L)$ is an $L{\times}L$ normalized diagonal matrix whose columns are eigenvectors of R. Then

$$R^{-1} = V\Lambda^{-1}V^T \tag{10}$$

λ_i ever larger, R^{-1} ever smaller, then the smaller of the value, the ability of **CEM** detecting small targets get worse, while the ability of detecting large targets increased.

λ_i ever small, $\boldsymbol{R}^{-1}$ ever larger, then the bigger of the value, the ability of **CEM** detecting small targets increased, while the ability of detecting large targets get worse.

In conclusion, if all eigenvalues of $\boldsymbol{R}$ are adopted, that is, eigenvector corresponding to small eigen-values are included into detector, then CEM will get good detection results.

2.3 Revision of CEM Tetection Large Targets

By observing the format of CEM, the performance of the CEM technique for target detection is strongly dependent on the structure of the sample correlation matrix $\boldsymbol{R}$. The ability to calculate the inverse of $\boldsymbol{R}$ accurately is a critically important consideration when applying the CEM operator.

The eigen-decomposition of $\boldsymbol{R}$ is given by

$$\boldsymbol{R} = \boldsymbol{V \Lambda V}^T \tag{11}$$

where $\Lambda = diag(\lambda_1, \cdots, \lambda_i, \cdots, \lambda_L)$ is an $L{\times}L$ diagonal matrix containing the eigenvalues of $\boldsymbol{R}$, and $V = (v_1, \cdots, v_2, \cdots, v_L)$ is an $L{\times}L$ unitary matrix whose columns are the eigenvectors of $\boldsymbol{R}$. Eigenvalue λ represents the size of target projection to feature space, eigenvector is the direction of target project on feature space (Shen F.L. 2001). According to the above analysis, in order to detect large target, a good estimate of the correlation matrix can be obtained by only considering the contribution of the first significant eigen vectors where p is an estimate of the intrinsic dimensionality of the hyperspectral data. Considering the contribution of only the first p eigenvectors, the correlation matrix estimate is given by

$$\hat{\boldsymbol{R}} = \tilde{\boldsymbol{V}}\tilde{\Lambda}\tilde{\boldsymbol{V}}^T \tag{12}$$

Where $\tilde{V} = (v_1, \cdots, v_i, \cdots, v_p)$ is the $L{\times}p$ matrix whose columns are the significant eigenvectors and $\tilde{\Lambda} = diag(\lambda_1, \cdots, \lambda_i, \cdots, \lambda_p)$ is a $p{\times}p$ diagonal matrix.. An estimation of $\boldsymbol{R}^{-1}$ is given by

$$\hat{\boldsymbol{R}}^{-1} = \tilde{\boldsymbol{V}}\tilde{\Lambda}^{-1}\tilde{\boldsymbol{V}}^T \tag{13}$$

where $\tilde{\Lambda}^{-1} = diag(\lambda_1^{-1}, \cdots, \lambda_i^{-1}, \cdots, \lambda_p^{-1})$. Replacement of $\boldsymbol{R}^{-1}$ in CEM target detection operator with $\hat{\boldsymbol{R}}^{-1}$ will provide the desired results for large targets.

After eigen decomposition, the decomposed matrix is not only helpful to understand geometry constructions of different classes, but also helpful to understand classification of high dimension space.

Because big eigenvalue represents big targets, autocorrelation $\hat{\boldsymbol{R}}$ are composed by several big eigenvalues and eigenvectors, large targets can be detected better.

2.4   Experiment Simulation

2.4.1   Experiment Image

The experiment image is 126 bands HYMAP hyperspectral remote sensing image. The data showed part of Indiana remote sensing test region in U.S. in September 1999, it contains agriculture and forest, etc. the characteristic of HYMAP data showed in table1. , a part of data was used in this experiment. Pseudo image of band 32、16、8 used as RGB channel, 4 interest targets was selected, P1 and P2 are small targets, P3 and P4 are big targets.

Table 1. Characteristic of HYMAP data used in the experiment

Characteristic	Parameter
Image size	151×266 pixel
Fly altitude	1500 meter
Resolution	3×3 square
Pixel depth	16 bits
Band number	126
Spectral domain	0.45-2.5 μm , 3-5 μm ☐8-12 μm
Spectral resolution	VIS-SWIR:10-20nm, TIR: 100-200nm

2.4.2   Experiment Results

Correlation matrix **R** of all the pixels in HYMAP image was decomposed in this experiment. According to big to small, eigenvectors corresponding to all eigenvalues: first 100, first 50, first 10, first 5 respectively are used to construct approximate $\hat{R}$, the detection results showed in Fig.2~Fig.6. From the detection results, we can see, when all eigenvectors are used, CEM detector can get good detecting results for small targets, while worse detection results for large target; along with the reduction of eigenvectors, that is, eigenvectors reduction proposed in this paper, CEM target detector get worse results for small targets, while better results for large targets.

Fig. 1. Pseudo image of band 32，16，8 and interest targets selected

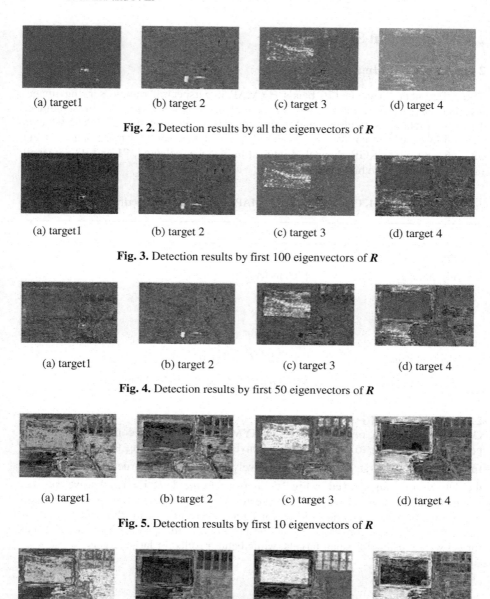

(a) target1   (b) target 2   (c) target 3   (d) target 4

Fig. 2. Detection results by all the eigenvectors of R

(a) target1   (b) target 2   (c) target 3   (d) target 4

Fig. 3. Detection results by first 100 eigenvectors of R

(a) target1   (b) target 2   (c) target 3   (d) target 4

Fig. 4. Detection results by first 50 eigenvectors of R

(a) target1   (b) target 2   (c) target 3   (d) target 4

Fig. 5. Detection results by first 10 eigenvectors of R

(a) target1   (b) target 2   (c) target 3   (d) target 4

Fig. 6. Detection results by first 5 eigenvectors of R

3 Conclusion

An eigenvector reduction method was proposed in this paper, and HYMAP hyper-spectral remote sensing data was used to test this new method. Autocorrelation matrix R was decomposed into eigenvalues and eigenvectors, then eigenvalues were selected from big to small, and corresponding eigenvectors were composed into new autocor-relation matrix. By computer simulations, three conclusions were got:

(1) When all eigenvalues and eigenvectors are used, CEM remains its primary characteristic, that is detecting small targets well;

(2) When half eigenvectors corresponding to larger eigenvalues were used, both small and large targets can be detected well;

(3) When seldom eigenvectors corresponding to larger eigenvalues were used, large targets can be detected well.

In conclusion, by using eigenvalue reduction method, the improved CEM method not only can detect large targets, but also can well detect large/small targets simulta-neously. When application, we can select all or part of eigenvectors flexibly in order to get good detection results for large or small targets.

References

Chang, C.I., Hsuan, R.: Linearly Constrained Minimun Variance Beamforming Approach to Target Detection and Classification for Hyperspectral Imagery. In: Proceedings of Geoscience and Remote Sensing Symposium, IGARSS 1999, vol. 2, pp. 1241–1243. IEEE, Los Alamitos (1999)

Chang, C.I., Heinz, D.C.: Constrained subpixel target detection for hyperspectral imagery. In: Proceedings of SPIE - The International Society for Optical Engineering, vol. 4048, pp. 35–45 (2000)

Geng, X.R.: Research of target detection and classification of hyperspectral remote sensing. Ph.D Dissertation, pp. 81–87 (2005)

Joseph, C.H.: Detection and Classification of Subpixel Spectral Signatures in Hyperspectral Image Sequenecs. Ph.D Dissertation, Maryland University, USA, pp. 84–89 (1993)

Shen, F.L., Ye, Z.F., Qian, Y.M.: Signal statistic analysis and processing, pp. 561–562. University of Science and Technology of China Press (2001)

Application of 3S Techniques in the Study of Wetland Environment of Dong Ting Lake

Baicheng Xie[1,*], Chunxia Zhang[2], Xiqiang Shuai[1], and Boliang Luo[1]

[1] Institute of Meteorology Science of Hunan Province ChangSha.Changsha Hunan province, P.R. China 410007
[2] Changsha Environmental Protetion College Changsha, Changsha Hunan province, P.R. China 430023,
Tel.: +86-731-85551337; Fax: +86-731-85551337
xbcyyhn@163.com

Abstract. Wetland is a multifunctional ecosystem in the earth, wetlands possess irreplaceable and enormous ecological functions and serve as an essential life supporting system. how to apply the new and high technology to the study of wetland has become the key point in wetland study area in our country. RS, GIS and GPS technology provide a new method for resource and ecological environment monitor. In this paper, the problem and development trend of 3S technology are discussed. Application of 3S technique in the study of wetland environment of Dongting Lake are described. The wetland monitor ways and technique route are introduced. The beach and water area information were gained.the dynamic changes of water were evaluated objective and quantitative. Finally, we point out the difficulties of wetland monitor of Dongting Lake.

Keywords: Remote sensing system,Geographic information system,global position system Monitor of wetland, Dongting Lake.

1 Introduction

Wetland is a multifunctional ecosystem in the earth, In addition to providing extremely rich resources for human production and living, wetlands possess irreplaceable and enormous ecological functions and serve as an essential life supporting system. It is surrounded and converged biosphere; hydrosphere, geosphere and atmosphere. At the same time, all kinds of energy and substance are changed here. It is praised "the kidney of the earth" and "the heaven of avian". It is important for us to protect environment and biodiversity.

Wetland is a kind of crucial resource, natural landscape and unique ecosystem offering physical contribution to human society such as maintenance of biodiversity and water quality attenuation of floods and mitigation of extreme weather etc. (Mitsch and Gosse Like, 1986; Lyon and Mccartby, 1995: Zhao etal, 2003). The wetland area in china occupies 10 percent of the whole earth, however, with the environment pollution becoming worse, our country wetland area reduces gradually in recent years, In

---

* Corresponding author.

D. Li and C. Zhao (Eds.): CCTA 2009, IFIP AICT 317, pp. 8–14, 2010.

addition, and it leads to serious ecology consequence. So far, how to apply the new and high technology to the study of wetland has become the key point in wetland study area in our country. The study mainly includes the swamp and seashore resource survey and wetland exploitation.

Geographic Information Systems (GIS), Remote Sensing (RS) and Global Positioning Systems (GPS), which are often abbreviated to Three—S technologies in China, play significant roles in many extensive and integrated research related to space and time, and are valuable techniques and tools in obtaining, storing, managing, analyzing and visualizing ecological, water resource and socio-economic data for effective and efficient inventory and optimal policy and decision making (He and Jiang 1995; Zhu etal 2002).

Nowadays the development of the spatial information technology, the advanced technology and ways were supported for monitoring wetland. Studies on wetlands range from the spatial structures and functions to succession of wetlands at different levels (Yi, 1995; Wang et al, 2001). Recently many researchers are interested in the study of landscape patterns of wetlands based on RS and GIS at the landscape scale (Liu, 2004; Kelly, 2001). However, few studies can be reportedly erected the wetland information database and wetland information system, and how to share wetland information and deliver information mechanism also became research topics.

Three-S technologies often were applied in land use planning, land resource surveying, crop yield evaluating, disaster assessing, flood monitoring, fire preventing, and desertification controlling by interrelated research institutes, universities, and government sectors. However, little research has been devoted to the used of Three-S technologies in the field of wetland. With the human being understanding wetland of remote sensing deeply, the remote sensing expert classification and decision model base were erected, and wetland information can be automatic classification extraction. Wetland of remote sensing application will become a developing trend. In this paper three-S technologies were applied to ecology resource survey and wetland draft. How to erect the ecology information database will be introduced. It will supply reasonable suggestion for wetland exploiting, utilizing, protecting and making a long-term developing plan.

2   Materials and Methods

2.1   The Situation of the Dong Ting Lake Wetland

Located in the north of Hunan Province and the south bank of the Yangtze River middle reaches, the Dongting Lake is one of the five largest freshwater lakes and of the seven most important landscapes of internationally important wetlands in China. The Dongting Lake, as a typical water level regulator of the Yangtze River, receives four main rivers (namely the Xiang, the Zi, the Yang, and the Li River) flowing through the Hunan Province, and the water of the three lake entrances of the Yangtze River. It is a buffers and plays a key role in aspects of tourism, irrigation, transportation, fishing, flood and drought controlling, climate regulation, drought controlling, beautifying environments and landscapes (Zhong, 1999).

There are 2740 km water area and 4080 km wetland area in the whole Dongting Lake region. It was separated into three parts, namely the South Dongting Lake, the East Dongting Lake and the West Dongting Lake, due to sedimentation and reclaiming lands in the lake. The Dongting Lake was selected as study area, which is located between 28°44´ −29°35´ N and 111°53´ − 113°05´ E. The main soil types are lake marshy soil and river marshy soil. The evaluation ranges from 35m to 55m. its climate belongs to the transitional zone from the middle subtropical to north subtropical, with the precipitation from 1200 to 1550mm. Due to the reclamation of wetland, grassland and forestland on a large scale in traditional agricultural practices (terracing, overgrazing, woodcutting cultivating fields, etc.), land use/cover had changed dramatically and the regional eco-environment had been rapidly deteriorating. The flood disaster raised awareness of the government and public for the importance of wetland conversation. Therefore, in 1994 the Chinese government signed the Ramsar Convention and designated Dongting Lake as wetland of International Importance. The Dongting Lake wetland is a national nature reserves in the study area, which are famous at home and abroad.

2.2 The Application of 3S Technology

The Three-S technologies were seldom employed in the research related to wetland in the 1980s, and they gained a wider appreciation in the late of 1990s. The Three-S technologies were used to the Wetland Resource Monitoring Centre in the first national wetland resource investigation. The Three-S technologies have great potentials in many kinds of research related to wetland. Land classification and changing detection techniques have been applying for mapping, investigation of wetland evolutionary process, landscape-change analyses, channel migration, flood, and wetlands resource monitoring. Spatial quantitative analyses and modeling are valuable methods in ecosystem service evaluation, wetland ecological process and risk assessment, wetland disease control, water quality monitoring and modeling, and wetlands hydrology.

Remote sensing can handle large amounts of input data and provide s an effective and efficient means to describe the characteristics of a wetland system (Koneff &Royle 2004) and its integral importance within watersheds and river basins. The Lake Institute in Nanjing used Landsate imagery to support the monitoring of the entry of sand and site into Poyang Lake (Zhao and Fu 2003) and Dongting Lake (Du et al.2001, Zhang et al. 1999) through analyzing the spectrum symbols, the landscape spectrum scenery and the living conditions of wetland floras in Sanjiang Plain, this paper found out the symbols of wetlands on remote sensing image, and established a series of classification models of remote sensing for different kinds of wetlands. The Guangdong wetland resources and environment information system (WREIS) was set up by the RS and GIS techniques (Huang, 1999). Areas inundated by floods and floodplains could be mapped effectively with remotely sensed data. The technology has been extensively used in the estimation of inundated area and the extent of flood damage along the Yellow River, Yangtza River, Huaihe River and many more basins (Zhang et al. 2003).

The wetland is rich in plant landscape diversity: there are about 40 families, 75 genus and 131 species of the hydrophytes, of which wetland plants are an important part of the vegetation (The Department of Wild Animals and Plants, State Forestry Bureau of China, 1996). It is especially important for us using the remote sensing technology to monitor great area wetland resource and eco-environment. Main applications are as follows: 1, using many wave bands is easier to distinguish land cover from vegetation style, namely, the land using and the style of wetland can be classified. 2, the soil surface moisture can be acquired from the shortwave infrared band data, thus, the boundary of scope and the change of area can be confirmed. 3, the vegetation cover information and the crop while is growing can be extracted from TM3 and TM4 band; the wetland style will be further classified. 4, the picture data can be overlaid the satellite images by vector superpose.

2.3 Methods and Technology Route

Wetlands are ecosystems typically found on the transition between terrestrial and aquatic systems. In order to be classified as a wetland, an area typically has at least one of the following three characteristics: first, there are a water table at or near the ground's surface during the growing season (including when the land is covered by shallow water); second, there are poorly drained or hydric soils; third, there is a unique diversity of wildlife and vegetation specifically adapted to thrive in wet environments. So far, there are a few researches in wetland bioloy diversity, wetland landscape pattern and how to define wetland boundry exactly. There are all kinds of wetland types in Dongting Lake, and the ecological environment is very complex, which is difficulty for us. First the rigions of wetland situation were suveryed, then

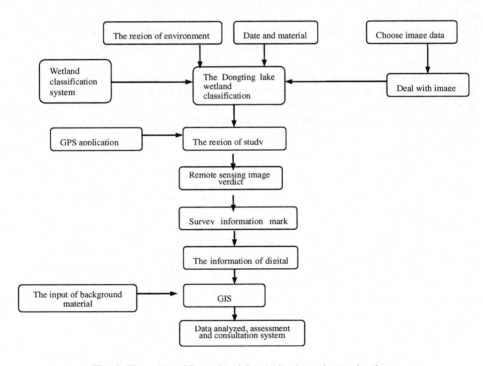

Fig. 1. The route of Dongting lake wetland monitor technology

classified systems of wetland are designed based on classification principal of international and investigation; second, we choose a proper waveband group, the image of TM being delt with, the style of wetland being determined; thrid, we use GPS to help locate in wetland investigation; Forth, the spots of image were determined, and drown; finally, the spots were digital; all of vector imformation will be imported from the GIS system.

We have collected the previous research ways about wetland remote monitor, and classified them. Finally, the wetland romote sensing monitor route will be established (Fig.1). We must get the wetland spectrum charactistic when the wetland are distinguised and classified. Because the substance spectrum characteritics are showed on the remote sensing image, at the same time, the spatial and time spectrum charactistic variety will be reavaled from it. The multi-temporal remote sensing information composition classification was chosen; such a way is mainly used to wetland spectral charactertic time effect. According to the vegetation living condition and the season difference, the multi-temporal images were compounded, then the wetland were divided into four types, such as, *Bulrush* wetland, *Phalarideae* wetland, *Carex Montana* wetland et al.

3 Results and Discussion

The different period remote sensing data were applied for monitoring Dongting Lake wetland , the results showed that the different seasons have different water level, from the Fig 2 we can see that rich water seasons mainly is from May to September , the low water mainly concentrate from October to April. The minimum water area is 520 km2 in october 15, 2006 (Fig3), the maximum water area is 2503km2 in September 15,2004 (Fig4).

Using the meteorological satellite to monitor the Dongting Lake water area, and caculate the water area, analytica the time and space changes Law.our government can make decides correctly by supplying the results when the flood is coming. Moreover, we can get great ecology environment law for monitoring the beach and water changes.

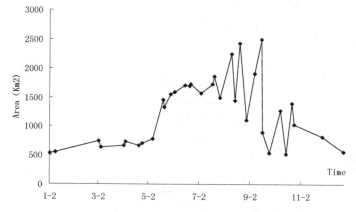

Fig. 2. The changes of water area in Dongting Lake from 2004 to 2006 yeatr

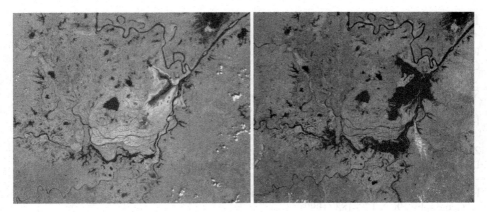

Fig. 3. The water area 520 km^2 in october 15,2006

Fig. 4. The water area 2503km^2 in September 15,2004

4 Problem and Prospect

The remote sensing technique and GIS technique have been applied extensively in wetland research, and it has solved many practical problems. But it cannot reach the sustainable development goal, many aspects need to be perfected and developedt further. First, there are obviously different wetlands spectral and other matter spectral on image, we can classify it. But some spectral character are similar to others types wetland spectral, it is very difficult for us to distinguish satellite image. Therefore, the wetland condition will be considered into wetland classification. Second, the water level of Dongting Lake has a big change in different seasons, which affects wetland dynamic monitor. Choosing the satellite image should consider the seasons because there are different results with abundant rainfall periods and low rainfall periods, it is possible to ensure the continuity of wetland variety. Third, The three-S technologies and ES will be integrated. The intelligent decision and support system can be erected. The utilization of people-machine-resource and the management integration, spatialization, intelligentization and visiblizaition can be realized; it is reference for reasonably making use of wetland resource. Forth, the many types of wetland information systems are erected based on Web system. Thus on the one hand it is likely to provide a large amount of wetland data, on the other hand, it will be provide people with the wetland science education for increasing people's consciousness of wetland protection.

Acknowledgements

Funding for this research was provided by china meteorological administration Department of Science and technique bureau (P. R. China). The first author is grateful to the whole member of project group for providing him with research ways at field survey in Dongting Lake.

14   B. Xie et al.

References

Kangqian, Y., Jinren, N.: Review of wetland studies 18(5), 539–546 (1998)
Guangyou, S.: Development and prospect of wetland science in china. Acta Electronica sinaca 15(6), 666–672 (2000)
Shuqing, Z., Chun, C., Enpu, W.: A Study on wetland classifying of remote sensing in Sanjiang Plain. Remote Sensing technology and application 14(1), 54–58 (1999)
Hongshen, D., Jiahu, J., Lake, D.: The Press of Science Technology University in China, HeHui, 29 p. (2000)
Zhengke, Y.: The background data and resource database in Dongting Lake, p. 7 (2001)
Lyon, J.G., Cartby, J.M.: Wetland and Environmental Applications of GIS. Lewis Publishers, NY (1995)
Huiping, H.: A Study on wetland resources and environment in Guangdong by GIS. Tropical Geography 19(2), 178–183 (1999)
Aihua, W., Shuqing, Z., Bai, Z.: Application of Remote Sensing and Geographical Information System Technology to Wetland Research. Remote Sensing technology and application 16(3), 200–204 (2001)
Bai, Z.: Application of Remote Sensing and Geographical Information System Technology to Wetland Research. Remote Sensing technology and application 11(1), 67–71 (1996)
Jinliang, H.: The Area Change and Succession of Dongtinghu Wetland. Geographical Research 18(3), 297–304 (1999)

A Classification of Remote Sensing Image Based on Improved Compound Kernels of Svm

Jianing Zhao[1], Wanlin Gao[1,4,*], Zili Liu[1], Guifen Mou[2], Lin Lu[3], and Lina Yu[1]

[1] College of Information and Electrical Engineering, China Agricultural University, Beijing, P.R. China 100083
[2] Yunnan Xin Nan Nan Agricultural Technology Co., Ltd. Yunnan Province, P.R. China 650214
[3] Kunming Agriculture machinery research institute, Yunnan Province, P. R. China 650034
[4] College of Information and Electrical Engineering, China Agricultural University, No. 17, Qinghua Dong Lu, Haidian, Beijing 100083, P.R. China, Tel.: +86-010-62736755; Fax: +86-010-62736755
gaowlin@cau.edu.cn

Abstract. The accuracy of RS classification based on SVM which is developed from statistical learning theory is high under small number of train samples, which results in satisfaction of classification on RS using SVM methods. The traditional RS classification method combines visual interpretation with computer classification. The accuracy of the RS classification, however, is improved a lot based on SVM method, because it saves much labor and time which is used to interpret images and collect training samples. Kernel functions play an important part in the SVM algorithm. It uses improved compound kernel function and therefore has a higher accuracy of classification on RS images. Moreover, compound kernel improves the generalization and learning ability of the kernel.

Keywords: compound kernel, remote sensing, image classification, support vector machine.

1 Introduction

Classification on remote sensing images is very important in the field of remote sensing processing. Among the classification algorithms, the accuracy of SVM (C. J. Burges, 1998) classification is one of the highest methods. SVM is short for Support Vector Machine which is a machine learning method proposed by Vapnik according to statistical leaning theory (Vapnik, 1995; Vapnik, 1998), it integrates many techniques including optimal hyperplane (Cristianini, 2005), mercer kernel, slave variable, convex quadratic programming and so on. Support Vector Machine successfully solves many practical problems such as small number of samples, non linear, multi dimensions, local minimum and so on. In several challenging applications, SVM achieves the best performance so far (ZHANG Xue Gong, 2000). In this paper, a

---

* Corresponding author.

D. Li and C. Zhao (Eds.): CCTA 2009, IFIP AICT 317, pp. 15–20, 2010.

compound kernel is proposed which is better than a single kernel to improve generalization and learning ability.

2 SVM Method

2.1 SVM Basis

Support Vector Machine is a machine learning algorithm based on statistical learning theory, using the principle of structural risk minimization, minimizing the errors of sample while shrinking the upperbound of generalization error of the model, therefore, improving the generalization of the model. Compared with other machine learning algorithms which are based on the principle of empirical risk minimization, statistical learning theory proposes a new strategy: the mechanism of SVM: find a optimal classification haperplane which satisfies the requirement of classification; separate the two classes as much as possible and maximize the margin of both sides of the hyperplane, namely, make the separated data farthest from the hyperplane. A training sample can be separated by different hyperplanes. When the margin of the hyperplane is largest, the hyperplane is the optimal hyperplane of separability (Cristianini et al. 2005).

2.1.1 Linear Separability
A two-class classification problem can be stated in following way: N training samples can be represented as a set of pairs (x_i, y_i), i=1,2…n with y_i the label of the class which can be set to values of ± 1 and $x \in R^d$ stands for feature vector with d components. The hyperplane is defined as $g(x) = w \cdot x + b = 0$.

Find the optimal hyperplane which leads to maximization of the margin. The optimal question is then translated to seek the minimization of the following function (1) and (2):

$$\Phi(w,\varepsilon) = \frac{1}{2}(w \cdot w) + C(\sum_{i=1}^{n} \varepsilon_i) \tag{1}$$

$$y_i[w \cdot x_i + b] - 1 + \varepsilon_i \geq 0 \ i = 1,\ldots,n \tag{2}$$

Where: w is normal to the hyperplane, C is a regularisation parameter, b is the offset.

The dual problem of the above problem is searching the maximization of the following function:

$$Q(\alpha) = \sum_{i=1}^{n} \alpha_i - \frac{1}{2}\sum_{j=1}^{n} \alpha_i \alpha_j y_i y_j (x_i \cdot x_j) \tag{3}$$

which is constrained by

$$\sum_{i=1}^{n} y_i \alpha_i = 0, 0 \leq \alpha_i \leq C, i = 1,2,\cdots,n \tag{4}$$

The classification rule based on optimal hyperplane is to find the optimal classification function:

$$f(x) = \text{sgn}\{(w^* \cdot x) + b^*\} = \text{sgn}\{\sum_{i=1}^{n} \alpha_i^* y_i (x_i \cdot x) + b^*\} \tag{5}$$

through solving the above problems, the coefficient α_i^* of a non support vector is zero.

2.1.2 Non-linear Separability

For a non-linear problem, we use kernel functions which satisfy Mercer's condition to project data onto higher dimensions where the data are considered to be linear separable. With kernel functions introduced, non-linear algorithm can be implemented without increasing the complexity of the algorithm. If we use inner products $K(x, x') = < \varphi(x), \varphi(x') >$ to replace dot products in the optimal hyperplane, which equals to convert the original feature space to a new feature space, therefore, majorized function of function (3) turns to:

$$Q(\alpha) = \sum_{i=1}^{n} \alpha_i - \frac{1}{2} \sum_{j=1}^{n} \alpha_i \alpha_j y_i y_j K(x_i, x_j) \tag{6}$$

And the corresponding discrimination (5) turns to:

$$f(x) = \text{sgn}\{\sum_{i=1}^{n} \alpha_i^* y_i K(x_i \cdot x) + b^*\} \tag{7}$$

2.2 Kernel Function

Kernel functions have properties as follows:

Property 1: If K_1, K_2 are two kernels, and a_1, a_2 are two positive real numbers, then K(u , v) = a_1* K_1 (u , v) + a_2 * K_2 (u, v) is also a kernel which satisfies Mercer's condition.

Property 2: If K_1, K_2 are two kernels, then K(u , v) = K_1 (u , v) * K_2 (u , v) is also a kernel which satisfies Mercer's condition.

Property 3: If K_1 is a kernel, the exponent of K_1 is also a kernel, that is, K(u , v) = exp (K1 (u , v)) (XIA Hongxia, 2009).

There are 4 types of kernels which are often used: linear kernels, polynomial kernels, Gauss RBF kernels, sigmoid kernels.

(1)linear kernels : $K(x_i \cdot x_j) = x_i^T x_j$	
(2)polynomial kernels : $K(x_i \cdot x_j) = (\gamma x_i^T x_j + r)^d$	
(3)RBF kernels $K(x_i \cdot x_j) = \exp(-\gamma \|x_i - x_j\|^2)$	
(4)sigmoid kernel: $K(x_i \cdot x_j) = \tanh(\gamma x_i^T x_j + r)$	

3 Compound Kernels

Currently, there are so many kernels each of which has individual characteristic. But they can be classified into two main types, that is local kernel and global kernels (Smits et al. 2002).

(1)local kernels

Only the data whose values approach each other have an influence on the kernel values. Basically, all kernels based on a distance function are local kernels. Typical local kernels are:

$$\text{RBF: } K(x_i \cdot x_j) = \exp(-\gamma \|x_i - x_j\|^2)$$

(2)global kernels

Samples that are far away from each others still have an influence on the kernel value. All kernels based on the dot-product are global:

$$\text{Linear: } K(x_i \cdot x_j) = x_i^T x_j$$

$$\text{Polynomial: } K(x_i \cdot x_j) = (\gamma x_i^T x_j + r)^d$$

$$\text{sigmoid: } K(x_i \cdot x_j) = \tanh(\gamma x_i^T x_j + r)$$

The upperbound of the expected risk of SVM is the proportion of the average number of support vector in the training sample to the total number of training samples: E[P(error)]≤E[number of support vector]/(total number of support vector in train samples-1)

We can get that if the number of support vector is reduced, the ability of generalization of SVM can be improved. Therefore, Gauss kernels can be improved as follows:

$$K(x_i \cdot x_j) = a * \exp(-\gamma \|x_i - x_j\|^2)\gamma > 0 \tag{8}$$

Through adding a coefficient a which is a real number greater than 1, the absolute value of the coefficient of the quadratic term of quadratic programming function in equation (8) is increased. Hence, the optimal value of α is reduced, and the number of support vector is reduced, the ability of generalization therefore is improved.

If the total number of train samples is fixed, the error rate of classification can be reduced by decreasing the number of support vectors.

If kernel functions satisfy Mercer's condition, the linear combination of them are eligible for kernels. Examples are:

$$K(x_i \cdot x_j) = a * K_1(x_i \cdot x_j) + b * K_2(x_i \cdot x_j) \tag{9}$$

Where both a and b are real numbers greater than 1, K_1, K_2 can be any kernel. Global kernels have good generalization, while local kernels have good learning ability. Hence, combining the two kernels will make full use of their merits, which achieves good learning ability and generalization.

For a compound kernel, four parameters need to be confirmed. The values of the parameters have great effect of the accuracy of classification. Optimal (C, α, a, b) is needed for the compound kernel.

4 Experimetal Results and Discussion

In this paper, a remote sensing image which resolution is 30 meters of rice paddy in Guangdong province is selected to test the results, the program of classification is written based on libsvm. The results are showed in table 1 as follows:

Table 1. The classification accuracy using different kernels

Number of pixels	kernel	Number of sv	Accuracy
350	RBF C=1 γ=0.1	310	91.2%
350	linear	210	85.3%
350	Compound of Linear and RBF C=2 γ=0.5 a=1 b=3	270	93.1%

Table 1. shows result of the classification of the remote sensing image. The classification method of compound kernel has the highest accuracy, and needs less number of sv.

From the result, we can get that the compound kernel has good ability of generalization and learning. Compared to the RBF kernel, the compound kernel has a higher accuracy of classification but the number of support vector is lower than it. Hence, the compound kernel achieves good generalization ability. In the test, the value of (C, α, a, b) has a great effect on the accuracy of classification. For different remote sensing image and different, different compound kernels and (C, α, a, b) should be selected. The optimal set of compound and value of (C, α, a, b) should be fixed through repeated trails.

5 Conclusion

In conclusion, the compound kernels yield better results than the single kernel. The compound kernels need less number of support vectors, which means that the kernels have good generalization ability and achieve higher classification accuracy than single kernels.

References

Burges, C.J.: A tutorial on support vector machines for pattern recognition. In: Fayyad, U. (ed.) Data mining and knowledge discovery, pp. 1–43. Kluwer Academic, Dordrecht (1998)
Cristianini, N., Shawe-Taylor, J.: An introduction to support vector machines and other Kernel-based learning methods: House of Electronics Industry (2005)

Smits, G., Jordaan, E.: Improved SVM regression using mixtures of kernels. In: IJCNN (2002)

Vapnik, V.N.: The nature of statistical learning theory. Springer, New York (1995)

Vapnik, V.N.: Statistical Learning theory. Wiley, New York (1998)

Xia, H., Ding, Z., Li, Z., Guo, C., Song, H.: A Adaptive Compound Kernel Function of Support Vector Machines. Journal of Wuhan University of Technology, 2 (2009)

Zhang, X.G.: Introduction statistical learning theory and support vectormachines. Act Automatica Sinica 1(26), 32–42 (2000)

Monitoring Method of Cow Anthrax Based on Gis and Spatial Statistical Analysis

Lin Li[1], Yong Yang[1,3,*], Hongbin Wang[2], Jing Dong[1], Yujun Zhao[1], Jianbin He[1], and Honggang Fan[2]

[1] Faculty of Animal Husbandry and Veterinary Medicine, Shenyang Agricultural University, 110161 Shenyang, P.R. China
lilin619619@yahoo.com.cn
[2] Faculty of anima Medicine l, Northeast Agricultural University, 150030 Harbin, P.R. China
[3] Faculty of Information and Electric Engineering, Shenyang Agricultural University, 110161 Shenyang, P.R. China
yangsyau@163.com

Abstract. Geographic information system (GIS) is a computer application system, which possesses the ability of manipulating spatial information and has been used in many fields related with the spatial information management. Many methods and models have been established for analyzing animal diseases distribution models and temporal-spatial transmission models. Great benefits have been gained from the application of GIS in animal disease epidemiology. GIS is now a very important tool in animal disease epidemiological research. Spatial analysis function of GIS can be widened and strengthened by using spatial statistical analysis, allowing for the deeper exploration, analysis, manipulation and interpretation of spatial pattern and spatial correlation of the animal disease. In this paper, we analyzed the cow anthrax spatial distribution characteristics in the target district A (due to the secret of epidemic data we call it district A) based on the established GIS of the cow anthrax in this district in combination of spatial statistical analysis and GIS. The Cow anthrax is biogeochemical disease, and its geographical distribution is related closely to the environmental factors of habitats and has some spatial characteristics, and therefore the correct analysis of the spatial distribution of anthrax cow for monitoring and the prevention and control of anthrax has a very important role. However, the application of classic statistical methods in some areas is very difficult because of the pastoral nomadic context. The high mobility of livestock and the lack of enough suitable sampling for the some of the difficulties in monitoring currently make it nearly impossible to apply rigorous random sampling methods. It is thus necessary to develop an alternative sampling method, which could overcome the lack of sampling and meet the requirements for randomness. The GIS computer application software ArcGIS9.1 was used to overcome the lack of data of sampling sites.Using ArcGIS 9.1 and GEODA to analyze the cow anthrax spatial distribution of district A. we gained some conclusions about cow anthrax' density: (1) there is a spatial clustering model. (2) there is an intensely spatial autocorrelation. We established a prediction model to estimate the anthrax distribution based on the spatial characteristic of the density of cow

---

* Corresponding author.

D. Li and C. Zhao (Eds.): CCTA 2009, IFIP AICT 317, pp. 21–26, 2010.

anthrax. Comparing with the true distribution, the prediction model has a well coincidence and is feasible to the application. The method using a GIS tool facilitates can be implemented significantly in the cow anthrax monitoring and investigation, and the space statistics - related prediction model provides a fundamental use for other study on space-related animal diseases.

Keywords: GIS, spatial statistic analysis, animal epidemiology, cow anthrax.

1 Introduction

Geographic information system (GIS) is powerful automated system for the capture, storage, retrieval, analysis, and display of spatial data. The system offers new and expanding opportunities for epidemiology because it allows an informed user to choose between options when geographic distributions are part of the problem. Even when used minimally, this system allows a spatial perspective on disease. Used to their optimum level, as tools for analysis and decision making, they are indeed a new information management vehicle with a rich potential for animal epidemiological research. Epidemiologists have traditionally used maps when analyzing associations between location, environment, and the disease (Busgeeth, K. et al., 2004). GIS is particularly well suited for studying these associations because of its spatial analysis and display capabilities. Recently GIS has been used in the surveillance and monitoring of vector-borne diseases (Glass GE, et al., 1995; Beck LR, et al., 1994; Richards FO, et al.,1993; Clarke KC, et al., 1991), water borne diseases (Braddock M, et al.,1994), in environmental health (Barnes S and Peck A. 1994; Wartenberg D, et al., 1993; Wartenberg D. et al.,1992), and the analysis of disease policy and planning (Marilyn O Ruiz, et al.,2004).Spatial analysis function of GIS can be widened and strengthened by using spatial statistical analysis, allowing for the deeper exploration, analysis, manipulation and interpretation of spatial pattern and spatial correlation of the animal disease.

Anthrax is a disease of warm-blooded animals, including humans, most livestock and some wildlife. It is caused by the spore-forming bacteria Bacillus anthracis. Herbivorous animals are highly susceptible to anthrax, For cow, the disease usually is acute, resulting in death in one to three days. By the time cow displays signs of disease, including staggering, trembling, convulsions, or bleeding from body openings, death usually follows quickly. Cow anthrax is biogeochemical disease and primarily occurs in alkaline soils with high nitrogen levels caused by decaying vegetation, alternating periods of rain and drought. Its geographical distribution is related closely to the environmental factors of habitats and has some spatial characteristics, and therefore the correct analysis of the spatial distribution of anthrax cow for monitoring and the prevention and control of anthrax has a very important role.

However, the application of classic statistical methods in some areas is very difficult because of the pastoral nomadic context. The high mobility of livestock and the lack of enough suitable sampling for the some of the difficulties in monitoring currently make it nearly impossible to apply rigorous random sampling methods. It is thus necessary to develop an alternative sampling method, which could overcome the lack of sampling and meet the requirements for randomness.

2   Materials and Methods

In this paper, we analyzed the cow anthrax spatial distribution characteristics in the target district A (due to the secret of epidemic data we call it district A) based on the established GIS of the cow anthrax in this district in combination of spatial statistical analysis and GIS. We established a prediction model to estimate the anthrax distribution based on the spatial characteristic of the density of cow anthrax. A two-stage random sampling was used. It consisted in generating sampling sites at random and then drawing the required number of cow from the nearest herd to each point, and this, for practical reasons, within a specified radius. In this sampling method, the primary unit was defined as a settlement, watering point or grazing area where animals were expected to be found. Due to a lack of an exhaustive list of these locations in some areas and because of the high mobility of pastoral herds, the unpredictability of their movement, the classical random selection of sites was not feasible. Therefore, there was the need to develop an alternative method that allows random selection of the first sampling units. To that effect, GIS computer application software (ArcGIS) was used.

2.1   Cow Anthrax Data Sampling Based on ArcGIS

ArcGIS Spatial analyst is an extension to ArcGIS Desktop that provides powerful tools for comprehensive, raster-based spatial modeling and analysis. Using ArcGIS Spatial Analyst, you can derive new information from your existing data, analyze spatial relationships, build spatial models, and perform complex raster operations. With its extension to generate random coordinates, ArcGIS is able to generate at random the required number of sites within the area where sampling needs to take place, be it at zone, country, region or even district level, simultaneously allowing the application of weight factors to different sub-units of the surveyed area. Each site is identified by its longitude and latitude coordinates. In reality, a generated site is materialized only as a point on the map and therefore a fixed radius is defined around each point in order to determine a geographical area within which the sampling of cow anthrax is carried out. The determination of length of the radius takes into consideration the cow anthrax density. A radius below 10 km in a dry area may result in too many sampling sites with no animals, whereas in a densely populated area around permanent watering points and grazing areas, a relatively shorter radius suffices. A surplus of the total number of sites is generated at random to replace target sites that could not be accessed due to natural obstacles or sites with no animals within the defined circle.

2.2   Cow Anthrax Density Spatial Cluster Analysis

Spatial cluster analysis plays an important role in quantifying geographic variation patterns. It is commonly used in disease surveillance, spatial epidemiology, population genetics, landscape ecology, crime analysis and many other fields, but the underlying principles are the same. Spatial clustering analysis is the main research field of spatial data mining. We analysis cow anthrax data spatial cluster using High/Low Clustering tool in ArcGIS which can measures the degree of clustering for either high values or low values. The tool calculates the value of Z score for a given input feature class. A z-score is a measure of the divergence of an individual experimental result from the most probable result, the mean. Z is expressed in terms of the number of standard deviations from the mean value.

$$z = \frac{x - \mu}{\sigma}$$

where: x is ExperimentalValue, μ is Mean, and σ is StandardDeviation.

This z-value or z score expresses the divergence of the experimental result x from the most probable result μ as a number of standard deviations σ The larger the value of z, the less probable the experimental result is due to chance. For this tool, the null hypothesis states that the values associated with features are randomly distributed. The higher (or lower) the Z score, the stronger the intensity of the clustering. A Z score near zero indicates no apparent clustering within the study area. A positive Z score indicates clustering of high values. A negative Z score indicates clustering of low values.

2.3   Cow Anthrax Density Spatial Autocorrelation Analysis

Spatial Autocorrelation is correlation of a variable with itself through space. If there is any systematic pattern in the spatial distribution of a variable, it is said to be spatially autocorrelation. If nearby or neighboring areas are more alike, this is positive spatial autocorrelation, Negative autocorrelation describes patterns in which neighboring areas are unlike. Random patterns exhibit no spatial autocorrelation. Spatial autocorrelation is important, because most statistics are based on the assumption that the values of observations in each sample are independent of one another. Positive spatial autocorrelation may violate this, if the samples were taken from nearby areas .Goals of spatial autocorrelation Measure is the strength of spatial autocorrelation in a map and test the assumption of independence or randomness. Spatial autocorrelation is, conceptually as well as empirically, the two-dimensional equivalent of redundancy. It measures the extent to which the occurrence of an event in an area unit constrains, or makes more probable, the occurrence of an event in a neighboring area unit. We analysis cow anthrax data Spatial Autocorrelation using Global Moran's I tool in GEODA. Moran's I is a measure of spatial autocorrelation developed by Patrick A.P. Moran. Moran's I is defined as:

$$I = \frac{N}{\sum_i \sum_j w_{ij}} \frac{\sum_i \sum_j w_{ij}(X_i - \bar{X})(X_j - \bar{X})}{\sum_i (X_i - \bar{X})^2}$$

where: N is the number of spatial units indexed by i and j, x is the variable of interest, $\bar{X}$ is the mean of x, and wji is a matrix of spatial weights.

Negative (positive) values indicate negative (positive) spatial autocorrelation. Values range from −1 (indicating perfect dispersion) to +1 (perfect correlation). A zero values indicate a random spatial pattern. For statistical hypothesis testing, Moran's I values can be transformed to Z-scores in which values greater than 1.96 or smaller than −1.96 indicate spatial autocorrelation that is significant at the 5% level.

3   Results and Discussion

The results of cow anthrax density spatial cluster analysis showed that the density of dairy cattle anthrax distribution in space with a high concentration of the Z Score = 5.6, P < 0.01, which means that there is a spatial clustering model. The results of Cow anthrax density spatial Autocorrelation analysis showed Moran's I = 0.4358, which means there is an intensely spatial autocorrelation. Based on the spatial characteristic of the density of cow anthrax we established a prediction model to estimate the anthrax distribution by Ordinary Kriging tool in ArcGIS. Ordinary kriging (OK) is a geostatistical approach to modeling. Instead of weighting nearby data points by some power of their inverted distance, OK relies on the spatial correlation structure of the data to determine the weighting values. Ordinary Kriging is a spatial estimation method where the error variance is minimized. Superimposing map of predict map (Fig1) and distribution of cow anthrax epidemic focus in district A (Fig2). Comparing with the true distribution, the prediction model has a well coincidence and is feasible to the application.

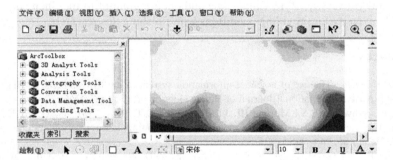

Fig. 1. Predict map of cow anthrax neural distribution in district A

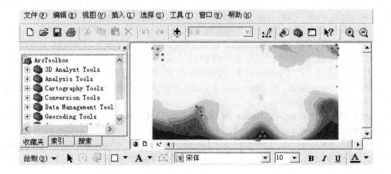

Fig. 2. Superimposing map of predict and the true distribution of cow anthrax neural

4 Conclusion

GIS are powerful computerized systems with capabilities for inputting, storage, mapping, analysis and display of spatial data associated with a location on the Earth's surface. GIS have tremendously enhanced ecological epizootiology, the study of diseases in relation to their ecosystems. They have found increasing application for surveillance and monitoring studies, identification and location of environmental risk factors as well as disease prediction, disease policy planning, prevention and control. In this article we discuss the application of GIS to veterinary and medical research and the monitoring method of cow anthrax based on GIS and spatial statistical analysis. The method using a GIS tool facilitates is implemented significantly in the cow anthrax monitoring and investigation, and the space statistics - related prediction model provides a fundamental use for other study on space-related animal diseases.

Acknowledgements

Funding for this research was in part provided by the postdoctoral fund of Shenyang Agricultural University, P. R. China. The authors are grateful to the Shenyang Agricultural University for providing conditions with finishing this research.

References

Busgeeth, K., et al.: The use of a spatial information system in the management of HIV/AIDS in South Africa. International Journal of Health Geographics 3, 13 (2004)

Glass, G.E., Schwartz, B.S., Morgan III, J.M., Johnson, D.T., Noy, P.M., Israel, E.: Environmental risk factors for Lyme disease identified with geographic information systems. Am. J. Public Health 85, 944–948 (1995)

Beck, L.R., Rodrigues, M.H., Dister, S.W., Rodrigues, A.D., Rejmankova, E., Ulloa, A., et al.: Remote sensing as a landscape epidemiologic tool to identify villages at high risk for malaria transmission. Am. J. Trop. Med. Hyg. 51, 271–280 (1994)

Richards Jr., F.O.: Use of geographic information systems in control programs for onchocerciasis in Guatemala. Bull. Pan. Am. Health Organ. 27, 52–55 (1993)

Clarke, K.C., Osleeb, J.R., Sherry, J.M., Meert, J.P., Larsson, R.W.: The use of remote sensing and geographic information systems in UNICEF's dracunculiasis (Guinea worm) eradication effort. Prev. Vet. Med. 11, 229–235 (1991)

Braddock, M., Lapidus, G., Cromley, E., Cromley, R., Burke, G., Branco, L.: Using a geographic information system to understand child pedestrian injury. Am. J. Public Health 84, 1158–1161 (1994)

Barnes, S., Peck, A.: Mapping the future of health care: GIS applications in Health care analysis. Geographic Information systems 4, 31–33 (1994)

Wartenberg, D., Greenberg, M., Lathrop, R.: Identification and characterization of populations living near high-voltage transmission lines: a pilot study. Environ. Health Perspect 101, 626–632 (1993)

Wartenberg, D.: Screening for lead exposure using a geographic information system. Environ. Res. Dec. 59, 310–317 (1992)

Ruiz, M.O., Tedesco, C., McTighe, T.J.: Connie Austin and Uriel Kitron. International Journal of Health Geographics 3, 8 (2004)

The Research on Natural Vegetation's Response to Agriculture in Tarim River Basin in Recent 50 Years Using Multi-Source Remote Sensing Data[*]

Xiaohua Wang[1], Shudong Wang[2], Zhenyu Cai[1], and Jianli Ding[3,**]

[1] School of Economics and management, Hebei University of Engineering,
Handan, Hebei, 056038, China
[2] State Key Laboratory of Remote Sensing Science;
Beijing Key Laboratory for Remote Sensing of Environment and Digital Cities;
Center for Remote Sensing and GIS, School of Geography; Beijing Normal University,
19 Xinjiekou Wai Street, Beijing, 100875, China
[3] Jianli Ding, College of Resources and Environment Science,
Xinjiang University. Urumqi, Xinjiang, 830046, China

Abstract. Excessive water is used by farm field in Tarim river basin, it is very important to deep study on the relation between area of farm field and natural vegetation in recent 50 years. Collecting and processing various decades of remote sensing data, and got area of farm field, grass, forest, town and non-serviceable field. These results indict that: Because of the high flow period between 1950 and 1972, a part of grassland converts forest. From 1972 to 1990, because of increase of farm field, limited water isn't enough to grassland and forest, and a fraction of forest area converts grassland again. With rapid increase of area of farm field, more water need from the river and underground, which causes water level depression, and natural ecology becomes deteriorative from 1990 to 2000. Comprehensive analysis indicts that agriculture control implemented by government is necessary to recovery of ecology, or farm field also would be lost eventually with destruction of natural ecology.

Keywords: relation between natural vegetation and farm field multi-source remote sensing data inner cause analysis Tarim river basin.

1 Introduction

Tarim river is called mother's river, which is the main source of farm field and ecology. However, unrestricted water consumption causes negative influences such as considerable deforestation, desertification, and increased soil salinity [1-5]. The ecological environment in the Tarim River basin is extremely vulnerable [6-9]. The local and

---

[*] The paper is supported by the project : the National Key Technology R&D Program of China （2006BAB07） and Open foundation of Key Laboratory of Oasis Ecology (Xinjiang University) Ministry of Education.
[**] Corresponding author.

D. Li and C. Zhao (Eds.): CCTA 2009, IFIP AICT 317, pp. 27–31, 2010.

center government have paid great attention to deterioration of the ecology and environment. In an attempt to restore the ecological system, some significant improvements have been made in regulating the stream flow of the Tarim River for flood control and ecological water releases from Boston Lake through Daxihaizi Reservoir were carried out several times from 2000. It is very important to deep study on the relation between area of farm field and natural vegetation in recent 50 years. Collecting and processing various decades of remote sensing data, we got area of farm field, grass, forest, town and non-serviceable field using multi-source remote sensing data such as MSS, Landsat-TM, aerial photograph, time series MODIS products NDVI (250m) [10-14].

2 Historical Data Processing and Analysis

2.1 Data Processing

To study the relation between farm field and area of vegetation, it is hard to analyze the tendency using short time series of data. We select the time from 1950s to 2000s and

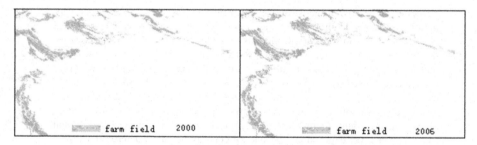

Fig. 1. Farm field area in 2000 and 2006

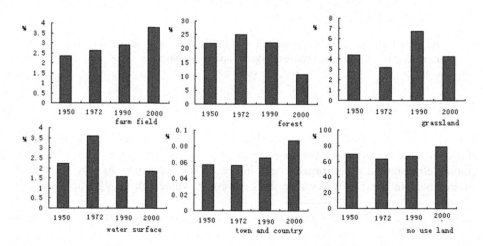

Fig. 2. Land use and cover change analysis form 1950s to 2000s

the remote sensing data of various sources which include Land sat MSS, Landsat TM and so forth, and the data processing methods include the transition from raster to vector data, supervise classification and so forth. For example, time series of MODIS data were used to extract farm field information using variance method (Fig.1). Finally, we got statistical data of various land use and cover (Fig.2).

2.2 Data Analysis

The analysis from Land use and cover change figure of Tarim river basin shows that the land use and cover always changes from 1950s:

(1) From 1950 to 1972, area of farm field has a tiny increasing rate of 2.35-2.73% (percentage of statistical area), at the same time, forest area had a little increase of 22.12%-25.13%, but area of grassland had a little decrease of 4.51% - 3.13%. (2) From 1972 to 1990, area of farm field still has a little increase rate of 2.73-2.92% compared with the changes from 1950 to 1972. At the same time, forest area decrease to about 22% again, but area of grassland has a rapid decrease from 3.13% to 7.02%. (3) From 1990 to 2000, area of farm field has a rapid increase rate of 2.92-3.82% compared with the changes from 1970 to 1990, but area of forest and grassland has a sharp increase trend, and that of forest rapidly decreases from about 23% to about 10% and that of grassland from 7.02% to 4.48%. (4) From 2000 to 2006, area of the farm field keeps stable, but the area of forest and grassland increase obviously using method of variance of time serious MODIS data products. As for area of town and City, there is a continued increase from 1970, and which becomes obvious from 1972 to 2000.

Comprehensive analysis indicts that agriculture control implemented by government is necessary to recovery of ecology, or farm field also would be lost eventually with destruction of natural ecology. According to historical material and above results, because of the high flow period between 1950 and 1972, a part of grassland converts forest. From 1972 to 1990, because of increase of farm field, limited water isn't enough to grassland and forest, and a fraction of forest area converts grassland again. With rapid increase of area of farm field, more water need from the river and underground, which causes water level depression, and natural ecology becomes deteriorative from 1990 to 2000. Water releases from Boston lake for restoration of the ecosystem agriculture control implemented by government help recovery of natural vegetation after 2000.

3 The Relation Analysis between Farm Field and Natural Vegetation

We analyzed the relation curve of farm field and vegetation from 1950s to 2000s, and the result indicts that the correlation coefficient between natural vegetation and farm field be 0.8775. It is higher than that of vegetation (natural vegetation and farm field) and farm field. It indicted that more water need from the river and underground, which causes water level depression, and natural ecology become deteriorative , with rapid increase of area of farm field.

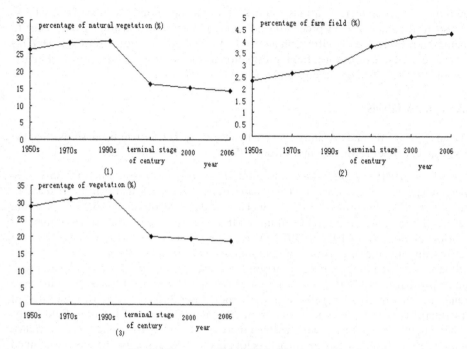

Fig. 3. Changes of farm field and vegetation

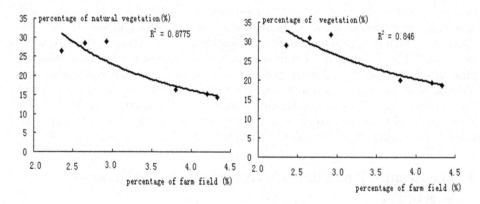

Fig. 4. Correlation coefficient between farm field and natural vegetation

4 Conclusion

Currently, Some significant improvements to restore the ecological system have been made in regulating the stream flow of the Tarim River for flood control and ecological water releases from Boston Lake through Daxihaizi Reservoir were carried out several times from 2000. Water releases from Boston lake for restoration of the ecosystem agriculture control implemented by government help recovery of natural vegetation after 2000. As a result, the natural vegetation begin to restore.

References

Anderson, G.L., Hanson, J.D., Haas, R.H.: Evaluating landsat thematic mapper derived vegetation Indices for estimating above-ground biomass on semiarid rangelands. Remote Sensing of Environment 45(2), 165–175 (1993)

Dozier, A.G.: EOS Science Strategy for the Earth Observing System, pp. 1–119. VASA AIP Press, Bekerly (1994)

Baret, F., Jacquemoud, S., Hanocq, F.: The soil line concept in remote sensing. Remote Sensing Environment 7, 6582 (1993)

Song, C.: Spectral mixture analysis for subpixel vegetation fractions in the ureban environment: How to incorporate endmember variability? Remote Sens. Environ. 95, 248–263 (2005)

Dymond, J.R., Stephens, P.R., Newsome, P.F., et al.: Percent vegetation cover of a degrading rangeland from SPOT. International Journal of Remote Sensing 13(11), 1999–2007 (1992)

Ferenc, C.: Spectral Band Selection for the Characterization of Salinity Status of Soils. Remote Sensing of Environment 43, 231–242 (1993)

Gillies, R.R., Carlson, T.N., Kustas, W.P.: A verification of the 'tri-angle' method for obtaining surface soil water content and energy fluxes from remote measurements of the Normalized Difference Vegetation Index (NDVI) and surface radiant temperature. International Journal of Remote Sensing 18(15), 3145–3166 (1997)

Goetz, S.J.: Muti-sensor analysis of NDVI, surface temperature and biophysical variables at a mixed grassland site. International Journal of Remote Sensing 18(1), 71–94 (1997)

Huete, A.R., Jackson, R.D.: Suitability of spectral indices for evaluating vegetation characteristics on arid rangelands. Remote Sensing of Environment 23, 213–232 (1987)

Jackson, R.D., Reginato, R.J., Idso, S.B.: Wheat canopy temperature: a practical tool for evaluating water requirements. Water Resour. Res. 13, 651–656 (1977)

Leonard, B., Chen, J.M., et al.: A shortwave Infrared Modification to the Simple Ratio for LAI Retriel in Boreal Forest-An Image and Model Analysis. Remote Sensing of Environ. 71(1), 16–25 (2000)

Leprieur, C., Verstraete, M.M., Pinty, B.: Evaluation of the performance of various vegetation indices to retrieve vegetation cover from AVHRR data. Remote Sensing Review 10, 265–284 (1994)

Liang, S., Fang, H., Chen, M., et al.: Validating MODIS land surface reflectance and albedo products: Methods and preliminary results. Remote Sensing of Environment 83, 149–162 (2002)

Liang, K.Y., Liu, P.J.: Resource and environment study on banks of Tarim River with RS. Science Press, Beijing (1990)

The Service Architecture of Agricultural Informatization

Zhiyong He[*], Lecai Cai, Hongchan Li, and Jujia Xu

Institute of Computer Application, Sichuan University of Science and Engineering,
Zigong Sichuan, P.R. China 643000,
Tel.: +86-13778520180; Fax: +86-813-5505966
hzy@suse.edu.cn

Abstract. Agricultural informatization is the basis for the modernization of agriculture, at present there are many different ways of construction and models in agricultural information systems. Agricultural informatization for the Theory and Application of the status quo, a comparative analysis of a typical agricultural information system, agricultural information service system architecture is advanced, discussed in detail the operating mechanism of service, standards and application of norms, information platform architecture and other major issues, for the construction of agricultural informatization technology standards and norms of the project has a reference value.

Keywords: agricultural informatization, service architecture, framework.

1 Introduction

Agricultural informatization is the strategy of rejuvenating the implementation and prospering rural market economy, promote solve problems of "agriculture, countryside and farmers", promoting the construction of rural well-off, agricultural science and technology service system is the important part of the construction. It is the vital significance to construct the agriculture information service system of agricultural informationization, and to improve the level of agricultural modernization and agricultural comprehensive competitiveness.

At present, the agriculture information service system has no unifying definition. Some articles which are related to agricultural information service system aims at agricultural informationization, of development; provides agricultural information service all kinds of main agricultural information service as the core, consist the organic system according to certain rules and system(Yingbo Li et al., 2005). Rural information service system is an integral component which it is engages in the organization of the rural information service, personnel, information infrastructure, information resources and the necessary information technology (Liang Xue et al., 1998). An organic whole is integrated by the agricultural information service system, it has agricultural information resource, agriculture information service participants, policies

---

[*] Corresponding author.

D. Li and C. Zhao (Eds.): CCTA 2009, IFIP AICT 317, pp. 32–39, 2010.

and regulations, operation mechanism, information infrastructure, and information technology, etc, agricultural information collection and processing, the spread and application are realized, its purpose is to provide technology and organization of agricultural informatization and security.

2 Domestic and Overseas Status

2.1 Overseas Status

The agriculture information service system as an important support system in the market economy developed countries. The market information service system construction is strengthened, the agriculture information legislation is noticed, International cooperation and agricultural market information service are strengthened, and the forceful measures construction of economic restructuring is promoted (Jing Zhao et al., 2007).

A huge market information network and information network of perfect function have been established by The United States, the policy of the development of agricultural information system is stringier (Xiangyu Zhang et al., 2005). The Japanese government attaches great importance to build a perfect agricultural market information service system, plays a folk in provide market information, and the role of computer in the popularization and application of rural (Yi Yang et al., 2005), the information service system of agricultural market as one of the key service system construction, Japanese agricultural informatization development in these successful experience is worth our using for reference. French formed government, folk, university diversified and comprehensive information service pattern and diversity of information service main body, which include the national agricultural subject, agricultural chamber, agricultural scientific research and teaching units, various agricultural industry organizations and professional associations, folk information media and various agricultural production cooperatives and mutual (Zuoyu Guo et al., 2000), etc. Europe officials are the subject of agricultural information service (Li Bai. et al., 2006), in addition to the official, agriculture information service industry association (including organization, agricultural chamber) and futures market and insurance institutions to provide information service also occupied very big one part, play an important role in the agricultural risk prevention. In addition, South Korea established level from agricultural service system, the collection scientific research, promotion and training trinity, unified in agricultural service system, process is simplified and concentrated, promote the new leadership of rural sports.

The common features are formed when all the agriculture information service system are constructed, namely government organizations from macroscopically management and coordination in the executive departments, and clear responsibilities of collaboration; Second, the information service, information service, diversification of diversification, such as France; Third, pay attention to strengthen service environment from scientific research, education to ensure service system.

2.2 China's Status

Compared with the developed countries, the Chinese rural information service system construction starts relatively late, construction of agricultural information system has made significant achievements. Since 1995, China agriculture proposed involving agricultural informatization construction "JinNong project", based on Countries with large agricultural products market, emphasis, main agricultural education scientific research units and the agricultural professional societies, associations, agricultural and rural informatization are accelerated and promoted, agriculture comprehensive management and service information system" is established. The agricultural basic database, the national agricultural monitoring, forecast and warning, the macroeconomic regulation and control and decision-making service application system, agricultural production situation and crop production forecast system for agricultural development are established and perfection, and play an important role in recent 5 years. At present, the provincial department of agriculture, 97% of the prefecture (city) and 80% of the county agriculture department has information management and service, 64 percent of the information service, the township set up more than 20 million people in the development of rural feedback team, preliminary build from central to local agricultural information system.

Information technology development and application achieve positive progress. the information application system of the administrative examination and approval and the government office network office automation, scheduling NongQing scheduling, satellite communication, remote sensing, agriculture and animal epidemic information use the information technology are developed by the agriculture department. Department of agriculture, according to the management and service requirement, information system is developed which is related soil testing formula, pest control, basic farmland management. It effectively promotes the agricultural department of e-government, for accelerating the development of modern agriculture, the narrowing of "digital divide" plays a positive role in promoting (Ministry of Agriculture PRC., 2008).

At present, information service work of agriculture, countryside and Peasant is developing, it covers market monitoring warning system of the main agricultural products, and provides strong basis for the leadership, macro decision ; Information release which provides timely information industry of agricultural economy for farmers by the ministry of economic information standard calendar; It plays more and more major role in the promoting agricultural, the rural economy development, and the construction of new countryside by coordinating the agriculture information service work. There are 4 million the industrialization of agriculture leading enterprise,17 million rural cooperative intermediary organizations, 61 million administrative villages, 95 million agricultural production and operation, and 240 million rural agent in our country, they can accept the department of agriculture information service by information networks and other forms.

3 The Current Problems

Agricultural informatization construction of our country is still in the initial, but compared service ability of the information service system with government, enterprises and the demand of agriculture information service, compared modern agriculture and the construction of socialist new countryside with the overall requirements of information work, compared with the developed countries in market economy, there is still a relatively large gap. Current outstanding problems mainly include:

1) The agricultural informatization regulations system is not perfect, the agriculture information standardization is lag;

2) Weak infrastructure and farmers in developing the information service of the "last mile" problems have not been solved;

3) Information collection and collection channels are not standardization, information resource is shortage, information content of practicality is weak, quality need to be promoted, and information resource integration is difficult;

4) The share degree of Existing hardware resources and information resources is low, and does not form the effective mechanism of sharing;

5) The function of agricultural information, for example Radio, television, newspapers, telecommunications, media communication, has not been fully developed and integration, support ability of information platform technology is not adequately;

6) Peasants culture quality is low, awareness of information is weak, and information ability is not strong;

7) The percentage of The agricultural population of Internet users is low, agricultural sites are less, practical contents are also less, unwrought depth, using low degree, application service ability is bad;

8) The construction of agricultural team is lag, the demand of work can not satisfy by the personnel quantity, and service ability;

9) The degree of marketing is not high, the agricultural product market of electronic business affairs still not completely, commodities and services network level is low;

10) Agricultural information service system is not perfect, the agricultural informatization development level of regional of the middle, east and west is serious imbalance, and differences of regional development are very serious.

In view of the above problems, regional agricultural comprehensive information public service platform model and application research is put forward in this paper, and the agriculture information service of the "last mile" problem of agricultural information platform, and integration of information resources sharing, and agricultural informatization service standard lag issues are solved, it is an important realistic significance for perfecting the agriculture information service system.

4   Development of Dynamic Analysis

The tendency of our agricultural, rural informatization construction is following:

1) The government's leading tendency. Public information service is provided as the important functions of the government, the government must attach great importance to, overall planning, and investment, speed up the construction on which the government must play a leading role

2) The complicated trend of demand. Different departments, various industries, of all kinds' enterprise, cooperative organizations and farmers have different subjects such information needs. On the demand of information quality, effective customized information becomes main stream. On the demand of information on the subject, the strong demand of farmers, various enterprises and intermediary organizations are also increasingly urgent needs.

3) The trend of the diverse channels. In the information services, multi-channel, more comprehensive application, modes means are emerge. "SanDian unity" information service in the background is the one of the effective way.

4) The trend of work. Agricultural information system construction includes agricultural production, processing, distribution and scientific education, promoting technology, consumption and other aspects, enrich the connotation of work, the extension of agricultural industries, and combining the closer, interweave propulsion trend.

5) The development trend of socialization. In recent years, the development of government leading practice showed that, the social forces in agricultural information system construction is the inevitable choice.

6) Precision trend of Production. Precision agriculture based on the global positioning system (GPS) and geographic information system (GIS), information collection and management system of farmland, intelligent decision support system of agricultural machinery system and system integration technology, becomes the core of agricultural informatization development.

5 The Research of Agricultural Information Construction Framework

5.1 The Nature of Agricultural Information Construction Framework

Agricultural information construction framework is a whole of many elements, which integrates information collection, possessing, communication, application and so on; the purpose is to provide technology and organization of agricultural informatization and security. Its main function is to collect information on the rural economy, processing, storage and transmission by feedback. Firstly, the governments and the relevant departments for the agricultural development plan and guide to provide the scientific basis as farmers, merchants and agricultural market. Secondly, it provides timely and accurate information service for farmers, merchants and agricultural market, accelerates the process of marketing of rural economy. Rural information service system provides information environment of the order transparent, information network of perfect accessibility and information service of high efficient, it is significant and profound influence on rural economic development.

5.2 The Mode of Agricultural Information Construction Framework

Notice of The national agricultural and rural informatization construction framework (2007-2015) by the issuance of agriculture of March 2008(Ministry of Agriculture PRC., 2008),pointed out: basic framework of our agricultural and rural informatization construction consists mainly in information infrastructure, information resources, talents, service and application system, and the development of the system, operation mechanism, whose rural economic, political, cultural and social fields, etc. study and formulate relevant software and hardware technology standards, data acquisition and processing standards, information, formulate standards, pay attention on information acquisition, storage, processing, processing standards and norms, accelerate information services, agricultural information classification and coding standard can be formulated quickly.

Now, there is not standard of agricultural information construction framework in China, through the research and analysis standard of e-government standard (rules) and the ministry of agricultural and rural informatization construction, the general framework of agricultural information service system frame model is put forward, as shown in Fig1.

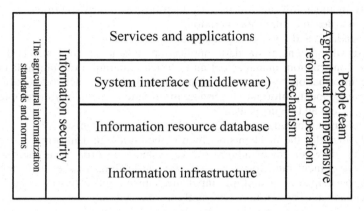

Fig. 1. The mode of agricultural information construction framework

Agricultural informatization basic framework consists mainly in rural economic, political, cultural and social fields of information infrastructure, information resource database, system interface, service and application system, talent team, agricultural comprehensive reform and the operation mechanism of agricultural informatization, information security and standard and standard etc.

1) The information infrastructure

It mainly includes network infrastructure, information technology equipment. Network infrastructure mainly has the computer network, the communication network and broadcast television, newspapers, magazines and billboards. Information technology basic equipment mainly refers to the development of information technology and popularization and application of the necessary facilities.

2) System interface layer

System interface layer is a system which the agricultural informatization construction to meet different levels of real-time information collection, transmission and exchange and sharing, service.

3) Information resource database

It is mainly includes science, technology, markets, policies, regulations, culture, education, health, and other information resources database. It serves the participants of production and living, and other social activities in the agriculture and rural.

4) People team

A practical and efficient service team is built, the information service mode innovation is strengthened, the applicability and effectiveness and scientific of the rural information service is improved.

5) Services and applications system

It mainly includes service and information technology application. Information service combines with rural economy, social development of the actual, the development of modern agriculture, basic conditions of the existing information, the industrialization of agriculture and rural economic, the need of cooperation organization.

6) Information security

It mainly includes information security facilities, technical solutions, systems and equipment in order to ensure safe operation.

7) The standards and norms of agricultural informatization

It consists of the general standard, application standards, safety standards and infrastructure construction, management and service integration of standard; including agriculture information resources development and sharing, network (station) construction management, information service, information technology development and application, safety and security, etc.

8) Agricultural comprehensive reform and operation mechanism

It mainly includes the mechanism which is the agricultural informatization development, deepening reform of administrative system and promoting each other and common development; and forms the government-leading, market and other social forces joint participation of multiple input mechanism, and forms sustainable development mechanism of agricultural and rural informatization.

6   Conclusion

Information infrastructure is the basic conditions of agricultural informatization construction. System interface layer is the data exchanging platform of agricultural informatizition. It implements the data exchange function between the different applications. Information resource databases are the important content of agricultural informatization construction. Service and application system is the starting point and the foothold of agricultural informatization construction. Information security is the running guarantee for agricultural informatization construction. Agricultural informatization standardization is the key link of agricultural informatization construction and development. Agricultural comprehensive reform and the operation mechanism of agricultural informatization is the fundamental guarantee.

Agricultural information service system is the basic work of agricultural informatization construction, from the theory and application situation of agricultural informatization, compares and analyses the typical agricultural information system, the agricultural informatization service system structure is put forward, relevant laws and regulations of agricultural informationization construction is accelerate formulate, the perfect work system is established, agricultural and rural informatization construction standardization and institutionalized are promoted, it is an important meaning to construct the agricultural informatization engineering standards and norms.

References

Li, Y.: Study of agricultural information service system, China Agricultural University, PhD thesis, pp. 34–35 (2005) (in Chinese)

Xue, L., Fang, Y.: Agricultural Informatization, vol. 2, pp. 218–230. Jinghua Press, Beijing (1998) (in Chinese)

Zhao, J., Wang, Y.: Summary Research on Agricultural Informatization in and Abroad. Document, Information & Knowledge 6, 80–85 (2007) (in Chinese)

Zhang, X.-y.: Study on strengthening the construction of agricultural information system. Agricultural economic problems 1, 22–24 (2003) (in Chinese)

Yang, Y.: Concerning the development and enlightenment of Japanese agricultural informatization. Modern Japan's economy 6, 62–64 (2005) (in Chinese)

Guo, Z.: See the agriculture information service in the agriculture information service network construction from France—training report of the agriculture information service network in France. Feed wide-angle 4, 28–29 (2000) (in Chinese)

Bai, L., et al.: Mode and enlightenment of Euro agriculture information service. World Agriculture 1, 33–35 (2006) (in Chinese)

Ministry of Agriculture PRC. Construction Planning of The 11th five-year plan period of National Agricultural Information System by the ministry of agriculture (2008) (in Chinese), http://www.agri.gov.cn/jhgb/t20080321_1029961.htm

Ministry of Agriculture PRC. The notices about National Agriculture and Rural Informatization Construction Framework (2007-2015) by the ministry of agriculture (2008) (in Chinese), http://www.agri.gov.cn/jhgb/t20080321_1029943.htm

A Forecast Model of Agricultural and Livestock Products Price

Wensheng Zhang[1,*], Hongfu Chen[1], and Mingsheng Wang[2]

[1] College of Civil Engineering, Shijiazhuang Railway Institute, Shijiazhuang, Hebei,
China 050043,
Tel.: +86311-87939151
xyns@hotmail.com
[2] College of Traffic Engineering, Shijiazhuang Railway Institute, Shijiazhuang, Hebei,
China 050043

Abstract. The frequent and excessive fluctuation of agricultural and livestock products price is not only harmful to residents' living, but also affects CPI (Consumer Price Index) values, and even leads to social crisis, which influences social stability. Therefore it is important to forecast the price of agriculture and livestock products. As a result, we make a research on the factors affecting agricultural and livestock products price, establish a forecasted model of agricultural and livestock products price, and develop its early-warning system which is suitable to China. Considering the direct relationship between the price and the output, multiple linear regression method is adopted to study this problem. The model is composed of three sub-models. This paper puts forward the concept of price equilibrium coefficient C_0, which describes the degree to which people accepting the forecasted price. With the establishment of the standard for the influence of price fluctuation, the influence of price fluctuation is measured. Each range of the C_0 value corresponds with a specific result, which may informs the government with the danger of price fluctuation. As a result, the model can early-warn the price rising caused by crop reduction due to sudden natural disaster, which may induce social turmoil and crisis. If the forecasted price rises heavily, the government should take measures to avoid crisis. This paper offers the method to control future price. At last, a forecasted model of pork price is calculated with simulated data. The forecasted result is in good agreement with actual situation.

Keywords: multiple linear regression, forecast model, price equilibrium coefficient, RS, GIS.

1 Introduction

Agricultural and livestock products are closely related with residents' life, and the stability of products' price is important to social stability. In 2008, the reduction of rice in the world's major grain-producing countries led to sharp drop in their exports

---

* Corresponding author.

D. Li and C. Zhao (Eds.): CCTA 2009, IFIP AICT 317, pp. 40–48, 2010.
© IFIP International Federation for Information Processing 2010

of agricultural products, which resulted in skyrocketing food prices, violence and even bloodshed in some areas of food-importing countries. From January to October in 2007, some cities in southern China saw pig fever, which led to less pigs farrowed, high mortality and sharp reduction of pig population. Pig population reduction triggered rising meat price at 22-26 Yuan per kilogram in October, which then quickly spread nationwide. Many ordinary Chinese residents missed a meal of pork and complained a lot. Rising price of agricultural products promotes the CPI to a large degree. The National Bureau of Statistics CPI data showed that in February 2008 CPI rose by 8.7% over the same period last year, and rose by 2.6 percent at ring growth. Among them, food prices rose by 23.3%, non-food prices rose by 1.6%. The above information shows the inadequate preparation for agricultural and livestock price rise led to a crisis.

Markov Chain and multistage fuzzy comprehensive estimation are applied in forecasting of the agro-products price (Yu pingfu, 2005). Time serial model is used to forecast pork price in Henan Province, China (Sheng Jianhua, et al., 2008).

At present, the forecast method for the price of agricultural and livestock products mainly based on the past products price and work experience, which can't forecast products price accurately and timely in some situation. Therefore, it is necessary to propose the agricultural and livestock products price prediction model in case of emergent agriculture, animal and husbandry disaster, so that decision-making department can foretell and adopt a positive response.

2   To Monitor and Analyze the Information of Agricultural Products Growth Based on RS and GIS

By using the correct high-resolution MODIS remote sensing images and professional remote sensing analysis software Erdas Imagine we can analyze the related information of farming land, such as the health state of growth, crop acreage etc. Establishing geographic information systems of the prices of agricultural products based on Map X 5.0, we can make analysis in the system and produce thematic price maps of agricultural products, measure the volume of growth area of local crops, and achieve database connection.

3   The Forecast Model of Agricultural and Livestock Products Price

In fact, a phenomenon is often associated with a number of factors, and it is more effective and realistic to predict or estimate a common dependent variable by a number of the optimal combination of independent variables rather than just one variable. In the regression analysis, if there are two or more independent variables, and each independent variable and dependent variable only has linear relationship, it is known as multiple linear regression. The prices of agricultural products are affected by many factors. Interactions between various factors are interrelated. Thus, the prediction method applies multiple linear regression method. Through the establishment of multi-linear equations for each of the influential factors as explanatory variables, forecast prices will be interpreted as a variable. Model is structured as follows:

$$Y = \beta_1 + \beta_2 X_1 + \cdots + \beta_k X_k$$

When prices of agricultural products are forecasted, the characteristics of various influential factors are taken into full account. The influential factors are divided into two categories: external factors and internal factors. External factors refer to the external environmental factors, such as population, transportation conditions and other factors. Internal factors refer to the reasons of agricultural products, such as agricultural yield and other factors. External factors and internal factors have different influences on the prices of agricultural products. Therefore, we use internal factors and external factors respectively to predict the prices of agricultural products. Finally, using the weighted method to integrate their predictive equation into the recent agricultural prices, the agricultural price prediction equation will be achieved. All the coefficient data of predictive equation will be integrated into equation set based on the recent data, and corresponding coefficient is achieved.

3.1 The Establishment of the Model

(1) Model I
Analysis the external factors that impact the prices of agricultural products. A number of these factors are used as explanatory variables, and forecasted prices will be interpreted as variables, price forecasts multivariate linear regression equation is constructed.

$$Y_1 = Q_1 + Q_2 X_1 + \cdots + Q_i X_i \qquad (1\text{-}1)$$

Where: Q_1, Q_2 and Q_i are the coefficients of the forecasting formula, X_1 and X_i are the external factors, Y_1 is the forecasting price of agricultural and livestock's products by external factors.

(2) Model II
Analysis the internal factors that impact the prices of agricultural products.
A number of these factors are used as explanatory variables, and forecasted prices will be interpreted as variables, price forecast multivariate linear regression equation is constructed.

$$Y_2 = P_1 + P_2 X_{i+1} + \cdots + P_j X_j \qquad (1\text{-}2)$$

Where: P_1, P_2 and P_i are the coefficients of the forecasting formula, X_{i+1} and X_j are the internal factors, Y_2 is the forecasting price of agricultural and livestock's products by internal factors.

(3) Model III
The factors of recent price of agricultural products are added to external factor prediction model as explanatory variables, and forecasted prices will be interpreted as variables, price forecast multivariate linear regression equation is constructed.

$$Y_3 = M_1 + M_2 X_1 + \cdots + M_i X_i + M_{j+1} X_{j+1} \qquad (1\text{-}3)$$

Where: M_1, M_2, M_3 and M_{j+1} are the coefficients of the forecasting formula, X_{j+1} is recent price of agricultural and livestock's products, Y_3 is the forecasting price of agricultural and livestock's products.

(4) Model IV

Based on the above three model structure, using the weighted method, the integrated prediction model is formed.

$$Y = K_1(Q_1 + Q_2 X_1 + \cdots + Q_i X_i) + K_2(P_1 + P_2 X_{i+1} + \cdots + P_j X_j)$$
$$K_3(M_1 + M_2 X_1 + \cdots + M_i X_i + M_{j+1} X_{j+1}) \qquad (1\text{-}4)$$

Where: Y is the finally forecasting price, K_1, K_2, and K_3 are the coefficients of each model. $K_1+K_2+K_3=1$. The selection of K value can be flexible according to the actual situation, focusing on the factors that need to be expressed.

In order to increase the accuracy of the forecasting result, we suggest the value of K_3 should be great than 0.33.

3.1.1 Model Applicability

The model can be applied to some local areas such as counties and townships, but also can be applied to other large areas such as big cities. As to agricultural products with a longer growth cycle and deeper influence, it can be forecasted within large areas (with province as a unit), which has great practical significance. Forecast time can have weeks, months, and years so on as units. Of course there is no practical significance if the forecasting time span is too short or too long. In addition, before the forecast, it is needed to determine the coefficients of the model structure based on the actual data.

3.2 Price Equilibrium Coefficient

In order to evaluate the prices of agricultural products, the concept of price equilibrium coefficient is proposed. The concept of price equilibrium coefficient C_0 refers to the extent of the prices of agricultural products that are acceptable to the residents. It is expressed through dividing the remainder of forecasted price subtracted by the pre-agricultural prices by the percentage of pre-agricultural prices expressed in absolute terms. The formula is as follows:

$$C_0 = \left| \frac{Y-P_1}{P_1} \times 100\% \right| \qquad (1\text{-}5)$$

$$C_0 = \left| \frac{Y-P_1*(1-3*C)}{P_1*(1-3*C)} \times 100\% \right| \qquad (1\text{-}6)$$

Where: Y is the forecasting price of the agricultural and livestock's products, P_1 is the recent price of the agricultural and livestock's products, C is current coefficient of CPI.

(1) When CPI<3%, formula (1-5) is applied.
(2) When CPI>3%, formula (1-6) is applied.

3.2.1 The Significance of C_0

According to large amounts of data I have read and my actual analysis, the significance of the actual values of C_0 is given, as shown in Table 1.

Table 1. The range and significance of value of C_0

The range of value of C_0	The significance of value of C_0
0~8%	The price fluctuation is reasonable
8~17%	The price fluctuation problem is minor
17~30%	The price fluctuation problem is major
30~55%	The price fluctuation problem is serious
More than 55%	The price fluctuation problem is amazingly serious

When confronted with the actual situation, for agricultural products are different from one another, the value of C_0 is possibly different. If necessary, C_0 can be multiplied by the values ranging between 0.2 and 0.7, so that a more accurate price equilibrium exponent will be achieved.

3.3 The Regulatory Function of the Forecast Model of Agricultural Products Price

The government policy makers can make corresponding macro regulations based on evaluated value of price equilibrium coefficient and actual condition, so that the price equilibrium coefficient can return to safe scope. The steps are as follows:

(1) The reasonable price is got by abductive reasoning, based on reasonable price range of price equilibrium coefficient.
(2) Aided by integrated evaluation model, we can get the factor-adjustable quantum by analyzing easily adjusted factors and adjusting single factor.
(3) Re-predict the price of agricultural products and conduct price evaluation.

For instance: When prediction shows the price equilibrium exponent of marketable rice ranges from 17% to 30%, the price fluctuation problem is comparatively serious over a period in the future, we can calculate the rice quantity we need to add to market by abductive reasoning, re-predict and evaluate the rice price based on reasonable price level and aided by integrated model structure.

4   The Forecast Model of Pork Price

From year 2008 to 2009, pork price in China had been fluctuated obviously. The pork price is not only related to the amount of pork production, but it is also affected by transport costs, population and economic ability, etc. Here we can apply the forecasting model of agricultural and livestock products price to forecast the pork price.

In order to establish the model, we need to consider many factors which affect the pork price, such as the volume of pig slaughter, the volume of pig stock, transportation cost, local non-agricultural population, per capita GDP, the amount of per capita consumption on pork, food price for pigs, recent pork price and so on.

SPSS (Statistical Package for the Social Science) is one of well-known statistical analysis software packages for social sciences in the world. It can carry out multiple linear regression analysis. Applying SPSS software analysis, we can get the following models.

(1) Model I

This model reflects the relationship between pork price and external factors, such as non-agricultural population, GDP, amount of per capita consumption on pork and so on.

$$Y_1 = -0.11X_1 + 0.068X_2 - 0.02X_3 + 9.174 \tag{1-7}$$

Where: X_1 is non-agricultural population in forecasted region (unit: 10,000 people), X_2 is per-capita GDP (unit: 10,000 Yuan), X_3 is the amount of per-capita consumption on pork (unit: Yuan).

(2) Model II

This model reflects the relationship between pork price and internal factors, which include the volume of pig stock, the volume of pig slaughter, food price for pigs and transportation cost.

$$Y_2 = -0.467X_4 - 0.251X_5 + 1.254X_6 + 0.374X_7 + 18.529 \tag{1-8}$$

Where: X_4 is the volume of pig slaughter (unit: 10,000 pigs), X_5 is the volume of pig stock (unit: 10,000 pigs), X_6 is the price of food for pigs per kilogram (unit: Yuan), X_7 is the cost of transportation per pig (unit: Yuan).

(3) Model III

By introducing the factor of recent pork price, then we got the following model III based on model I.

$$Y_3 = 0.002X_1 - 0.12X_3 + 0.771X_8 + 2.75 \tag{1-9}$$

Where: X_8 is recent pork price per 500 grams (unit: Yuan).

(4) Integrated Model IV

Based on the above three models, we got the following integrated model IV.

$$Y = K_1*(0.11X_1 + 0.068X_2 - 0.02X_3 + 9.174) + K_2*(-0.467X_4$$
$$-0.251X_5 + 1.254X_6 + 0.374X_7 + 18.529) + K_3*(0.002X_1 - 0.12X_3$$
$$+0.771X_8 + 2.75) \tag{1-10}$$

According to the simulated data shown in Table 2, we can get the forecasting price which is 10.91 Yuan per 500 grams by the integrated model IV. Because C is less than 3%, we can use formula 1-5 to calculate price equilibrium coefficient C_0. As a result, the C_0 value is 9.1%, indicating the pork price fluctuation will cause a small problem in the next period in the city. If no measures are taken, price will increase, thus giving a fillip to CPI index.

Table 2. Simulated Data used in Model

Factor Name	Factor Value	Factor Name	Factor Value
non-agricultural population in city	1,500,000	price of food for pigs	2.5Yuan/kg
present amount of pig stock	300,000	transportation cost	12 Yuan per pig
present amount of pig slaughter	70,000	present pork price	10 Yuan per 500g
current CPI value	2.5%	K1	0.25
K2	0.35	K3	0.4

4.1 The Regulation Function of the Forecast Model of Pork Price

The factors which influence the price of pork are as follows: the volume of pig slaughter, the volume of pig stock, etc. The population of pigs and slaughtered ones are macro-controlled through comprehensive analysis of these adjustable factors.

(1) Through the abductive reasoning of formula C_0, reasonable pork price is 10.8 Yuan per 500 grams.

(2) Based on formula 1-10, we calculate that 3000 slaughtered pigs need to be increased, and pig stock population increases by 8000.

(3) The re-forecasted price is 10.79 Yuan per 500 grams, C_0 exponent is 7.9%, and the price is a reasonable price.

Through the forecast and adjustment process mentioned above, a potential crisis is resolved successfully. This not only provides convenience to the government, but also brings about tangible benefits to the residents.

5 Conclusion and Discussions

Through the study on the forecasting model of agricultural and livestock products price and its application in pork price with simulated data, we can get conclusions as follows:

(1) With multiple linear regression method we can predict pork price. The results in SPSS show that there is a high correlation between pork price and influential factors. Considering that agricultural and livestock products price is correlated with its own output, we can apply the present method to the price prediction of other products.

(2) The study on agricultural and livestock's products price fluctuation evaluation can evaluate the predicted price. According to the price fluctuation evaluation, the predicted products price can be evaluated so that price warning can be made.

(3) The adjustment of agricultural and livestock products price can play a very important role in the government's management on agricultural and livestock products price. This paper provides the process how pork price is adjusted with the help of the model. The government can predict the time when the price would change and control its change by regulating its output or other related factors.

More practical experiments are needed to enhance the model's accuracy. In addition, the prediction model is set up based on the simulated data in Hebei province, China. For the sake of the model's wider application, the real data of related factors is needed when it is applied to predict agricultural and livestock products price in some specific areas.

Acknowledgements

Great thanks go to Professor Stefano Bocchi from the University of Milan for his guidance on research methods and ideas. Besides, thank a lot the Hebei Academy of Agriculture and Forestry Sciences for the help provided in the application.

References

Marion, B.W., Geithman, F.E.: Concentration-price relations in regional fed cattle markets. Review of Industrial Organization 10(1), 1–19 (1995)

Yue, E., Jianbing, Z., Yeping, Z., Kaimeng, S.: Agriculture economy information service system based on embedded GIS. Computer Engineering 34(23), 269–271 (2008) (in Chinese)

Kanis, E., Groen, A.F., De Greef, K.H.: Societal Concerns about Pork and Pork Production and Their Relationships to the Production System. Journal of Agricultural and Environmental Ethics 16(2), 137–162 (2003)

Ehui, S., Polson, R.: A review of the economic and ecological constraints on animal draft cultivation in sub-Saharan Africa. Soil Tillage Res. 27, 197–210 (1993)

Juan, G., Huimin, L.: An Analysis of the Trend of Impact of Agricultural Product Price Fluctuation in China. Social Science Journal of Shenyang Agricultural University 9(2), 144–147 (2007) (in Chinese)

Kiers, H.A.L., Smilde, A.K.: A comparison of various methods for multivariate regression with highly collinear variables. Statistical Methods and Applications 16(2), 193–228 (2007)

Apergis, N., Rezitis, A.: Mean spillover effects in agricultural prices: Evidence from changes in policy regimes. International Advances in Economic Research 9(1), 69–78 (2003)

Chunling, P.: Analysis and suggestion: thirty years of fluctuation and development of live-pig production of Liaoning province. Social Science Journal of Shenyang Agricultural University 10(6), 649–653 (2008) (in Chinese)

Chakrabarty, S., Chopin, M., Darrat, A.: Predicting Future Buyer Behavior with Consumers' Confidence and Sentiment Indexes. Marketing Letters 9(4), 349–360 (1998)

Jianhua, S., Gongpeng, C., Xixin, J., Guangzhi, R., Tengyun, G., Xiaofeng, L.: Study on the change rule and forecast of pork price in Henan Province. Journal of Henan Agricultural University 42(6), 672–676 (2008) (in Chinese)

Jagtap, S., Amissah-Arthur, A.: Stratification and synthesis of crop-livestock production system using GIS. GeoJournal 47, 573–582 (1999)

48   W. Zhang, H. Chen, and M. Wang

Bocchi, S., Annamaria, Castrignanò: Identification of different potential production areas for corn in Italy through multitemporal yield map analysis. Field Crops Research 102(3), 185–197 (2007)

Pingfu, Y.: Study on the Construction and Application of Model Prediction System of the Agro-products. Journal of Anhui Agricultural Sciences 33(7), 1284–1286 (2005) (in Chinese)

Zanias, G.P.: Inflation, Agricultural Prices and Economic Convergence in Greece. European Review of Agricultural Economics 25, 19–29 (1998)

A Network-Based Management Information System for Animal Husbandry in Farms

Jing Han[1] and Xi Wang[2,*]

[1] College of Information Technology, Heilongjiang August First Land Reclamation University,
Daqing, Heilongjiang Province, P.R. China 163319
[2] College of Engineering, Heilongjiang August First Land Reclamation University,
Daqing, Heilongjiang Province, P.R. China 163319,
Tel.: +86-459-6819224; Fax: +86-459-6819224
hanj1202@163.com, ndwangxi@163.com

Abstract. It has already been a trend for the management of animal husbandry in farms to employ the advanced management system(MIS)and the software-developing platform to scientifically manage the information about animal husbandry in farms through the network. This is a network system that combines B/S structure and ASP techniques, and the method adopts both the computer network technology and database technology to implement system and data integration. Based on the characteristics of the system, the security of the management information system is studied. The security strategy model of the MIS is proposed, and the technical strategy of the MIS's security is discussed in the level of the application software and the database. And the security management method of the B/S structure is introduced briefly.

Keywords: management information system, ASP, animal husbandry information, network technology.

Currently, the economy of agriculture and rural areas of China is now developed to a new historical phase, and animal husbandry has gradually evolved into the preponderant and predominant industry in the economic development of rural areas (Kou Zhanying). In the agricultural production of the reclamation areas in Heilongjiang Province, animal husbandry accounts for a significant proportion. The problems of traditional method of animal husbandry information management have inevitably come up, such as inadequate in information, low in efficiency and slow in feedback. The informationization of animal husbandry is a way to spur the modernization of animal husbandry through informationization, combining information technology, network technology, artificial intelligence technology, computer technology and modern management theory. And this will have far-reaching significance to promoting the economy of rural areas and agricultural development. Establishing a management information system of animal husbandry (MISAH) is beneficial to raising the efficiency of information work in animal husbandry and quickens the progress of animal husbandry informationization. And setting up such a system in Heilongjiang Province is

---

* Corresponding author.

D. Li and C. Zhao (Eds.): CCTA 2009, IFIP AICT 317, pp. 49–54, 2010.
© IFIP International Federation for Information Processing 2010

also conducive to increasing the degree of local husbandry information management, and making it more normalized and routinized, thus providing reliable foundation for the informationization of animal husbandry in the overall reclamation areas. Building up such a system can make full use of the resource of animal husbandry in Heilongjiang province, steering it into creating greater value for production of animal husbandry, and it can also meet the needs of developing modern animal husbandry, thus exerting considerable influence on the economic development of the country.

1   The Analysis of the System and Overall Design

1.1   The Concept of the Design of the System

The system has as its developing object the Red Star farm in the reclamation area of Heilongjiang province. According to its functions, the management information system (MIS) can be divided into four comparatively independent modules. For the individual module, its needs are easy to realize. Therefore, based on the above-mentioned analysis and comparison and based on the realistic situation of the Red Star Farm, it's decided that the development of the system will utilize the method of prototype and that of structural life period. The former analyzes the needs, with the modified and identified prototype system as the basis for developing the system, and the latter is used to design. Standard datasheets are adopted in the process of system-designing, the transmission of data will be realized through the system of data exchange, and the standard of statistics in the datasheets will be normalized, to ensure the normalization of the data and the safety of transmission.

The reclamation area in Heilongjiang province is vast in its extent, and naturally the economic situations and natural environments are diverse, so the information system platforms of animal husbandry to be developed should have different focus. The system is designed according to the characteristics of the Red Star Farm, and the following principles are followed in designing and developing the integrated platform of MISAH.

(1) The commonality should be taken into account in designing the system platform, and its content should be combined with the characteristics of the local place, giving prominence to its direction, thus making it convenient to use for the animal husbandry department and people engaged in animal husbandry.

(2)The operation of the project should be simplified, functionalized, giving full play to the role of information system platform.

(3) The system can analyze, sort out and mobilize the information on animal husbandry in the Red Star Farm.

(4)The system upholds the graphical user interface designing principle, the interface is direct-viewing and transparent to users.

(5)When users use the software, they can fully understand the functions of the interface, and without much training they can conveniently use the system.

The system can be applicable to the other farms with simple modification.

1.2   The Analysis of the Functions of the System

According to investigation, analysis and categorization of the needs of MISAH, it can be divided into the subsystems of animal husbandry production management, of

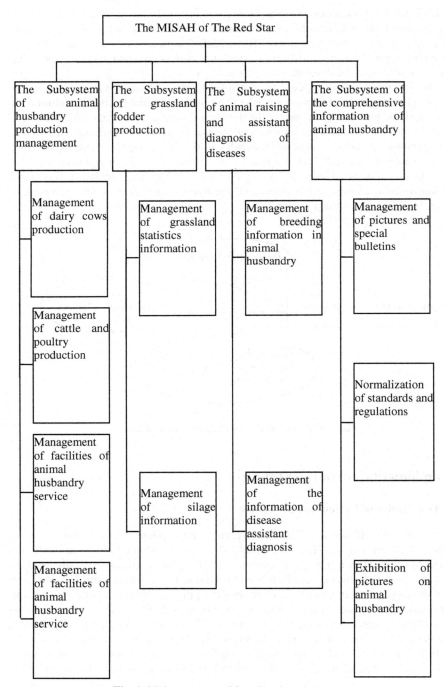

Fig. 1. Main structure of functional module

grassland fodder production, of the animal-raising and the assistant diagnosis of diseases, and of the comprehensive information of animal husbandry, with every sub-system falling into several modules. Thus, such functions can be realized such as information input, search, modification, back-up, collection, data analysis and infor-mation classification. Its diagram is shown in Fig.1.

The system designate three different roles to various users: system administra-tors, the administrators of subsystems (the personnel in animal husbandry administra-tion of farms), and common users (the farmers engaged in animal husbandry and plantation in the subordinate institutions of farms). System administrators can be held by the persons in charge of animal husbandry in the farm, and their main function is managing animal husbandry information of farms and the limits of rights of the other users. Administrators of subsystems mainly add, delete and modify information. Common users can browse and inquire all the open information of the management system. Its diagram is shown in Fig.2.

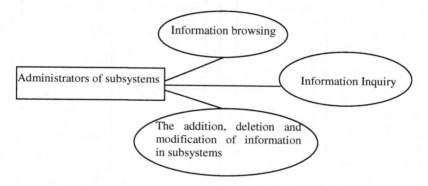

Fig. 2. Case Diagram of subsystem administrators

2 Implementation of the System

2.1 The Choice of Systematic Structure

The structural development of MIS has experienced the phases of Client/Server mode and Browser/Server mode. The traditional C/S structure adopted the open mode, but the openness is only limited to development of the system. Because it failed to pro-vide users with the real expected open environment, the software with C/S structure had to develop software of different versions to apply to different operating systems. In addition, its efficiency is low while the price is high. In B/S structure, users' work-ing interface is realized though WWW browsers. The main working logic is realized in the Server, with little of it done in Browser, thus forming the so-called 3-tier struc-ture. In this way, the load in the client is simplified, decreasing the cost and working load of system maintenance and ultimately the TCO of users (Liu Huaquan et al., 2000).

If the system is successfully developed, the users basically need not to be trained. Their needs can be satisfied though the universal browsers and by clicking the mouse. The structure has low requirement for clients' hardware, with high degree of information resource sharing and of expansibility. And lots of users have access to wide area network. Based on these features, the system has developed and adopted B/S structure.

2.2 The Choice of the Working Plat Form for the System

Based on the analysis of the functions and roles of the system, many modules can be established, which are both inter-independent and interrelated. The system working platform adopted Windows 2003 Server, and background database utilized network database SQL Sever 2000 with relation model. The designing method used one of the standard database designing methods: New Orleans designing mode, which classifies designs of database into four stages: analysis of needs, conceptual design, logical design and physical design. And after the database is designed, there are two other stages: that of implementation and operation and that of operation and maintenance.

Conceptual structure design is the key to the overall database design; it refers to the process of transforming the application need got from the analysis of needs into information structure or conceptual model, and optimizing the model. The entities designed according to the above-mentioned principles include: Diary Cows Production Monthly, diary cows statistics in the end of the year, families raising more diary cows than the others, vets and technicians, facilities of animal husbandry service, information of grassland statistics, the kinds of silage and their planting methods, the plantation and harvest of silage and the plantation of silage at the end of the year and the statistics of its storage.

To guarantee the inner links between different modules and the consistency of data, in designing database, all the datasheets in the system will be managed in one database file. In the process of designing database, the fields in every datasheet have been fully optimized and assembled, so that the redundancy of the data can be reduced and the efficiency of data-utilizing can be improved.

3 The Security of the System

The safety management includes the safety of the operating system of the network server, the safety of the application program and the safety of database. The following aspects should be considered in designing.

3.1 Safety Designing of Infrastructure

(1) The back-up system: The system or the database can break down in the process of operation, and the common breakdown includes working breakdown and system breakdown. To prevent the appearance of incorrect data in the database, or to renovate the destroyed data in the database so that the database can restore to the condition of consistency, the system must be backed up regularly.

(2)To install anti-virus software and firewall to prevent virus: Installing anti-virus can monitor the operating situation of the program in the computer and can find out and wipe out the virus program, thus protecting the safety of the computer system and the information. To install firewall can avoid the visits of information from unauthorized users and programs.

3.2 Safety Design of the system

(1)Optimizing programming to avoid the illegal invasion of Web application program. Web page parameters or data shouldn't be transmitted directly, instead, they should be transmitted through component technology to achieve the purpose of conceal the parameters.

(2)The classification of the users being considered , the limits of rights should be allocated according to different users, meanwhile different operating interface should be set up to control them. To realize authority differentiation in visits, user's name and password are to be set up for subsystem administrators and their superiors, and the method of access control and identity verification should be adopted. Besides, every user should not exceed his authority to visit the contents unauthorized.

(3)To prevent the data from being accepted illegally during transmission, the system used the data encryption technology (Zhenli GU, 2005).

4 Conclusion

The MIS was developed according to the practical needs of the Red Star Farm in Heilongjiang province, and has its practical significance and value. The system adopted B/S structure, constructing a open comprehensive management platform of information resource for the Red Star Farm, greatly improving the efficiency and level of animal husbandry management in the farm. In designing, effective method of database technology solved the sophisticated data-processing process. The ultimate goal to develop MIS is to provide subsidiary decision-making information. The system has collected a large amount of object data in a planned way, and the data of the need for subsidiary decision-making has been abundant. Then the developing emphasis of the system would be to utilize the technologies of analysis and of data processing such as data mining technology to find out the hidden and inner valuable information, and to emphasize its analytical function of data and its roles of guiding the work and assisting decision-making, so that a more advanced decision-making and expert system can be developed.

References

Zhanying, K.: Review and Forecast about Animal husbandry of China,
    http://www.china-av.net
Huaquan, L., et al.: The Development of 3-tiers based B/S structure database management system. Microcomputer & Its Applications 19(3), 43–45 (2000)
Gu, Z.: Analysis and Implementation of SQL Server Database Technology for Remote Access. Computer and Modernization 8, 56–58 (2005)
Songtao, Z.: Development of Dynamic web site based ASP. Publishing House of Electronics Industry, Beijing (2006)
Ruijun, G., et al.: Examples of Database developed. Publishing House of Electronics Industry, Beijing (2005)

Research of Animal Disease Information System Based on GIS Technology

Hongbin Wang[*], Lin Li, Jing Dong, Danning Xu, and Jing Li

College of Veterinary Medicine, NorthEast Agricultural University,
Harbin,Heilongjiang Province, P.R. China, 150030,
Tel.: +86-451-55191940; Fax: +86-451-55190470
neau1940@yahoo.com.cn

Abstract. It is necessary to build animal disease information system in order to dynamically monitor and predict diseases distribution, and propagation, such as HPAI and FMD taking place, Powerful functions of data management analysis and display with the GIS can be applied oriented to animal disease information. This disease information platform can be applied to dynamiclly monitor and predict disease, as well as to support the disease prevention decision making, in order to decrease the loss.

Keywords: animal disease; GIS; information system.

1 Introduction

In recent years, HPAI and FMD diseases as well as other kind of animal diseases have caused disastrous result by disease outbreak and propagation in some areas of China, . With HPAI for example, in 2005 there were 13 areas having avian influenza outbreak. More than 22 million live birds were slaughtered, while 7 persons were infected in China with bird flu, including five deaths. These are seriously hampered the development of livestock and poultry farming. As a result, it is most important to construct an animal disease information system for dynamic monitoring disease distribution and spread. Animal medicine is a subject with multi-branches involving micro-structure as well as macro-system, of which a great amount of data in relation to the macro system are blessed with the features of spatial distribution. It is evident that the occurrence and epidemic of infectious diseases, the distribution and etiology of endemic, the frequently-occurring disease with regional feature, and the animal husbandry and veterinary health institutions are all closely related to the distribution of spatial information Animal disease data relating to the spatial feature is the prerequisite condition for the application of GIS (Jiang Chenghua, 2004).

GIS (Geographic Information System) is a computer aided spatial data information management system. Generally speaking, the application of GIS enables the spatial data management (collection, storage maintenance), implements the analysis (statistics spatial modeling) and graphics display (figures and maps) (Xu yong, 2006). GIS

---

[*] Corresponding author.

D. Li and C. Zhao (Eds.): CCTA 2009, IFIP AICT 317, pp. 55–59, 2010.

integrate the unique map visual effect with the geographical analysis and database operation. It is therefore the important means to realize the decision support management system (Norstrom M, 2001). In the field of animal hygiene, many state agricultural departments are aware of animal hygiene database and the importance of the construction of GIS in order to display, summary and analyze data of animal's epidemic situation. For instance, in 1996 USA established, the "State Animal Health Report System", at the same time Canada, Australia, New Zealand, Thailand, Malaysia, and other countries have operated respectively the similar system with perfective functions. Our Government has also awared of this requirement, and, established the National Animal Epidemiology Research Center in 2000(Teng Xiangyan, 2005). With a special fund, a national animal health information system has been established integrating health animal health information management and animal health GIS. However in the provinces and cities of our country the livestock diseases geographic information system is hardly established and put into application. Domestic and international GIS application in health research indicates that the technology animal diseases geographic information system based on GIS is an inevitable trend for animal disease information management with scientific and automation feature(Pferiffer D U,2002).

2   Materials and Methods

2.1 System Target

(1)Integrating correlated data information about livestock and poultry, establish a multi-source databases covering whole province which includes the data of graph,, the data of image and the data of statistic.

(2)Establish and develop a livestock epidemic monitoring information system according with the state standards, and applying GIS technology with web technology. Provide technical support for the construction of provincial units with no provisions for animal epidemic.

(3)Provide the multi-level information service about related information to the distribution, occurrence and prediction of disease for the provincial animal husbandry, veterinary department, vet workers and farmers.

(4)Provide decision support for the provincial animal husbandry and veterinary department.

2.2 System Framework and Basic Functions

(1)GIS platform

The GIS support platform is established with ESRI company's groupware GIS package Mapobjccts (hereinafter referred to as the MO). The application system is an user-oriented program by Visual Basic language 6, based on the Windows operating system.

(2) The spatial database and attribute database drawing the corresponding thematic maps by using GIS, based on provincial 1:1 million electronic map, and the spatial database and attribute database which established by the obtained epidemic monitoring information. The information of spatial database digital marks according to existing standard map. Including the foundation geographical data such as administrative

divisions, highway, railway, rivers, lakes and special maps such as animal population spatial distribution,nursery geographical distribution, space scatter diagram after animal epidemics occurrence, the space distribution of transmission intermediary, etc. Attribute database, including soil water, traffic condition, the density of livestock and avian, genius morbi, epidemic disease characteristics, the size of veterinary hospital, etc. Data is available for management by Access database which may link through the SQL.

(3)System structure

Animal diseases information system diagram is shown in figure 1

System is divided into two major modules: daily management module and decision support module, and divided into three subsystems, and each has its characteristics and functions.

Disease monitoring and management subsystems: which can deal with databases, such as natural resources, animal resource information database, disease database, and make them visualization, scientific, process optimization.

Diseases of comprehensive evaluation subsystems: determine the optimal breeding structure and the breeding density, enact reasonable prevention and control strategies and scheme in key prevention and control areas, through the space analysis and space statistics of relevant data of animal diseases.

Emergency decision of Epidemic outbreak subsystem: scientific and reasonable take emergency measures, reduce the cost, improve the decision-making response speed and accuracy by epidemic situation, GIS technology and many analysis method.

(4)Introduction of functions of the system

①Realizing data acquisition, comprehensive inquiry, statistics analysis, and visualization of management

Visual management of livestock and poultry resources, disease spatial databases and attribute data, visualization with epidemic areas data and the reports: information of animal epidemic disease and no provisions for animal epidemic (e.g., husbandry statistical data: veterinary data, immune data etc.)When the map clicked by the attributes tools, this area highlighted and a new form showed with all relevant information of livestock and avian. The data can be printed and forming statements.

Be involved in the management of information and data types and format, data security, data transfer, statements statistics and infuse information system, through the Internet, realize the real-time data acquisition and the outbreak of information flow of information transmission, epidemic disease timely pass the text and images site information, and can be distributed laohutai monitoring and information processing the livestock and poultry. Through the management and livestock data report, realizing the function of data input and report to ensure accurate data and corresponding relationship.

Related information of the livestock can be comprehensive inquired. When the searching information of livestock finished, the results will come from some area that the number of livestock is bigger than a certain data. These areas are presented on map by highlight, this is also applicable to the animal diseases number query a value greater than the query. Make the special map which about total incidence of animals, total loss of animals by the reports of any period of epidemic disease. The information of livestock such as variety distribution, personnel distribution, but also provides the technical support for future outbreaks decryption release.

②System management of animal husbandry related laws and beforehand library, provide reference for instituting policies and emergency decision-making

Collect the information that about the relevant laws, regulations and standards of animal sanitation which issued of all levels Chinese governments and the relevant departments since 1949, andand various laws and regulations issued of international organizations and other governments. Computerized management of the information systematically and comprehensively. Provide reference for instituting policies.

③Analyze the resource of livestock and avian and the space-time information of disease. Provides the basis for livestock production, macro-control of disease control and prevention.

By using spatial analysis function of the GIS (stack analysis, density analysis, cluster analysis, and the dynamic analysis), the human resources and the distribution of resources for livestock and avian can be overall assessment reported, predict distribution of disease for the whole animal diseases regional according to the disease data has been collected and report. Determine the scope of the services by the distance factors, consider the position of existing hygienic medical institutions and radius of services for a variety of factors, and allocate the limited health resources reasonable. Considering the geographical environment, the distribution of existing institution, potential distribution of requirement and other factors, select location scientific by superposition, density and reclassification and other analysis method, provide evidence for selecting the address of new breeding base and veterinarian institutions, etc.

④Realizing sudden outbreak positioning, diffusion trend visual simulation, urgent resources optimization ration

Obtaining the information of epidemic-stricken area promptly, accurately, exhaustively confirming livestock epidemic areas it can be divided into several levels showing visual effect of the epidemic area the rational resources and personnel deployment through the hierarchy of epidemic allocated more accurately rational allocating resources

Search the adjoin area of epidemic disease spot Simulate the spread trend of the epidemic disease It may indicate Anti-epidemic station surrounding areas well prepared for prevention and quarantine and the spread of disease Simulate the spread trend of the epidemic disease by some factors such as the choice of livestock epidemic disease in quantity make for prophylaxis

Provide decision support to control epidemic disease by rationing the urgent resources optimally and analyzing epidemic resource information provide scientific control measures for occurred the disease Calculate the distance from the genesis of disease to the epidemic prevention station and provides the optimal path to the epidemic area of infectious disease by network analysis method. Through calculation of the distance, Estimated that veterinary officers to the time the disease occurred, A timely manner in order to control the epidemic. Select the optimal route to the destination, saving the emergency cost.

3 Summary

Concerning with the development of livestock influenced seriously by the major disease, poultry aquaculture, this research work has implemented dynamic monitoring for livestock diseases distribution and spread based on GIS animal diseases information system,, applying GIS spatial analysis function with network technology..

By combining the GIS technology with network technology, the telemetry data, and the management information from municipal (county) stations are integrated. It is therefore able to monitor the livestock epidemic disease, and to supervise epidemic-prevention, and livestock supervisory information in a regular and systematic statistical base. It is also able to grasp the dynamic trend and make data resources visualization. The historical disease condition, spatial distribution rule, etiological analysis of ecological environment, the prevention of disease and reasonable allocation of resources functions may then be analyzed by synthesis. Through the analysis of the deeply used information, the macro management and scientifically decisions of livestock diseases is strengthened. The government administration departments, various monitoring institutions, and scientific research units understand livestock diseases more deeply, and start livestock diseases prevention work with scientific basis.

This system not only provides technology for an animal disease management for the provincial animal husbandry and veterinary department, but also provides good application and technical reference for construction of our animal diseases geographic information system.

Acknowledgements

The authors thank the financial support from National Project of Scientific and Technical Supporting of China during the 11th Five-year Plan 2006BAD10A02-04.

References

Chenghua, J., Yang, D.: The application of geography in medicine geography research. Overseas medicine geography fascicl 25(4), 182–184 (2004)

Yong, X., Siqing, Z., Yang, Y.: Research and application of geographic information systems for Prophylaxis and control disease. Disease monitor 21(1), 45–47 (2006)

Norstrom, M.: Geographic information system(GIS) as a tool in surveillance and monitoring of animal disease. Acta veterinaria Scandinavica Supplement 94, 79–85 (2001)

Xiangyan, T., Baoxu, H., Xueguang, Z.: The application of GIS in animal health domain. Journal of Chinese veterinarian 41(6), 58–60 (2005)

Pferiffer, D.U., Hugh-Jones, M.J.: Geographical information systems as a tool in epidemiologicl assessment and wildlife disease management. Rev. Sci. Tech. Off. Int. Epiz 21(1), 91–102 (2002)

Greenough, G., McGeehin, M., Bernard, S.M.: The potential impacts of climate variability and change on health impacts of extreme weather events in the United States. Environ. Health Perspect. 109(suppl. 2), 191–198 (2001)

Modern Agricultural Digital Management Network Information System of Heilongjiang Reclamation Area Farm

Xi Wang, Chun Wang*, Wei Dong Zhuang, and Hui Yang

Engineering Collage, Heilongjiang August the First Reclamation Land University,
Daqing, Heilongjiang Province, P.R. China 163319,
Tel.: +86-459-6819224; Fax: +86-459-6819224
wangchun1963@126.com

Abstract. To meet the need of agriculture management modernization of Heilongjiang reclamation area, further boost large-scale integration level of modern agriculture production and boost management level of agriculture production.On Red Farm, we have established the digital management network information system with the remote sensor technology, GIS technology, GPS technology, database technology, network technology, agriculture intelligent technology, multimedia technology, information auto acquired technology and control technology applied in the system. Modern agriculture digital information system of Red Star farm is composed of base construction of agricultural digital information, digital management system construction of agricultural production, digital technological equipment of agriculture etc. The digital and network management of agriculture can offer all management department the best convenient to master management information in time and boost the technological level of agricultural production, form the digital technological system of farmland, explore new way of agricultural production in information age and seek new production way of high efficiency, high production, high quality, low consumption.

Keywords: agricultural production, agricultural information, network system, geography information.

1 Preface

At present, Heilongjiang reclamation area has already established the largest scale national farmland group with the highest degree of mechanization. Heilongjiang reclamation area takes modern agricultural equipment as guidance and development modern agriculture firmly. According to the need of agricultural modernization of Heilongjiang reclamation area, to further boost large-scale integration level of modern agriculture production, boost management level of agriculture production, establish the digital management network information system. The digital and network

---

* Corresponding author.

D. Li and C. Zhao (Eds.): CCTA 2009, IFIP AICT 317, pp. 60–64, 2010.

management of agriculture can offer all management department the best convenient to master management information in time and boost the technological level of agricultural production.

2 Farmland Information System of Agricultural Production Management

Farmland information system of agricultural production management is composed of management schedule center of agricultural production, remote sensor material of farmland, GIS, crop farming digital information system, digital management information of agricultural machine, irrigation digital information system of farmland, forestry digital information system of farmland, stock raising digital information system of farmland, weather information system etc (Zhuang Weidong et al., 2005).

2.1 Management and Schedule Center of Agricultural Production

The management and schedule center is cored by management information system of agricultural production, 3S (GPS, GIS, RS) technology and network technology. The management information of agricultural production and video monitor information is displayed on the 126 inches plasma board. The screen constitutes 3*3 42 inches plasma screen and the size of the screen is 2880 * 1670 *176 mm. The screen is equipped with a RGB matrix, a video and audio matrix, plasma screen control computer and operation display computer of agricultural production management schedule.

2.2 Remote Sensing Information System of Farm Blocks

Purchase the satellite remote sensing images covered the whole farmland with 2.5 meter resolution ratio and measure at the certain point on the ground. After geometry adjustment, the ground position space information of farmland including farm blocks, road, irrigation, forestry, reservoir, residence area is gained to transfer the remote sensing material to system server. Through GeoBeans network Web GIS software, by using network browser, the operation including zoom in, zoom out, move, distance measurement, land block area measurement can be practiced to facilitate the overall planning and production management of farm.

2.3 Geography Information System of Farmland

Establish the farmland time and space database through deciphering the remote sensing images, adjusting the demonstration area space data, and acquiring the crop planting production information in relation with space data. Introduce GeoBeans Web GIS software developed by Remote Sensing Institute of Chinese Academic of Science and the function including zoom in, zoom out, move, farm block inquiring, the longitude and latitude of current position, distance measurement, land blocks area measurement can be practice. The agricultural information management and space information visualization can be realized. Input the farmland blocks map and previous land number to computer, and base condition and geography position of certain number block can be inquired on the computer.

2.4   Network Information System of Agricultural Machine Management

Network information system of agricultural machine management is composed of agricultural machine management database, GPS dynamic tracing and scheduling system, remote network video monitoring system, working machines wireless remote video monitoring system, GSM short message group sending system, working progress statistic system, single working machine assignment statistic system, agriculture machines pervious statistic data and agriculture machine usage experience exchange system(Li Qiang et al., 2007).

2.5   Crop Farming Digital Management Information System

The system includes geography information system of farmland, geography information system of soil nutrient (N, P, K, PH value, organic matter), soil nutrient database and land blocks file database information. The computer management and network support of agriculture data information and data resource sharing provide reliable information and data support for decision analysis of agriculture production management. Design crop farming digital management web pages for Red Star which include brief introduction of crop farming, policy and law for agriculture production management, agriculture resource photos, agriculture production management photos, agriculture production propaganda photos, announcement and news distribution, contact ways and responsibilities of agriculture production management department.

2.6   Digital Management Information System for Organic Agriculture Production

The system includes inquiry of organic production executive standard information, inquiry of organic crop production operation regulation, inquiry of organic crop disease prevention and cure, traceable system of organic production etc.

2.7   Digital Management Information System for Farmland Irrigation

On the base of investigation and research to production management and water resources on Red Star farm,we established specialized database and resource database and form the digital management information system of farmland irrigation for Red Star farmland. The system includes farm irrigation geography information system and natural irrigation resource information database. Establish relative database information management system through accumulation of scheme and technology step by step. The computer management and network support of irrigation data information and data resource sharing provide reliable information and data support for irrigation management decision analysis.

2.8   Digital Management Information System for Forestry

Digital management information system for forestry is cored by "3S" technology which is data sharing information system supported by network technology and database technology. The system includes forestry GIS, information database of forestry resource information, forestry ecological management and protection, forestry production management, forestry fire prevention, afforestation and policy and law information. Establish relative database information management system through accumulation of scheme and technology step by step to achieve the computer management and network support and data resource sharing. The system can provide reliable information and data support for forestry management decision analysis through inquiry of forestry network database.

2.9 Digital Management Information System for Stock Raising

Stock raising information industry is the development direction of modern stock raising and it is the only way of sustainable development. Establish stock raising information which can drive the modernization of stock raising. The system includes stock raising information system, stock disease assistant diagnose system, milk cattle management and standard specification and laws which are significant for agriculture and country economic development.

2.10 Weather Information System

There is a close relationship between weather and agriculture and the master degree of weather information can affect the agriculture production efficiency directly. The weather information system can help farmers and manager to master the local weather information and make scientific production decision. The system includes weather forecast, agriculture weather, artificial weather information and weather magazine.

3   Usage Effects of the System

Modern agriculture development center of Red Star farm in Heilongjiang reclamation area takes "stand on digital agriculture and develop modern agriculture" as construction logos. Save resources and gain profit to the largest extent through GPS technology, GIS technology, RS technology and high technology investment and management. Agriculture production schedule and command and calculation by use of network information system transform agriculture management from original distributive and extensive management to present integrative and network management. The system provides a convenient way for producers and management staffs to learn working situation and income and expenses and daily dynamic information of agriculture production about the whole farm on the web.

4   Conclusion

Red Star farm in Heilongjiang reclamation area can take advantage of large scale farmland modern agriculture production and boost agriculture production management level and bring the farm agriculture production to network, digital and information age through the modern agriculture digital management information system. The system can make agriculture production more scientific, standard, quantitive and more efficient and force the optimization of agriculture economic structure and transformation of agriculture growth. The system opened the first modern agriculture production management mode for Heilongjiang reclamation area under new system and new situation.

References

Weidong, Z., Chun, W., Xi, W.: Design and development of precision agriculture web of Heilongjiang Reclamation area. Research on Agriculture Machine 14(4), 251–252 (2005)

Qiang, L., Xi, W., Weidong, Z.: Research on Agriculture Machine Information Management System based on Network. Agriculture Network Information 15(11), 44–46 (2007)

Case Analysis of Farm Agriculture Machinery Informatization Management Network System

Hui Yang*, Xi Wang, and Weidong Zhuang

Institute of Engineering, Heilongjiang August First Land Reclamation University,
Daqing, 163319, Heilongjiang Province, P.R. China,
Tel.: +86-459-6819224; Fax: +86-459-6819210
hicleni@gmail.com

Abstract. In the process of China's agricultural modernization, especially agricultural machinery modernization, in terms of equipment, we've chose the way that foreign imports (and domestic research) with the combination of self-developed, in the software, it is difficult to fully apply this approach, the specific reasons are: the modernization of China's agriculture development model is diversified, it is difficult to find a unified management model, even in the scale of operations of the representative state-owned farms and the abroad farms are also very different management models. Due to various types of growth models of biological complexity, diverse climatic and geographical environment factors, coupled with the characteristics such as long cycle of agricultural production, high input, high-risk, and decentralized management, industrial management mode it is very difficult to apply. Moreover, the application of modern management tools is also difficult to quantify the benefits, leading to the current research and application are in a state of comparatively dropped behind.

Combination the development of "Farm agricultural machinery informatization management system in The network integration system of digital agriculture of Heilongjiang province", Heilongjiang August First Land Reclamation University Digital Agriculture Project Team carried out case studies and experiments in a few more advanced farms (Red Star Farms, Seven Stars Farm, etc.) in Heilongjiang Province. In order to facilitate the unified management of the farm machinery operation, the agricultural machinery operation management system was achieved through a form of a web site design. With the support of precision agricultural technique, which include information management, farm machinery job scheduling, agricultural machinery performance standards and the accounting standard of the workload, the supply of fuel and spare parts, maintenance organizations and reminders, statistics analysis of the technical and economic indicators, financial accounting, as well as the necessary network meeting support systems, etc.

To integrate the analysis and design of the informatization management network system of a farm, the efficient path of the farm machinery management modernization of a farm should be researched, in this paper, some details were discussed in the research and constructive foundation of agricultural machinery

---

* Corresponding author.

D. Li and C. Zhao (Eds.): CCTA 2009, IFIP AICT 317, pp. 65–76, 2010.

management modernization, the roles and actions of farm machinery informatization management network system in agricultural modernization, description of the issue of the farm machinery informatization management network system, the design of internal modular structure, external network system interface, open source questions and the improvement path to follow, etc.

Keywords: agricultural; agricultural machinery; farm machinery; informatization; management modernization; network system; job scheduling; performance standards; maintenance organizations.

1   Project Background

In the process of China's agricultural modernization, especially agricultural machinery modernization, in terms of equipment, we've chose the way that foreign imports (and domestic research) with the combination of self-developed, in the software, it is difficult to fully apply this approach, the specific reasons are: the modernization of China's agriculture development model is diversified, it is difficult to find a unified management model, even in the scale of operations of the representative state-owned farms and the abroad farms are also very different management models. Due to various types of growth models of biological complexity, diverse climatic and geographical environment factors, coupled with the characteristics such as long cycle of agricultural production, high input, high-risk, and decentralized management, industrial management mode it is very difficult to apply. Moreover, the application of modern management tools is also difficult to quantify the benefits, leading to the current research and application are in a state of comparatively dropped behind.

Combination the development of "Farm agricultural machinery informatization management system in The network integration system of digital agriculture of Heilongjiang province", Heilongjiang August First Land Reclamation University Digital Agriculture Project Team carried out case studies and experiments in a few more advanced farms (Red Star Farms, Seven Stars Farm, etc.) in Heilongjiang Province.

To integrate the analysis and design of the informatization management network system of a farm, the efficient path of the farm machinery management modernization of a farm should be researched, in this paper, some details were discussed in the research and constructive foundation of agricultural machinery management modernization, the roles and actions of farm machinery informatization management network system in agricultural modernization, description of the issue of the farm machinery informatization management network system, the design of internal modular structure, external network system interface, open source questions and the improvement path to follow, etc.

2   System Overview and Description of Structure and Function

In order to facilitate the unified management of the farm machinery operation, the agricultural machinery operation management system was achieved through a form of a web site design. With the support of precision agricultural technique, which include information management, farm machinery job scheduling, agricultural machinery performance standards and the accounting standard of the workload, the supply of fuel and spare parts, maintenance organizations and reminders, statistics analysis of the technical and economic indicators, financial accounting, as well as the necessary network meeting support systems, etc.

Discussed below in light of the actual cases.

2.1 Build Information Management and Publishing Platform

Broadly speaking, project of farm agricultural machinery informatization management system in the farm agricultural machinery informatization management network system in fact covers the modules shown in Figure 1.

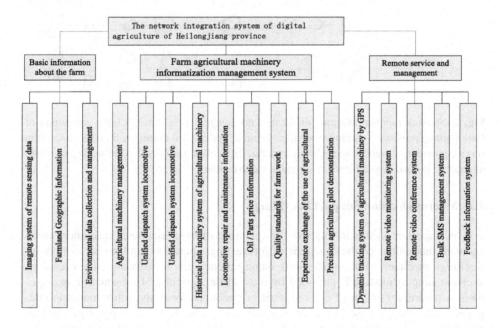

Fig. 1. Farm agricultural machinery Informatization management system HIPO Chart

About web site design we chose the Browser/Server structure, behind the web server, we also need some map engine and some database server for basic information and realtime information, etc.

Network topology diagram of the network integration system of digital agriculture of Heilongjiang province shown in Figure 2.

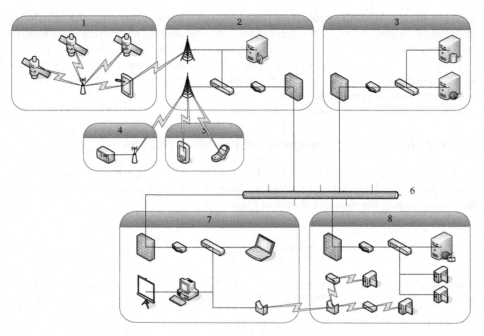

Fig. 2. Network topology diagram of the network integration system of digital agriculture of Heilongjiang province①dynamic tracking system of agricultural machiney by GPS; ②telecom service provider; ③web server; ④environmental data collection; ⑤bulk SMS management; ⑥ internet; ⑦unified dispatch center; ⑧remote video capture

2.2 Macros Decision Support Providing

It is an important means to improve the scientific of the agricultural machinery management that enhanced the functions of forecasting, operations and decisions in a agricultural mechanization information management system software, the necessary decision support module is provided by most of the agricultural information management software, a number of conventional or unconventional algorithm model is integrated, serve the purpose of forecasting, operations and decisions support as much as possible. For example, the forecast function in the development of agricultural machinery is achieved by use of least square method, BP neural networks, gray-Markov chain model; the combination forecasting function in the development of agricultural machinery is achieved by use of rough set theory; the decision function in the development of agricultural machinery is achieved by use of operations research model such as AHP(Xinjie Yu, 2004).

The intelligent decision support function modules of soil testing fertilization is included in the system design of "The digitalization of farm agricultural machinery management system in the farm agricultural machinery informatization management network system", it is still under research and development, and match with the historical database of land information of precision agricultural and ancillary equipment such as variable rate fertilization device have been entered the implementation phase. The forecast, decision function modules referred above have not been included in this system since the study focused on different purpose.

2.3 Depth in the Direction of the Core Management Modules Such as the Enterprise Resource Planning and Scheduling Management

At present, a series of core information management function modules have been get implemented such as the enterprise resource planning and scheduling management, it has been fully applied in industries such as coal, automobiles, hydraulic and electric power, etc. Due to some special reasons, the research and application of the process of agriculture production is lagging behind the general level, we now need to devote more energy to study.

As an example, studies has been done on agricultural machines optimization and allocation and service system construction in Shanghai farm by Jun Shi, the management system played an important role in following aspect: uniform implementation of standardization of technical measures on agricultural operations, uniform management of agriculture equipment, uniform scheduling mechanical operations, uniform maintenance and optimum combination, achieved with minimal input and the lowest cost to maximize machinery operation to meet demand for agricultural production, In order to achieve large-scale agricultural machinery management, agricultural equipment played the economies of scale, implement the standardization of agricultural machinery management, strengthen the total quality management of agricultural production process, to ensure safety in agricultural production with agricultural machinery, etc. but in fact they did not put the negative effects of the planned economy. It is in a matter of ongoing exploration that how to resolve keeping the above-mentioned advantages of the original management system under the premise in both, also contradictory to adapt with the construction of agricultural mechanization service system against the backdrop of market-oriented economy in agriculture. A highly efficient and flexible information technology itself, characterized by open and transparent, make it more suitable for the role of supplying a scientific planned guidance for a low cost of the non-mandatory advice and services for the agricultural market, It retains the advantages of guiding plans, also with a very flexible response capability on the rapidly changing market economy. With the opportunity for a comprehensive science and technology, information technology and biotechnology as the leading process of agricultural modernization. It is not impossible to provide a way out to explore the solutions for these conflicts status quo(Jun Shi, 2006).

The system attempt to cover the agricultural machinery job scheduling, agricultural machinery operations standards and the workload accounting, fuel and spare parts supply, maintenance organizations and reminding, report forms analysis and statistics of the techno-economic indicators, financial accounting and other typical management problems of agricultural machinery management, while it is primary, however, this information on behalf of the further request of the farm management informationize.

A simple working procedure of farm station locomotive work looks like figure 3:

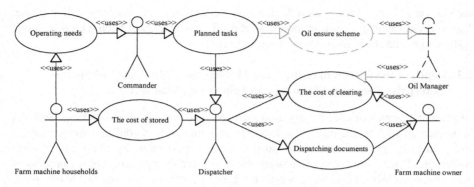

Fig. 3. A simple working procedure use case of farm station locomotive work

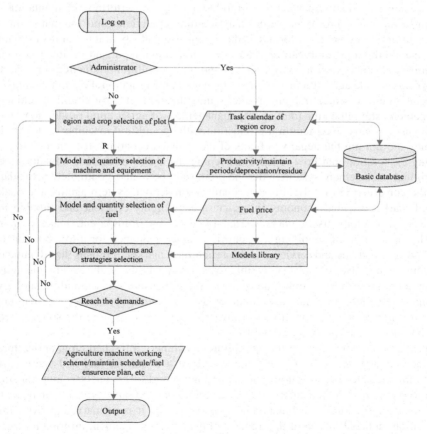

Fig. 4. Agriculture machine job scheduling constitute breviary flow figure

From the figure 3, the key sector of farm workstation working flow is the constitution and carrying into execution of agriculture machine job scheduling. The scheduling can be divided into long term, middle and short term scheme, and a profile of the agriculture machine job scheduling constitution flow figure looks like figure 4.

3 Review and Analysis of System Development

3.1 Construction Repeats Problem

This includes the concept overlap with other agricultural information systems, the duplication of software R & D issues with similar areas of domain, the sharing and cooperating with other relevant public information, and so on.

According to the authors know, farm-level managers have become increasingly aware of the important role of information technology in farm management, However, Specific operation in the process, may be biting off more than for large, aggressive committed errors, sometimes there is an obvious problem of redundant construction. Although the contribution rate of the informatization of agriculture is not very easy to calculate accurate, I suggest that at least the value of the project evaluation should be done by the concerned departments to carry out.

Farms now have a certain autonomy, but also the guide department of the government in charge of the development of farms, duplication should be avoided by coordination between the farm and the government, to break the technical barriers and identify the common domain. To resolve this problem I advocate component-based technology and open source development model (See part 4).

3.2 Accurate Positioning of the System, Evaluation and Feedback

Accurate positioning of the system needs to re-examine the meaning of agricultural management, agricultural mechanization management is the general term of the integrated use of management about the main theories, the mind, the ways and the means, the organizations and the systems and so on, on the management of agricultural mechanization, of main body of the management of agricultural mechanization. The traditional management concepts of agricultural mechanization is the management, supply, manufacture, maintenance and repair, scientific research, technical support, training and the use of such work of agricultural machinery, unified management by the agricultural sector of the government. With the development of agricultural mechanization, agricultural mechanization means more detailed management functions, an increase of agricultural technology popularization and application, agricultural machinery safety supervision, identification of agricultural products and quality supervision, identification of vocational skills of skilled workers, agricultural mechanization information and propaganda (Qicai Li, 1999). As well as the penetration is expected in the functions of the general plan, organizational, command, control, coordination, feedback and so on (Xingguo Li, 2006).

General agricultural information management system usually contains most of the information service functions, forecasting, decision support functions and internal information management functions, such as conventional modules. To determine the degree of importance and urgency of a system function, we should look at the impact of its economic effectiveness of enterprises or social benefits, for example, early warning of nature factors in agriculture, the process management of agricultural mechanization production and so on.

Mingyan Hu carried the research on demand analysis and resource configuration of the regional manufacturing informatization, and in her paper the efficiency of financial resources and the economic growth rate of informatization is calculated(Mingyan Hu, 2004). Wen Ru carried the research on construction strategy and comprehensive evaluation of China's enterprise informatization, the calculation method of the tangible business benefits and the evaluation method of the intangible benefits of enterprise is talked about(Wen Ru, 2003). A specific analysis for the focus of the project should be carried through the use of these algorithms of quantitative analysis and methods of qualitative analysis for guiding the direction of development to fetch the desired effects.

4   To Explore the Evolution of System Software

Combine the understanding of the development and construction of the system mentioned above, if the entire informatization can be divided into two parts: hardware and software, the following text try to talk about some of the details of the whole from the perspective of software evolution.

4.1   The Development, Reuse and Sharing of Network Components

According to the software product line concepts and ideas pointed out by scholar Fuqing Yang (JadeBird engineering put forward the concept during "75" period): the software production process can be divided into components production, framework production, components and framework reusing. Software components / framework technology is the core of software production, organically linked the domain engineering and application engineering through the aspects of component management and re-engineering. Combined with project management, organizational management, such as management issues, form a complete software production process (Fuqing Yang, 1996; Fuqing Yang, 2005).

For such a software role of agricultural information management network system, it is used primarily as an integrated production of application development, it should play the role of components reusing in the software product line. Once the farm want to consider a further upgrade of the system to a deeper level of the enterprise management, the current system is going to be integrated into a larger system as a subsystem, or attempting to take certain level of component for reuse, decomposition of component is to support or face-out, is a very tough choice. Therefore, it is very necessary to consider the picture of the whole, to do a good job at the beginning of the project on the specific historical background and forward-looking long-term planning.

Scholar Fuqing Yang believe that network-oriented computing environment is currently moving from Information Web to Software Web, software componentware should be formed as internetware. That is, to adapt to the open and dynamic external networks environment, as well as the users use the personalized request, the adjustment of the static and dynamic evolution should be carried out in accordance with the function, performance indicators, such as the indicator of the credibility, so that the highest possible reliability software patterns for users can be formed.

From the perspective of micro, internet will enable the research focus of the development of system software and supporting platform shifted from the operating system to the new middleware platform, breakthrough in internetware theory, methods and techniques will lead to a breakthrough in establishing a new pieces of innovative technology platform of new type of intermediate. The current software technology follow the law of the combination of software and hardware (microelectronics), the combination of application and systems development. To develop systems and products application-oriented, integration, for personal, reflecting individual. The overall development trends of software technology can be summarized as follows: software platform for network-based, object-based approach, system component, products of family, and the development of engineering, process standardization and production scale, and international competition (Fuqing Yang, 2002).

Based on such a background, the design of farm agriculture machinery informatization management network system should be subjected to the network-based, component-based development path. Decompose the system components for the suitable size, full research argues and give priority to the use of existing component library products. Sharing the good reusable components that are not inconsistent with the premise of the TRIPS Agreement, increasing the flexibility and robustness structures of components by the perspective of a comprehensive plan from the using of design patterns.

(1)Perhaps the existing system can borrow ideas from the integration patterns of the SNS which known as the more popular examples: Widget is used to increase the capacity of cross-domain interaction, as far as possible the original system changes slightly;

(2)A new design can be adopted for the newly developed system, the whole of the components fully taken into account, to use the mature design patterns as far as possible, design reusable components;

4.2   Try the Applicability of the Open Source Technology System on the Informatization of Agricultural

In view of the difference of China's agricultural enterprises and the ordinary industrial enterprises or the tertiary industry, to start a number of government-funded open-source projects, I guess, will have a multiplier effect. At present, open source technology system have shown strong vitality to flourish in many areas, the revitalization of the software industry of our nation are also depends on the healthy development of the open-source technology system, I think that might be a good starting point to seek cross-point in the agricultural areas and the open source areas that the demand and the supply both are unique.

International original open-source resources web site platform such as http://sourceforge.net, http://apache.org/, updating very frequently, domestic http://cosoft.org.cn/ is also a good open source resource site. The GNU organization, Free Software Foundation (FSF), and the various open-source licenses represented by GPL, as well as the wealth products of the platform and a large number of workers, provide a broad space for the development of open-source system. It can be said that all the software in the field of closed source systems are provided corresponding products with the open source side by side, although they are not necessarily the same level, in fact each other high. There is no reason for agricultural information technology to turn a blind eye to these resources, which is a magic weapon for development the software industry by leaps and bounds, which is the only way to break the technical barriers, rapid approach international advanced level and access to the independent intellectual property rights.

I envisaged, we can distribute information technology component or code fragments through open-source license, allowing users to assemble their own open-source product, which will play a powerful role in adding fuel to the flames on cultivating a large number of agriculture technicist familiar with open source products, improve the overall quality of the agriculture technical personnel of grass-roots level, speed up the process of informatization. of course, it is not realistic for the grass-roots level technical personnel to learn to count on the complexity, as the leader, it requires our universities and research institutions to have dedication, to launch open-source team, give their time and energy.

Once the open source components are capable of covering the basic necessary component set of the farm agriculture machinery informatization products, when it is feasible for a agriculture grass-roots level to build their own applications of information technology, open source technology system will certainly be recognized by everyone, then the benefits of it will be multi-faceted, not just the software products directly own low-cost, high efficiency, will also train a large number of technical staff, bring whole Industry a qualitative progress. The great social benefits, reactive in contemporary, benefit future generations.

Limited to the subject matter hereof, on the open-source technology systems, business models are not discussed, it can be said, the resources-related projects and documents, academic papers are voluminous and can not be enumerated one by one (Haifang Chang, 2007; Guangfeng Wang, 2008).

4.3 A Phased Implementation Strategy

Hebei Agricultural University, Dr. Zhang Honggang studied on the informatization development strategy of the county economic(Honggang Zhang, 2008), a comprehensive summary of the macro level of the county agricultural development measures is talked about. The development and countermeasures of China's enterprise informatization is talked about by Lili Zhang of Taiyuan University of Technology (Lili Zhang, 2003), in the paper, it referred several correlative aspect, such as, to improve staff quality, it need to be considered people-oriented when determine the construction and implementation strategies of informatization of China's enterprise; planning objectives and step by step; emphasis the support of on core business and the core competitiveness; to do a good job in corporate strategic data planning; to avoid the risk of informatization; and combine with the management innovation; focused, at different levels and progressive implementation; to do a good job in the normalization and standardization of enterprise information management; to deal with external parties and corporate relations; to do a good job in the evaluation and inspection, in maintenance and expansion of enterprise informatization; to pay attention to the industry experts, co-operation with the third-party consulting firm, etc.

Accordance with the positioning of system development in part 3.3, in order to achieve optimal distribution and positive roll of informatization resources, it is necessary to carry out a overall planning for informatization process. This is also researched and discussed very much. Concerned about the title in terms of the farm agriculture machinery informatization management, the author advocates the development of an incremental process. Focused on the system, carry a gradual implementation of a hierarchical, step by step into the support level of enterprise core business and core competitiveness, push the evolution down-to-earth from the digitization to the comprehensive informatization, in order to circumvent the risks of informatization, so that the process of informatization can be realy implemented.

References

Yang, F.Q.: Thinking on the development of software engineering technology. Journal of Software 16(1), 1–7 (2005)

Yang, F.Q.: The present and development of JadeBird engineering—discussion on development approach of national software industry. In: Yang, F.Q., He, X.G. (eds.) Proc. of the 6th National Software Engineering Academic Conf. Tsinghua University Press, Beijing (1996) (in Chinese with English abstract)

Yang, F.Q., Mei, H., Lü, J., Jin, Z.: Some discussion on the development of software technology. Acta Electronica Sinica 30(12A), 1901–1906 (2002) (in Chinese with English abstract)

Chang, H.: The research and implementation of the OSS based administration system of telecom integrated OA, Beijing University of Posts and Telecommunications, Master degree thesis, 3 (2007)

Wang, G.: Competition between open source and proprietary software: research based on the system software market, Liaoning University, Doctoral dissertation, 4 (2008)

Zhang, H.: Research on the Strategies in the Development of the County Economic Informatization. Management Science, Hebei Agricultural University, Doctoral dissertation, 6 (2008)

Zhang, L.: Present situation and countermeasure research of enterprise information of our country, Management Science and Engineering, Taiyuan University of Technology. Master's thesis, 4 (2003)

Yu, X.: The research on the MIS for agricultural mechanization based on Web GIS, agricultural mechanization engineering, Zhejiang University, Master's thesis, 7 (2004)

Shi, J.: Studies on agricultural machines optimization and allocation and service System construction in Shanghai farm, Cultivation, Yangzhou University, Master's thesis, 10 (2006)

Jia, Y.: Study on agricultural macroscopic decision-making support system based on GIS, Cartography and Geographic Information System, Beijing Forestry University, Master's thesis, vol. 5, p. 21 (2007)

Li, X.: Environment analysis and study of evaluation index on agricultural mechanization of China, Agricultural Mechanization Engineering, Hebei Agricultural University, Master's thesis, 3 (2006)

Li, Q., Xiao, Y., Zhang, X., Li, R.: Research on management innovation of agricultural mechanization. Journal of Agricultural Mechanization Research 8(3), 23–41 (1999)

Hu, M.: The research on demand analysis and resource configuration of the regional manufacturing informatization, Mechanical design and theory, Tianjin University, Master's thesis, 1 (2004)

Ru, W.: Research on informatization construction strategy and comprehensive evaluation of China's enterprise, Southwest Petroleum Institute, Master's thesis, 4 (2003)

Study on Retrieval Technique of Content-Based Agricultural Scientech Multimedia Data[*,**]

Xiaorong Yang[1,2] and Wensheng Wang[1,2]

[1] Agriculture Information Institute, Chinese Academy of Agriculture sciences,
No.12 Zhongguancun South St., Haidian District, Beijing 100081, P.R. China
[2] Key Laboratory of Digital Agricultural Early-warning Technology (2006-2010),
Ministry of Agriculture, The People's Republic of China

Abstract. Traditional text information–based management and utilization methods can not satisfy application demands of implicit and unstructured agriculture multimedia data. For providing a readable-friendly agriculture science & technology information service, this paper proposes a relation database model oriented multimedia data method, which implements content-based multimedia data search by indexing every kind of multimedia data.

Keywords: Multimedia, Data Matrix, Content-Based Retrieval.

1 Introduction

With the development of rural informatization in China, agriculture scientech information service for agriculture management departments, agriculture scientific workers, popularizing agricultural techniques workers and formers is an important content of socialist new countryside construction. Because traditional text information–based agriculture scientech information service can not satisfy users' demands, multimedia information such as pictures and videos becomes more popular due to its good intuitionism and readability. But multimedia information takes up mass storage space and hasn't semantic interpretation, text information–based management and utilization methods can be not used for the management of the implicit and unstructured agriculture multimedia data. So multimedia technology should be used to store and manage all kinds of multimedia data.

2 Management Method of Content-Based Agricultural Scientech Multimedia Data

Multimedia data include three kinds: The first kind is multimedia primitive objects such as static or active images, audio files and video files. The second kind is metadata which are description information about multimedia objects. The third kind is the

---

[*] 国家科技支撑计划课题"农业资源利用与管理信息化技术研究与应用"
(2006BAD10A06).
[**] 科研院所技术开发研究专项"基于智能检索的西藏科技资源共享技术".

D. Li and C. Zhao (Eds.): CCTA 2009, IFIP AICT 317, pp. 77–83, 2010.

correlation methods between multimedia objects and their metadata. On account of multimedia data are composed of different multimedia objects, different kind of multimedia data should adopt different representation format and access methods.

2.1 The Storage Model of Object-Oriented Relational Data

Because multimedia data are high-capacity, none explanatory and nonstructural, Relational Database Management System (RDBMS) based on relational data model can not satisfy the storage demands of multimedia data. We analyzed three common data models which were object-oriented relational data model, pure object-oriented data model and relational&object-oriented data model. Being improved based on relational data model and object-oriented data model, they have limitations in managing multimedia data. On one hand, the method based on pure object-oriented data model is very difficult and lacks uniform data model and strong theory foundation. And it has poor compatibility and transplantability and doesn't support SQL statements. Besides it is poor in query optimization and view function. On the other hand, though it integrates the mature relational data model and object-oriented data model, the method based on relational& object-oriented data model can not support some object-oriented semantic. It gives up some function of object-oriented database to preserve the storage structure of RDBMS so that it reduces overall efficiency of database. And different database which are developed by different manufacturers adopt different object-oriented mechanism (W.L.Geroski et al. 1998). So we selected the object-oriented relational data model which combines object-oriented technology and database technology and applied object-oriented idea to complicated kind of data. We adopted the model and used C++ to develop application software to fulfill these functions including of data storage, management and retrieval. The storage address information of multimedia objects in a server was stored in RDBMS. For example, a real multimedia object was stored in the specified directory of a server according to some naming rules and its keywords, text title, the address information and name were stored in RDBMS as index. We established the connection between the storage directory of multimedia objects and their keywords so that users could access multimedia objects. High-capacity multimedia objects were stored as outside data files so the optimization of database wasn't reduced. But multimedia objects could not get protection from database. So data backup should be done regularly to protect them (Hao Ying 2004).

2.2 Agricultural Scientech Multimedia Metadata

Because agricultural scientech multimedia objects don't have interpretation, we must establish their metadata. Users can access multimedia objects which have specified features or certain contents according to the corresponding relation between multimedia objects and their metadata. Metadata are used to describe multimedia objects' content, structure, semantics, size, type, created time and so on. We created two levels metadata which were lower level vision features and higher level semantic features.

Metadata about lower level vision features are those related to content. They were extracted automatically by using the feature extraction function. These metadata included the information about colors, texture, contour, shape, volume, spatial relation, movement, deformation(for example object's bending), vision object's source and features(for example source object, source event, source property, event, event property and so on) and model. For example, one person's facial features information (such as his eyes&nose's type and his hair's color) or the photography movement information(such as panning the camera and changing the focus) can be gotten. And Metadata about higher level semantic features were marked by people. Information editor described the features according to his impression such as facial expression (angry or happy) and the information related to multimedia objects such as photographer's name as background information.

We created different metadata according to different multimedia objects as follows:

(1) Text Metadata

Text metadata were described by using text information language SGML. They were about text content description, text data presentation (text formatting, coding and compression technique), text history (modifying date), text positioning (text storage location) and so on.

(2) Images Metadata

Images metadata were extracted by using generating algorithm which divided an image into several areas or objects to locate the objects of the image according to the image's property such as color and texture and so on. The technique of areas growth was used to separate an image's objects. It separated an image into a group of single pixels or areas of a group of pixels whose internal parts were growing up to their boundary according with the object's boundary. Every area's property information such as grayness, color and texture was identified and assigned a value. These values formed the image's metadata by merging set operation of sets.

(3) Video Metadata

A video is an unstructured, 2-dimensional image flow sequences. It has dynamic such as the change of the lens movement, the moving object's size and the trajectory path of video object besides common still images' features. Video data are described with act, scene, lens, frame and so on. An act is composed of a series of correlative scenes. A scene is composed of some lens. A lens is composed of a series of continuous frames. And a frame is a still image and the smallest video unit (Xu Jiuling et al. 2003).

We created three kinds of video metadata. The first kind was those related to content which were used to describe a video's primitive features such as the camera movement, light and dark of light, the object trajectory path of video frame sequences and the color chromatography curve of independent video frame. The second kind was those about content which were used to describe a video's content such as the lens' distance, the camera angle, object kind in lens and the frame's brightness, colors, texture, object kind in a frame of independent video frame. The third kind was those unrelated to content which were used to describe the whole video's features such as its created date and director's name (Yang Zhiping et al. 2002).

3 Retrieval System of Content-Based Agricultural Scientech Multimedia Data

3.1 The Architecture of Content-Based Agricultural Scientech Multimedia Data System

The retrieval system of content-based agricultural scientech multimedia data adopted this kind of method which was different from that of traditional database and approximate matching technique. Fig.1 is the system architecture map. The system includes 4 layers. The first layer is the expression Layer. Users access the portal to get agricultural scientech multimedia data. The second layer is the application layer. The application system provides retrieval service for users. After having dealt with

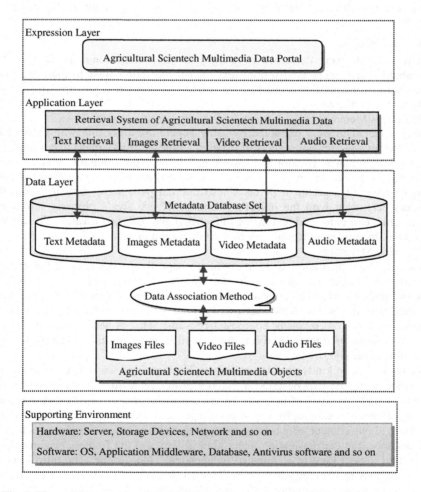

Fig. 1. The architecture of content-based agriculture scientech multimedia data system

users' request, it accesses the database and returns the retrieval result to users. The third layer is the data layer. There are metadata database set and agricultural scientech multimedia objects. The data association method is adopted to create the corresponding relation between multimedia objects and their metadata. The system finds the multimedia objects according to its metadata. The forth layer is the supporting environment. It includes hardware such as server, storage devices, network and software such as OS, application middleware, database, antivirus software and so on.

3.2 Retrieval Method of Content-Based Agricultural Scientech Multimedia Data

Content-based retrieval method is to create multimedia object's index by adopting its metadata. It is a kind of similarity retrieval method about multimedia objects. The retrieval result is gotten with successive refinement by adopting the method of approximate or partial matching. The retrieval process is as following: firstly, the system analyzes the real content of a multimedia object and evaluates the indicated predicate. The retrieval object matches with query standards approximately. A corresponding characteristic vector is given to a multimedia object based on approximate concept. That is to say, every multimedia object is mapped to a point in attribute space. The query range around the query point expands according to a specified amount of deviation. Multi-dimension index structure is used to query the object corresponding with the query point in query range. The query results probably include the objects which don't satisfy the condition but certainly include all objects which satisfy the condition. Then two-stage solving technology is used to continue querying in the query results gotten by rough query (Zhang Zhigang et al. 2007).

Different kind of multimedia object adopts different content-based retrieval method.

(1) Content-based text retrieval method

Content-based Text retrieval method adopts full text retrieval technique and probabilistic model namely probabilistic dependency among entrys or between entry and document.

(2) Content-based images retrieval method

By combining pattern recognizing technology and human-machine interactive technology, feature index is created and stored in the feature-database according to colors, texture, shape and spatial relation o f an image. When searching, the system analyzes an image's features to find the needed image. The features include colors' distribution, correlation, component, texture's structure, direction, symmetric relation, contour's shape and size, sub-images' relationship, number and attribute, the image's approximate description.

The retrieval system based on color features pre-treats quantitatively image color specifications firstly. An image is indexed according to its overall color distribution or color collection vector (CCV). When searching, firstly the system compares the similarity of two images by using color histogram. Then the system calculates every color's pixel. At last the image which has the same color content can be searched.

An image is composed of different texture areas. Some images are irregular in local areas but regular as a whole. When searching, the system mainly measures an image's texture features such as roughness, regularity, line similarity, contrast, direction and so on and then selects the needed texture from samples.

The retrieval system based on shape features can search an image in relation to its contour line by using template matching method. Users can make an image's rough shape or contour line by hand or mapping instrument and the system matches it with the images stored in system (Luo Deyong et al. 2003, He Weilie 2007).

(3) Content-based video retrieval method

The content-based video retrieval system firstly treats video information in order to restore motion information when searching. It analyzes a video's structure and calculates frame difference by detecting lens border and different lens' color and brightness, and then divides a video into some basic lens units. The system selects the key frames to represent complicated video by using frame average method and Histogram average method. By extracting every lens' features it creates index to form feature space, then depends on feature space to compare lens' content. At last according to these features it combines the lenses which have similar contents. When searching, the system matches feature according to user's certain requirements and searches again according to user's feedback until user gets satisfying results. This is a refinement process step by step (Wang Weihua et al. 2007).

4   Conclusion

Multimedia object takes up mass storage space and hasn't semantic interpretation, traditional text information–based management and utilization methods can be not used for the management of the implicit and unstructured agriculture multimedia data. The object-oriented relational data model was adopted to manage multimedia data efficiently. High-capacity multimedia objects were stored as outside data files so the optimization of database didn't be reduced. Agricultural scientech multimedia metadata were created to relate multimedia object and its index so that the system can access the needed data. Retrieval system of agricultural scientech multimedia data was developed to fulfill the storage and access of content-based multimedia data efficiently.

With the rapid development of digital communication network and increasing capacity of computer processing, it is practicable and economic to produce and utilize multimedia data. Transmitting the agricultural scientech multimedia data gives new content for the agricultural scientech information service and promotes the development of rural informatization in China.

Acknowledgements

The work is supported by the Academy of Science and Technology for Development fund project "intelligent search-based Tibet science & technology information resource sharing technology", National Science and Technology Support Program (Grant No. 2006BAD10A06), special project of the Ministry of Science and Technology "TD-SCDMA based application development and demonstration validation in agriculture informationization", and the special fund project for Basic Science Research Business Fee, AII (No.2009J-06).

References

Geroski, W.L., et al.: Management technical manual of multimedia information. Cambridge University Press, Cambridge (1998)

Ying, H.: Treatment of multimedia data in ORDBMS. Journal of Beijing Institute of Civil Engineering and Architecture 17(2), 77–79 (2001)

Jiuling, X., Huizhen, X.: Study on metadata. Journal of Information Theory and Practice 26(2), 163–166 (2003)

Zhiping, Y., Sumei, F.: Storage and retrieval of metadata in multimedia database. Journal of Chongqing Teachers College 19(1), 14–19 (2002)

Zhigang, Z.: Study on multimedia database and its content-based retrieval method. Journal of Information Technology and the Informationization 2, 61–62 (2007)

Deyong, L., Hai, M.: Study on content-based multimedia data retrieval method in digital library. Journal of Information Exploring (1), 21–23 (2003)

Weilie, H.: Study on content-based multimedia data retrieval technology. Journal of TV University Engineering 2, 59–61 (2007)

Weihua, W., Weiqing, W.: Content-based multimedia data retrieval technology. Journal of Computer Project and Design 10, 2373–2375 (2007)

A Design of an Experimental System for Trapping Pressure of Agricultural Gear Pump

Xiangdong Zhu[*], Kejiang Zang, Guoling Niu, and Xiaohai Li

College of Mechanical Engineering, Jiamusi University, Jiamusi,
Heilongjiang Province, P.R. China 154007,
Tel.: 0454-8763218
zhuxiangdong_jms@163.com

Abstract. The trapping phenomenon stems from the principle of gear pumps, and it affects directly the performance and life of gear pump. Reducing pressure in trapping area is an important aspect to improve working performance of gear pump, and how to test the pressure of trapping area is one of the most important problems. According to such requirements, an experimental system has been designed, and then relevant experiments are conducted on the designed experimental system. The experimental results indicate that the system can fulfill the requirements commendably.

Keywords: gear pump, trapping, oil pressure, testing system.

1 Introduction

In order to make gear teeth of gear pumps mesh well, and make inlet chamber and outlet chamber seal tightly for exporting working medium evenly and continuously, the overlap coefficient must be more than one $(\varepsilon > 1)$. In other words, before one pair of teeth is about to stop meshing, another pair of teeth has meshed. More than one pair of teeth mesh simultaneously. Generally, there is a sealed space formed by two pairs of gear teeth, and the oil can be trapped in the sealed space. The sealed space is called trapping chamber. But the sealed space will turn smaller and smaller along with the pump work continuously. When the two meshing nodes locate the symmetrical position of two pairs of gears' notes, the sealed space will be reduced to the smallest cubage. Since the oil compressibility property is small, the oil pressure turns high rapidly for the decrease of the cubage of the sealed space. When the oil pressure exceeds far the outlet chamber pressure, the oil is extruded from the both sides of the adjoining plane parts. So the shafts and bearings of the gear pump endure heavy shock load, and make power loss increase, medium heat, vibration and noise produce. Thus the work stability and the life of the pump will be decreased. As gears continue to mesh, the trapping chamber cubage turns bigger and bigger, so the vacuum will be formed. Bubbles, cavitations, vibration, noise and other hazards appear subsequently. This impact load change periodically along with alteration of the trapping space

---

[*] Corresponding author.

D. Li and C. Zhao (Eds.): CCTA 2009, IFIP AICT 317, pp. 84–89, 2010.
© IFIP International Federation for Information Processing 2010

volume in turn. And moreover the original stack load will be superimposed. Although the trapping phenomenon is happened shortly, its frequency is very high. So the alternating load has a periodic property.

The above is a general description of the trapping phenomenon. In fact, under the environment of the micro-clearance gear meshing or anti-backlash gear meshing, when a pair of teeth meshes, the trapping phenomenon will also occur. There are two trapping chambers are formed under a pair teeth meshing, and three trapping chambers are formed under two pairs teeth meshing. The trapping phenomenon may affect the life of pump and the working performance badly.

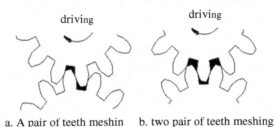

a. A pair of teeth meshin   b. two pair of teeth meshing

Fig. 1. Trapping area

Reducing the trapping pressure of gear pump is one of the most important aspects to improve the pump's work performance. Home and abroad scholars have done large amount of job on it. Among them, unloading grooves is the most effective method and it is applied broadly. So many theoretical analyses have done on the trapping pressure but the problem that how to test the pressure is rarely reported. A test method that has mentioned in the document (Zhao Liang, et al., 2004) is difficult to carry out. In recent years, the author studies on how to reduce the trapping pressure of gear pump and designs an experiment test system to test the trapping pressure change. Following content introduce this experiment system only for reference.

2 The Principle of Unload-Pressure-Relief Groove

(1) work surface and non-work surface

Gear tooth has two figure surfaces. One is meshing surface for the sake of transferring power, the other is towards it for the sake of holding balanced transferring. In order to narrate easily, we called the former "work surface" and the latter "non-work surface".

(2) unload-pressure-relief groove

The unload-pressure-relief groove is a groove on the not work surface, show as Fig.2. As a result of radial groove, it can not destroy meshing and intensity of the work surface, and stability is invariable.

(3) Principle

The principle of the unload-pressure-relief groove is show as Fig.3, which not setting pressure unloading grooves, supposing meshing non-profile clearance. From

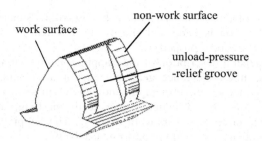

Fig. 2. Gear tooth with the unload-pressure-relief groove

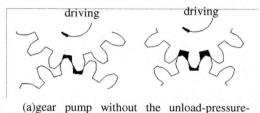

(a)gear pump without the unload-pressure-relief groove

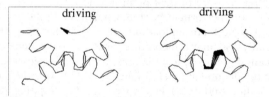

(b) gear pump with the unload-pressure-relief groove

■: the trapping volume without the unload-pressure-relief groove
▓: the additive trapping volume with the unload-pressure-relief groove

Fig. 3. Relation of gear pump without the unload-pressure-relief groove and with the unload-pressure-relief groove

the Fig.3, we can see that a pair of gear teeth is meshing, then two trapping chambers are formed when the unload-pressure-relief grooves are not set; while there isn't trapping chamber when unload-pressure-relief grooves are set. Two pairs of gear teeth meshing, there are three trapping chambers, when the unload-pressure-relief grooves are not set; while there are two trapping chambers when the unload-pressure-relief grooves are set, moreover they are connected. In general, the numbers of trapping chambers are reduced after the unload-pressure-relief grooves are set.

3 Experiment Test System

The experiment test system is shown in Fig.4. A volume flow control circuit is built with a volume adjustable hydraulic pump 3 and a single-way constant displacement motor 6. The tested pump 7 is driven by the single-way constant displacement motor. The tested pump's speed can be adjusted by changing volume adjustable hydraulic pump during the test. Bypass valve 5 is used as a safety valve and throttle valve 9 is used for adjusting the tested gear pump's work pressure. Motor drives the tested pump through a coupler. The apparatus 8 is speed transducer, which measure the tested pump's operating rotational speed, two pressure transducers are installed at the two ends of tested pump for testing the trapping pressure change. Datum getting from the pressure transducers and speed transducers is collected by the datum acquiring meter 10 and written into the recording equipment. Finally, through soft ware in the computer disposes the datum to obtain the data and the graph that needed (H Yanada, et al., 2006).

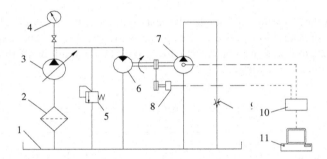

1. tank 2. filter 3. variable displacement pump 4. pressure meter 5. relief valve 6. fixed displacement motor 7. tested gear pump 8. speed transducer 9.check valve 10. datum collecting instrument 11. recording equipment (computer)

Fig. 4. Experiment test system sketch

4 The Design of the Tested Gear Pump

The aim of this experiment is to test the pressure alternation of the pump's trapping pressure for studying the effects of different measures, which are taken to decrease the pressure of trapping chamber, so the structure of tested pump is three pieces types structure, which is shown in Fig.5. The tested pump mainly consists of the back cover 1, pump body 4, front cover 6, the driving gear 7 and driven gear 5. Inlet port and outlet port are installed in the back cover. The structure of the tested pump is similar with CB-B gear pump. The driven gear's design is a key part. In order to get the pressure signals from the ends of the driven gear, so the driven gear and driven shaft are designed one body. At the end of shaft the trapping area pressure signals are got from axial holes and radial holes, in the driven gear and driven shaft and are tested by the pressure transducer 3. The sketch of gears' structure and assembly is shown in Fig.5.

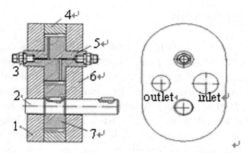

1. back cover 2. driver shaft 3. pressure transducer 4. pump body 5. driven gear 6. front cover
7. driver gear

Fig. 5. Structure of the tested gear pump

5 Test Experiment Example

According to the above test principle, the test equipment had been developed to test
the trapping pressure in a tested gear pump show as Fig.6. The experiment of
measuring trapping pressure change is carried out. The conditions of the tested gear
pump are shown as following.

Table 1. Structural parameters of gear pump

parameter	modulus	Tooth number	press angle	tooth width	tooth top parameter	tooth top clearance parameter
	mm		deg	mm		
Value	5	13	20	35	1	0.25

The pressure transducer model is BPR-2, which is connected with datum collecting
instrument whose model is WS-U20116 [B] by an electric resistance strain
instrument. The pressure value of trapping area is collected continuously, and the
datum collected can be put into computer directly and disposed by the collecting
instrument datum handling system. Fig.4 shows trail curve of pressure (n=200r/min).
Since the datum are measured continuously, so the pressure curves reflect not only the
trapping pressure alteration but also the alteration of the inlet chamber pressure, the
transition region pressure, the outlet chamber pressure. Here the peak value area of
the curve expresses trapping area's pressure.

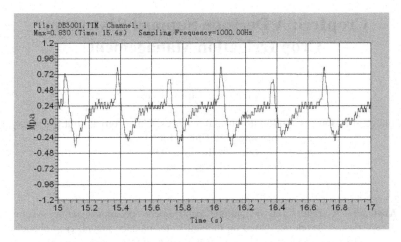

Fig. 6. Trail curve of pressure/time (n=200r/min)

6   Conclusions

In this experimental test system, the pressure signals were obtained from the driven shaft's end. The test points following with the driven gear traveled inlet chamber, the transition region, and outlet chamber. The pressure curves which we obtained from the points not only included the trapping pressure alteration but also obtained the inlet chamber pressure, the transition region pressure, the outlet chamber pressure. So gear pump's working performance can be studied from the pressure curves.

Acknowledgements

The authors would like to thank the project (10541223) supported by Scientific Research Fund of Heilongjiang Province Education Department.

References

Liang, Z., Xiyan, R., Dong-ping, W.: Analysis of gear pump for trapping in a very small clearence between gears. Chinese Journal of Mechanical Engineering 35(6), 77–80 (2005) (in Chinese)

Yanada, H., Ichikawa, T., Itsuji, Y.: Study on the phenomenon oil—trap stems of gear pump. Chinese Hydraulics & Pneumatics (3), 47–52 (2006) (in Chinese)

CropIrri: A Decision Support System for Crop Irrigation Management

Yi Zhang and Liping Feng*

College of Resources and Environmental Sciences, China Agricultural University,
Beijing, P.R. China, 100193,
Tel.: +86-10-62733939
fenglp@cau.edu.cn

Abstract. A field crop irrigation management decision-making system (CropIrri) was developed based on the soil water balance model, crop phenology model, root growth model, crop water production function, and irrigation management model. The irrigation plan is made through predicating of soil water content in root zone and daily crop water requirement using historical and forecasting weather data, measured real time soil moisture data. CropIrri provided four decision modes of non-limiting irrigation, water-saving irrigation, irrigation with experience and user custom irrigation. The main function of CropIrri includes: pre-sowing and real-time irrigation management decision-making support, simulation of soil water dynamics in the root zone, evaluation of the effect of certain irrigation plan on crop yield reduction, and database management. A case study of wheat crop irrigation management by CropIrri showed its practical value and benefit. It could be an objectives-oriented, multi users-oriented and practical irrigation management decision-making tool.

Keywords: Field crop, Irrigation, Decision support, Model.

1 Introduction

Field crop irrigation scheduling is a major part in crop production management, it is important in using of water resources rationally and increasing crop water productivity. In recent years, studies on the establishment of optimal irrigation methods and irrigation decision support system have obtained important achievements (J. A. de Juan, 1996; J.-E. Bergez et al, 2001; Zhu et al, 2003; Zhu et al, 2005; Zhang et al, 2006). These studies helped to improve crop water management and irrigation decision-making level, but there still exist many problems, such as limited to certain regions, difficult to determine growth period accurately, complex model parameters or large database, and so on.

This paper highlights the following aspects to improve and to overcome the traditional weaknesses of agricultural irrigation systems, then to build the field crop irrigation management decision support system, CropIrri. (1) Using the multi-annual mean meteorological data to make irrigation schedule before sowing, and use the forecast

---

* Corresponding author.

D. Li and C. Zhao (Eds.): CCTA 2009, IFIP AICT 317, pp. 90–97, 2010.
© IFIP International Federation for Information Processing 2010

weather data to carry out the real-time irrigation management; (2) Using simulation model for crop phenology to determine the adaptability of different varieties in different regions, and to simulate the length of growth stages, which is important to enhance the accuracy of parameters at different stages; (3) Using root growth model to simulate the root growth and elongation, then system can compute soil water content in the root zone more accurately; (4) Set custom irrigation schedules for senior user, call the crop water production function to evaluate yield losses in different stages for certain irrigation schedule; (5) simplify some input parameters to ensure system running. When input parameters are short, users can select several kinds of parameters provided by system. This can expand its regional serviceability and the farmer's usability.

2 System Design and Principle

Through the analysis of basic relation and quantitative algorithm between soil water deficit and crop water consumption, management decision-making levels and types of crop variety, environmental factors and production level, the field crop irrigation management decision support system, CropIrri, was established by taking account of the soil moisture prediction model, crop phenology model and irrigation decision-making model. The flowchart of CropIrri is shown in figure 1.

The CropIrri system is developed by using Visual Studio.NET 2005 language and run on the Windows XP platform.

ETmi is maximum crop evapotranspiration on day i, SWBM is soil water balance model, CWPFM is crop water production function model.

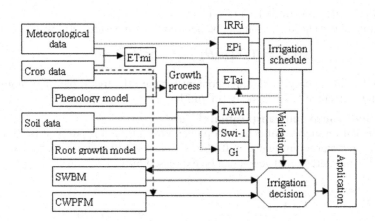

Fig. 1. Irrigation flowchart of CropIrri

2.1 Main Function

CropIrri system is designed for dryland crops (wheat, maize and soybean) to provide a practical decision tool for irrigation management. The main functions include: (1) Irrigation decision services. To evaluate crop water requirements, and to make pre-sowing and the real-time irrigation plans based on the historical weather data and weather forecast information. (2) To simulate daily change of soil moisture content in the root zone. (3) To evaluate a given irrigation schedule, and to develop optimal irrigation schedule in addition. (4) To modify the planned results according to the measured actual soil moisture content during crop growth period to enhance the forecasting accuracy. (5) Database management capability.

2.2 Main Function Modules

CropIrri system combines environmental conditions like climate and soil with crop growth characteristics as a whole, and was established through soil water balance model, crop phenology model, root growth model, crop water production function, and irrigation decision-making model.

2.2.1 Soil Water Balance Module
Soil water balance model can reflect the dynamics of soil water content in root zone and can be expressed as flow equation (Richard G Allen et al., 1998):

$$SW_i = (ETa_i + RO_i + OP_i) - (EP_i + G_i + IRR_i) + SW_{i-1} \quad (1)$$

Where: SWi is soil water depletion in the root zone at end of the day i [mm], ETai is actual crop evapotranspiration on day i [mm], ROi is runoff from the soil surface on day i [mm], OPi is deep percolation on day i [mm], EPi is effective precipitation on day i [mm], Gi is capillary rise from the groundwater table on day i [mm], IRRi is net irrigation on day i [mm], SWi-1 is soil water depletion in the root zone at end of the previous day, i-1 [mm].

(1) Initial soil water depletion (SWi-1)
The initial soil water depletion can be derived from measured soil water content by:

$$SW_{i-1} = 1000 \times (\theta_{fc} - \theta_{i-1}) \times Zr_{i-1} \quad (2)$$

Where:θi-1 is the average soil water content for the effective root zone [m3/m3], θfc is the water content at field capacity [m3/m3].

(2) Actual crop evapotranspiration (ETai)
The calculation method of actual crop evapotranspiration is adopted from FAO-56. It equals crop water requirement multiplied by the soil water stress coefficient.
The soil water stress coefficient can be expressed by:

$$Ksi = \begin{cases} 1 & SWi < RAWi \\ \dfrac{TAWi - SWi}{TAWi - RAWi} & SWi > RAWi \end{cases} \quad (3)$$

$$TAWi = 1000 \times (\theta fc - \theta wd) \times Zri \qquad RAWi = Pi \times TAWi$$

Where: TAWi is total available soil water in the root zone on day i [mm], θwd is the water content at wilting point [m3/m3], RAWi is the readily available soil water in the root zone on day i [mm], Pi is fraction of TAW that a crop can extract from the root zone without suffering water stress.

(3) Effective precipitation (EPi)
Effective precipitation is the part of natural precipitation that actually added to the crop root layer soil moisture, it is expressed by:

$$EPi = a \times TPi \qquad a = \begin{cases} 0 & TPi < 5mm \\ 1 \sim 0.8 & 5mm \leq TPi \leq 50mm \\ 0.7 \sim 0.8 & TPi > 50mm \end{cases} \qquad (4)$$

Where: TPi is the forecast precipitation or natural precipitation on day i [mm], a is the rainfall recharge coefficient, Its value is related to rainfall amount, rainfall intensity, duration, soil properties, ground cover, landform and so on.

(4) Capillary rise from the groundwater table (Gi)
Capillary rise depends mainly on soil type, the depth of the water table and moisture of the root zone. General Gi can be assumed to be zero when the water table is more than about 1 m below the bottom of the root zone. Capillary rise from the groundwater table is given by:

$$Gi = ETai \times e^{-\sigma Ho} \qquad (5)$$

Where: σ is experience coefficient (Sand=2.1, loam=2.0, clay=1.9), Ho is the depth of water table.

2.2.2 Crop Phenology Module
The predicating of crop development is the key to determine the date of irrigation. The crop phenology model was adopted from the general crop phenological theory model (CPTM) (Feng L. et al., 1999). Based on multi-annual mean meteorological data in crop growing region, the length of growth stages with different sowing date could be simulated, which is important to enhance the accuracy of parameters at different development stages.

2.2.3 Root Growth Module
Root growth model is used to calculate soil water content in the root zone. The planting depth (generally 0.03-0.05m) is considered as the initial crop rooting depth, maximum rooting depth (soybean is 0.6-1.3m and maize is 1-1.7m) was adopted (China's agricultural encyclopedia Agrometeorological volume Editorial Committee, 1986). The daily rooting depth of soybean and maize was interpolated by the initial and maximum rooting depth. Wheat root growth model was adopted from the flowing equation (Feng et al, 1998):

$$Zr_i = Zr \times \left(0.005628 + 2.3501 * tr - 4.5548 \times tr^2 + 3.2148 tr^3\right) \qquad (6)$$

Where: Zri is the rooting depth on day i [m], Zr is the maximum rooting depth [m], tr is relative time, that means the ratio of days after sowing (on day i) and the number of days that roots reached the maximum rooting depth, flowering stage of winter wheat reached the maximum rooting depth.

2.2.4 Crop Water Production Function Module

The module is used for evaluating the impact of irrigation schedule on crop yield. Yield reduction is expressed by the ratio that the difference between the highest yield and the actual yield to the highest yield (highest yield means the output under non-limiting irrigation schedule). Water shortage in certain stage not only affects this period, but also affects on the whole development period. Jensen model is a high precision mathematical model in evaluating the impact of water shortage in each growth stage on crop yield under limited water supply conditions (Ge et al, 2003). Jensen model can be expressed as:

$$\frac{Ya}{Ym} = \prod_{j=1}^{n} \left(\frac{TETa}{TETm} \right)_{j}^{\lambda j} \tag{7}$$

Where: Ya is actual yield of crop [kg/ hm2], Ym is maximum yield of crop [kg/ hm2], TETa is actual crop evapotranspiration on stage i [m3/hm2], TETm is maximum crop evapotranspiration on stage j [m3/hm2], λj is yield response factor on stage j, j is divided the whole growth period for j stages.

2.3 Modes of Irrigation Scheduling

Irrigation decision-making concerns the date and amount of irrigation, as well as the impact of selected irrigation schedule on crop yield. CropIrri supplies four modes of irrigation scheduling as follows.

2.3.1 Non-limiting Irrigation Schedule

Non-limiting irrigation is to meet the need of water requirements and to obtain maximum crop production. By comparing the daily soil moisture deficit with readily available moisture in the soil profile, when soil moisture deficit approaches readily available moisture, water stress is occurred and irrigation is made. Soil water content equaling to 80% of field capacity as suitable irrigation index was used to avoid leakage caused by deep water losses.

2.3.2 Water-Saving Irrigation Schedule

The goal of water-saving irrigation is to obtain highest yield with highest water utilization efficiency (WUE). In this case, actual crop evapotranspiration is less than potential evapotranspiration. When soil moisture content in the root zone reached 85% ~ 90% of field capacity, it is appropriate for crop growth ; When soil moisture content in the root zone below 60% of field capacity, it affects the normal growth and output of crop. In this research, the suitable soil water content in root zone ranges from 70% to 60% of field capacity in non-critical periods of of water requirement, and ranges from 75% to 65% of field capacity in critical period of water requirement. The critical period of water requirement is booting stage for wheat, flowering stage for soybean. The critical period of water requirement is from flowering stage to milk stage for maize.

When the soil moisture content in the root zone is below the appropriate low-limited water content, irrigation schedule is made to irrigate to the appropriate upper-limited water content.

2.3.3 Irrigation Schedule with Experience

Irrigation schedule is made by taking account into irrigation experience. In order to ensure crop emergence, priority should be given to sowing irrigation; then to consider the importance of the crop water requirement to determine irrigation plan. Taking wheat as an example, if one irrigation, irrigation should be at booting period. If two irrigations, irrigation should be at turning green stage and booting stage for the situation of irrigation at sowing, and at winter stage and booting stage for the situation of non-irrigation at sowing. If three times, irrigation should be at winter stage, turning green stage, and booting stage.

Each irrigation amount should reach the soil water content as 80% of field capacity.

2.3.4 Advanced (User Custom) Irrigation Schedule

The mode of advanced irrigation schedule is for researchers and technicians. Users can custom the date and amount of irrigation for different purpose, such as periodic irrigation with certain amount of water, for example, irrigation with 50 mm of water or soil water content reaching to field capacity at soil moisture content decreasing to 60% of field capacity, or irrigation with 100 mm at fixed interval of 30 days. So that we can understand the change of soil moisture content and crop water consumption. This could support and assist scientific researches in crop water relation.

3   Case Study for Wheat Crop

The case study was conducted for irrigation management by CropIrri system during the winter wheat growing season at Quzhou experiment station, Hebei, China in 2007-2008. The historical weather data of 30 years and the measured soil data in Quzhou Experiment Station were used. Wheat variety was Han 6172, a mid-maturing wheat cultivar. The soil water content was adequate for planting wheat due to heavy rainfall before sowing. Wheat sowed on 23 October with 330×104/ha of basic seedlings and 3 cm of sowing depth. The measured mature date was June 3.

Table 1. Report of pre-sowing irrigation plan made by CropIrri for wheat

Irrigation plans	Growth stages	IRD	IRR	YRR
		(d-m)	(mm)	(%)
Non-limiting irrigation	Overwintering-turning green	8-Dec	39	0
	Turning green-jointing	25-Feb	44	
	Booting-flowering	18-Apr	68	
	Flowering- filling	29-Apr	61	
Water-saving irrigation	Overwintering-turning green	18-Dec	43	10.83
	Jointing-Booting	8-Apr	71	
	Booting-flowering	21-Apr	37	

IRD is irrigation date, YRR is yield reduction rate.

The pre-sowing decision-making report for winter wheat under non-limiting irrigation and water-saving irrigation is shown in Table1. Comparing the two irrigation plans, both of them irrigated in winter, which played a role in water storage to a certain extent to meet the need of soil water for winter wheat in turning green stage. Irrigation didn't apply at turning green stage in water-saving irrigation schedule, which could cause some water stress. It was a similar irrigation schedule during the jointing-flowering stages which is rapid increase in water consumption. Irrigation at this stage is conducive to yield increase. Two heavy rainfalls during wheat growth stage on April 20 and May 3 was 35 mm and 60 mm respectively. The actual irrigation under water-saving irrigation plan on the April 21 was zero.

The experiment results and measured water data is shown in table2. Water utilization efficiency (WUE) increased 3.3% under water-saving irrigation schedule. Nearly double amount of water was saved. The final crop yield reduction was only 9.3%, which approached the predictive value of 10.83% (table1). This may be caused by the difference between multi-annual mean rainfall data with 0 mm and actual rainfall data with 60 mm on May 3. The actual rainfall helped to increase yield under water-saving irrigation plan. Over-irrigation in non-limiting irrigation plan at late growth period couldn't be all used by wheat and might waste water. Study also showed that over-irrigation decreased water use efficiency and resulted in the waste of water resource (Xu et al, 2003). The water-saving irrigation schedule had a better performance in Quzhou region and was favorable to water-saving and high yield.

Table 2. Water consumption, yield and water utilization efficiency for winter wheat under different treatments in Quzhou

Irrigation plans	TWC	RAIN	IRR	Yield	WUE
	mm	mm	mm	kg/ha	kg/(ha.mm)
Non-limiting irrigation	472	127	212	7245.4	15.35
Water-saving irrigation	364	127	114	6570.7	18.05

TWC is total water consumption, WUE is water utilization efficiency, AAI is the actual irrigation depth.

4   Conclusion and Discussion

A field crop irrigation management decision-making support system was developed based on the soil water balance model, crop phenology model, root growth model, crop water production function, and irrigation management model. CropIrri system could be used in pre-sowing and real-time irrigation management decision-making support, simulation of soil water dynamics in the root zone, evaluation of the effect of certain irrigation plan on crop yield reduction, and database management. It is developed for the dryland crops of wheat, maize and soybean.

The major characteristics are to provide the different irrigation management schedules for different level of users. It not only has default irrigation schedule for the common user, also has the custom irrigation schedule that are suitable for the senior user. Through embedding crop phenology module, it could predict crop development and to support to determine irrigation date more accuracy. CropIrri system could be an objective-oriented, multi user-oriented and practical irrigation management decision-making tool.

CropIrri could allow the single crop for management decision at present. The further study is to include varying cropping patterns, such as intercropping cultivation to enhance its function.

Acknowledgements

The authors acknowledge the financial support provided by the National High Technology Research and Development Program of China (2006AA10Z224), and the National Key Technologies R&D Program (2006BAD10A12).

References

Allen, R.G., Pereira, L.S., Raes, D., Smith, M.: Guidelines for computing crop water requirements. FAO Irrigation and Drainage 56 (1998)

China's agricultural encyclopedia agrometeorological volume editorial committee. Encyclopedia of China's agriculture - agricultural meteorology volume. Agricultural Press, Beijing (1986)

Feng, G.L., Liu, C.M.: Analysis of root system growth in relation to soil water extraction pattern by winter wheat under water-limiting conditions. Journal of Natural Resources 13(3), 234–241 (1998)

Feng, L.P., Gao, L.: A general crop phenological theory model. Journal of China Agricultural University 4(suppl.), 16–19 (1999)

Ge, Y., Zhou, L.H., Zhang, G.Y., Zhang, P., Xin, G., Jiang, X.M.: Water production function and sensitivity index to water deficit for winter wheat in Shenyang Region. Journal of Shenyang Agricultural University 34(2), 131–134 (2003)

de Juan, J.A., Tarjuelo, J.M., Valiente, M., Garcia, P.: Model for optimal cropping patterns within the farm based on crop water production functions and irrigation uniformity I: Development of a decision model. Agricultural Water Management 31, 115–143 (1996)

Bergez, J.-E., Debaeke, P., Deumier, J.-M., Lacroix, B., Leenhardt, D., Leroy, P., Wallach, D.: MODERATO: an object-oriented decision tool for designing maize irrigation schedules. Ecological Modelling 137, 43–60 (2001)

Smith, M.: CROPWAT, A computer program for irrigation planning and management. FAO Irrigation and Drainage 46 (1992)

Xu, Z.Z., Zhou, G.S.: Agricultural water use efficiency and its response to environments and managing activities. Journal of Natural Resources 18(3), 294–303 (2003)

Zhang, B., Yuan, S.Q., Li, H., Cong, X.Q., Zhao, B.J.: Optimized irrigation-yield model for winter wheat based on genetic algorithm. Journal of Agricultural Engineering 22(8), 12–15 (2006)

Zhu, C.L., Peng, S.Z., Sun, J.S.: Reseach on optimal irrigation scheduling of winter what with water-saving and high efficiency. Journal of Irrigation and Drainage 22(5), 77–80 (2003)

Zhu, Y., Hu, J.C., Cao, W.X., Zhang, J.B.: Decision support system for field water management based on crop growth model. Journal of Soil and Water Conservation 19(2), 160–162 (2005)

Collection of Group Characteristics of Pleurotus Eryngii Using Machine Vision

Yunsheng Wang[1], Changzhao Wan[1], Juan Yang[1], Jianlin Chen[1], Tao Yuan[1], and Jingyin Zhao[1,2,*]

[1] Technology & Engineering Research Center for Digital Agriculture,
Shanghai Academy of Agriculture Sciences, Shanghai, P.R. China 201106
[2] No.2901 Beidi Road, Shanghai, 201106, P.R. China,
Tel.: +86-21-62204989
wys188@163.com

Abstract. An information collection system which was used to group characteristics of pleurotus eryngii was introduced. The group characteristics of pleurotus eryngii were quantified using machine vision in order to inspect and control the pleurotus eryngii house environment by an automated system. Its main contents include the following: collection of pleurotus eryngii image; image processing and pattern recognition. Finally, by analysing pleurotus eryngii image, the systems for group characteristics of pleurotus eryngii are proved to be greatly effective.

Keywords: Pleurotus eryngii, Group characteristics, Machine vision, Collection.

1 Introduction

World's edible fungi output reached 13 million tons at 2006, China is the largest producer of edible fungi, edible fungi production in China is the world's edible fungi production for more than 70%. Agricultural production in China, the sixth production of edible fungi, edible fungi have become the pillar industries in the agricultural economy (Huang Chienchun, 2006; Lu Min, 2006).

At present, the small-scale, extensive management production of edible fungi in China is still the main form, although this mode of production and low cost initial investment, but subject to natural risks and market risks is very weak, it is difficult to guarantee product quality, production efficiency is very low. Therefore, the industrialization and standardization of edible fungi production has become the development direction of edible fungi industry.

In the early 1990s, China has introduced several edible fungi production line, so that productivity increased a hundredfold, after nearly 20 years of practice and exploration, China has gradually developed a model of development of mushroom industry.

Edible fungi production in factories, we find that the edible fungi cultivation of industrial technology as the core, "Mushroom house climate control system" continues to experience-based control, far from being done in accordance with the edible fungi production automation features of the Tracking control. Edible Fungi groups

---

* Corresponding author.

D. Li and C. Zhao (Eds.): CCTA 2009, IFIP AICT 317, pp. 98–103, 2010.
© IFIP International Federation for Information Processing 2010

uniformity and the uniform shape and other characteristics of indicators, industrial cultivation of edible fungi has been the main basis for control measures. For the current status of these indicators of edible fungi in the diagnosis of a major in mind the number of measurements and visual way, there is a slow investigation, the charges, subjective strong, and the error of defects, and some features were also difficult to describe in considerable Restrictions on the extent of the industrial production of edible fungi in time during the strain of scientific management, industrial production of edible fungi is present on the "bottleneck" problem. Therefore, exploration accurate and rapid collection of edible fungi groups feature of the new methods, carefully observe the growth of edible fungi minor changes, summed up the seasons, days and nights change on the growth of edible mushrooms, edible fungi to establish the best model of environmental control So that the quality of edible mushroom cultivation factory stability, cultivation and management of fully automated intelligent, and machine vision technology in the industrial production of edible fungi in the show great potential for application.

Machine vision research using computers to simulate the visual function of things from an objective image to extract information, be understood to be processed and eventually used for the actual detection, measurement and control. Machine vision technologies, including image acquisition, image processing and pattern recognition, machine vision can be simulated by the human eye to the edible mushrooms to the close-up photography visible spectrum, and then use artificial intelligence, digital image processing techniques to analyze the image information, access to research Object required information.

2 Related Research

The use of machine vision technology feature extraction Mushroom has undertaken a number of studies, the representativeness of the research results:

(Van De Voo ren et al., 1992) The use of machine vision technology to determine morphological characteristics of a variety of mushrooms, Circularity, Bending Energy, Sphericity and Eccentricity and so on to describe the the the shape of the characteristics of mushrooms; (Yu GaoHong et al., 2005) Identification of the use of machine vision algorithms for contour description of the individual mushroom; (Vízhányó et al., 1998) color image analysis of edible fungi of the mechanical damage in the color and different color sick to to maximize the commercial value; (Tünde, et al., 2000) The use of image enhancement algorithm to distinguish between diseases of the color difference between edible fungi; (Heinemannp et al., 1994) The use of machine vision technology for classification of edible fungi; (NJ Kusabs et al., 2006), such as the use of machinery vision technology from a number of morphological parameters for grading of edible fungi; (Van Loon et al., 1996) The use of machine vision to detect the mushroom the size of individual developmental stages; (Tillett et al., 1989) in the edible mushroom harvesting machines use machine vision to locate Mushroom; (Tillett et al., 1991) of the edible fungi mushroom bed positioning algorithm; (Reed et al., 1995) The use of machine vision to identify the mushroom mushroom harvest size.

Study on the application of machine vision are mostly limited to simple, idealized monomer, the complex problem of overlapping research groups small, and the growth of the state of the monomer can not accurately reflect the actual population growth of the law can not be used to guide production management practices. Characteristics of the identifiable group because of its image with a lot of interference factors for image segmentation and feature extraction very difficult, very difficult to identify, and thus need to study more efficient segmentation algorithm based on feature extraction, as the demand for the development of applications depth, machine vision technology will be gradually applied to the recognition of group identities.

3   System Introduction

3.1   System Design

Precise control through the small environment the production of pilot experiments with the factory-based, as the leading machine vision technology, neural networks, machine learning as the link for example, integrated expert system and a variety of mushroom production technology and methods, and make full use of mushroom production expertise to achieve the characteristics of the smart machine groups mushroom identification and diagnosis of mushroom growth of the state of the technological breakthrough.

The overall technology map as shown in Fig.1

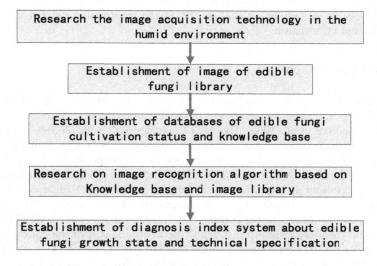

Fig. 1. Overall technology map

3.2   Research Methods and the Implementation of the Program

The implementation of the program in Fig.2

(1) Study the characteristics of pleurotus eryngii, the establishment of a wealth of examples of environmental test samples of small rooms, as a group of image characteristics and growth of information, data access to sources of information. Controlled trial of small environment, including different temperature, humidity, CO2

concentration to deal with. Industrial production of samples, including groups of as many as possible samples of the edible fungi. Through a rich mushroom habitat type and sample number of learning samples, to improve the accuracy of the machine to identify and enhance the applicability of the future to build systems.

(2) Pleurotus eryngii groups to carry out multi-dimensional space-time image data acquisition. pleurotus eryngii with cameras in key growth period, from the top, middle and base in different parts of using down, up, at eye level groups depending on the perspective of image; used traditional method homogeneous simultaneous determination of edible fungi and pleurotus eryngii covered neatly color, strength and other characteristics of growth examples of learning data, and invited experts in edible fungi pleurotus eryngii growth of the state assessment; the establishment of pleurotus eryngii growth of the state information database.

(3) Image feature extraction, image feature database to establish. To obtain images of pleurotus eryngii pretreatment to remove noise; to degradation caused by the phenomenon of recovery. The use of morphological operators and Hough transform techniques for the separation characteristics. Pleurotus eryngii growing in view of different culture production is based on different indicators, the growth of edible fungi in the group stage of the growth characteristics and state indicators to be classified by level of importance to determine the main objectives. Comprehensive application of

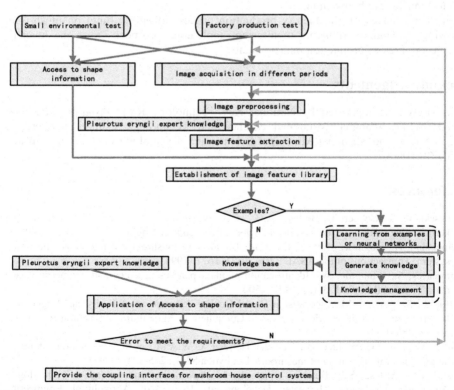

Fig. 2. Groups the characteristics of acquisition and pleurotus eryngii growth of the state of rapid diagnostic technology map

image recognition, neural networks, machine learning from examples, such as fuzzy mathematical algorithm to be addressed, and through comparison, the screening, to determine the optimal algorithm. Difficult for some of the characteristics of image recognition, will take full advantage of the experience of mushroom experts, in order to relax the requirements of machine vision technology.

(4) Summary of mushroom factory production management knowledge and experience to create the rules, the establishment of image recognition applications knowledge.

(5) Characteristics of groups in the pleurotus eryngii image acquisition, intelligent identification on the basis of the establishment of sets of machine vision technology in pleurotus eryngii growth of the state of diagnosis index system and technical specifications.

4   System Characteristics and Prospects

System is to be used in machine vision technology to identify intelligence mushroom groups and growth characteristics of the state of the practice of rapid diagnosis of an attempt to change the traditional mushroom factory groups characterized by the production of visual and manual collection method to enable automatic environmental control mushroom house intelligent.

Research based on the factory production process of edible fungi on mushroom growing demand for dynamic monitoring, application prospects, raise the level of mushroom factory production management.

Acknowledgements

This study was supported by National Key Technology R&D Program (NO.2006 BAD10A11), Shanghai Municipal Science and Technology Commission Project (073919103) and Shanghai Municipal Science and Technology Commission project (08DZ2210600).

References

Gunasekaran, S., Cooper, T., Berlage, A.G., et al.: Image processing for stress cracks in corn kernels. Transactions of the American Society of Agricultural Engineers (1), 266–271 (1987)

Heinemann, H., Hughesr, Morrow, C., et al.: TGrading of mushrooms using a machine vision system. Transactions of the ASAE 37(5), 1671–1677 (1994)

Humphries, S., Simonton, W.: Identification of plant pans using color and geometric image data. Trans of the ASAE 36(5), 1493–1500

Felfdi, J., Szepes, A.: Machine Vision Based Quality Assessment of Fruits and Vegetables. In: Proceedings of the World Congress of Computers in Agriculture and Natural Resources, pp. 42–48 (2002)

van de Vooren, J.G., Polder, G., Van der Heijden, G.W.A.M., et al.: Application of image analysis for variety testing of mushroom. Euphytica 57(3), 245–250 (1991)

Reeda, J.N., Milesa, S.J., Butlera., J., et al.: AE-Automation and Emerging Technologies Automatic Mushroom Harvester Development. Journal of Agricultural Engineering Research 78(1), 15–23 (2001)

Kacira, M., ling, P.P.: Design and development of an automated and non-contact sensing system for continuous monitoring of plant health and growth. Trans of the ASAE 44(4), 989–996 (2001)

Ling, P.P., Kuzhitsky, V.N.: Machine vision techniyucs for measuring the canopy of tomato seedling. Journal of Agricultural Enginccring Itescarch 65(2), 85–95 (1996)

Marchant, J.A., Ongungo, C.M., Street, M.J.: Computer vision for potato inspection without singulation. Computers and Electronics in Agriculture (4), 235–244 (1990)

Morimoto, T., Hashimoto, Y.: AI approaches to identification and control of total plant production for SPA&SFA to environmental control. In: 3rdIFAC/CIGR workshop on artificial intelligence in Agriculture, pp. 1–20 (1998)

Kusabs, N.J., Trigg, L., Bollen, A.F., Holmes, G.: Objective measurement of mushroom quality relative to industry inspectors. International Journal of Postharvest Technology and Innovation 1(2), 189–201 (2006)

Reed, J.N., Crook, S., He, W.: Harvesting mushrooms by robot. In: Science and cultivation of edible fungi, pp. 385–391 (1995)

Seginer, I., Elster, R.T., Goodrum, J.W., et al.: Plantwilt detection by computer vision tracking of leaftips. Transactions of the ASAE 36(5), 1563–1567 (1992)

Shimizu, H., Hcins, R.P.: Computer vision based system for plant growth analysis. Transaction of the ASAE 38(3), 959–964 (1995)

Tarbcll, K.A., Rcid, J.F.: A computer vision system for characteriziug corn growth and development. Transaction of the ASAE 31(5), 2215–2255 (1991)

Tillett, R.D.: Image analysis for agricultural processes: a review of potential opportunities. J. Agric. Engng. Res. 50, 247–258 (1991)

Tillett, R.D., Batchelor, B.G.: An algorithm for locating mushrooms in a growing bed. Comp. Electron. Agric. 6(3), 191–200 (1991)

Tillett, R.: Locating mushrooms for robotic harvesting, SPIE Conference on J. Non-destructive crop mcasurentcnts by image processing for crop growth control. Journal of Agricultural Engineering Research 61(2), 97–105 (1995)

Van Loon, P.C.C.: Het bepalen van het ontwikkelingsstadium bij dechampignon met computer beeldanalyse. Champignoncultuur 40(9), 347–353 (1996)

Vízhányó, Tillett, R.D.: Analysis of mushroom spectral characteristics. In: Proceedings of AgEng 1998 International Conference on Agricultural Engineering, pp. 24–27 (1998)

Woebbeckc, D.M.M., Von Bargen, K., et al.: Shape features for identifying young weeds using image analysis. Transaction of the ASAE 38(1) (1995); Intelligent Robots and Computer Vision VIII, 260–269 (1989)

Trooien, T.P., Hermann, D.F.: Measurement and simulation of potato leaf area using image processing. Transactions of the ASAE 35(5), 1719–1721

Trooieu, T.P., Hermann, D.F.: Measurement and simulaticm of Ilotato leaf area using image processing (II) Model development. Transaction of the ASAE 35(5) (1992)

Tünde, Vízhányó, Felfldi, J., et al.: Enhancing colour differences in images of diseased mushrooms. Computers and Electronics in Agriculture 26(2), 187–198 (2000)

Van De Vooren, J. G., Polder, G., et al.: Identification of mushrooms cultivars using image analysis. Transactions of the ASAE 35(1), 347–350 (1992)

Wolfe, R.R., Swaminathan, M.D.: Determining orientation and shape of bell peppers of machine vision. Transactions of the American Society of Agricultural Engineers (6), 1853–1856 (1987)

GaoHong, Y., Yun, Z., Ge, L., et al.: Transactions of the Chinese Society of Agricultural Engineering 21(6), 101–104 (2005)

The Design of Flower Ecological Environment Monitoring System Based on ZigBee Technology

Xiaoqing Guo[1,*] and Xinjian Xiang[2]

[1] School of Biological and Chemical Engineering, Zhejiang University of Science and Technology, Hangzhou, Zhejiang Province, P.R. China 310023
Tel.: 0571-85070376
hzguoxiaoqing@163.com
[2] School of Automation and Electrical Engineering, Zhejiang University of Science and Technology, Hangzhou, Zhejiang Province, P.R. China 310023

Abstract. Ecological environment is the key point of improving the flower's quality and quantity. Due to China's flower production management at a lower level, there is no scientific method in real-time monitoring of the flower's ecological environment. In order to solve the problem such as high costs; poor monitoring point scalability, poor mobility and other issues in traditional flower basement's data acquisition system, this paper devises a wireless real-time system based on ZigBee technology for the monitoring of flower's ecological environment. By the analysis of ZigBee technology's characteristics, it focuses on the design of wireless gateway with S3C4510B; wireless sensor node control module AT89S51 and the communication module CC2430; analyses the Zigbee protocol stack network's formation and designs data acquisition and communication procedures. By monitoring every flower's ecological environment indicators in practice, this system can meet the needs of the real-time monitoring for flower's ecological environment.

Keywords: Flower basement; eco-environmental monitoring; Zigbee; wireless sensor network; data acquisition.

1 Introduction

China is the world's largest producer of flowers, flower production occupies an important share of the national economy, but in the current production management, there is no scientific method in real-time monitoring of the flower's ecological environment such as growth conditions, cultivation and prevention of pests and diseases. (Ji Qing et al., 2007) Wireless sensor network (WSN) is consist of a large number of low-cost micro-sensor node with communication, sensing and computing deployed in the monitoring region, through self-organization constitutes a "smart" monitoring and control network (Jiang Ting et al., 2006). University of California, Berkeley used Micromote nodes to deployed wireless sensor network in a 70M red cedar for monitoring

---

D. Li and C. Zhao (Eds.): CCTA 2009, IFIP AICT 317, pp. 104–108, 2010.
© IFIP International Federation for Information Processing 2010

its environment change. Intel used Crossbow's Mote nodes in the deployment of WSN in a vineyard to monitor small changes in environment (Li Xiaomin et al., 2007). These studies provide an effective reference to the real-time monitoring for flower's ecological environment.

Zigbee with star, mesh, tree-like network topology, can be used in the WSN network and other wireless applications. ZigBee works in the 2.4GHz license-free frequency band can accommodate up to 65,000 nodes. Those nodes with low-power consumption can work 6 to 24 months by two AA batteries (Liang Yufen et al., 2007). Besides, it has a higher degree of reliability and security. These advantages of ZigBee-based wireless sensor networks are widely used in industrial control, consumer electronics devices, automotive electronics, home and building automation, medical equipment control(Liu Yongqiang et al., 2007).

For the effective monitoring flower ecological environment, improve the production and management process, and promote high-yielding flowers and eugenics, this paper devises a wireless real-time system based on ZigBee technology for the monitoring of flower's ecological environment, realized the real-time monitoring of flower ecological factors such as temperature, humidity, light, as well as nutritional status, provided technical support for the best fostering strategy and scientific management.

2 System Design

At present, many companies have introduced their Zigbee wireless development system based on CC2430. Because of its low-cost, this paper choose protocol stack source code and development kit support Zigbee protocol standards provided by TI company. Flower ecological environment with complex environment variable need the network routing and data fusion capabilities, and any network structure are based on a simple star structure. This star structure is also studied in this paper. A number of Zigbee terminal nodes (RFD) and a Zigbee Coordinator (FFD) compose a star-WSN, RFD sensors collect environmental parameters (temperature, humidity and light, etc.), and send to the FFD through the network, finally feedback PC by the serial interface. WSN nodes including RFD and FFD, with MCU and RF chip mode to meet the complex monitoring stability and applicability of the ecological environment. Each node is consisting of data acquisition sensors, processing module, wireless communication module and power module.

2.1 Data Acquisition Module

Temperature, moisture and light are important factor in the monitoring region. This paper take temperature for example, introduced the WSN data acquisition module. DS18B20 digital temperature sensor with 12-bit Celsius temperature measurements, it communicates over a 1-Wire bus and has an operating temperature range of -55℃ to +125℃. DS18B20 includes 3-pin grounded (GND), data input / output pins (DQ) and power line (VDD), as well as two power supply mode parasitic capacitance and external +5V power. In the paper, access from the VDD power supply pin make the DQ does not need to strengthen the pull, so the bus controller does not keep high in the temperature conversion and allow the data exchanges, GND pin can not be left vacant.

2.2 Microprocessor and RF Module

Microprocessor as the key unit finishes data collection, processing and delivery with other units. Zigbee protocol stack on the system require 8-bit microprocessor; full-function device (FFD) node of the protocol stack, ROM <32k; simple function device (RFD) node protocol stack, ROM about 6k; RFD also need to have sufficient RAM to save node binding table, found table and routing table. AT89S51 as microprocessor and CC2430 as RF chip had been selected.

3 Software of Wireless Data Acquisition Node

3.1 Data Transmission Networks

FFD and RFD initialization procedure both include hardware initialization, Zigbee network initialization and serial communications initialization, call the stack function ZigBee_Init () and Console_Init () to complete Zigbee agreement and serial communications initialization, at the same time call application-layer function Hardware_Init () to complete hardware initialization. Define the current Primitive to track network communication primitives, at the beginning of program running, current Primitive mean NO-PRIMITIVE, when it finish other primitives, it mean NO-PRIMITIVE again.

3.1.1 RFD Nodes Software Flowchart

Top layer send NLME-NETWORK-FORMATION request to network protocol layer for establish a new network. After the establishment of network, network layer will ask MAC layer for the energy detection scan to the channel that protocol or physical layer provides. Successfully received the results of energy detection scans, it will sort the channel with increased energy and discard those channel beyond their energy range, finally choose the allowed channel. If the network establish success, report to the Top layer through NLME-NETWORK-FORMATION confirm, else RFD will connect MAC layer device's network. MAC layer using the primitive MLME ASSO-CIATE indication initialization, if the network allows devices to join, MLME – AS-SOCIATE response its primitive state. At last, MAC layer send a successful response to network layer, and network layer report to RFD. RFD sends inquiries key information to determine whether there is any terminal node to send information.

3.1.2 FFD Nodes Software Flowchart

After initialization, first of all to determine whether the equipment has joined the network. If FFD already joined the network, send a message directly; If have not yet joined the network, it will be treated as a solitary point to join the original network, or the network layer rescan the new network; if joined a new network, the application layer send NLME-NETWORK-DISCOVERY request primitive to bottom layer. The primitive includes scan channel parameter and time parameter, network layer receive the primitive and ask MAC layer to run scan program, feedback the scan result to application layer, then application layer will connect one of those networks and send NLME-JOIN request to the network layer. Network layer find o a suitable father de-vice from its neighbor table, and send MLME-ASSOCIATION request to MAC layer for a connection. The result will back through MLME-ASSOCITION confirm primitive. If FFD join network as a solitary point, send NLME-JOIN request to network layer for a connection. If already joined network, it would start the environment parameter conversion and send information such as temperature and humidity to RFD which ask for data query.

3.2 Temperature Acquisition Software Flowchart

DS18B20 has a 1-Wire bus architecture, temperature conversion and read/write need follow basic rules by writing command in ROM (write 44H mean start the temperature conversion once; write BEh mean read the register). The data acquisition process is function row_reset () as sensor reset and initialization; function write_byte () write command to the ROM for next temperature conversion or read/write; function read_byte () read the register which save the temperature value and send the value to wireless protocol stack.

4 System Testing

After finish the program write to the RFD and FFD, connect to computer by RS-232, using serial debugger software as screen output on computer. The software baud set at 19200bps with 8-bit data bit, 1 stop bit, no parity-tested and data flow control bit. Then power the RFD devices, they will search the channel to establish network, if done, display 4 bit data with 16-band, which means the new network number. Now RFD allows FFD join the new network. Power the FFD devices and wait until the network address display on the screen, which means the whole network establish successfully.

Once the network established, push button at the RFD device to send the inquire command, when the FFD device receive this command, it start the temperature conversion and send the temperature value back to RFD device, which output: "Received 25.0" on the computer by RS-232. "25.0" is the region temperature where set those RFD and FFD devices, the wireless temperature acquisition finished. Other devices in the star network work as the same way finally send their region's environment parameters and complete the multi-point real-time acquisition.

5 Conclusion

Modern agricultural production needs the support from science and technology. In this paper, AT89s51 MCU and CC2430 RF chip complete the wireless sensor networks with star topology. Zigbee protocol stack complete the data exchange and network management by the primitive used in each layers. RFD sensors collect flower basement's environment parameters and send from application layer to physical layer by primitives, through wireless physical channel to the FFD device, completed the wireless real-time acquisition. The design provides a simple low-cost solution for wireless data acquisition, an effective method for monitoring flower environment parameters and flower growth, a great support for flower cultivation in decision-making and production management. It's an agricultural automation system example, with the agricultural scale development, this agricultural automation system based on wireless network technology has a very broad application prospects.

Acknowledgements

This material is based upon work funded by Zhejiang Provincial Natural Science Foundation of China under Grant No. Y108268.

References

Qing, J., Peiyong, D.: Study and implementation of a wireless sensor network based on zigbee technology. Sensor World (10), 30–35 (2007)

Ting, J., Chenglin, Z.: Zigbee technology and application, pp. 167–237. Press of Beijing University of Posts and Telecommunications (2006)

Xiaomin, L., Zhihong, Z., Zhi, G., et al.: Study and experiment on zigbee wireless sensor network. Electronic measurement technology 30(6), 133–137 (2007)

Yufen, L., Deyun, G., Yanchao, N., et al.: Review of the application system of WSN. Application of electronic technology (9), 3–9 (2007)

Yongqiang, L., Bing, Z.: Environmental monitoring for the design of wireless sensor network node. Popular Science & Technology 99(5), 46–47 (2007)

Zigbee Wireless Sensor Network Nodes Deployment Strategy for Digital Agricultural Data Acquisition

Xinjian Xiang[1,*] and Xiaoqing Guo[2]

[1] School of Automation and Electrical Engineering, Zhejiang University of Science and Technology, Hangzhou, Zhejiang Province, P.R. China 310023
Tel.: 0571-85070268
hzxxj@sina.com
[2] School of Biological and Chemical Engineering, Zhejiang University of Science and Technology, Hangzhou, Zhejiang Province, P.R. China 310023

Abstract. ZigBee is an emerging wireless network technology, according to china's digital agricultural feature such as remote, dispersion, variability and diversity, the ZigBee-based wireless sensor network for digital agricultural data acquisition is one of the best ways to build the system. In this paper, based on ZigBee wireless sensor network deployment planning principles and the status of our digital agriculture, we study several ZigBee wireless sensor network nodes deployment program for different condition and calculated their largest network capacity, network latency and other parameters, finally proposed a wireless sensor network nodes deployment strategy for digital agricultural data acquisition. This strategy is no specific requirements to the wireless sensor network topology structure, and can support the ZigBee wireless sensor network by random and manual deployment.

Keywords: Digital agricultural data acquisition, zigbee wireless sensor network, node deployment network planning; capacity calculation.

1 Introduction

China's agriculture has characters such as geographically dispersed, diverse objects, biological variation, and uncertain environmental factors, also the most obvious areas affect by environment, it's essential for digital agricultural data acquisition (Wei Quan et al., 2008). Existing data acquisition system use manual or pre-wiring cable acquisition. Those methods increase the work and can not keep real-time effective data acquisition; they have obvious limitations affected by geographical location, physical lines and the complexity of environmental factors. Modern wireless network technology and computer application develop fast in several years, it is necessary to combine the last technology with the area of agriculture. According to china's digital agricultural feature such as remote, dispersion, variability and diversity, the ZigBee-based wireless sensor network for digital agricultural data acquisition is one of the best ways to build the system (Lu Zhao quan et al., 2008).

---

* Corresponding author.

D. Li and C. Zhao (Eds.): CCTA 2009, IFIP AICT 317, pp. 109–113, 2010.

Wireless sensor network is consist of many micro-sensor nodes deployed in the monitoring region. Through wireless communication, a form of self-organizing multi-hop network, which aims at collecting objects' information without topographical constraints and send to control center (Tian Wang-lan et al., 2008). Zigbee wireless technology with a small size, low power feature, therefore, the establishment of a ZigBee-based wireless sensor network technology can meet these requirements. In this paper, based on ZigBee wireless sensor network deployment planning principles and the status of our digital agriculture, we study several ZigBee wireless sensor network nodes deployment program for different condition and calculated their largest network capacity, network latency and other parameters, finally proposed a wireless sensor network nodes deployment strategy for digital agricultural data acquisition.

2 ZigBee Technology Introduction

ZigBee is a new kind of short-distance, low-power, low data transfer rate, low cost, low complexity wireless network technology, is a communication solution between wireless markup technology and Bluetooth technology, mainly used for close wireless connections. Zigbee connect and communicate among thousands of tiny sensor. These sensors require very little energy to send data from one sensor to another sensor through radio waves in a relay way, and the communication efficiency is very high (Ban Yanli et al., 2007).

The foundation of ZigBee is the IEEE 802.15.4 agreement, which is consist of PHY and MAC layer based on IEEE 802.15.4 and the network and application support layer of ZigBee. The network system has low power consumption, low cost, short delay time, high capacity, safety features. IEEE 802.15.4 agreement clearly defines the three kinds of topologies: star structure, cluster-like structure and network structure. Agreement defines two kinds of physical equipment used mutually, named the full function equipment (FFD) and simplify function equipment (RFD):

The full function equipment support any kind of topology structure, and can communicate with any equipment; simplified function equipment only support star structure, and can only communicate with the full function equipment (FFD). IEEE 802.15.4 network need one full function equipment as network consultation, and terminal node generally use simplified function equipment (RFD) to reduce the system cost and improve power battery life. In ZigBee network layer, the distributed address scheme is used to distribute network address. This scheme distributes a limited network address for the equipment, and these addresses are the unique in a particular network. ZigBee Coordinator is a full-function device, which decided the greatest number of sub-equipment to connect its network.

3 Zigbee Wireless Sensor Network Node Deployment in Digital Agricultural Data Acquisition

Before the design, it is necessary to know product application environment and Zigbee characteristic well. Application environment includes data acquisition structure, wireless channel and transmission environment, and product features include RF output power and receiver sensitivity.

Network capacity and latency should be considered in general node deployment. Standard ZigBee network capacity can support up to 65,000 network nodes, each two adjacent nodes need 15ms to complete a communication. However, in practical applications need to consider network coverage and response time. Node capacity is proportional to network coverage, and system doesn't with long response time. This requires design a different network topology structure for different application environment. Design of network capacity and node deployment in different network topology structure under ideal conditions as follows.

3.1 Space Structure of Digital Agriculture Data Acquisition

China's agriculture has characters such as geographically dispersed, diverse objects, biological variation, and uncertain environmental factors. According to China's topography, climate truck characteristics and the regional agricultural economic development in different ways, agricultural production area can be plain, high slope, hills, grasslands, mountains, lakes, wetlands and so on, corresponding to the respective sites of agricultural production is farmland, orchards, gardens, pastures, fish ponds, fisheries, etc. Based on the principle of ZigBee networking, the structure of digital agricultural data acquisition as followed:

(1) Linear structure. Such as small rectangular greenhouses, slender shape fish ponds.
(2) Planar network structure. Such as a square canopy, large-scale farmland, pasture, etc...
(3) Space network structure. Such as orchards, gardens in the mountains and so on.
(4) Mixed structure. It is a cyberspace with linear, planar network and space network structure. Such as large-scale farm include fish ponds, farms, and orchards.

3.2 Zigbee Wireless Sensor Network Nodes Deployment with Linear Structure

Linear structure network is a simply network can be used in rectangular greenhouses' digital agricultural data acquisition. The entire network, only a single path, which determines the number of nodes equal to network layers, or hop (Hop). The central node located in the center of the entire network, network scanning cycle (the time central node collecting network data form all the backbone nodes) directly depend on the network hops, or backbone nodes. Every communication cycle for the calculation of 15ms, then the entire network scanning cycle T can be expressed as:

$$T=15*2*(1+2+3+\cdots n) \tag{1}$$

Where: T for the entire network scan cycle (ms); n for the network layer, and the network nodes is 2n. When T = 30s, the calculation of available n = 43, that is, linear structure network has maximum capacity of 86 nodes in the 30s scanning cycle. Communication distance of 100m (0.1km) calculated, then the coverage of the entire network for (n-1) * 2 * 100 = 8.4km long linear region. Based on the above formula, when n = 22 , T = 5s, that is, network can be divided into six slip in the 30s scanning cycle, each slip has maximum capacity of 44 nodes, this will allow network coverage up to 14km2(approximate $\pi r2 = 13.85$), the entire network capacity also increased to 264 nodes.

Analysis of Conclusions: a single slip of linear network has maximum capacity of 86 nodes in the 30s scanning cycle, and to minimize hops help increase network capacity.

3.3 Zigbee Wireless Sensor Network Nodes Deployment with Planar Network Structure

Planar network structure such as large-scale farmland, pasture is complex, due to the multi-path network; scanning time analysis is also more complicated, this paper takes square instead of round for a brief analysis.

ZigBee planar network structure diagram as shown in Fig.1, assuming set a node at any two straight lines' cross-point, and central node at the center of network, there are 8 nodes(which around the central node) can communicate directly (1 hop), 16 nodes for 2 hop and 24 nodes for 3 hop. The whole network scan time can be expressed as:

$$T=15ms*8* (1+4+9+\cdots+n *n) \tag{2}$$

$$N=8*(1+2+3+\cdots+ n) \tag{3}$$

Where: n for the network layer; N for the network nodes' capacity.

When T = 30s, the calculation of available n = 8, N = 268, that is, planar network has maximum capacity of 268 nodes in the 30s scanning cycle. Communication distance of 100m (0.1km) calculated, and then the coverage of the entire network is 1.6 * 1.6 = 2.56 km2 area. If the central node at the edge of the entire network topology, it is clear that the network will increase the number of layers, extend the scanning time and reduce the network nodes' capacity.

Analysis of Conclusions: The central node should set at the center of the network topology as much as possible, the more close to the edge, the longer scanning time, and in the limited scan time, the whole capacity of the network will become even smaller.

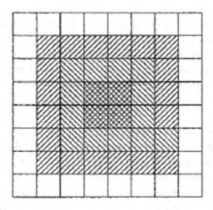

Fig. 1. ZigBee planar network structure

3.4 Zigbee Wireless Sensor Network Nodes Deployment with Space Network Structure

Space network structure is more complex. It is conceivable that in the N-storey orchard, each storey with n blocks of terrace set one node. In this structure, the whole network scans time as follows:

$$T = N*t + n*15 \text{ ms } (1+2+3+\cdots+N-1) \tag{4}$$

Where: T for the entire network scanning cycle; t for single-storey scan cycle; n is the number of nodes on each plane; N for the space layers. Based on the above formula, each plane can be treated as a planar network structure, and all of these make the space network structure. In 30s scanning cycle, it is no more than 11 layers when each layer has 25 nodes. If treat each plane as a two-branch linear network structure, it is no more than 8 layers when each layer has 24 nodes.

Analysis of Conclusions: The system performance of space network structure is relate to each layers' node number. In order to increase the network capacity, every layer's node number should be limited.

4 Conclusion

Zigbee network nodes deployment strategy is very important in the process of build digital agricultural data acquisition network. In order to reduce the workload of optimize network deployment, and build flexible network, the network topology structure should be decided by the environment. Network deployment is a systematic project, need for a wealth experience of wireless network planning and accumulation.

Acknowledgements

This material is based upon work funded by Zhejiang Provincial Natural Science Foundation of China under Grant No. Y108268.

References

Rui, W.Q.H., Qiang, J.: Industrial Control Network Model Based on ZigBee. Mathematics In Practice and Theory 15, 174–177 (2008)

Zhao quan, L., et al.: Greenhouse Planting Temperature Monitoring System Based on ZigBee Wireless Network. Journal of Anhui Agri. Sci. 36(13), 5682–5684 (2008)

Wang-lan, T.: Studying of the Factor that Affects the Energy Consume in the Wireless Sensor Network. Computer Knowledge and Technology 4, 742–744 (2008)

Yanli, B., Qiaolin, C., Chen, W.: ZigBee tree routing algorithm based on energy balance. Computer Applications 11, 2791–2794 (2008)

Research on 3G Technologies-Based Agricultural Information Resource Integration and Service

Nengfu Xie[1,3,*] and Wensheng Wang[2]

[1] Agricultural Information Institute, The Chinese Academy of Agricultural Sciences,
Beijing, China, 100081
nf.xieg@caas.net.cn
[2] Key Laboratory of Digital Agricultural Early-warning Technology, Ministry of Agriculture,
The People's Republic of China
[3] Agricultural Information Institute, No.12 Zhongguancun South St.,
Haidian District Beijing 100081, P.R. China
Tel.: +86-10-82109819; Fax: +86-10-82109819

Abstract. With the urgent demand of agricultural information for farmers and quick development of telecommunications industry in china, the convenience, quickness and validity of agricultural information service are becoming more important. Now, the problem of the digital divide between rural and urban areas comes from the shortage of effective information transmission means. But the agricultural information services limited by network bandwidth and geography will be changed completely with the mobile phone network coverage to rural areas, especially TD-SCDMA network into rural areas in china. The paper will introduce the development of the 3G technologies-based effect on the agriculture informationization, and discuss the problems of agricultural information resource integration and service adapted to 3G technologies in detail, based on which a 3G technologies-based agricultural information resource application mobile service hierarchy (3GAIMSH) is provided.

Keywords: Agricultural Information, 3G, TD-SCDMA, service design, integration.

1 Introduction

With the urgent demand of agricultural information for farmers and quick development of telecommunications industry in china, the convenience, quickness and validity of agricultural information service are becoming more important. The shortage of agricultural information, poor telecommunications networks and uncoordinated information resources are crucial problems facing agricultural information sources in china. How to integrate the agricultural information coming from different agricultural information sources and provide information service for farmers in rural areas, especially in many poor rural areas is the current research focus of information service because human factors are still the main means of collecting and disseminating

---

* Corresponding author.

D. Li and C. Zhao (Eds.): CCTA 2009, IFIP AICT 317, pp. 114–120, 2010.

information in many rural areas. In the near future, wireless networks will appear with greater ubiquity, from public access networks offering connectivity in agricultural information services, to telecoms operated 3G and 4G networks, especially TD-SCDMA network into rural areas in china, which allows 9 hundred million of farmers access to services as important and varied as information acquisition, health care, education, and financial and governmental services. The mobile phone has been described as the most likely modern digital device to support economic development in developing nations (The real digital divide, 2005). Tapan presented CAM - a framework for developing and deploying mobile applications in the rural developing world. Supporting minimal, paper-based navigation, a simple scripted programming model and offline multimedia interaction, CAM is uniquely adapted to rural user, application and infrastructure constraints (Parikh etc., 2006). The agricultural information services limited by network bandwidth and geography will be changed completely with the mobile phone network coverage to rural area, which will help solve the problem of "last kilometer" in agricultural informationization.

Third generation mobile network (3G) is the latest advancement in the field of mobile technology. Providing high bandwidth communication of 8kbit/s-2Mbit/s and a revolutionary introduction of multimedia services over mobile communication, it aims to make mobile devices into versatile mobile user terminals. TD-CDMA (Time Division CDMA) wireless technology in china can meet the rapidly growing demand for mobile broadband services in agricultural information systems. The mobile information services will make it easier and quicker to provide agricultural information services in today's more complex web environment.

Now, the problem of the digital divide between rural and urban areas comes from the shortage of effective information transmission means. But the agricultural information services limited by network bandwidth and geography will be changed completely with the mobile phone network coverage to rural areas, especially TD-SCDMA network into rural areas in china. The paper will introduce the development of the 3G technologies-based effect on the agriculture informationization, and discuss the problems of agricultural information resource integration and service adapted to 3G technologies in detail, based on which a 3G technologies-based agricultural information resource application mobile service hierarchy is provided.

2 The Problems of Agriculture Information Services

As the amount and complexity of web agricultural information continues to grow, quick and efficient information retrieval & intelligent information services are becoming a critical focus for research and development, especially for multiple heterogeneous information sources. The information sources are distributed at different locations, most of which are belonged to many research or commercial unites' private proprietary information. Such this situation makes public users or farms impossible to access the information. On the other hand, the usage rate of the distributed information is very low. In fact, these sources can be combined and integrated into a unified environment, which provides a united interface for users, requiring information, who will not need to reach each information and seeks the information that they need.

2.1 Digital Divide

Currently, many information resources in a certain unit can not be very limited or no access at all. Though, Accessing the Internet will bring a wealth of information to all agriculture farmers in rural areas and will help in overcoming the digital divide. As most farmers in china have no hands-on experience or access to digital networks, and most information on the web is very confused that may bring many problems in agriculture production process and agriculture production market. On the other hand, most useful and practical information are coming from agriculture government and Research Institutes. But the access of such information may be limited. So we an open and integrated agricultural information system should be encouraged, which help farmers access a lot of useful information if each country tries to develop contents in the language people are using.

2.2 Lack of Rapid and Timely Information Distribution

With special characteristic such as region, domain and timeness, information required cannot reach users in many rural areas. Rural areas have spatially intermittent connectivity. By requiring online usage of an application, villagers would have to travel to a connected location to access it. This can be very inconvenient, possibly taking an entire day. One of the most innovative new tools for rural development might be in your pocket.

Over the last few years cell phone use has exploded in china, with more uses than anyone could have imagined. Using communication technologies for agricultural development was the theme of the World Conference on Agricultural Information and IT held in Japan last week. Over a billion people use the internet, but more than three times as many use cell phones. Especially with the 3D technology development, agricultural information required is no longer a problem for the most isolated rural area, but the big need is for good content.

2.3 Limitations of Current Mobile Software Platforms

While mobile phone hardware is well-suited to rural conditions, the same is not true for software. Mobile web interfaces are notoriously difficult to use, even for developed world users (George etc., 2001). Typing URLs with a numeric keypad is slow and painful. Therefore, users must rely on a portal or set of bookmarks to access web sites. Most web pages are designed for large screen resolutions, making navigation within a page also problematic (Parikh & Lazowska, 2006). Currently mobile software platforms in agricultural domain have been established in many rural areas in China. However, how to push the right content to users in a right way is still a big problem because there is not a content-abundant integrated environment that has many applications and tools suitable 3G technology-based information process and distribution. Especially, with the emergence of 3G technology, information access and distribution are becoming very easily, and available for users in many rural areas.

3 The 3GAIMSH Hierarchy

At present, the mobile service of agricultural information had been conducted in many provinces in China, especially TD-SCDMA based network into rural areas in china recently. 3G mobile technology will make video telephony or video call in areas within a network coverage and use a wider range of value added services. Such mutimedia-base content will help farmers to understand agricultural information more easily. By 3G mobile network, the agricultural production direct, news, maket information and etc. will be transmitted to farmer's 3G mobile phones, which are available for viewing and exporting retrieved information.

In 3GAIMSH hierarchy, we provide a environment that integrate related different agricultural information sources into the agricultural information center by data adapter. The source can be deleted, added and indexed dynamatically. The center will also provide Update Tool guarantees that all data are always upto- date. The 3G-based information distribution service ensures that the information requested will be pushed farmer's 3G mobile phones quickly.

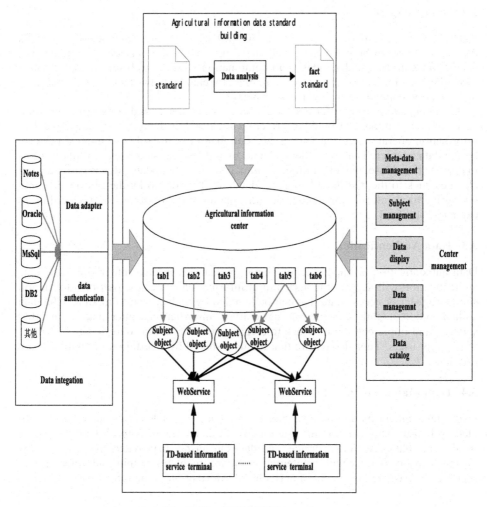

Fig. 1. The 3GAIMSH hierarchy

3.1 Agricultural Information Sharing Standard Building

In order to let distributed information sources join the information center, the related agricultural information sharing standard must be built, including access standard, service standard, and exchange standard. In building these standards, firstly agricultural information model must be studied and get an application-suitable information standard. In the access standard, the criterion of information access interface must be defined. We have created a web service interface supporting the construction of distributed agricultural access interfaces. For information exchange, we have defined XML-based agriculture domain information exchange standard. To improve interoperability of Web Services, service standard will consider SOAP 1.1, UDDI 2, ... to help the deployment of profile compliant Web Services.

3.2 Data Adapter

Data adapter is an important part of the 3GAIMSH hierarchy. Data adapter provides the communication between the distributed sources and the agricultural information center. We can use the data adapter in combination with the web service-based interface. That is these two interfaces combine to enable both data access and data manipulation capabilities using access standard.

The data adapter can perform select, insert, update, delete and conversion information operations in the distributed information sources. For the insert, update and delete information operations, we are using the continuation of the select command perform by the data adapter. That is the data adapter uses the select statements to fill an information Set and use the other three commands (insert, update, delete) to transmit changes back to the distributed sources' database. In order to let the information into the agricultural information center, the information must be transferred according to exchange standard.

3.3 Data Authentication

Data authentication is important for many applications in data integration. The data authentication will keep in the agricultural information center the part of information source register & access data that can be handled by authentication module. The data authentication module can provide agricultural information center data from other sources, and implement an interface to create new information sources, which will be sharing for users. This also ensure that the data shared by a certain users with a given privilege.

3.4 Data Management

Data Management module is designed for the data add, delete, and query in distributed agricultural information integration environment. It will control the data quality of different databases, data service, and data clustering. It provides the management on the distributed information integration environment for helping administrator to regulate the basic management of agricultural information in the information center.

3.5 Metadata-Based Information Category and Service

Metadata information category & service will help users find the information they required easily. The information meta-data is building an object-oriented repository technology that is integrated with agricultural meta data management that process metadata, which will provide a united category that organize the information in information center, and a standard of the information' store, and support exchange of model data. The metadata-based information service will provide a united interface for users to access the information by metadata-based information category.

3.6 Agricultural Information Distribution

Agricultural information distribution is about assuring that the right agricultural information is available to the right agricultural users. In order to distribute the information to the users timely, we will build 3G-based information distribution service, by which the information will be accessed by the 3G-based mobile phone. In the means, the information distribution will help the communication between the users and the information center, which can accept the request and send answer to the users. The information Distribution also makes a plan, and pushes the information to the users' 3G mobile phone by 3G network.

4   Conclusion

In the paper, we have presented the research, design and analysis of 3GAIMSH hierarchy -- a novel 3G technologies-based agricultural information resource application mobile service hierarchy for rural information services in china. The hierarchy-based application will help integrate distributed agricultural information source into a united environment to be reuse and sharing, and provide the right information to the farmers or other users. In the future, more 3GAIMSH-based applications and tools will be developed, and the paper will study ontology-based agricultural information organization for providing users more intelligent agricultural information service.

Acknowledgements

The work is supported by the Academy of Science and Technology for Development fund project "intelligent search-based Tibet science & technology resource sharing technology", Special Project on of The National Department of Science and Technology "TD-SCDMA based application development and demonstration validation in agriculture informationization", Special fund project for Basic Science Research Business Fee, AIIS (No.2009J-06) and the National Science and Technology Program (Grant No. 2006BAD10A06).

References

Parikh, T.S.: Using Mobile Phones for Secure, Distributed Document Processing in the Developing World. IEEE Pervasive Computing, 4(2), 74–81 (2005)
The real digital divide. The Economist (March 2005)

Buchanan, G., Farrant, S., Jones, M., Thimbleby, H., Marsden, G., Pazzani, M.: Improving mobile Internet usability. In: WWW 2001: Proceedings of the 10th international conference on World Wide Web, pp. 673–680. ACM Press, New York (2001)

Parikh, T.S., Lazowska, E.D.: Designing an architecture for delivering mobile information services to the rural developing world. In: Proc. WWW 2006 (2006)

Research on Agricultural Surveillance Video of Intelligent Tracking

Lecai Cai*, Jijia Xu, Jin Liangping, and Zhiyong He

Institute of Computer Application, Sichuan University of Science and Engineering,
Zigong Sichuan, P.R. China 643000,
Tel.: +86-13508170118; Fax: +86-813-5505966
clc@suse.edu.cn

Abstract. Intelligent video tracking technology is the digital video processing and analysis of an important field of application in the civilian and military defense have a wide range of applications. In this paper, a systematic study on the surveillance video of the Smart in the agricultural tracking, particularly in target detection and tracking problem of the study, respectively for the static background of the video sequences of moving targets detection and tracking algorithm, the goal of agricultural production for rapid detection and tracking algorithm and Mean Shift-based translation and rotation of the target tracking algorithm. Experimental results show that the system can effectively and accurately track the target in the surveillance video. Therefore, in agriculture for the intelligent video surveillance tracking study, whether it is from the environmental protection or social security, economic efficiency point of view, are very meaningful.

Keywords: video tracking, Mean-Shift algorithm, Epanechnikov core, density estimation.

1 Introduction

China's agricultural research, although the application of information technology is springing up in recent years, but the development is very fast. During the period of 1970s to 1980s, the State Science and Technology Commission, the Ministry of Agriculture has supported some crops expert system, decision support system, and agricultural production management system, Scientific and technological information networks, databases and other systems of research making some important results, many of them have been applied. Since 1996, the national "863" Project carried out. The application of intelligent agricultural information technology demonstration projects, research and development of a series of technical content was high, with independent intellectual property rights of agricultural technology should be service platform and a variety of practical tools, has established more than 100 agriculture applications. Intelligent expert system has a wide range in the domestic demonstration area from the original four provinces have developed to the 20 provinces. Therefore the research on agricultural surveillance video of intelligent tracking will bring us significant economic benefits.

---

* Corresponding author.

D. Li and C. Zhao (Eds.): CCTA 2009, IFIP AICT 317, pp. 121–125, 2010.

The first concept of Mean Shift (Fukunaga et al,1975) is Fukunaga et. in 1975 in a density gradient with respect to the probability estimation function proposed by its original meaning, as its name implies, is to shift the mean vector, where is the Mean Shift a noun, it refers to a vector, however, with the development of Mean Shift Theory, the meaning of Mean Shift has changed, if we say that Mean Shift algorithm, generally refers to an iterative approach, that is calculated to offset the current point average, to move the point to its partial Mean-shift, and then as a new starting point and continue to move until the end to meet certain conditions.

However, after a long period of time has not Mean Shift unnoticed until 20 years later, that is, in 1995(Yizong Cheng et al, 1995), and Mean Shift on one of the important documents was published. In this important literature, Yizong Cheng of the basic Mean Shift Algorithm (Tan Tieniu et al, 2002) in the following two have done a promotion, first of all, the definition of a family of Yizong Cheng kernel, making shift (Liu Ruizhen et al, 2007;Yu Shiqi et al, 2007) with the sample points with different distance, the offset of the mean shift vector contribution (Wang Liang et al, 2002; Hu Weiming et al, 2002; Tan Tieniu et al, 2002) are also different, followed by Yizong Cheng also set a weight coefficient, making the importance of different sample points are not the same, which greatly expanded the scope of application of the Mean Shift.

Behind, this article will detail the basic idea of Mean Shift and its expansion, as well as the algorithm steps. Finally, this article will also be given in the video.

Mean Shift tracking of specific application (Liu Fuqiang et al, 2003; Li Lianye et al, 2005; Yan Hui et al, 2005; Zhu Weidong et al, 2005).

2 Density Estimation

Density estimation (Silverman B W et al, 1986) is from a group of unknown probability density distribution of the observations to estimate the probability density of its distribution to meet. There are usually two methods: parametric and non-parameter method. Parameter method is based on the assumption of data points is determined by our known distribution (such as Gaussian distribution) arising from, and then by the known distribution to approximate the requirements of the distribution; not known to find the distribution of its similar distribution, so that the probability estimation more accurate density distribution. Non-parametric density estimation method of many, such as the histogram method, the nearest method, kernel density estimation (Philippe Van Kerm et al, 2003.) methods, etc., which is the kernel density estimation of the most widely used technology, the following kernel density estimation is given the formula. In d dimensional space Rd in a given n data points (xi) i = 1,2 ,..., n, with K core density is estimated that the width of the window h. There are

$$f(x) = \frac{1}{nh^d} \sum_{i=1}^{n} K(\frac{x - x_i}{h}) \tag{1}$$

Which the kernel function K (x) must satisfy two conditions: 1, K as a symmetric probability density function (such as Gaussian density);

2, $\int_0^\infty K(x)dx = 1$。

For K (x), there are several different types of kernel functions to choose from, including Epanechnikov core(Scott D W et al, 1992) in the sense of minimum mean-variance optimal, one of the kernel function: In this paper, the experiment is carried out using core Epanechnikov density estimate of.

$$K_e(x) = \begin{cases} \dfrac{1}{2} c_d^{-1} (d+2)(1 - \|x\|^2) & \text{if } \|x\| < 1 \\ 0 & \text{otherwise} \end{cases} \tag{2}$$

Of which: cd as a unit the size of d-dimensional sphere, for example, c1 = 2, c2 =, c3 = apparently, according to type (1) data can be a given probability density function estimation.

3 Mean Shift Algorithm

Mean Shift algorithm(Dorin Comaniciu et al;Peter Meer et al,1999) gradient method is iterative calculation using the probability density function of extreme points. Data have been obtained in accordance with the probability density function, its gradient is:

$$\nabla f(x) = \frac{1}{nh^d} \sum_{i=1}^{n} \nabla K(\frac{x - x_i}{h}) \tag{3}$$

Can be further transformed into:

$$\nabla f(x) = \frac{n_x}{n(h^d c_d)} \frac{d+2}{h^2} (\frac{1}{n_x} \sum_{x_i \in S_h(x)} [x_i - x]) \tag{4}$$

The scope of which Sh (x) for the radius of spherical super-h, volume of hdcd, their centers for x and include data points nx.

Mean Shift Vector Mh (x) is defined as:

$$M_h(x) = \frac{1}{n_x} \sum_{x_i \in S_h(x)} [x_i - x] = \frac{1}{n_x} \sum_{x_i \in S_h(x)} x_i - x \tag{5}$$

The same type (4) can be written:

$$M_h(x) = \frac{h^2}{d+2} \frac{\nabla f(x)}{f(x)} \tag{6}$$

Style (5) and type (6) show that the local Mean Shift vector mean and the difference between window centers, point to the probability density function of the direction of the peak or valley points. The calculation steps are as follows:

First of all, to jump out of a cycle threshold e, from x = xi, i = 1 ,..., n of departure: 1, calculating the Mean Shift Vector Mh (x); 2, will move the window by Mh (x) the size and the direction of movement; 3, if Mh (x) than e, and then jump back to step (1) repeat; 4, to the point of convergence Department reservations.

It can be seen, Mean Shift to speed up the increase in variable step size (down), and ultimately converge to the probability density function of the peak (valley) points.

4 Mean Shift Algorithm to Realize Thought a Experiment

Mean Shift-based video tracking algorithm in the experiment, mainly through the tracking target on the current frame compared with the adjacent frames to determine the next frame the candidate for the current module or modules. Overall thinking is as follows:

Initialize the location of the current frame, and for the current Bhattacharyya coefficient (hereinafter referred to as bc coefficient) values. Calculate the location of new goals, new objectives and then update the location of the distribution, the experiment used to carry out core Epanechnikov kernel density estimation.

Then bc obtained updated coefficient, calculated the distance between two vectors, that is relatively current and updated bc coefficient, so as to come up with the movement more similar to the frame. Frame and set the current frame, so again the next candidate for the current module and the comparison module, until the track before the experiment to reach the frame default value, end of the code used to define the variables. Manually in the first frame (input file type) are identified, automatic target tracking. The experiments also support mouse control, which can be used in tracking the course of the mouse to control the suspension, termination or continuation of such.

In order to verify the previous chapter described the principle of Mean Shift tracking algorithm, this chapter will be in the video file in a cat on the track movement experiment. This video is downloaded from ww.snuffx.com. In this experiment the cat is a moving target, jumping from one side to the other side, moving target small and fast application of the agricultural surveillance video of intelligent tracking have a strong persuasive. Sequence in each frame size of 320*240 in the first frame to be determined manually to determine the objectives of tracking, the algorithm in VC + +6.0 under development. Take the following video of the cat as an example, once every 10 frames as follows:

Tracking at the 4th frame

Tracking at the 14th frame

Tracking at the 24th frame

Tracking at the 34th frame

5 Conclusion

The agricultural tracking intelligent video surveillance is a new research topic, this article through experiments. Especially for the tracking algorithm analysis, Mean Shift Algorithm for the traditional model in the entire follow-up due to always remain the same cycle, positioning is not accurate enough, especially in the experiment to track moving objects in video action on the substantial difficulties in tracking moving target, this experiment a little Mean Shift Algorithm improved. Generally speaking, the more successful experiments and experiments show that the method of a small amount of real-time high, able to effectively track the moving target video.

Acknowledgements

Funding for this research was provided by Department of Science and Technology of Sichuan Province (P.R. China).

References

Silverman, B.W., et al.: The Estimation of the Gradient of a Density Function, with Applications in Pattern Recognition 48(9), 1022–1023 (1975)

Tieniu, T.: Intelligent Video Surveillance Technology. In: The First National Conference on intelligent visual surveillance, Beijing, vol. 38(11), pp. 1007–1011 (2002)

Scott, D.W.: Multivariate Density Estimation, vol. 56(11), pp. 1037–1044. Wiley, New York (1992)

Ruizhen, L., Shiqi, Y.: OpenCV Tutorial, 1st edn., pp. 357–392. Beijing University of Aeronautics and Astronautics Press, Beijing (2007)

Liu, F.-q.: Mathematics Video Surveillance System Development and Application, pp. 112–117. Mechanical Industry Press, Beijing (2003)

Lianye, L., Yan-hui, Z.W.: TV monitoring system to prevent invasion of design and construction technology, vol. 13(2), pp. 138–149. Publishing House of Electronics Industry, Beijing (2005)

The Estimation of the Gradient of a Density Function, with Applications in Pattern Recognition 38(5), 478–481 (1975)

Liang, W., Ming, H., Tieniu, T.: Visual analysis of human movement. A synthesis of Chinese Journal of Computers 25(3), 225–237 (2002)

Silverman, B.W.: Density Estimation for statistics and Data Analysis, vol. 48(9), pp. 698–701. Chapman and Hall, London (1986)

Van Kerm, P.: Adaptive kernel density estimation. In: 9th UK Stata Users meeting. In: 9th UK Stata Users meeting,Royal Statistical Society, London, May 19-20, vol. 26(10), pp. 147–151 (2003)

Comaniciu, D., Meer, P.: Mean Shift Analysis and Applications. In: Proceedings of the International Conference on Computer Vision, vol. 27(12), pp. 1191–1197 (1999)

Expert Control Based on Neural Networks for Controlling Greenhouse Environment

Le Du

Beijing Institute of Technology, Beijing, 100081, China

Abstract. Depending upon the nonlinear feature between neural units in artificial neural networks (ANN), artificial neural network was used to develop a model for greenhouse inside air temperature management. Data was collected and processed for training and simulation of temperature model. The design of the network structure and transfer function was also discussed. Because heuristic logic of greenhouse environment exists, a simple ANN controller does not perfect complex task, the algorithm of ANN with heuristic logic is developed. Experimental result indicated that the performance of the new control scheme meet the requirement of greenhouse environment control.

Keywords: complex system, neural network, greenhouse environment.

1 Introduction

The greenhouse environment belongs to the complex system (G. van Straten et al., 2000; N. Sigrimis et al.,2000; M.Trigui et al., 2000), it is difficult to control due to high nonlinear, and uncertain. System configuration is shown in Fig. 1.

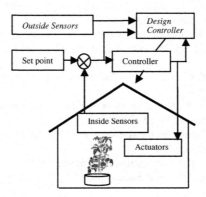

Fig. 1. Greenhouse Environment Control Architecture

The main purpose of greenhouse is to improve the environmental conditions in which plants are growth .The aim of greenhouse environmental control is to provide means to further improve these conditions in order to optimize the plant production process .The greenhouse climate is influenced by many factors, for example inside air

D. Li and C. Zhao (Eds.): CCTA 2009, IFIP AICT 317, pp. 126–132, 2010.
© IFIP International Federation for Information Processing 2010

temperature, the actuators, outside weather and crop. Inside air temperature is the most important factor affecting crop growth in greenhouse climate. Temperature exerted significant effects on photosynthesis. The net photosynthetic rate (Pn) varied in day time, it was double peak curve ,there was obvious midday depression of photosynthesis, and 21-33℃ was the suitable temperature for photosynthesis in summer. Methods aimed at efficiently controlling the greenhouse temperature must take these influences into account, and that is achieved by the use of models. This model is of capital interests in the study of the dynamics of the greenhouse temperature under different climate conditions, aimed at producing inside temperature regularity which helps to avoid extreme situations (high temperature or low temperature, etc.) and to optimize crop production by achieving adequate temperature integrals while reducing pollution and energy consumption.

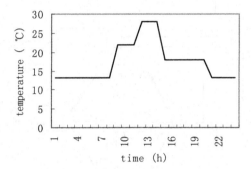

Fig. 2. Chart of variable – temperature

In this control system for plant growth, the key problems include control system input so called set point and control algorithm. The traditional set point is shown in Fig. 2. Because set point changes abruptly, that wastes a large of energy sources. So we must seek a method to transform a segment of steps chart into a smooth cure for saving energy.

If we control a variable – temperature, the traditional method is to use a controller (PID) to generate control output u(t). However, the control u(t) relates many actuators, so optimal method is used to decide what actuator to act. Since the traditional method is difficult to control this system, Artificial neural networks and symbolic description methods are introduced.

Artificial neural networks are widely applied in greenhouse environmental control to perform some type of non-linear system mapping. In the fields of identification and modeling of non-linear systems their universal approximation property is exploited. Previous work on this ANN modeling task relied in an input output model structure selected from (Cunha et al., 1996) in the context of dynamic temperature models identifications. This structure was selected by means of the (Akaike 1974) information criterion where several hypothesis were tested and the best one chosen. The application of RBF NN to greenhouse inside air temperature modeling has been investigated (Ferelra, P.M. Ruano, A.E., 2002). These models always arise like non-stability, non-regularity, which result in the complete loss of energy. This paper uses artificial neural network theory and method to control the range of temperature change and discuss the ANN model with heuristic logic.

2 Algorithm Design of Neural Networks Direct Self-tuning Control for Greenhouses with Heuristic Logic

Direct self-tuning control with an artificial neural network is shown in Fig.3 which includes three basic section:

(1) Feedback loop consists of self-tuning controller and controlled plant.
(2) The parameters of the controller are obtained by designing artificial neural networks identification and controller.
(3) Heuristic logic decision what actuators are operated .

It is obvious that the key technique of self-tuning controller is how to design artificial neural networks identification and controller well and give appropriate heuristic logic.

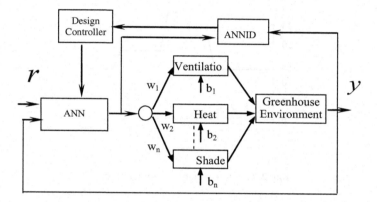

Fig. 3. Neural networks indirect self-tuning control Consider the nonlinear system, $w_1, w_2, \cdots, w_n$ are the weight value of n actuators, $b_1, b_2, \cdots, b_n$ are the valve value of n actuators.

$$y(k+1) = g[y(k),\cdots,y(k-n+1;u(k),\cdots,u(k-m+1)] + \varphi[y(k),\cdots,y(k-n+1);u(k),\cdots,u(k-m+1)]u(k) \tag{1}$$

Where u, y are the input and output of the system, and $g[\bullet], \varphi[\bullet]$ are nonzero function.

If $g[\bullet], \varphi[\bullet]$ are known functions, control algorithm of the controller according to certain equivalence principle is as follow:

$$u(k) = \frac{-g[\bullet]}{\varphi[\bullet]} + \frac{r(k+1)}{\varphi[\bullet]} \tag{2}$$

The control system attempts to make the output $y(k)$ match the reference input $r(k)$.

If $g[\bullet]$, $\varphi[\bullet]$ are unknown functions, the artificial neural networks identification is trained on line so that $g[\bullet]$, $\varphi[\bullet]$ approximate controlled plant. Using $Ng[\bullet], N\varphi[\bullet]$ substitute for $g[\bullet], \varphi[\bullet]$, control algorithm of the controller according to certain equivalence principle is as follow:

$$u(k) = \frac{-Ng[\bullet]}{N\varphi[\bullet]} + \frac{r(k+1)}{N\varphi[\bullet]} \qquad (3)$$

where $Ng[\bullet], N\varphi[\bullet]$ is nonlinear dynamic the artificial neural networks respectively.

2.1 The Artificial Neural Networks Identification

For simple problem, one order system is governed by the following equation:

$$y(k+1) = g[y(k)] + \varphi[y(k)]u(k) \qquad (4)$$

The artificial neural networks identification is shown in Fig. 3. The identification consists of two three-lays nonlinear DTNN, where number of order of $Ng[\bullet], N\varphi[\bullet]$ is equal to number of order of $g[\bullet], \varphi[\bullet]$. The input of the artificial neural networks is $\{y(k), u(k)\}$, and the output is as follows:

$$\hat{y}(k+1) = Ng[y(k); W(k)] + N\varphi[y(k); V(k)]u(k) \qquad (5)$$

where W, V are two neural networks weights which are given:

$$W(k) = [w_0, w_1(k), w_2(k), \cdots, w_{2p}(k)]$$
$$V(k) = [v_0, v_1(k), v_2(k), \cdots, v_{2p}(k)]$$

with p is hidden layer' node number

$$w_0 = Ng[0, W]; \quad v_0 = N\varphi[0, V]$$

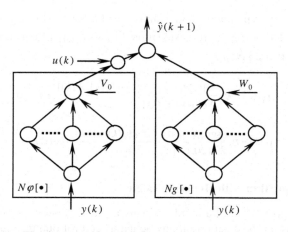

Fig. 4. The artificial neural networks identification

In Fig. 4, L is linear node, and H is nonlinear node.

substitute (4) into (5), then the output of the control system:

$$y(k+1) = g[y(k)] + \varphi[y(k)] \left\{ \frac{-Ng[\bullet]}{N\varphi[\bullet]} + \frac{r(k+1)}{N\varphi[\bullet]} \right\} \tag{6}$$

if $Ng[\bullet] \to g[\bullet]$ and $N\varphi[\bullet] \to \varphi[\bullet]$,

$y[k] \to r[k]$.

By minimizing the quadratic performance criterion, get weights tuning process thought training the artificial neural networks identification

$$E(k) = \frac{1}{2}[r(k+1) - y(k+1)]^2 = \frac{1}{2}e^2(k+1) \tag{7}$$

$$W(k+1) = W(k) + \Delta W(k); \quad V(k+1) = V(k) + \Delta V(k)$$

Using Bp learning algorithm (Ferelra et al., 2002; Tetsuo et al., 2000)

$$\Delta w_i(k) = -\eta_w \frac{\partial E(k)}{\partial w_i(k)}; \quad \Delta v_i(k) = -\eta_v \frac{\partial E(k)}{\partial v_i(k)}$$

Substitute (7) into (8), we get:

$$\Delta w_i(k) = -\eta_w \frac{\varphi[y(k)]}{N\varphi[y(k);V(k)]} \left\{ \frac{\partial Ng[y(k);W(k)]}{\partial w_i(k)} \right\} e(k+1) \tag{8}$$

In the some way, we get:

$$\Delta v_i(k) = -\eta_v \frac{\varphi[y(k)]}{N\varphi[y(k);V(k)]} \left\{ \frac{\partial N\varphi[y(k);V(k)]}{\partial v_i(k)} \right\} e(k+1)u(k) \tag{9}$$

where $\varphi[y(k)]$ is unknown and sign is known, which is marked as $\text{sgn}\{\varphi[y(k)]\}$. After $\text{sgn}\{\varphi[y(k)]\}$ is substituted to $\varphi[y(k)]$ of equation (9) and (10), $w_i(k+1) \cdot v_i(k+1)$ are given by

$$w_i(k+1) = w_i(k) - \eta_w \frac{\text{sgn}\{\varphi[y(k)]\}}{N\varphi[y(k);W(k)]} \left\{ \frac{\partial Ng[y(k);W(k)]}{\partial w_i(k)} \right\} e(k+1) \tag{10}$$

$$v_i(k+1) = v_i(k) - \eta_v \frac{\text{sgn}\{\varphi[y(k)]\}}{N\varphi[y(k);V(k)]} \left\{ \frac{\partial N\varphi[y(k);V(k)]}{\partial v_i(k)} \right\} e(k+1)u(k) \tag{11}$$

2.2 Control Algorithm with Heuristic Logic

As stated above, given control u involves many actuators operation, so control output u of fist step does not take action on any actuator, control output u must accomplish following mapping to yield real control output for actuator.

$$u(t) \cdot \omega_1 - b_1 = u_{r1}$$
$$u(t) \cdot \omega_2 - b_2 = u_{r2}$$
$$\vdots \tag{12}$$
$$u(t) \cdot \omega_n - b_n = u_{rm}$$

Where $w_1, w_2, \cdots, w_n$ and $b_1, b_2, \cdots, b_n$ result from heuristic logic, so the control algorithm possesses intelligence.

3 Experimental Results of Control

In this part, the effectiveness of the proposed control method will be demonstrated by a case study for winter heating system, many actuators are forbidden due to energy conservation, the actuator of heating system operates only. In this case, $w_2 = 1$, and b_2 is determined by prior knowledge. Fig 5 is experimental chart measured by the control system in Feb. 9, 2004.

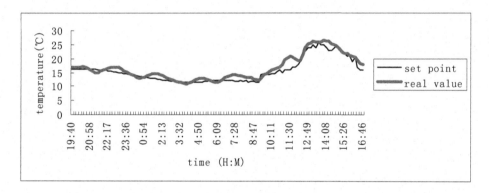

Fig. 5. Comparison chart of temperature between set point and real value

From Fig 5, it is seen that the chart of real control value tracts that of set point and light intensity increases in Fig 5 gradually at 7 to 16, provided energy for greenhouse decrease gradually. Energy chart omitted.

4 Conclusion

The presented the algorithm of ANN with heuristic logic proved to be very effective in meeting formal requirements for greenhouse control such as set point and tracking set point. ANN modeling set point is feasible. The key technique for greenhouse environment control uses ANN and heuristic logic reasonably to create a good algorithm. The proposed control algorithm is currently implemented in based CAN bus control system for greenhouse environment control.

References

van Straten, G., Challa, H., Buwalda, F.: Towards user accepted optimal control of greenhouse climate. Comp. Electronics Agric. 26(3), 221–238 (2000)

Sigrimis, N., Anastasiou, A., Rerras, N.: Energy saving in greenhouses using temperature integration: a simulation survey. Comp. Electronics Agric. 26(3), 321–341 (2000)

Trigui, M., et al.: A strategy for greenhouse climate control. part I: model development. J. Agric. Engng. Res. 78(4), 407–413 (2000)

Cunha, J., Boaventura, A.E., Farria, E.A.: Dynamic temperature models of a soil-less greenhouse. In: Proceedings of the 2nd Portuguese Conference on Automatic Control, Portuguese Association of Automatic Control, Porto, Portugal, vol. 1, pp. 77–81 (1996)

Akaike, H.: A new look at the statistical model identification. IEEE transactions on Automatic Control 19, 716–723 (1974)

Ferelra, P.M., Ruano, A.E.: Choice of RBF model structure for predicting greenhouse inside air temperature. In: 15th Triennial World Confress, IFAC, Barcelona, Spain (2002)

Morimoto, T., Hashimoto, Y.: An intelligent control for greenhouse automation, oriented by the concepts of SPA and SFA —— an application to a post-harvest process. Comp. Electronics Agric. 29(2), 3–20 (2000)

A Review of Non-destructive Detection for Fruit Quality

Haisheng Gao, Fengmei Zhu, and Jinxing Cai

Department of Food Engineering, Hebei Normal University of Science & Technology,
Changli, Qinhuangdao, Hebei Province, P.R. China 066600

Abstract. An overview of non-destructive detection in quality of post-harvest fruit was presented in this paper, and the research and application were discussed. This paper elaborated the fruit quality detection methods which were based on one of the following properties: optical properties, sonic vibration, machine vision technique, nuclear magnetic resonance (NMR), electronic noses, electrical properties, computed tomography. At last, the main problems of non-destructive detection in application were also explained.

Keywords: fruit quality, non-destructive detection, research, application.

1 Introduction

Fruit commercialization is adopting the scientific method and detecting, grading, packaging the fruits on the basis of comprehending the physiology metabolism law, protecting and improving the quality, and achieving the change from the elementary raw material to high added-value commodity. Nowadays, reducing the consumption of post-harvest fruit is the most concerned question for the world agricultural trade. It was reported that the consumption of post-harvest fruit in developed countries accounted for the 15-20% of the total amount. China is the world's largest fruits and vegetables production country. The breeding, culturing, and pest control was paid much attention, however, the post-harvest processing technology was neglected, the question of detecting, grading, transporting, preservation was not solved, so the lost of post-harvest fruits and vegetables in circulation was huge, the loss ratio was 30%~40% every year(Gao Haisheng. 2003, Li Lite et al. 2003, Lu Lixin et al.2004).

With the rapid development of science and technology and computer vision technique to the development of agricultural field, new methods of non-destructive detection for fruit quality were provided. The main methods included optical properties, sonic vibration, nuclear magnetic resonance (NMR), machine vision technique, electrical properties detection, computed tomography and electronic noses technique and so on.

2 Detection of Fruit Quality Using Optical Properties

Due to the difference of fruits internal components, fruits have different absorption and reflection properties in the light of different ray. The non-destructive measurement for fruits quality was achieving in the combination of these properties and optical detection device.

D. Li and C. Zhao (Eds.): CCTA 2009, IFIP AICT 317, pp. 133–140, 2010.
© IFIP International Federation for Information Processing 2010

In the fruit interior quality inspecting systems and their detecting principle, Ying yibin et al. (2003) analyzed three different kinds of measuring methods, such as regular reflection, transmittance and diffuse reflection, and elaborated the research and application technology for sugar content, acidity and firmness of fruits. Liu yande et al. (2003) studied the principle of fiber sensing technique, designed a detection system for fruit quality, and investigated a fiber sensing technique in reflectance, interactance and transmittance mode. The results showed diffuse reflection was the best method for fruits internal components detection.They also established the non-destructive method using infrared spectroscopy diffuse reflection technique for determination of the sugar content in apple and intact honey peach, and attained the good results (Liu Yande et al. 2004). At the same time, Fu xiaping et al. (2004) carried out the related experiments and obtained the good results too.

Han donghai et al. (2003) demonstrated that the absorbencies of the bruised and normal parts had the same variational tendency by analyzing on color, tissue and near infrared spectrum characteristics of the Fuji apple for their normal and bruised parts, and concluded Fuji apple had the least surface color difference in 1 hour compared to Qinguan and "Golden Delicious" apples. The range of wavelength from 760 nm to 960 nm could be used to detect the superficial damaged apple. The absorbencies increased with time according to the equation Y= aXb.

Tian haiqing et al. (2007) developed a measuring system for soluble solids content (SSC) of watermelon based on near infrared transmittance technique. The system mainly consisted of an available spectrometer, optic fiber transmittance accessories, data sampling card and light source. Prediction test for SSC of 50 'Qilin' watermelons were carried out by the system. The calibration modes between the spectra (the original spectra, its first derivative, and its second derivative) and SSC were established using partial least square and principle component regression methods. The results showed that the second derivative spectra with partial least square method provided the best prediction of the SSC of watermelons with the correlation coefficient of 0.951, the root mean square error of calibration of 0.347 and the root mean square error of prediction of 0.302. The correlation coefficient between predicted and measured values was 0.910.

Some scholars in Japan developed the sensor of detection the pear and apple maturity by visible light and infrared spectroscopy, then they developed the selecting fruit device which select fruit maturity and color quickly, and applied this technology to auto-selecting fruits production line, linked the maturity, color sensor, auto-grading, package production line, achieved a highly automatized non-destructive grading fruits (He Dongjian et al. 2001).

Liu xinxin et al. in China Agricultural University discussed the non-destructive measurement of water core in the storage process of apples. They established the equation of light intensity and weight with different time by detecting the change of light intensity with self-made differential instrument. In the storage, the light intensity of water core apples decreased more sharply than apples with no water core. In the later stage of storage, the symptoms of diseased fruit with not serious disease disappeared. Meantime, the morbidity of apples with larger weight had high incidence (Han Donghai et al. 2004). In addition, Han donghai et al. (2006) detected the internal breakdowns of apples by visible-near infrared spectroscopy (650 nm~900 nm), analyzed the spectra of the apples, and selected three wave numbers, 715 nm, 750 nm, and 810 nm as the character wave numbers. The results showed that the correct probability of classification was 95.65% by using the above-mentioned three wave numbers.

As a word, the detection for fruits using optical properties is one of the most practical and most successful technique in non-destructive measurement. It has the following characteristics: high-sensitivity detection, good adaptability, lightweight equipment, flexible usage, not harmful to humans. This technique has gradually applied to the practical stage on abroad.

3 Detection of Fruit Quality Using Sonic Vibration

It was reported that energy absorption and sonic and ultrasonic vibration reflected the status of internal damage, but these two techniques were more suitable for assessing bruise susceptibility. Compared with application of fluorescence and delayed light emission to chlorophyll containing fruits and vegetables, optical absorbance to all fresh produce was widely tested to evaluate damage. However, these methods developed were essential for detection of physical damage. For researchers and industry, an exciting prospect arises from a better understanding of internal damage of fruits and vegetables during post-harvest processing and circulation. X-ray analysis, magnetic resonance imaging, and laser inspection can now be used to detect the internal damage in some limited application; but not practical for routine damage testing because the equipment is expensive. Like all other technologies, the cost of sonic vibration detection method was reduced sharply, and detection capability was improved highly (Jiang Yueming et al. 2002).

4 Detection of Fruit Quality Using Machine Vision Technique

Machine vision technique had an earlier application in agriculture to identify plant species. With the rapid development of image processing technology and computer software and hardware, machine vision system had fast development for fruits quality auto-detection and grading. The study indicated that black-and-white image processing technology had already applied to apples surface bruise detection in U.S.A.

Ying yibin et al. (2004) explored a methodology for the maturity inspection of citrus with machine vision technology, and used the surface color information and the ratio of total soluble solid to titratable acid (TSS/TA) as maturity indexes of citrus. The results stated that at the wavelength of 700 nm, the green surface and saffron surface of citrus were of higher spectral reflection, the difference between them reached the maximum, and the image acquired at this wavelength could be of much color information for the maturity inspection. The test results showed that the identification accuracy was 91%.

A tendency of non-destructive detecting technology for quality and safety of agricultural and poultry products is a hyper-spectral imaging system, which possesses the advantages of both computer vision and spectroscopy inspecting technologies. Hyper-spectral imaging technology had higher wavelength resolution than multi-spectral imaging, the precision was 2∼3 nm. The inspecting technology of interior quality of kiwifruit by means of hyper-spectral imaging was discussed. The new process methodology of data for hyper-spectral imaging was analyzed and future research aspects were pointed out. The range of wavelength was 650~1100nm. The predict mode of SSC was established by PLS. The results showed that the predict of SSC by near-infrared spectra technique had high precision, predict error was 1.2Brix. The injury, decay, bruise and detection of soil pollution were studied by hyper-spectral imaging. Second uneven difference algorithm was designed to Separate the defects and contamination in apples. The defects and contamination regions were differentiated (Liu Muhua et al. 2005).

5   Detection of Fruit Quality Using Nuclear Magnetic Resonance (NMR)

Nuclear magnetic resonance (NMR) is a technique which detects the concentration of hydrogen nuclei and is sensitive to variations in the binding state. The researchers found that the mobility of water, oil and sugar hydrogen nuclei would change with the change of content in the maturation process of fruits. In addition, the concentration and mobility of water, oil and sugar related to mechanical injury, tissue degeneration, over maturity, decay, insect damage and frost injury. Thus different quality parameters of fruits would be detected by measuring the concentration and mobility based on these properties.

Researchers could measure many parameters of fruits quality using NMR image technique as non-destructive method. The relationship of NMR parameters and fruits quality parameters could be obtained easily, and the development of NMR technique was improved. Although NMR technique had already applied to detecting tumor and other medical fields commercially, the potential of testing fruits defects and other qualities was not totally developed. Therefore this technique has not been reported yet in China (Lu Lixin et al. 2004).

6   Detection of Fruit Quality Using Electrical Properties

With the drop of fruits' freshness, the equivalent impedance of spoiled or damaged apples is less than that of intact apples, but the measurement results may be affected by the excursion of frequency. The dielectric constant of spoiled or damaged apples was more than that of intact apples. Some experts studied the relationship of frequency properties of fruits electrical properties constant and fruits quality characteristics from 0.1 kHz to 100 kHz in frequency, taking apple and pear as experiment object. The results showed that frequency properties of fruits electrical properties constant and fruits quality characteristics had close relationship. The base of non-destructive detection and auto-grading for fruits was established (Zhang Libin et al. 2000).

Guo wenchuan et al. (2007) investigated electrical and physiological properties of peaches in order to understand electrical properties of post-harvest fruitss and to explore new quality sensing methods based on electrical properties. It was observed that the relative dielectric constant varied with cosine law roughly and loss tangent decreases as peaches' aging. The maximum relative dielectric constant appeared at peak of respiration. The reasons why electrical parameters change were analyzed. Furthermore, BP neural network technology was used to identify freshness of peaches when relative dielectric constant and loss tangent were selected as input characteristic parameters. Results showed the average distinguishing rate was 82%.

7   Detection of Fruit Quality Using Computed Tomography

Computed tomography (CT) is a method of examining body organs by scanning them with X rays and using a computer to construct a series of cross-sectional scans along a single axis. Xu shumin et al. used picked Fuji apples as the experiment objects. The layer of X-ray and computer scan was applied to detect the CT value of apples drop from different heights. On the same scan layer, the CT value of destructive apples decreased with increased of storage time, and the more destruction was made, the lower the CT value was. With the increasing the thickness of scan layer, the CT value of non-destructive apples decreased, while the CT value of destructive apples increased. By changing the storage time, the relation between CT value and destruction was also different (Xu Shumin et al. 2006). Zhang jingping et al. scanned the red Fuji apples with CT technology and analyzed the CT image properties of apples, obtained the significant linear relationship of sugar content of a point on fruits and CT value. Thus the sugar content would be attained from CT value, a new method of using CT image to detect sugar content distribution on line was obtained. The mode of linear relationship of sugar content of a point on fruits and CT value was established and verified by non-destructive method. The results showed that the average error rate of this mode was only 4.36%(Zhang Jingping et al. 2007). Meanwhile, the CT technique was used as the input of neural network to forecast the major components of Fuji apple. The results showed that the average forecast errors of moisture content, sugar content and acid content are 1.75%, 5.81% and 0.72% respectively, and the precision of this method could meet the requirements of Fuji apples non-destructive measurement(Zhang Jingping et al. 2008).

8   Detection of Fruit Quality Using Electronic Noses

Electronic noses have been developed as systems for the automated detection and classification of odors, vapors, and gases since the end of last century. An electronic nose is generally composed of a chemical sensing system and a pattern recognition system. There was a kind of portable non-destructive detector called Sakata fruits detector in Japan. It could detect the immature, mature and spoiled fruits with 99% accuracy. Zhang libin et al. (2000) developed a set of electronic nose system which composed of metal oxide semiconductor gas sensor array by simulating the functioning of the olfactory system, and analyzed the samples with neural networks. The detection accuracy was 80%.

Zou xiaobo et al. (2005) gave a new method to classify apples by the odor of apples, and developed an electronic nose equipment to classify apples. Fifty good apples and fifty bad apples was classed. Five feature parameters were developed from every data curve of sensor arrays, and all the feature parameters were called input vectors. Principal component analysis and genetic algorithm radial based function (GA-RBF) neural network were used to combine the optimum feature parameters. Good separation among the gases of different apples was obtained using principal component analysis but a bit was overlapped. The recognition probability of the GA-RBF to the learning samples and the testing samples were 100% and 96.4%.

9 Conclusion

The non-destructive detection methods including optical properties, sonic vibration, machine vision technique, nuclear magnetic resonance(NMR) for fruit quality had distinctive advantage compared to other instrumental analysis and chemical analysis methods, and had broad application prospects and development potential. Traditional chemical analysis methods had following disadvantage: time-consuming, hard sledding, high cost. Many analytical instruments had large volume and weight and high price. These instruments only were used in laboratory. It is necessarily for agricultural sector, quality inspection sector, market management sector to put in use portable non-destructive detection instruments for fruits and vegetables sampling detection. Non-destructive detection was also a development for researchers combine the research results and production practice. However, the non-destructive detection was only for one product and one item in China. The comprehensive detection methods for many internal qualities of fruits were deficient. With the development of non-destructive detection technique, data-processing technology, automatic control technology and computer technology will play an important role in non-destructive measurement for fruit quality.

The non-destructive detection methods have their strengths and weaknesses. For instance, the method of optical properties can detect the injury of fruits surface, but can not detect the internal quality. Sonic properties, NMR detection technique are still in laboratory experiment, not applying to commercial fields. The application of computer vision technique for post-harvest fruits commercialization is in practical stage at home and abroad. At the same time, with the rapid development of computer technology and specialization in many fields, the automation of fruits measurement will come true. Therefore it is essential that simple, quick, accurate and comprehensive detection methods apply to the research and application in the future. In addition, the combination of machinery and optics technology, multi-spectrum technology and machine vision technology should be strengthened. Last but not lease, the combination of independent research and develop and introduction from abroad should be adopted. New techniques apply to post-harvest fruits processing, the competitiveness of China's fruits in international market will be enhanced.

References

Haisheng, G.: China's Fruit and Vegetable Storage Industry to be a Breakthrough, 7.7(A4). China Food News, Beijing (2003) (in Chinese)

Lite, L., Jie, W., Yang, D., et al.: State and New Technology on Storage of Fruits and Vegetables in China. Journal of Wuxi University of Light Industry 22(3), 106–109 (2003) (in Chinese)

Lixin, L., Zhiwei, W.: Study of Mechanisms of Mechanical Damage and Transport Packaging in Fruits Transportation. Packaging Engineering 25(4), 131–134 (2004) (in Chinese)

Yibin, Y., Yande, L.: Study and application of optical properties for nondestructive interior quality inspection of fruit. Journal of Zhejiang Agricultural University (Agric.& Life Sci.) 29(2), 125–129 (2003) (in Chinese)

Yande, L., Yibin, Y.: A Study on Fiber Sensing Technique Used for Fruit Interior Quality Inspection. Journal of Transcluction Technology (2), 170–174 (2003) (in Chinese)

Yande, L., Yibin, Y.: Determination of Sugar Content and Valid Acidity in Honey Peach by Infrared Reflectance. Acta Nutrimenta Sinica 26(5), 400–402 (2004) (in Chinese)

Xiaping, F., Yande, L., Yibin, Y.: Application of Near Infrared Spectra Technique for Nondestructive Measurement of Fruit Internal Qualities. Journal of Agricultural Mechanization Research (2), 201–203 (2004) (in Chinese)

Donghai, H., Xinxin, L., Lili, Z., et al.: Color, Tissue and Near-infrared Spectrum Characteristics of Bruised Apples. Transactions of The Chinese Society of Agricultural Machinery 34(6), 112–115 (2003) (in Chinese)

Haiqing, T., Yibin, Y., Huirong, X., et al.: Near-infrared Transmittance Measuring Technique for Soluble Solids Content of Watermelon. Transactions of The Chinese Society of Agricultural Machinery 38(5), 111–113 (2007) (in Chinese)

Dongjian, H., Maekawa, T., Morishima, H.: Detecting Device for on Line Detection of Internal Quality of Fruits Using Near Infrared Spectroscopy and the Related Experiments. Transactions of The Chinese Society of Agricultural Engineering 17(1), 146–148 (2001) (in Chinese)

Donghai, H., Xinxin, L., Lili, Z., et al.: Research of Nondestructive Detection of Apple Watercore by Optical Means. Transactions of The Chinese Society of Agricultural Machinery 35(5), 143–146 (2004) (in Chinese)

Donghai, H., Xinxin, L., Chao, L., et al.: Study on Optical-nondestructive Detection of Breakdown Apples. Transactions of The Chinese Society of Agricultural Machinery 37(6), 86–89 (2006) (in Chinese)

Yueming, J., Shiina, T.: Advances in Evaluations of Damage of Postharvest Fruits and Vegetables. Transactions of The Chinese Society of Agricultural Engineering 18(5), 8–12 (2002)

Yibin, Y., Xiuqin, R., Junfu, M., et al.: Methodology for nondestructive inspection of citrus maturity with machine vision. Transactions of The Chinese Society of Agricultural Engineering 20(2), 144–147 (2004) (in Chinese)

Muhua, L., Jiewen, Z., Jianhong, Z., et al.: Review of Hyperspectral Imaging in Quality and Safety Inspections of Agricultural and Poultry Products. Transactions of The Chinese Society of Agricultural Machinery 36(9), 139–143 (2005) (in Chinese)

Libin, Z., fangJia, X., Canchun, et al.: Nondestructive Measurement of Internal Quality of Apples by Dielectric Properties. Transactions of The Chinese Society of Agricultural Engineering 16(3), 104–107 (2000) (in Chinese)

Wenchuan, G., Xinhua, Z., Kangquan, G., et al.: Electrical Properties of Peaches and Its Application in Sensing Freshness. Transactions of The Chinese Society of Agricultural Machinery 38(1), 112–115 (2007) (in Chinese)

Shumin, X., Yong, Y., Jun, W.: Study on CT Value of Damaged Apple. Transactions of The Chinese Society of Agricultural Machinery 37(6), 83–85 (2006) (in Chinese)

Jingping, Z., Hui, W., Zheng, P.: Study on Relationship Analysis of Apple Profile CT Value and sugar content distribution. Transactions of The Chinese Society of Agricultural Machinery 38(3), 197–199 (2007) (in Chinese)

Jingping, Z., Hua, Z., Hui, W.: Non-destructive Test of Fuji Apple's Major Components by CT. Transactions of The Chinese Society of Agricultural Machinery 39(7), 99–102 (2008) (in Chinese)

Xiaobo, Z., Jiewen, Z., Yinfei, P., et al.: Quality Evaluation of Apples Using Electronic Nose Based on GA-RBF Network. Transactions of The Chinese Society of Agricultural Machinery 36(1), 61–64 (2005) (in Chinese)

Automatic Grading of the Post-Harvest Fruit: A Review

Haisheng Gao*, Jinxing Cai, and Xiufeng Liu

Department of Food Engineering, Hebei Normal University of Science & Technology,
Changli 066600, Hebei Province, P.R. China
Tel.: +86-0335-2039374; Fax: +86-0335-2039374
spxghs@163.com

Abstract. Mechanical fruit grading and automatic fruit grading have been detailed in this paper. The studies and applications of mechanical fruit grading, and computer visual and automatic fruit grading were also particularized. Computer vision technology for detecting fruit size, color, bruise and surface defects and evaluation of fruit overall quality were discussed. The primary problems and development in the future in application of automatic fruit grading in China were pointed out in the end.

Keywords: fruit, grading, mechanization, automatization.

1 Introduction

Fruit commercialization is the main purpose of its grading. Fruit in the same tree differ in quality such as feature, flavour because their growth was effected by many environmental factors. Especially, fruit from different orchards differ significantly in size and quality. Grading may not only standardize fruit product but also promote management of the fruit tree in orchard and product quality (Haisheng Gao, 2003; Lite Li et al., 2003).

Fruit and vegetable are very difficult to grade exactly and rapidly because of their significant difference in feature such as size, shape and color as a result of changeable conditions of nature environment and manual factors. Grading of fruit and vegetable was performed primarily by visual inspecting in many countries. Manual grading is lack of objectivity, accuracy and has lower efficiency because there is individual difference in visual inspecting which is affected by human healthy condition, psychological condition, lightness, fatigue and so on. But levels of mechanical and automatic grading of fruit and vegetable go higher and higher with the enhancement of mechanization, automatization and application of computer technology.

2 Mechanical Fruit Grading

Mechanical fruit graders are classified by their working principles as size grader, weight grader, fruit color grader and fruit color and weight grader which classifies fruit by their size and color.

---

* Corresponding author.

D. Li and C. Zhao (Eds.): CCTA 2009, IFIP AICT 317, pp. 141–146, 2010.

Fruit grading by their size and color is the advanced technology of the post-harvest fruit handling nowadays in the world. Its working principle is a combination of automatic size grading and color grading. At first, size grading of fruit is done on conveyer belt with changeable apertures. Fruit down from the belt are irradiated by lamps under conveyer belt and yield reflex. Then the reflex signal was passed to computer and different grades of fruit such as the whole green, the half-green and half-red and the red under every conveyer belt were obtained according to different reflex signals. Finally, the graded fruit were transported by different conveyer belts. The capacity of the apple grading product line might reach 15 to 20 tons per hour (Haisheng Gao et al., 2002, 2003; Miller B. K., 1989).

The near-infrared (NIR) transmittance spectrum of peach from different areas in wavelength range from 730 to 900 nm was studied to estimate peach ripeness in terms of sugar content and firmness. The accuracy of the estimation reached 82.5% (Carlomagno et al. 2004).

The grading of "Jonagold" apple was studied at the most effective wavelength at spectral range from visible light to near infrared light (450nm to 1050nm) (Kleynen et al., 2003). Apple image information was gathered using CCD (Charge Coupled Device) and apple were sorted on the basis of their sizes, color and defects on surface. The results showed the high efficiency of apple grading was obtained at wavelengths of 450, 500, 750 and 800nm.

A grading criterion for cucumber was put forwards (Zengchan Zhao et al., 2001, 2003). Picking robots and conveyors based on machine vision system were developed and a set of hardware and software system for cucumber grading was designed.

3   Computer Vision and Automatic Grading of Fruit

Computer vision system can simulate human vision to perceive the three-dimensional feature of spatial objects and has partial function of human brain. The system will transfer, translate, abstract, and identify the perceived information, and consequently work out a decision and then send a command to carry out expectant task. The simple computer vision system consists of illuminating chamber, CCD camera, image collecting card and computer. The chamber maintains an optimal work condition for the camera, namely, keeping a symmetrical and identical illumination in CCD vision area. CCD camera is an image sensor for capturing image. The image collecting card abstracts the image and translate video signal into digital image signal. The computer handles and identifies the digital signal to work out a conclusion and explain.

3.1   Detection of the Fruit Size

Fruit images can be captured by computer vision system. And then the detection of their edge was executed. The measuring direction of the fruit size would be oriented according to the fruit symmetry, and then the fruit size would be measured. The method of detecting the fruit edge is very suitable for processing blurry image. The method has not only high processing speed, but also does not need the further process such as fining and sequencing. The fruit axis is selected by its symmetry in detecting of the fruit axis direction, which has good universality and higher detecting accuracy.

In two groups of fruit experiments, the correctness of detecting of axis direction reaches 94.4% and the maximum absolute error of measuring the fruit size is 3 mm. It can meet producing need and the axis direction detected by this method agreed with the manual one. The method accords with international standard (Bin Fen, 2003).

On the basis of overall analysis of the fruit shape, the shape was described with 6 feature parameters, namely, radius index, continuity index, curvature index, symmetry of radius index, symmetry of continuity index and symmetry of curvature index. The reference shape method was firstly employed in the analysis of apple shape feature and the artificial neural network was used in the evaluation and sorting of the fruit shape. The results showed that the average consistency between computer vision grading and manual grading is over 93% in terms of obtained feature parameters and the fruit shape analysis technology (Jing Zhao et al., 2001).

3.2 Detection of the Fruit Color

Color is extrinsic reflection of intrinsic quality of fruit and vegetable. Consequently, it becomes an important study object and a basis of grading in computer vision system. Some color models should be adopted for evaluating color feature of the fruit surface in color discrimination (Dongjian He et al., 1998). Many special color models had been set up in some relevant studies. RGB and HIS model were often used in computer vision system to describe color, which is more similar to the manner of human vision. The HIS model includes three elements: hue, saturation and intensity. The color threshold values for discriminating different color grades were chosen according to the findings of color study and relative criteria. The accumulative frequency relative to the threshold value was obtained and then color grade would be done. HIS color system is very suitable for color evaluating and image processing. Meanwhile, the correctness of grading was more than 90% if indicating color feature with hue histogram and multi-variables identifying technology were used in detecting color of potato and apple (Shuwen Wang et al., 2001).

Tomatoes were classified into six maturity stages by computer vision system according to the USDA standard classification: Green, Breakers, Turning, Pink, Light Red, and Red (Choi et al., 1995). The classification results agreed with manual grading in 77% of the tested tomatoes and all samples were classified within one maturity stage difference. The grading correctness of bell peppers was 96% using computer vision system to grade fresh fruit and vegetable (Shearer et al., 1990).

3.3 Detection of the Fruit Bruise and Defects on Its Surface

The bruise and surface defects of fruit have severe effect on the intrinsic and extrinsic quality of fruit. Removing bruised or surface defective fruit not only is requirement of grading, high quality high price, but also is an important process for preventing fruit from rotting and deteriorating. Detecting bruise and defects of fruit is yet an obstacle to implementing the real-time grading of fruit.

Some findings showed that there is a different spectral reflectivity in the range of visible light for the bruised or/and the defective region on fruit surface compared to the normal one. Hereby, surface defects can be detected in wavelength range of visible light. In addition, fruit bruises often occurred at random in the process of picking, loading, unloading, and transportation. Grade criteria in terms of the number of the fruit bruises and the area of each bruise were set up in some countries. So they become criteria for grading in a computer vision system.

There was a color difference in the joint region of the defective and non-defective regions of Huanghua pear (Yibin Ying et al., 1999). The light values of R (red) and G (green) were used to distinguish the defective region from the non-defective region. The whole defective region was found by means of region growing method. Finally, the area of the defective region was calculated. The automatic detecting of pear bruise was studied (Tailin Zhang et al., 1999). The bruise signal was separated from background image and pear normal tissue image by employing many image pretreatment techniques. And then different bruised regions were separated each other with the aid of denoting different bruised regions with different gray values. A detecting criterion was set up according to the national pear grading standard in order to use computer detecting system in practice. A mathematic model for calculating bruise area was put forward according to pear shape and the characteristics of bruise. Measuring relative errors can be controlled with in 10%.

3.4 Overall Quality Evaluation

The fruit quality is a concept of overall quality. Bigger but off-color fruit and smaller but colourful fruit were often found in production. The fruit feature such as shape, size, and color was dependent on the inherent character of its variety. Fruit will have good feature if they normally grow in suitable areas and are harvested at the ripe time and vice versa. In a word, shape, size, color and bruise of fruit altogether decide its quality. The purpose of grading is a quality classification. The index measuring is only a means.

A real-time fruit sorting mechanism and its controller, which are the key parts of a robot for the fruit quality inspecting and sorting, were developed in Zhejiang University. The sorting mechanism consists of a feeding and upturning system, a computer vision inspecting system, and a grading system. Double cone-shaped roller for the fruit feeding and upturning keeps the fruit forward feeding at a certain speed and rotating at random on a parallel axis to make the whole surface of fruit being detected by inspecting system and get adequate images. The computer vision inspecting system processes these images to encode an order about the grade of every fruit and its real position in the sorting mechanism and then pass it to the grading system through the controller of inspecting system. Fruit grading will be done (Yibin Ying et al., 2004).

A type of controller for the fruit synchronous tracking and automatic grading was developed (Yonglin Huang et al., 2002). Results of image processing were sent to a parallel port of a computer from a shift register. Certain pulse signals from sensors were taken as the shift signals, and the positions of processing results in shift register keep the same pace with that of every fruit on the grading line. In this way, the synchronous tracking of fruit was realized. Meanwhile, an effective method to control the pulse distribution and correlative action of step motors, which were used to drive the grading mechanism, was also developed.

4 Conclusion

The manufacture of the fruit grading equipments in China is still limited in the field of mechanical grading equipments which mainly include size grader and weight grader at present. The mechanical and automatic graders are mainly imported and aren't used widely. Model 6GF-1.0 size grader for fruit, developed in China nowadays, was designed by the advanced grading principle according to the clearance between roller and belt. The working principle is: (1) the grading roller rotates at uniform velocity; (2) the conveyer belt moves in line; (3) fruit will fall down from the clearance into their receiver if their sizes are less than the clearance; (4) fruit grading will be done. Model GXJ-W series horizontal sorters for fruit and vegetable were manufactured in Xixia Maoyuan Machinery & Equipment Factory in Shandong province, which are fit for sorting spheroidal fruit or vegetable (e.g. pear, apple, persimmon, peach, lemon, guava, tomato, orange and potato). They are high efficiency sorters (Huan Liu, 2006).

The study on dynamic properties of agricultural materials was not put a premium in China because of the limitation of basic techniques. The development of automatic grading equipments in China is still in the stage of laboratory testing. Therefore, China should keep pace with international, newest developments in these fields, make the best of findings acquired abroad, and seek after new theories and methods to greatly develop new-type automatic grading equipments and promote the grading handling capacity to a great extent.

The great progress in post-harvest fruit grading technology will occur with developments of computer technology, mechanical technology and electronic technology, and their combination. The synchronous grading device with functions of detecting the fruit feature and evaluating its quality is hopeful to be manufactured in the future. Different species of fruit need different mechanical performance in industry of the fruit tree in China, for example, anti-bruising function for white pear, color detecting for yellow pear, shape detecting for particular shape pear (e.g. 'Ya' pear, 'Kuerle' sweet pear), detecting blackheart for blackheart-suffering pear.

In conclusion, commercialization of the post-harvest fruit and vegetable is a systems engineering. It consists of series of correlative and matched technologies. It is an integration of technologies in many different fields.

References

Gao, H.: Pursuing breakthrough in storage of fruit and vegetables in China. China Food Newspaper 7.7(A4) (2003) (in Chinese)

Li, L., Wang, J., Dan, Y., et al.: State and new technology on storage of fruit and vegetables in China. Journal of Wuxi University of Light Industry 22(3), 106–109 (2003) (in Chinese)

Gao, H., Sun, H.: Mechanization and automatization of harvested fruit disposing. Journal of World Agriculture 9, 36–38 (2002) (in Chinese)

Gao, H., Li, H., Zhang, H.: Automatization of post-harvested fruit grading. Machinery for cereals, oils and food processing 2, 34–35 (2002) (in Chinese)

Miller, B.K., Delwiche, M.J.: A color vision system for peach grading. Transactions of the ASAE 34(6), 2509–2515 (1989)

Carlomagno, G., Capozzo, L., Attolico, G., et al.: Nondestructive grading of peaches by near-infrared spectrometry. Journal of Infrared Physics & Technology 46, 23–29 (2004)

Kleynen, O., Leemans, V., Destain, M.F.: Selection of the most efficient wavelength bands for "Jonagold" apple sorting. Journal of Post harvest Biology and Technology 30, 221–232 (2003)

Zhou, Z., Zhang, X., Wu, J., et al.: Development of automated grading system for cucumbers. Transactions of the Chinese Society of Agricultural Engineering 19(5), 118–121 (2003) (in Chinese)

Zhou, Z., Bontsema, J., Vankollenburg-Crisan, L.: Development of cucumber harvesting robot in Netherlands. Transactions of the Chinese Society of Agricultural Engineering 17(6), 77–80 (2001) (in Chinese)

Feng, B., Wang, M.: Detecting method of fruit size based on computer vision. Transactions of the Chinese Society for Agricultural Machinery 34(1), 73–75 (2003) (in Chinese)

Zhao, J., He, D.: Studies on technique of computer recognition of fruit shape. Transactions of the Chinese Society of Agricultural Engineering 17(2), 165–167 (2001) (in Chinese)

He, D., Yang, Q., Xu, S., et al.: Computer vision for color sorting of fresh fruit. Transactions of the Chinese Society of Agricultural Engineering 14(3), 202–205 (1998) (in Chinese)

Wang, S., Pan, W., Zhang, C.: Farm produce inspection and machinery vision technology in its processing. Journal of Research of Agricultural Mechanization 3, 103–105 (2001) (in Chinese)

Choi, K., Lee, G., Han, R.J.: Tomato maturity evaluation using color image analysis. Transactions of the ASAE 38(1), 171–176 (1995)

Shearer, S.A., Payne, F.A.: Machine vision sorting with bell peppers. In: Proc. of the 1990 conf. on food processing and automation, vol. 3, pp. 289–300 (1990)

Ying, Y., Jing, H., Ma, J., et al.: Application of machine vision to detecting size and surface defect of huanghua pear. Transactions of the Chinese Society of Agricultural Engineering 15(1), 197–200 (1999) (in Chinese)

Zhang, T., Deng, J.: Application of computer vision to the detection of pear's bruising. Transactions of the Chinese Society of Agricultural Engineering 15(1), 205–209 (1995) (in Chinese)

Ying, Y., Rao, X., Huang, Y., et al.: Controller for real-time sorting mechanism of fruit. Transactions of the Chinese Society for Agricultural Machinery 35(5), 117–121 (2004) (in Chinese)

Huang, Y., Ying, Y.: Controller for fruit synchronous tracking and auto-classification used in real-time fruit grading system. Transactions of the Chinese Society of Agricultural Engineering 18(4), 163–164 (2002) (in Chinese)

Liu, H.: Fruit grading pioneer—intelligent fruit grading product line. Journal of Modern Agricultural Equipment 3, 36–37 (2006) (in Chinese)

A Voice Processing Technology for Rural Specific Context

Zhiyong He*, Zhengguang Zhang, and Chunshen Zhao

Institute of Computer Application, Sichuan University of Science and Engineering,
Zigong Sichuan, P.R. China 643000,
Tel.: +86-13778520180; Fax: +86-813-5505966
hzy@suse.edu.cn

Abstract. Durian the promotion and applications of rural information, different geographical dialect voice interaction is a very complex issue. Through in-depth analysis of TTS core technologies, this paper presents the methods of intelligent segmentation, word segmentation algorithm and intelligent voice thesaurus construction in the different dialects context. And then COM based development methodology for specific context voice processing system implementation and programming method. The method has a certain reference value for the rural dialect and voice processing applications.

Keywords: voice processing, COM, specific context, rural informatization.

1 Introduction

Intelligent voice processing technology is to resolve how to make computers understand natural language of mankind and can be output of natural language fluently, so that the computer has the capacity of human language. It is mainly divided into the speech synthesis technology, voice recognition technology, and voice evaluation and voice coding techniques. In the popularization of agricultural informational, China has large rural population, the larger regional language differences. People have put forward higher requirements for speech technology.

In this paper, the problem is therefore in the formation of the use of COM technology, based on the specific context and mixed multilingual text as the background voice processing, according to the characteristics of TTS technology, a library based on the establishment of specific context voice processing technology and intelligent voice processing rules system, in order to address the specific context of multi-lingual text and mixed-voice processing problems.

2 Development and Application of TTS Technology in Specific Context

Specific context refers to time, place, occasion, object and the use of objective factors such as language, identity, ideology, personality, occupation, self-cultivation, the

---

* Corresponding author.

D. Li and C. Zhao (Eds.): CCTA 2009, IFIP AICT 317, pp. 147–152, 2010.

situation, feelings and other subjective factors posed by the use of the language environment (Yibiao Yu et al., 2002; Bo Yang et al., 2005). Specific context of speech synthesis research, including the realization of multi-lingual speech synthesis, such as: minority languages, the local dialect speech synthesis, Chinese and foreign language. TTS that is, "from text to voice." It is the use of linguistics and psychology, in the built-in support chips, and smart text to flow into a natural voice. It can convert a text file in real-time, converting a short period of time can be seconds. In its unique role in the smart controller voice, the voice of the text output to achieve smooth temperament, making the listener feel when listening to nature, there is no voice output machines and jerky sense of indifference. Common TTS system mainly includes text analysis, prosodic processing, and speech synthesis. Text analysis of the input text linguistics is analyzed, the sentence in vocabulary, grammar and semantic analysis to determine the sentence structure and every word of the phonemes. Dealing with synthetic sound quality is not only a rhythm, but also making the sound quality of voice to speak close to the voice synthesis to ensure the natural tone of words, consistency. Speech synthesis is a good means to deal with the text of the corresponding word or phrase from the speech synthesis library extracted from the linguistic description into a speech waveform. The TTS core of this intelligent speech processing system used this technical route (Weijun Shen et al., 2000). Its technical route diagram is shown in Fig.1.

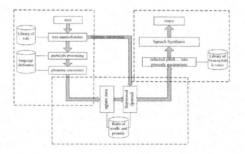

Fig. 1. Technical route of TTS

For the specific context of the complexity of the situation, Clustering Algorithm uses in text analysis for word segmentation. First, all possible results are available in the stage of segmentation, part of the lexical analysis level ambiguous rule excluded. Second, excluded from the remaining ambiguous areas on the Chinese understanding of the follow-up stage, through a combination of ontological knowledge base for semantic analysis in order to rule out the ambiguity, and then the right does not understand the results of a new thesaurus word segmentation feedback to the thesaurus management system, at the same time record the frequency of its emergence, when the frequency of a standard to meet when it joined the segmentation thesaurus in order to achieve the recall of unknown words and improve the efficiency of system operation. At the same time in the understanding of understanding of the final stage is finished, before the quasi-automatic model segmentation results of the feedback module clustering words based on statistical training systems, clustering results at the same time feedback to the thesaurus management system, to improve the next segmentation accuracy and efficiency (Qiang Li et al., 2000).

Processing stage in the rhythm uses feedback forms of 3-layer structure self-organizing network (Chen Zhao et al., 2002). (1) Enter the relevant context information, the language of information along a statement (sentence) → prosodic phrase (phrase) → syllable (syllable) step-by-step breakdown of information to find syllable C [consonant type C1, vowel type C2, tone type C3, in the words of the location of C4, with the former syllable coupling C5, and after the coupling syllable C6]; (2) adjacent to the former syllable Information L [vowel type L1, tone type L2]; (3) adjacent to the former syllable information N [consonant type N1, tone type N2]; (4) syllable of the prosodic phrase information where W [number of syllables W1, the location of the sentence W2, accent type W3, stress from the previous distance W4, stress from a distance after the W5, stress and stress the distance between W6]; (5) statement of information S [statement type S1 and the number of prosodic phrases S2]. More than the prosodic features (17 acoustic parameters) as the neural network input, through a multi-sample competitive template, the template matching with the established best or with the most similar to natural sounds as the output template.

TTS voice processing module is a key link. PSOLA optimization algorithm is used that is time-frequency distribution algorithm. The algorithm adjusts voice pitch and time Len of the original voice splice units in frequency domain and time domain. First the voice processing module synchronous analysis and tags keynote when the rhythm parameters arrival, and then the pitch of short-term analysis of signal changed in frequency domain (Min Han, et al., 2004).

3   Intelligent Voice Processing System in Specific Context

Intelligent voice system uses hierarchically structured design. The system is divided into that presentation layer, business layer, data access layer and data layer. Which the business layer and data access layer contains the recognition engine to develop a special voice interface and database interface, the data layer with special voice library can be easily loaded under various voice module, to facilitate the system's compatibility, the core processing module is designed to voice control command recognition and voice processing. Its system structure diagram is shown in Fig.2.

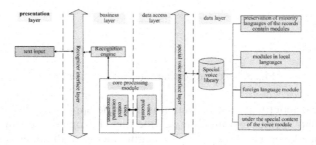

Fig. 2. System structure of intelligent voice processing

The presentation layer is a simple application layer which collects text data. Recognition engine is composed of the Recognizer interface layer and the recognition engine modules. It captures packets and decodes, and then puts the results to a place designated in the data structure. The core processing module primary recognizes the data which sent from recognition engine, link data, distinct and process. These procedures include further word processing, determine errors, and call voice module. The core processing module uses voice-processing algorithms and rule base to match the rules of voice processing, find a voice module for synthesis. The special voice interface layer uses correlation analysis and sequence analysis to match and find a new voice module, and then sends the matched voice library to the core processing module by COM components methods. Special voice library contains the records kept by minority languages modules, local languages modules, foreign language module, and other specific voice context modules. It support upgrade a variety of voice expansion modules.

4   The Key Technology of Intelligent Voice Processing Technology

4.1   Voice Processing Control Command Recognition

Voice processing control command recognition is the core aspect of intelligent information processing. First of all, the text of the application to identify the command interface program designed to identify and initialize, then the statement of the interface objects required creating a shared text recognition engine. Second, we must create a specific context of the context of speech synthesis interface, the realization of multi-lingual speech synthesis engine function. Once the correct order of operations to be identified, the text sent to the host program identification information. A speech synthesis module interface is created, loaded and activated for application, when the application has been received information from incident immediately (Flanagan，J.L et al., 1972). Its process diagram is shown in Fig.3.

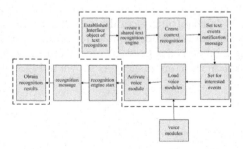

Fig. 3. Voice processing command recognition

The main text input text data collection is a simple application layer. Including a variety of formats such as text input: text,. Doc files. Recognition engine by the Recognizer interface layer and the recognition engine modules, the interface is crawling information packet capture and decode analysis, and the results of the analysis to a place designated in the data structure. Is mainly responsible for the core processing engine will be sent to identify the data, connect data identification, handling. Include:

further word processing, to determine errors, such as voice calling module. Core processing uses voice processing algorithms and rule base to match the rules of voice processing, direct find a voice module for synthesis. Treasury special voice interface layer through the use of correlation analysis and sequence analysis of the match to find a new voice module, method of use of COM components to match the voice to the core processing module library. Special voice library: preservation of minority languages of the records contains modules, modules in local languages, foreign language module, under the specific context of the voice module, to support the expansion of a variety of voice module upgrades. (Randy Abernethy et al., 2000; Ash Rofail et al., 2005)

4.2 The Development of COM Components

COM (Component Object Model) (Microsoft Co., 2009) is a new software development technology; it is helpful to improve computer industry's software manufacture more in line with the behavior of human beings. Under COM framework, various kinds of components with specific function will be developed, and the components could be combined together as needed to compose a complex application system. Application or component system could be formed by the combination of multi COM objects. At the same time, component could be unloaded or replaced during its running, there is no need to re-link or compile the application. (Flanagan，J.L et al., 1972).

When the application started, the voice processing system creates a process to execute new application. Component program is realized as DLL (Dynamic Link Library). When there was language text input, the core processing module calls component program, loads the required component program to its process, and then connects to specific voice module in special voice library through the special voice library interface layer. After the establishment of communication between processing requirements and component program, the interface pointer received by core processing program will point to component program's vtable directly. The vtable pointed by demand module interface pointer includes member function's address, and then business layer could call service provided by interface directly. For the component outside of process, component program and presentation layer are not in the same process space, the communication between component program and client program must pass through the process border, therefore the component outside of process must be processed with dynamic DLL firstly, then the parameters and other call information will be assembled into a packet and passed to component program. When the component program received the data packet, it will unpack the packet, read out the parameter information, call the actual interface and send the result collected back to the core processing program at last. Then a function all has been complete.

5 Concluding Remarks

The design proposed in this paper, combined COM with voice intelligent processing technology. It is helpful to develop new ideas in future in many fields, such as multi language interaction, translation, search engine and etc. it could promoted information technology to develop better and faster.

Voice processing technology for specific context could be used in various fields, such as keyboard input, optical scanning, handwriting recognition, web-based database, PDA, home appliances, digital products and etc. it is helpful to solve the shortcomings of traditional applications and overcome the communication obstacles between human and machines. It would play a greater role in various fields in the future. The adoption of this technology will play a positive role in the rural information promotion at the same time.

References

Yu, Y., Duan, K., Shi, R.: Prosodic Control for Speech Synthesis of Wu-Dialect Chinese text to speech system. Communications Technology (10), 87–90 (2002) (in Chinese)

Yang, B., Jia, Y., Li, Y., Yu, H.: Researches on prosody control technique and applications in Tibetan TTS. Journal of Northwest University for Nationalities (Natural Science) 26(1), 69–71 (2005) (in Chinese)

Shen, W., et al.: A chinese Text-to-Speech system. Computer Engineering and Applications 36(9), 76–80 (2000) (in Chinese)

Li, Q., Liu, Y., Zhu, X.: An Algorithm of Pitch Prediction. Journal of University of Electronic Science and Technology of China 29(5), 495–498 (2000) (in Chinese)

Zhao, C., Tao, J., Cai, L.: Rule-learning Based Prosodic Structure Prediction. Journal of Chinese Information Processing 16(5), 30–37 (2002) (in Chinese)

Han, M., Tian, L.: Chinese prosody stepwise synthesis based on FD & TD PSOLA. Journal of Shandong University (Engineering Science) 34(6), 35–37 (2004) (in Chinese)

http://www.microsoft.com/com/default.mspx (2009)

Rofail, A., Shohoud, Y.: Translator: Chung, P.: COM and COM + from entry to the master, pp. 136–165. Publishing House of Electronics Industry, Beijing (2005) (in Chinese)

Abernethy, R.: Translator: Wang, H.: COM/DCOM technology insider, pp. 56–70. Publishing House of Electronics Industry, Beijing (2000) (in Chinese)

A New On-Line Detecting Apparatus of the Residual Chlorine in Disinfectant for Fresh-Cut Vegetables

Chao Hu[1], Shu-qiang Su[2], Bao-guo Li[1,*], and Meng-fang Liu[1]

[1] Institute of Food and Biotechnology, University of Shanghai for Science and Technology,
Shanghai 200093, P.R. China
lbaoguo@126.com
[2] Shanghai General Cooling Technology Co., Ltd, shanghai 201201

Abstract. With the fast development of modern food and beverage industry, fresh-cut vegetables have wider application than before. During the process of sterilization in fresh-cut vegetables, the concentration of chloric disinfectant is usually so high that the common sensor can't be used directly on the product line. In order to solve this problem, we have invented a new detecting apparatus which could detect high concentration of chloric disinfectant directly. In this paper, the working principle, main monitor indicators, application and technical creations of the on-line apparatus have been discussed, and we also carried on the experimental analysis for its performance. The actual demands in factory could be met when the detecting flux is 2L/min, the dilution ratio is 15 and input amount of the disinfectant is 200ml per time, the max of the detecting deviation achieves ±4.8ppm(mg/L). The main detecting range of residual chlorine is 0~300ppm.

Keywords: Fresh-cut Vegetable; Residual Chlorine; Dilution Ratio; Sensor.

1 Introduction

China is a large agricultural nation. Vegetables are rich in resources and varieties with high quality and low price. In recent years, the volume of the world trade on vegetables have increased significantly. In China, the loss ratio after vegetables' picking reaches as high as 40%~50%, the same on the commodity surpasses 30%, resulting in a huge economic loss [1]. Fresh-cut vegetables are also called semi-processed vegetables, cooking vegetables, minimally processed vegetables, which are kind of instant vegetable products based on the fresh vegetable as raw material. They are processed by cleaning, peeling, cutting, trimming and packing, then entering the supermarket after the refrigeration and transportation for sale. They are fresh in quality, edible conveniently, nutritious, clean etc [2]. Moreover, they can extend the shelf life effectively and reduce loss markedly.

Through the vegetable's cutting process, the internal tissue is vulnerable to damage because of the microbial contamination. So it usually uses chloric disinfectant for

---

* Corresponding author, Tel: +86-21-55271117, Email:lbaoguo@126.com

D. Li and C. Zhao (Eds.): CCTA 2009, IFIP AICT 317, pp. 153–160, 2010.
© IFIP International Federation for Information Processing 2010

cleaning and sterilizing process to extend their shelf life and keep the vegetables safety [3].

At present, commonly used as chloric disinfectant is chlorine molecule, hypochlorous acid or calcium hypochlorite. Chloric disinfectant hydrolysis in water to form hypochlorous acid(As shown in formula(1)(2)(3)), and hypochlorous acid further decompose to form intense-oxidation oxygen [O], which has so highly oxidizing capability towards bacteria, viruse and pathogenic micro-organisms, it also can cause the protein denaturation.

$$Cl_2 + H_2O \rightarrow HClO + HCl \tag{1}$$

$$NaClO + H_2O \rightarrow HClO + NaOH \tag{2}$$

$$Ca(ClO)_2 + 2H_2O \rightarrow 2HClO + Ca(OH)_2 \tag{3}$$

The hypochlorous acid will decompose the hydrogen ion and the hypochlorite ion in water(4):

$$HClO \leftrightharpoons H^+ + ClO^- \tag{4}$$

The bactericidal effect of disinfectant has been proportioned with the concentration of hypochlorous acid. The bactericidal effect of HClO in disinfectant is about a hundred times stronger than ClO^-, HClO 's ratio increases along with the pH value reducing. Therefore, in the disinfection we may judge its bactericidal effect correctly through the content of hypochlorous acid. Pirvani abroad carried lettuce for sterilization experiment and he found that there existed bactericidal effect when the actual concentration of residual chlorine is 50~150mg/L[4].Domestically, Professor Lu Zhaoxin in Nanjing Agricultural University studied that sodium hypochlorite solution kill the E. coli most effective when the actual concentration of residual chlorine is 125mg/L [5].

Vegetable processing enterprises use chloric disinfectant whose actual concentration of residual chlorine is mainly 75~200mg/L in the production of the fresh-cut vegetables. Whereas in the current market, detector on chloric disinfectant focus on 0~5mg/L, so it can not be directly used on the product line of fresh-cut vegetables for detecting high concentration of chloric disinfectant [6]. In this article, it uses the method of dilution ratio, developing on-line detecting apparatus of the high concentration of residual chlorine to solve the problem above.

2 Working Principle

2.1 Principle of Residual Chlorine Sensor

The residual chlorine sensor used in apparatus is a galvanometer sensor based on the electrochemistry. It consists of the cathode, anode and electrolyte, the cathode is covered by a layer of membrane with gas permeability. Residual chlorine in measured fluid spread to the cathode through the membrane, then the cathode and anode polarize between appropriate voltage. Residual chlorine is reduced in the cathode when producing electrical current and thus measure the concentration of residual chlorine. The current is directly proportional with the chemical reaction in the sensor of residual chlorine from solution.

2.2   Principle of the On-Line Detecting Apparatus

In view of common residual chlorine sensor, whose measuring scope is finite from 0~5,0~10 or 0~20ppm. We choose dilution in certain ration as the method of the detecting apparatus. As shown in Figure 1, the diluent water and a certain amount of disinfectant mixes in the A. B place controls the best examination current capacity scope by the transit discharge controlling element, then completes the examination in C place, finally the density value which will be examined through the D place where returns the concentration to original state and demonstrate the actual disinfection fluid density. Simultaneously, using the programmed logic controller guarantee that the disinfection in product line satisfies the sterilize request all the while.

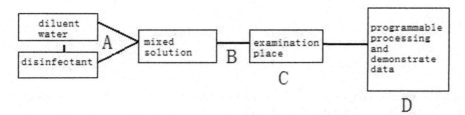

Fig. 1. operative principle of on-line detecting apparatus

3   The Main Monitoring Factors

3.1   Residual Chlorine

Residual chlorine is the total chlorine's content after the reaction with the deoxidizer when the disinfectant adding into the water to be possessed, including active chlorine, hypochlorous acid and organic chloride rather than chlorine ions. The bactericidal effectiveness of chloric disinfectant depends on the concentration of hypochlorous acid from residual chlorine. In the vegetable process, the concentration of hypochlorous acid is not only the key parameter measuring whether it has enough bactericidal effect, but also an important indicator of detection at the end of the cleaning the vegetables for wiping off the residual chlorine. The purpose of monitoring is to ensure the concentration of hypochlorous acid is in certain extent that have effective sterilization. Generally the concentration of hypochlorous acid is too high may increase the residual chlorine's content and lead to waste of resources; or too low would reduce the bactericidal effect.

3.2   pH

According to pH value in different water, the possible exist form of residual chlorine is: chlorine molecule dissolved in water(Cl_2), hypochlorous acid ($HClO$) or Hypochlorite ion ($ClO-$). Cl_2 and $HClO$ is in equilibrium in a certain temperature($K25°C= 4x10^{-4}$).In the same way, $HClO$ and $ClO-$ is also in equilibrium in a certain temperature($K25°C= 2.9x10-8$).

Figure 2 shows that in a typical drinking water, the pH value approximately equal to 7.5. Under this condition, both $HClO$ and $ClO-$ exist. The bactericidal effect of

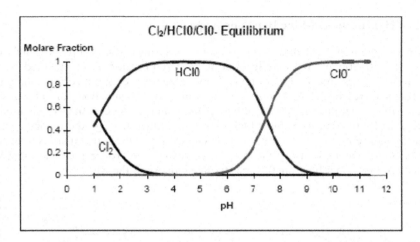

Fig. 2. The equilibrium carve between Cl2, HClO and ClO- in different pH

HClO in disinfectant is about a hundred times stronger than ClO-, so the effect of chloric disinfectant depends on its pH value.

3.3   Temperature

Effect of temperature on the residual chlorine measurement is divided into two parts: first, because the dynamic balance between HClO/ClO- depends on the temperature; Second, because the principle of sensor is based upon the electrochemical reaction, while the temperature will affect the measurement of hypochlorous acid by bringing fluctuations in electrochemical reaction. Therefore, accurate detection of concentration of chloric disinfectant needs monitoring temperature.

4   Results and Discussion

4.1   The Effect on the Apparatus by the Flux of Disinfectant

According to the principle of current sensor, detecting flux is very important for the accurate detection. As can be seen from Figure 3, when the detecting flux is between 20~80L/h, the concentration of HClO detected by the sensor is smaller than the titration value by iodometry. A possible reason is the speed that the testing disinfectant washout the probe head is too slow, causing the HClO in the testing fluid do not penetrate into tectoria membrane of the probe head completely. When the flux is greater than 120L/h, the residual value of detection of the sensor is greater than the titration value by iodometry, this possible reason is the velocity of testing disinfectant in the pipeline has speeded up and HClO amount enters tectoria membrana has increased in per unit time which heighten effective degree to washout electrode. At the meantime, the frequency of shifting electric charge speed up, thus causing the correspondence concentration value is high. When the detecting flux is 80~120L/h, both test results are basic consistent. So the design prototype selects 100L/h as the detecting flux.

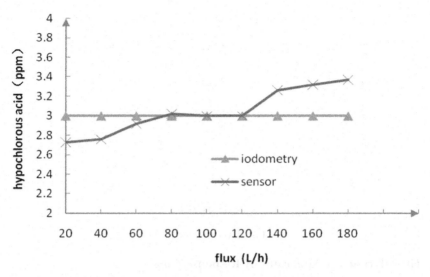

Fig. 3. Effect of detecting flux on the measurement of HClO

4.2   The Effect on the Apparatus by pH of Disinfectant

The bactericidal effect of disinfectant has been proportioned with the concentration of HClO, and HClO 's ratio to ClO- in the water increases along with the pH reducing. Figure 4a, 4b showed that when the pH = 7.5 (25 ℃), only 50% of residual chlorine turned to HClO; but at pH = 6.5, there was about 85% of HClO; at pH = 5.5, the residual chlorine at the form of 100 % HClO. In Figure 4a the original concentration of residual chlorine is 5ppm, Figure 4b is 10ppm. The changes are basically in accord with the trend curve of HClO in Figure 2. Thus, we can draw that the key point in the preparation of vegetable disinfectant is controlling its pH value, as far as possible make it in 6 below for maximum amount of HClO.

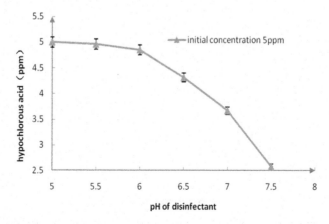

Fig. 4. a,b: Effect of pH of disinfectant on the measurement of HClO

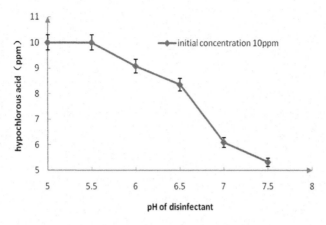

Fig. 4. (*Continued*)

4.3   The Effect on the Apparatus by Response Time

Known from Figure 5, with the response time extending, the apparatus displayed a gradual increase in concentration of HClO. When the time range was 5s to 30s, the concentration of HClO increased significantly, then the data on the screen came to basic balance after 30s. This is because the residual chlorine sensor needs certain time for accomplishing the electrochemical reaction. HClO was detected and consumed from the testing disinfectant when it entered tectoria membrana of the sensor, and non-stop supply of disinfectant to achieve a balance. From the Figure 5, the initial concentration of HClO is 5,10ppm respectively, and for the on-line detecting apparatus the best response time is 30~40 s.

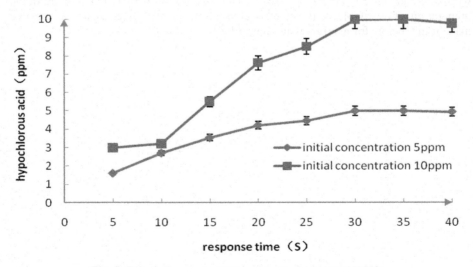

Fig. 5. Effect of response time on the measurement of HClO

4.4 The Test Experiment of Apparatus in Factory

Normally, the vegetable product line will prepare 150ppm of sodium hypochlorite as disinfectant. We took some sample for testing, the results showed in Figure 6: the pink carve describing the testing concentration of HClO by the apparatus, the result's maximum deviation from the actual data is ±4.8ppm. Sodium hypochlorite's own nature has decided the examination deviation. Because, the sodium hypochlorite very easy to decompose in the illumination and under the washout condition. Besides, in the clean process, in order to separate the microorganism from the vegetables' surface as soon as possible, air blower used to produce the air bubble which has the function to accelerate the separation. Involuntarily, the air bubble also intensifies the decomposition of the sodium hypochlorite. In addition, operational site's illumination has also created the condition for decomposition. Therefore, the actual concentration of HClO has the drop slightly. We also learned from the blue line which representing the response time of the on-line detecting apparatus in Figure 6, the average reaction time is 37s, which satisfies the actual working request.

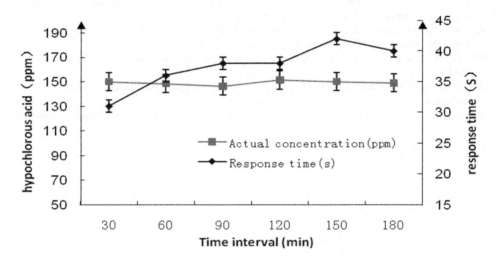

Fig. 6. Field-testing of the apparatus and its response time

5 Conclusion

The on-line detecting apparatus of residual chlorine in disinfectant for fresh-cut vegetables adopts the simple principle of proportion dilution, acquiring residual chlorine data by galvanometer sensor, and then processing the data through the microprocessor to realize the intelligent of machine. The results of measurement are basically consistent with the GB method. Compared with foreign devices, the relative deviation of measurement of domestic detector is higher. While the relative deviation of this detector is lower than the domestic similar products, and much closer to foreign analyzer in results. The main performance index also reached advanced level contrast with the overseas similar instruments. Furthermore its low price will reduce the cost of customers. The apparatus is controlled by programmable logic controller(PLC), so it can meet the requirements of the on-line measurement of disinfectant in food industry because of its convenient operating, good stability and high accuracy.

References

Qinsheng, L.: Our country vegetables processing and circulation present situation and trend of development. Machinery For Cereals,oil and Food Processing (3), 1–4 (2000)
http://www.fresh-cuts.org (International association of fresh-cuts product)
Xuejie, Z.: Our country vegetables processing industry present situation and faced with WTO opportunity and challenge. Science and Technology of Food Indusry (1), 82–85 (2001)
Pirovani, M., Piagentini, A., Güemes, D., et al.: Reduction of chlorine concentration and microbial load during washing-disinfection of shredded lettuce. Int. J. Food Sci. & Technol. (39), 341–347 (2004)
Likui, Z., Zhaoxin, L.: Study on predictive model for total coliform group reduction on fresh-cut lettuce treated with NaClO. Food Science (7), 67–71 (2004)
Sapers, G.M.: Efficacy of Washing and Sanitizing Methods for Disinfection of Fresh Fruit and Vegetable Products. J. Food Technol. (39), 305–311 (2001)
Shihao, M., Bo, L.: Hospital sewage disposal. Chemical industry publishing house, Beijing (2000)
EMERSON. Chlorine measurement by amperometric sensor. EMERSON Process Management Company
Shimano, T.: New design of geodesic lenses. In: MOC/GRIN 1989, Tokyo, pp. 130–135 (1989)

The Design of the Automatic Control System of the Gripping-Belt Speed in Long-Rootstalk Traditional Chinese Herbal Harvester

Jinxia Huang*, Junfa Wang, and Yonghong Yu

Institute of Mechanical Engineering, Jiamusi University, Jiamusi,
Heilongjiang Province, P.R. China 154007,
Tel.:13836641830
hjxlcj2006@sina.com

Abstract. This article aims to design a kind of gripping-belt speed automatic tracking system of traditional Chinese herbal harvester by AT89C52 single-chip micro computer as a core combined with fuzzy PID control algorithm. The system can adjust the gripping-belt speed in accordance with the variation of the machine's operation, so there is a perfect matching between the machine operation speed and the gripping-belt speed. The harvesting performance of the machine can be improved greatly. System design includes hardware and software.

Keywords: gripping-belt speed, traditional chinese herbal harvester, single-chip microcomputer, fuzzy PID.

1 Introduction

China is one of the most abundant and productive countries in the world. Nowdays, there are more than 500 kinds of Chinese traditional herbals which have already been gained successfully by man-made planting among 11146 kinds of medical plants. There are more than 1000 kinds of Chinese medicinal herbs which are popular in the market, about 25% of which are medicinal herbs of roots type (Yang JunJie et al., 2007). In recent years, the seven seas advocate "to return back into the nature", rising a green consumption upsurge, which makes a sharp demand for herbal medicine in the international market. In the global pharmaceutical market, the annual trading volume of the natural plant medicine is nearly $30 billions, and is increasing at an annual rate of 20% (Yang Yong et al., 2005).

2 The Actual Problems of the Roots Herbs Planting

Various herbs will be mainly produced by man-made planting because of the national disorderly digging forbidden and awareness for environment protecting. At present, long-rootstalk traditional chinese herbal' planting area has been over 250 million mu

---

* Corresponding author.

D. Li and C. Zhao (Eds.): CCTA 2009, IFIP AICT 317, pp. 161–168, 2010.

all over china including northeast, northwest, southwest ,etc. The actual problem of bottleneck herbs planting lies in harvesting which is mainly manual digging or manual picking after digging with deep loosening machines, so the disadvantages are high labor intensity, low efficiency and high loss of harvest. Besides, the time of chinese medicine herbal harvest is strictly required, or it will greatly influence the quality of the herbals and the planters' income. The problems above result in moratorium area of chinese traditional herbal planting. Therefore, the harvester development, especially long rootstalk chinese traditional herbals' harvester is an urgent task for us.

3   Hardware Design of Control System

Rubber gripping-belt is one of the core parts for long-rootstalk chinese traditional herbal harvester, as shown in Fig1. Whether the gripping-belt speed is matched with its machine's operation or not is the most important factor which affects the productivity and operating quality during the harvesting. At present, the gripping-belt speed regulation of harvester is controlled by operator's manual gearshift and

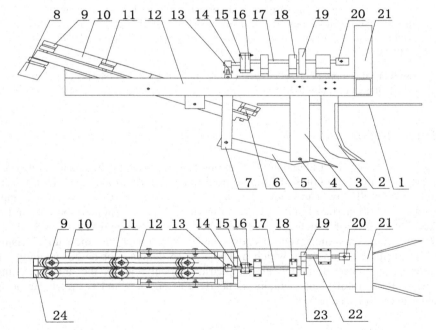

1 sub-seeding device 2 colter 3 fore-stock 4 bolt 5 vibration digging shovel 6 gripping-wheel 7 vibration rod 8 hydraulic motor 9 driver 10 gripping-belt 11 tensioning wheel 12 chassis 13 vibration block 14 back partial axle 15 back partial flange 16 front partial flange 17 front partial axle 18 front partial pedestal 19 chain wheel 20 gimbals 21 suspension bracket

Fig. 1. Long roots of chinese herbal medicines schematic harvester

adjusting hydraulic contin uously variable transmission according to the machine's operating speed, which will cause large speed-fluctuations and controlling inaccuracy as well as untimely operation. The system chooses single-chip microco mputer as the control center, and it is directed by fuzzy PID control, making actuators and gripping-belt speed into a closed-loop system, so it can achieve an automatic adjusting system in order to satisfy the requirements of the harvester operation.

The control system is used for measuring the current machine operation speed and gripping-belt speed. According to the fuzzy PID control algorithm, it can deal with the measured values, produce the corresponding control signal and drive electromagnetic regulating speed valve into a movement state, thus gripping-belt speed can be changed by tracking the machinery operation speed. The control system mainly takes an 8-bit AT89C52 single-chip microcomputer as a main control unit, periphery circuit includes input and output signal processing circuits, 4*4 keyboard and LED display and so on. The hardware system structure diagram is shown in Fig2:

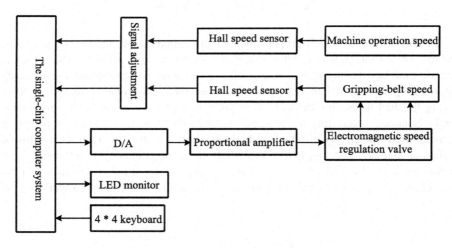

Fig. 2. Automatic control system hardware diagram

3.1 Detection Methods of Machine Operation Speed and Gripping-Belt Speed

Currently there are mainly three kinds of methods to measure rotative speed in industry: electromagnetic rotative speed sensor, incremental type optical coded disc sensors and hall switch speed sensors. But the first two kinds aren't suitable for its condition after analysis of the characteristics of the three sensors and poor working condition of herbal harvester. By contrast, the output signal's voltage amplitude of hall switch sensors is not influenced by rotative speed. It has high interference ability to prevent from electro magnetic wave, convenient installation and easy maintenance, simple struc ture, small volume, sensitive, long service life, high reliability, undertaking high impact resistance, resistance to shock. Therefore, hall switch speed sensors are more suitable for this kind of environment (Zhang Xin et al., 2002). Switch type hall speed sensor working principle is shown as Fig3.

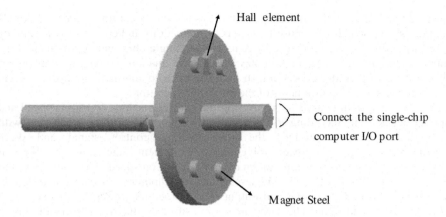

Fig. 3. The principles of hall sensors

The magnetic steel is well-proportioned conglutinated on the surface of the rotator, and hall switch integrated circuit is installed on a fixed location. When the rotator is running each lap, the circuit will output a pulse signal (or multiple pulse signals) each lap, the output pulse of hall switch circuit is directly proportional to its speed. After having finished measuring the sum of its pulse in unit time, the speed of this rotator can be obtained. According to the requirement of the system, the sensor's detecting elements are respe ctively installed on the transmission shafts which are located inside of the machine and gripping-belt's drive shaft, measuring speed. When transmis sion shafts turn a certain angle every time, hall sensor produces a pulse. According to the transmission ratio between transmission shaft and the drive shaft, we can work out the

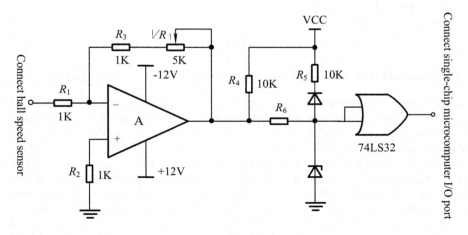

Fig. 4. Velocity measurement circuit

sum of pulse which produced by the drive shafts when they turn the same angle. After the detected pulse signal of hall sensor should be dealt by shaping circuit as shown in Fig4, we can obtain a standard pulse with its amplitude +5V. There's no need for A/D conversion, standard pulse can be counted relying on directly input the single-chip microco mputer's pins, which can work out machine operation speed and gripping-belt rotational speed.

3.2   The Single-Chip Microcomputer Application System

AT89C52 single-chip microcomputer is a kind of low voltage, high perfo rmance CMOS 8-bit type which is produced by American ATMEL company. It contains an 8K bytes read-only memory (ROM) which can be repeated erasing and writing and a 256 bytes random access memory (RAM), and it is produced by ATMEL company adopting the high density, nonvolatile mem ory technology (Wang ZhiGang, 2004).

3.3   DAC Converter

The actuator of this design is electromagnetic regulating speed valve. Its control voltage regulation is for 0V~ 10V.The signal which is output by single-chip microcomputer may control the electromagnetic valve after DAC convert and proportional amplifying. The system employ DAC0832 digital-to-analog converter for the advantages of low cost, simple interface and easy conversion control.

3.4   Input/Output Devices

Keyboard and display unit can be controlled by chip 8279. It can complete the function of dialogue between human and machine through the keyboard. The operator sends the control instruction to the chip and input necessary data. Display unit is a type of 4-bit LED display, using dynamic display. It can display machine operation speed and the gripping-belt current speed after analysis of the single-chip microcomputer's calculation.

3.5   The Software Design of Automatic Control System

The function of the main program is to initialize the working state of single-chip microcomputer and other chips, simultaneously organize and invoke every subprogram to complete control function according to predet ermined instruction demand. The following is the specific program functions:

(1) The single-chip microcomputer and other chips are initialized.
(2) Invoke subprogram of gripping-belt speed measuring.
(3) The speed information on LED display for driver reference.
(4) The data of the machine's operation speed enters into the main program after the measuring subprogram of machine operation.
(5) Examine the match extent between the machine operation speed and the gripping-belt speed. It invokes the subprogram which control electromagnetic valve adjusted correspondingly.

Language C is used in this program, which improves the design efficiency relatively to assembly language, source program's strong readability and easy maintenance. The main program flowchart is shown in Fig5:

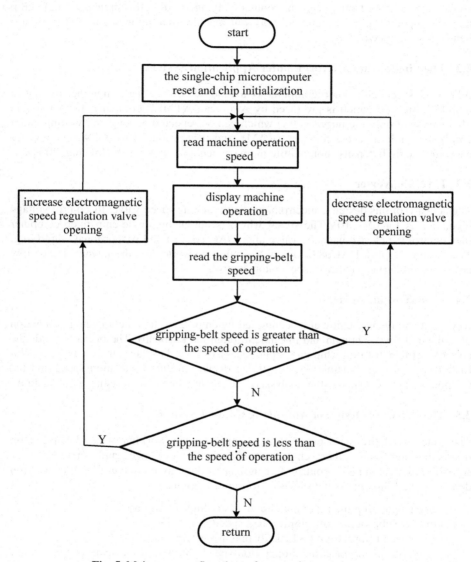

Fig. 5. Main program flowchart of automatic control system

4   The Algorithm Design of Automatic Control

Based on the characteristics of the system and the control requirements, the fuzzy PID control algorithm is adopted. The controller structure is shown in Fig6:

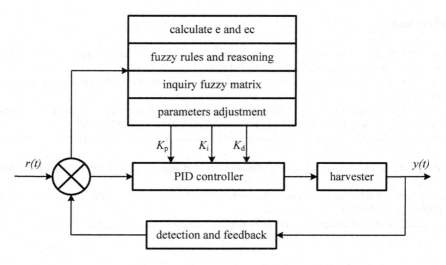

Fig. 6. Fuzzy PID control structure

Fuzzy PID controller can carry PID parameters self-regulating online for PID controller by fuzzy controller. The first task is to find out the fuzz relationship between three parameters (kd, ki and kp) of PID and deviation change rate (EC), and then, through constant detecting the deviation (E) and its change rate in operation. According to the fuzzy control theory, the parameters above can be modified online to meet the need of the different requirements of controller parameters when different deviation and change occur (Han JunFeng, 2003).

The deviation value E with real time between gripping-belt speed and machine speed, as well as its change rate EC both input variables into the fuzzy controller. Electromagnetic speed regulation valve can speed up or slow down by means of adjusting the output of fuzzy controller, thus a double input and single output fuzzy controller are established and speed control can be achieved.

5   Conclusion

Based on AT89C52 single-chip microcomputer, the control system is designed in accordance with the detected machine operating speed to control the speed of gripping-belt. It has a simple structure and be easy to be deal with. The control system of the chinese herbal medicine harvest can make the gripping-belt speed automatically match according to the change of machine's speed to achieve the perfect cooperation. Based on the debugging of each subsystem as well as the whole system, the results showed that the control system is stable in operation, simple operating and worthy of further research and development. The design purpose is achieved successfully.

References

JunJie, Y., Jian, Y., HaiXia, Y.: Henan herbs grow resource survey. Journal of modern medicine research and practice 21(5), 13–14 (2007)

Yong, Y., ShiMin, Y.: China, eu, Japan's medicinal plants. Norms and pharmaceutical 14(4), 14–17 (2005)

Xin, Z., ShenLong, C.: New hall-effect sensor characteristics and application in the measurement and control. Journal of university Daniel 21(10), 28–31 (2002)

ZhiGang, W.: MCU application technology and training. Tsinghua University Press, Beijing (2004)

JunFeng, H., et al.: The fuzzy control technology. Chongqing University Press, Chongqing (2003)

Science and Technology Project Supported by National the 11th five- year plan (2006BAD11A07-6)

Key Science and Technology Project Supported by Jiamusi University (LZJ2008-001)

A Virtual Prototyping Technology for
Design of Pressing Equipment of Dried Tofu

Mingji Huang[1,*] and Xiuping Dong[2]

[1] Department of Mechanical Engineering, University of Science and Technology Beijing,
Beijing, P.R. China 100083
Tel.: +86-010-62332538; Fax: +86-010-62329145
huangmingji@263.net
[2] Department of Mechanical Engineering, Beijing Technology and Business University,
Beijing, P.R. China 100048

Abstract. The industry of dried tofu products wants to achieve breakthrough, the key lies in developing efficient processing equipments. Pressing equipment of dried tofu has been mostly designed according to the line –"manual design (some with a CAD) - prototype production - prototype testing - design modification". However this way not only gets a high cost of research, but also a long cycle. It used virtual prototyping technology, at first, according to the design requirements, created three-dimensional modeling by using Pro/E software ; at last, made the feasibility analysis of kinematics and dynamics by using of ADAMS software. Through optimization design, analysis and control of cost, it had several characters as follows, simple machine, low cost, high productivity, it had great significance to mechanization and industrialization of pressing equipment of dried tofu.

Keywords: virtual prototyping, bean product, Pro/E, ADAMS.

1 Introduction

Nowadays, the production of soybean products, industrial automation degree of production is still relatively backward (Jiang Lianzhou et al., 2007; Wu Yuefang et al., 2006), especially dried tofu production; most manufacturers remain handicraft workshop. Mechanized processing and industrialized processing are slow and tortuous, expect the market factors, the weakness of technical force and the lack of advanced research and technology is a major obstacle. Although the production of dried tofu has a long history and accumulated rich production experience, some experience has not been theoretical and some phenomenon can not be explained by the theory. Pressing equipment of dried tofu has been mostly designed according to the line –"manual design (some with a CAD) - prototype production - prototype testing - design modification". However this way not only gets a high cost of research, but also a long cycle. Virtual prototyping technology provides a new means for the design of equipment. The paper is application research of virtual prototyping technology in the design of production equipment of dried tofu.

---

* Corresponding author.

D. Li and C. Zhao (Eds.): CCTA 2009, IFIP AICT 317, pp. 169–175, 2010.

2 System of Virtual Prototyping

It chooses Pro/E and ADAMS to build the research environment of virtual prototyping. In Mechanical/Pro modules, not only moving parts and sports relations can be defined, but also there is data conversion interface with ADAMS(Basset M et al., 2002), Although ADAMS has a powerful simulation of kinematics and dynamics, its 3D modeling capabilities and interoperability are weak. The system uses Pro/E to carry out 3D modeling and Mechanical/Pro module to definite constraint. Block diagram of virtual prototyping system is shown in Fig.1 (Zorriassatine F et al., 2003; Xiong Guangleng et al., 2001), the prototype is shown in Fig.2.

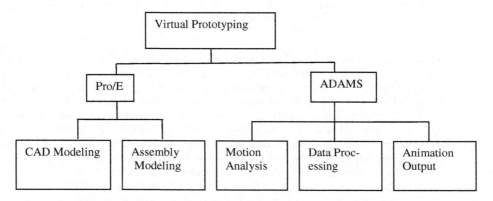

Fig. 1. Virtual prototyping system

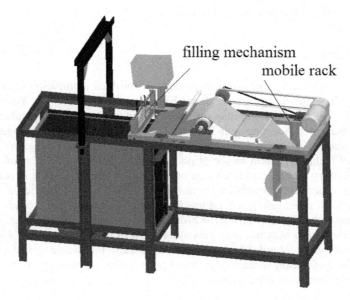

Fig. 2. Prototype of Pressing Equipment

3   Kinematics and Dynamics Analysis of Crucial Element

3.1   Kinematic Analysis

Even laying of mobile platforms is one of the key mechanical structure of pressing equipment of dried tofu, the carrying of the machine body, other key mechanical structure are placed on the mobile platform. Mobile platform is moving back and forth driven by the rocker, the rocker connects with the belt through the straight hinge, as shown in Fig.3.

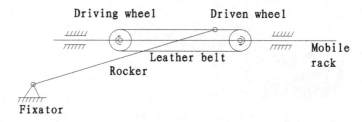

Fig. 3. Schematic drawing of motion principle of mobile rack

The velocity simulation of rocker centroid in the vertical plane is shown in Fig.4 using ADAMS, the interception of their 10 seconds of the motion diagrams, the abscissa is time(s), ordinate is speed(mm/s). Straight-line describes the level of the ground-rocker is relatively static, subfoveal and protruding are the part of movement, the highest and the lowest point is the point that the mobile platform changes the movement direction.

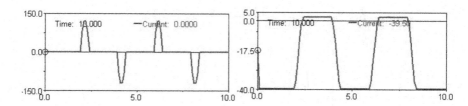

Fig. 4. velocity of the rocker centroid   **Fig. 5.** position of the rocker centroid

Rocker centroid position in the vertical plane is shown in Fig.5, abscissa is time(s), ordinate for the high(mm), because the upper surface of the mobile platform is defined as zero-plane in the assembly drawing, rocker is on the top of the mobile platform, its moving symmetrical line isn't at 0, Fig.5 shows the moving range of rocker centroid is 42mm, we can see the moving range of the rocker and the belt hinge's place is 84mm , the diameter of the pulley is 80mm, in addition hinge place 2mm, moving range of the rocker at the hinge place is just 84mm.

The upper horizon line in the Fig.5 shows the hinge point of the rocker and the belt is at loose side of belt, the lower horizon shows the hinge point of the rocker and the

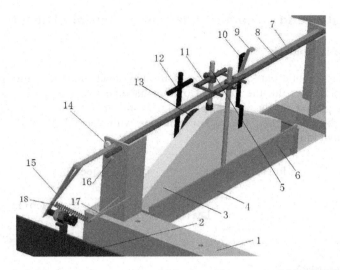

1. movable rack, 2. big frame, 3. upper material bucket, 4.lower material bucket, 5. upper material bucket valve, 6. lower material bucket valve, 7. rotary rod, 8.holding rod, 9.rotation hook, 10. material bucket hook,11.valve location rod, 12. portable rod, 13.feeding hole, 14.jack bolt, 15.impact tablets, 16.stop rotation tablets, 17.spring fixed flap, 18. tension spring。

Fig. 6. The overall drawing of filling mechanism

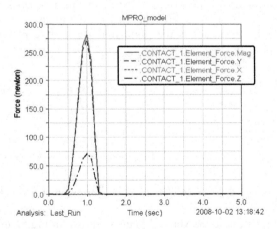

Fig. 7. Whereabouts high of drop-off material bucket and the opened valve

belt is at tight side of belt, the phase of the curve between upper and lower shows the centroid moving of the hinge point of rocker and belt on the pulley.

In the process of valve opening, the centroid trajectory of the drop-off material bucket and material bucket valve is shown in Fig.7, abscissa is time(s), ordinate is the height(mm), solid line is the centroid motion trajectories of the drop-off material bucket, the dashed line is the centroid motion trajectories of drop-off material bucket valve. the height of the valve to open is approximately 9mm from the figure.

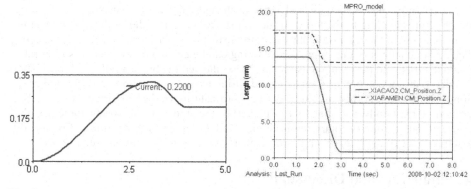

Fig. 8. rotation angle survey chart of rotary rod   **Fig. 9.** Impact force of impact tablet and rotary rod

Fig.8 is the rotation angle survey chart of rotary rod, abscissa is time (s), ordinate is the rotation angle (rad), rotating rod turns soon when it rotates to a certain point, at this time the rotation is drop-off material bucket has fallen off, the entire movable device moves in the opposite direction.

3.2 Kinetic Analysis

The simulation can measure the collision force of the rotary rod and the big frame, and Fig.9 shows the collision force of the collision tablets on the rotary rod by, solid line shows resultant force, long-dashed line shows Y force, and short dashed line shows X force , and the center line shows Z force. the greatest collisions force of is about 260N.

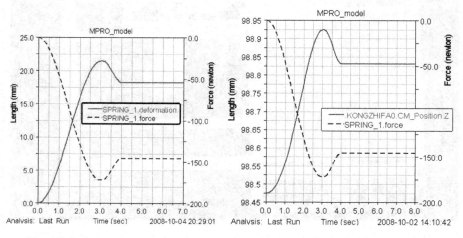

Fig. 10. Spring length and tension   **Fig. 11.** Position of rotary rod and spring tension

Fig.10 shows spring length and tension changes, the solid line shows the spring length, corresponding to the left ordinate (mm), dashed curve shows the spring tension, corresponding to the right ordinate (N). Spring stiffness coefficient is 800N/m, in accordance with changes in the length of the spring can be calculated the largest spring rally is about 170N, and the spring tension curve also showed the biggest rally is about 170N.

Fig.11 shows position of rotary rod and spring tension, the solid line shows the position of rotary rod, corresponding to the left ordinate(mm), dashed curve shows spring tension, corresponding to the right ordinate(N). the figure of the spring force and the figure of the position of rotary rod is similar ,and the figure of the position of rotary rod and the figure of the rotation angle curve chart of rotary rod is similar (see Fig. 7).

4 Simulation and Discussion

When we assemble in the Pro/E, certain order must be complied. The first installation of spare parts must be carefully chosen, because the other parts are based on the first, and the default location of the first parts can not be changed, if changing, the remaining parts will change accordingly. Because some parts should be inserted into other

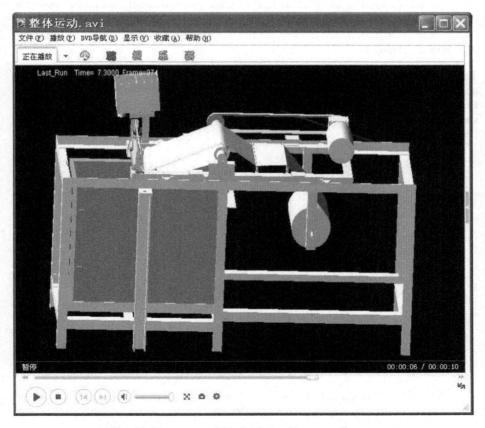

Fig. 12. Movement of dried tofu machine state diagram

parts of the location, we must pay attention to the part which is based on (the parent parts), the parts which is inserted are the sub-items, which should be moved with the parent parts, of course, there may be no parent parts when parts are assembled directly to the coordinates of drawing, so that even if changes happen in other parts, it will not be affected. The final assemble drawing in Pro/E is shown in Fig.2. Then the drawing is conversed to ADAMS by connecting software Mechanical/Pro. The various constraints and the power source can be directly loaded in Pro/E, and converted to the ADAMS or loaded directly in ADAMSADAMS/View.

In ADAMS/View, every part loaded has to be definite, if it does not affect the visual effects, animation and analysis, we should make the amount of parts to be the fewest in ADAMS/View. If there is no relative motion and no mechanical interference of two parts, we can connect the two parts into a part, so that the model can be simplified without affecting the results of simulation analysis.

It can make movement simulation and analysis after finishing a variety of constraints and drivers, ADAMS can output the results of the animation and analysis by post-processing module of ADAMS/PostProcessor. By comparing the test data and simulation results, ADAMS simulation can also verify the validity of the analysis, carry on mathematics operation to the data result and statistical analysis of the output. Fig.12 is the dummy specimen simulation animation which records from ADAMS/PostProcessor, The next work is to gathering test data, comparing to the simulation result.

5   Conclusion

It applies virtual prototyping technology in design of dried tofu production equipment, chooses Pro/E and ADAMS to build virtual prototyping, has overcome the insufficiency of ADAMS in 3D modeling, meanwhile it is highly effective to establishing virtual prototyping and accurate using Mechanical/Pro and the ADAMS data conversion, it has realized the entire virtual movement through the ADAMS simulation. The findings have the vital significance for design of pressing equipment of dried tofu, also has the reference value to similar products' development.

References

Basset, M., Pearson, R., Fleming, N., et al.: An approach to model-based design. In: The 2002 North American ADAMS Users Conference in Scottsdale, Arizona (2002)

Lianzhou, J., Shaoxin, H.: Status and Suggestions for the Development of Soybean Processing Industry in China. Journal of Agricultural Science and Technology 9(06), 22–27 (2007) (in Chinese)

Yuefang, W.: Development of Chinese Bean Food Trade. For the Food & Beverage Industry (04), 38–39 (2006) (in Chinese)

Zorriassatine, F., Wykes, C., Parkin, R., et al.: A survey of virtual prototyping techniques for mechanical product development. Proceedings of I Mech. E. Part B:J. Engineering Manufacture 217(1), 513–530 (2003)

Guangleng, X., Bohu, L., Xudong, C.: Virtual Prototyping Technology. Journal of System Simulation 13(1), 114–117 (2001) (in Chinese)

Headland Turning Control Method Simulation of Autonomous Agricultural Machine Based on Improved Pure Pursuit Model

Peichen Huang, Xiwen Luo[*], and Zhigang Zhang[3]

Key Laboratory of Key Technology on Agricultural Machine and Equipment of South China
Agricultural University, Ministry of Education, Guangzhou 510640, P.R. China,
Tel.: +86-020-38676975; Fax: +86-020-38676975
xwluo@scau.edu.cn

Abstract. According to the features of headland turning, new path planning and headland turning control algorithms for autonomous agricultural machine were presented in this paper. The turning path planning considered both the minimum turning radius and headland space was created by applying three straight lines. A path tracking algorithm based on the improved pure pursuit model was also proposed. This study used the BP neutral network to implement the dynamical look-ahead distance control for the improved pure pursuit model. Based on simplified bicycle kinematics model parameters, MATLAB/Simulink simulation results showed that the path planning algorithm were simple, occupied small headland space while still had a high tracking accuracy. The control method is feasible and practical.

Keywords: automatic guidance, agricultural machine, headland turning, path planning, pure pursuit algorithm, simulation.

1 Introduction

With the development of computer and sensor technologies, automatic guidance of agricultural machine has received deep research over many developed countries and regions (Toru Torii, 2000). The automatic guidance of agricultural machine technology is not only the basic platform for precision farming but also one of the research hotspots in fields of agricultural engineering. The property of headland turning is remarkably different from traveling on straight line. A proper headland turning not only improve the tracking accuracy when a machine transfers from the current working row to the next one, but also minimize the time spent in the headlands so as to increase the efficiency of farming operation (Zhu Zhongxiang et al., 2007).

A headland turning algorithm for rice transplanter was designed by Y. Nagasaka etc to guide the rice transplanter to move forward and backward during a turn

---

[*] Corresponding author.

D. Li and C. Zhao (Eds.): CCTA 2009, IFIP AICT 317, pp. 176–184, 2010.

(Yoshisada Nagasaka et al., 2004). This method could minimize the headland space but makes steering control complicated due to its backward motion. Kise. M et al (2001) developed two types of turning paths, namely forward turning and switch-back turning, which were created by applying a third-order spline function based on the minimum turning radius and maximum steering speeds. Computer simulation showed that the maximum tracking error was less than 0.2m. As the curve turning path is created by utilizing the third-order spline function, the headland turning for machine has to be implemented by curve tracking. The control difficulty will increase correspondingly. ZHU Zhongxiang (2007) etc proposed an optimal control algorithm. A time-minimum suboptimal control method was used to generate the turning path. A path-tracking controller consisting of both feedforward and feedback component elements was also proposed. This method implemented the optimal path with respect to traveling in minimum time, but the design of tracking controller is very complicated.

According to the features of headland turning and pure pursuit algorithm, agricultural machine headland turning control algorithm was presented in this paper. It works as follows: 1) applying simplified bicycle kinematics model, the path planning was produced by assembling three straight lines; 2) using the dynamical look-ahead distance which was implemented by the BP neutral network, an improved pure pursuit algorithm was applied in the headland turning.

2 Materials and Methods

2.1 Pure Pursuit Model

Pure Pursuit is a method for geometrically calculating the arc necessary for getting a vehicle onto a path (R.Craig Conlter, 1992; Vijay Subramanian et al., 2007). The method is simple, intuitive, easy to implement. The whole point of the algorithm is to choose a proper look-ahead distance. It is analogous to human driving in that humans look a certain distance ahead of the vehicle and steer such that the vehicle would reach a point at the look-ahead distance (Vijay Subramanian et al., 2007). The Pure Pursuit algorithm has been widely used in the field of path tracking. The algorithm is expressed as Fig.1 (All the parameters shown in Fig.1 are based on the machine's coordinate system without specification in the following part).

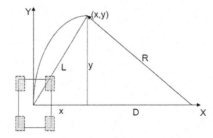

Fig. 1. Geometry of the Pure Pursuit Model

The x and y axis construct the machine's coordinate system. The point (x, y) is a point some distance ahead of the machine. The L is the length of the cord of the arc connecting the origin to the point (x, y). R is the radius of curvature of the arc. The relationship of x, L and R is as follows:

$$D + x = R \tag{1}$$

$$D^2 + y^2 = R^2 \tag{2}$$

$$x^2 + y^2 = L^2 \tag{3}$$

From Eq. (1), Eq. (2) and Eq. (3),

$$R^2 - 2Rx + x^2 + y^2 = R^2$$

$$R = \frac{L^2}{2x}$$

By choosing a look-ahead distance and calculating the path error x, the radius of the curvature required to get the machine on the required path can be calculated.

2.2 Headland Turning Control Method

2.2.1 Simplified Bicycle Kenimic Model

In this study, a simplified bicycle kenimic model is used to describe the machine motion. For the sake of simplified, it is assumed that the machine moves at a low constant speed over a flat surface with no wheel slippage and the wheels is considered as rigid wheels.

According to the kenimatic analysis (A.J.Kelly, 1994), the machine's motion equation is given as follows:

$$x'(t) = v(t)\cos(t)$$

$$y'(t) = v(t)\sin(t)$$

$$'(t) = v(t)\tan(t)/l$$

where l is the wheel base; δ is the front wheel steering angle; ϕ is the heading angle. As illustrated in Fig.2:

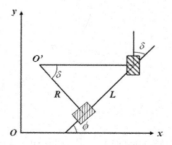

Fig. 2. The Kenimec Model of Simplified Bicycle

$$R = l/\tan \delta$$

where R is turning radius of the machine, l is wheel base, and δ is the front wheel steering angle. Combined with the pure pursuit equation derived above, the steering angle is represented by $\delta = \arctan(2lx/L^2)$. It indicates the relationship between the pure pursuit algorithm parameters and steering angle. This will help to lay a theoretical foundation of building the tracking control system.

2.2.2 Headland Turning Path Planning

Fig.3 shows the algorithm of path planning for headland turning in case of left turning. It is assumed that the minimum turning radius of a machine is $0.9m$ and the width M denotes $2R_{min}$, i.e. $1.8\ m$. The headland turning path shown in the Fig.3 is created by using three straight lines. In the established coordinate system, the three straight lines equations can be expressed as follows:

Path1: $x = 1.8$ $(0 \le y \le 3.5)$

Path 2: $y = 3.5$ $(0.9 \le x \le 1.8)$

Path 3: $x = 0$

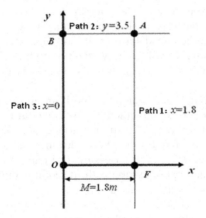

Fig. 3. Path planning of headland using three straight lines

As shown in Fig.3, the width M, i.e, the distance between path1 and path2, is two times equal to the machine's minimum turning radius, which not only fulfills the constrain of minimum turning radius, but also ensures the machine occupy the smallest width of headland. Path2, i.e $y = k$ is the important part in headland turning. The value k affects the turning space size and determines whether the machine is able to approach the next operation path precisely. The machine will have enough distance to adjust its position before approaching the next operation path while k is sufficient large, but meanwhile it occupies sizeable headland space. On the contrary, if k is smaller, it will reduce the space but the corresponding distance to adjust the machine's position will also be shortened.

After repeated simulations, a qualitative conclusion obtained from results is presented in this paper: During the headland path planning, the value k which can both make the machine get exactly enough distance to adjust its position and occupy the smallest space is determined by the width M and the machine velocity. In this paper, the width is $1.8m$ and machine velocity is set at $0.3 \ m/s$. When the value k is set at 3.5, it can obtain a favorable effect on both aspects. When the machine approaches at the end of the current operation path i.e. point F, it then starts to track Path1. When approaching the end point A of Path1, by using the path switcher, the machine switches to track the Path2. Likewise, the switch between Path2 and Path3 is to implement in the same way. Finally, the machine approaches at the start point O of the next operation path.

2.2.3 The Dynamical Look-Ahead Distance Control Based on Neural Network

There is one key parameter in the pure pursuit algorithm, the look-ahead distance. Its value has great effect on the tracking accuracy. Large look-ahead distances result in a gradual and smooth approaching of the path, but one which may take a considerable amount of time. Short look-ahead distances approaches the path quicker, but may result in oscillation about the path (R.Craig Conlter, 1992). Based on the analysis above, the large and short look-ahead distances both go against tracking effect. Many researchers may prefer to get a relative better look-ahead distance among large and small through repeated experiments. This method may has a relative better tracking effect on a certain degree but it is not the optimum as the look-ahead distance is unchanged during the whole headland turning. In addition, the velocity of a machine also affects the path tracking. Small velocity result in high tracking accuracy but also high time consumption, and vice versa so it is also necessary to consider the selection of the velocity.

According to the headland path planning algorithm above, the major task during the path1 and path2 tracking is to implement the machine turning, so a fixed moderate look-ahead distance can be chosen. Krešimir Petrinec, Zdenko Kovacic etc pointed out that a good choice of fixed look-ahead distance is around one wheelbase (Krešimir Petrinec et al., 2003) (in our case $1m$). The path3 tracking is related to the accuracy when entering the next operation path. It requires high accuracy, so the dynamical look-ahead distance control is used.

There will be a large offset at the initial stage when tracking path3 as the machine has to take some time to adjust the heading angle. Therefore a smaller look-ahead distance is necessary at this stage to make the machine regain the path quicker. When the offset becomes small enough (within assigning range), larger look-ahead is used to let the machine regain the path with less oscillation. Further considering the machine velocity, according to the Preview Follower Theory (Wang Jingqi et al., 2003), higher velocity requires larger look-ahead distance and vice versa. So different look-ahead distances should be used according to different velocities.

This paper utilized the neural network's seft-study and association memory ability to teach a neural network how to dynamically control the look-ahead distances. This network has two inputs (the machine velocity and the x coordinate) and one output (the desired look-ahead distances). After the comprehensively considering with the tracking accuracy and the consuming time, two velocities are selected in this paper. They are $0.3m/s$ and $0.4 \ m/s$. The headland was divided into three areas by two straight lines ($x = -0.5$ and $x = 0.5$). Different look-ahead distances in every individual area are used. After using different combination of velocities and look-ahead distances in the repeated simulations, it is determined that when the velocity is 0.3 *m/s*, 1 *m* is used as large look-ahead distance and $0.01m$ as small look-ahead distance; when the velocity is 0.4 *m/s*, 1 *m* is as large and 0.5 *m* as small. Before establishing the neural network, it is necessary to determine the training set. At each training sample, the *x* coordinates and velocity values were used as the inputs and the desired look-ahead distance *L* for the output, e.g, $[x = -1, v = 0.3, L = 0.01]$ is one training data, thus there were a total of 148 examples (cases) in the training set. Table1 shows the training samples.

Table 1. Sample Data for Training Neural Network

x	v	L	x	v	L
-1	0.3	0.01	-1	0.4	0.5
-0.99	0.3	0.01	-0.99	0.4	0.5
-0.98	0.3	0.01	-0.98	0.4	0.5
...	...	...	...	...	...
-0.52	0.3	0.01	-0.52	0.4	0.5
-0.51	0.3	0.01	-0.51	0.4	0.5
-0.5	0.3	1	-0.5	0.4	1
-0.4	0.3	1	-0.4	0.4	1
-0.3	0.3	1	-0.3	0.4	1
...	...	...	...	...	...
1.7	0.3	1	1.7	0.4	1
1.8	0.3	1	1.8	0.4	1

In Table 1, *x*, *v* and *L* represent the *x* coordinate, velocity (m/s) and look-ahead distance (*m*) respectively. After training with the training set, the network which memorized the look-ahead distances in different areas was established finally by using BP algorithm, which is commonly used in the engineering. The dynamical control of look-ahead distance for tracking can let the machine adapt to the changing of practical situation and occupy the small headland space as much as possible.

3 Simulation Results and Discussion

3.1 The Headland Turning Simulation Based on Pure Pursuit Model

The headland turning simulation based on the pure pursuit model was performed on MATLAB/Simulink platform. In this case, the parameters were initialized as follows:

$V = 0.3m/s$; $M=1.8m$; $R_{min} = 0.9$ m; the starting position was set at $(1, 0)$; the initial heading was set at 90^0. Two fixed look-ahead distances are simulated separately, the results are shown in Fig.4(a) and Fig.4(b). The simulation process and results of the case $V=0.4\,m/s$ is similar to the case $V=0.3\,m/s$, so this paper only present the latter case.

In the whole headland turning simulation results, the tracking trajectory of path3 is the most important as it can show if the machine can entry the next operation path accurately, so the results only show the desired path3 and machine tracking trajectory. The straight line represents the path3 and circles represent the machine trajectory.

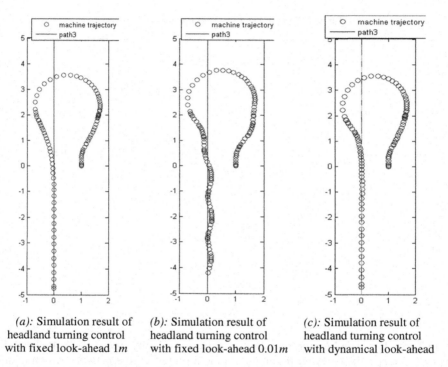

(a): Simulation result of headland turning control with fixed look-ahead $1m$

(b): Simulation result of headland turning control with fixed look-ahead $0.01m$

(c): Simulation result of headland turning control with dynamical look-ahead

Fig. 4. Simulation result of headland turning control with different look-ahead distance

Fig.4(a) and Fig.4(b) illustrate the simulation results with fixed look-ahead distances 1 m and 0.01 m. Same initial conditions were set at both simulations.

It was found that when using the smaller fixed look-ahead distance 0.01m, the machine trajectory is more oscillations as shown in Fig.4(b). The larger look-ahead distance 1m resulted in a gradual and smooth trajectory, but took longer time for the machine approaching to the path as shown in Fig.4(a).

3.2 The Headland Turning Simulation Based on the Improved Pure Pursuit Model —— BP Network Implement The Dynamical Look-Ahead Distance Control

The training of the network was performed with BP network function package provided by the MATLAB Neural Network Toolbox. First, function newff( ) was utilized to create a three layers' network (one input, three hidden, one output layer). Second, function train( ) was utilized to perform training. The training epochs were set at 100 while the mean square error (MSE) was 10^{-6} (reasonably low) after training. Finally, the network Simulink module generated by function gensim( ) was combined with the normalized and anti-normalized modules to create the dynamical look-ahead distance control module.

Fig.4(c) shows the simulation result after using the dynamical look-ahead distance control during the last desired path $x = 0$. When the x coordinate is large, small look-ahead distance (in our case $0.01m$) tends to converge to the path more quickly. When the x coordinate is small, large look-ahead distance (in our case $1\ m$) tends to regain the path with less oscillation.

It was indicated in Fig.4 that the dynamical look-ahead distance control could significantly improve the tracking effects. Not only the machine could converge the path more quicker, but also with less oscillation. This was favorable for the machine to entry the next operation accurately.

4 Conclusion

According to the pure pursuit algorithm which was originally devised as a method for mobile robot tracking the path, combined with the characteristics of agricultural machine, a new control method of headland turning based on the improved pure pursuit model for agricultural machine was presented in this paper. The MATLAB simulation results showed that this control method was feasible and effective.

The path planning algorithm, which is implemented by combining three straight lines, makes the path planning simple and easy to implement while the machine's tracking effect is good.

The dynamical look-ahead distance control was also proposed. This could let the machine adapt to the changing of practical situation. Compared to two fixed look-ahead distance simulation, the dynamical look-ahead distance control results show that it has the advantages of reducing time and space, more reliable and high accuracy during the headland turning.

In future work, attention must be paid to providing an accurate formula to quantitatively calculate the path2 position by utilizing the path1 and path3 positions so that the machine can obtain the best on both adjusted distance and tracking accuracy.

Acknowledgements

Research conducted at South China Agricultural University, funding from National Fund 863. Special thanks are extended to Professor Luo Xiwen and Mr.Zhang Zhigang for their technical support.

References

Kelly, A.J.: A Feedforward Control Approach to Local Navigation Problem for Autonoumous Vehicles. CMU Robotics Institute Technical Report, CMU-RI-TR-94-1 (1994)

Kise, M., Noguchi, N., Ishii, K., Terao, H.: Development of the agricultural autonomous tractor with an RTK-GPS and a FOG. In: Proceedings of the 4th IFAC symposium on Intelligent autonomous vehicles. IFAC, pp. 103–106 (2001)

Petrinec, K., Kovacic, Z., Marozin, A.: Simulator of Multi-AGV Robotic Industrial Environments. IEEE, Los Alamitos (2003)

Craig Conlter, R.: Implementation of the Pure Pursuit Path Tracking Algorithm. Camegie Mellon University (1992)

Torii, T.: Research In Autonomous Agriculture Vehicles In Japan. Computers and Electronics in Agriculture 25, 133–153 (2000)

Subramanian, V., Burks, T.F.: Autonomous Vehicle Turning In the Headlands of Citrus Groves. In: An ASABE Meeting Presentation, ASABE, p. 1015 (2007)

Jingqi, W., Huiyan, C., Pei, Z.: Application of Fuzzy Adaptive PID and Previewing Algorithm in Steering Control of Autonomous Land Vehicle. Automotive Engineering 25(4), 367–371 (2003) (in Chinese)

Nagasaka, Y., Umeda, N., Kanetai, Y., Taniwaki, K., Sasaki, Y.: Autonomous guidance for rice transplanting using global positioning and gyroscopes. Computer and Electronics in Agriculture (43), 223–234 (2004)

Zhongxiang, Z., Jun, C., Yoshida, T., Torisu, R., Song, Z.-h., Mao, E.-r.: Path Tracking Control of Autonomous Agricultural Mobile Robots. Journal of Zhejiang University SCIENCE A 8(10), 159–1603 (2007) (in Chinese)

Design of Plant Eco-physiology Monitoring System Based on Embedded Technology

Yunbing Li[1,2], Cheng Wang[1,*], Xiaojun Qiao[1], Yanfei Liu[1], and Xinlu Zhang[1]

[1] National Engineering Research Center for Information Technology in Agriculture,
Beijing, P.R. China 100097
[2] Engineering college, South China Agricultural University, Guangzhou, Guangdong Province,
P.R. China. 510642,
Tel.: +86-010-51503409
wangc@nercita.org.cn

Abstract. A real time system has been developed to collect plant's growth information comprehensively. Plant eco-physiological signals can be collected and analyzed effectively. The system adopted embedded technology: wireless sensors network collect the eco-physiological information. Touch screen and ARM microprocessor make the system work independently without PC. The system is versatile and all parameters can be set by the touch screen. Sensors' intelligent compensation can be realized in this system. Information can be displayed by either graphically or in table mode. The ARM microprocessor provides the interface to connect with the internet, so the system support remote monitoring and controlling. The system has advantages of friendly interface, flexible construction and extension. It's a good tool for plant's management.

Keywords: plant eco-physiology monitor system, embedded technology, MODBUS Protocol.

1 Introduction

Plant eco-physiology is the research of interaction between plant and the environment. Physiological parameters included such as stem sap flow, stem diameter variation, transpiration and leaf temperature, plant ecological parameters include the ambient environment of plant such as the total irradiances, density of CO_2, temperature and moisture of soil and so on. Plant eco-physiology signal can be monitored roundly and automatically by the system. Physiological parameters provide important information, especially if integrated into control systems or computer models in the so-called SPA (speaking plant approach) (Hashimoto et al., 1989). Lots of researches have showed that physiological parameters reflect the status of the plant growth and physiology demand directly and immediately (Meng Zhaojiang et al., 2005). The conventional routes of monitoring environmental parameters in greenhouse realize environment regulation achieve plant growth environment optimization (Yang Weizbong et al., 2006; Teng Guanghui et al.,2002). But plant growth's influence factors are

---

* Corresponding author.

D. Li and C. Zhao (Eds.): CCTA 2009, IFIP AICT 317, pp. 185–190, 2010.
© IFIP International Federation for Information Processing 2010

complex, different environment are required for different plant, depending upon the environment parameters purely cannot applied to many plant species widely(S.L. Speetjens et al., 2008).The continuous and automatic monitoring system for plant eco-physiology is able to feed back the information of plant needing, and it is used for environment regulation when mismatches appear between environmental status and requirements of plant. In developed countries, plant physiological parameters monitored prevail in some orchards (Tom Helmer et al., 2008). But its apparatus communication limited in a small area, and its extension was inflexible. What's more, it needs the computer to support work. But these apparatus have similar shortcomings: functional simple, poor scalability, and the need for PC support. With the development of embedded technology, people use microprocessor as a substitute for computer. In this paper embedded technology and wireless sensor network were adopted to design plant eco-physiology monitoring system. The wireless sensing network collect information for ARM microprocessor, and the microprocessor realized the whole system's control and predicts status of plant from the collected information. The plant eco-physiology monitor system can be adopted in plant's management, water management and environment monitoring, which has forecast and guide functions.

2   System Framework

The system is organized as three parts: ARM controller module, data acquisition module and long-distance communication module. Wireless sensor network (WSN) with ZigBee function is utilized in the data acquisition module. The WSN gathers various plants' eco-physiological information and transmits it to the ARM controller. The controller, in which ARM micro-processor as its core and μC/OS-II real time operation system as its developing platform, it can control the whole system and make tasks scheduled. Touch screen is used as an output device which can display the information as well as input device. System parameters can be set according to user's needs. The monitor system communicates with server through Ethernet.

3   System's Hardware Design

The ARM controller has many peripherals. The power source management module provides the multi-channel power sources for system models respectively. The WSN termination nodes connected with controller through RS232 interface. The Coordinator node of WSN exchange information with controller through ModBus protocol. Through the Internet system can exchange information with the long-distance server. People can browse information and realize the remote operation through the Internet.

3.1   Resource of ARM Controller

The core of embedded system is based on ARM7TDMI processor. The processor connects with the programmable memory flash, data storage flash, Ethernet controller chip CS8900A, USB-Host controller. The ARM interface is shown in Fig. 1. It contains a flash on the board, a standard extend RTC real-time clock, UART controller, IIC and SPI Bus. The flash can storage 1 GB data, A real-time operating system μC/OS-II is used in this system, which supports the TCP/IP protocol and FAT document management. Resources of the system help us to realize the long-distance network access and control. The system supports the CF card for extended storage and U-disk makes information transmitted and saved conveniently.

4 Software Design

We can execute task scheduling and management by the μC/OS-II platform. μC/OS-II system supports 64 tasks at most. In this system we use 4 tasks respectively: Man-machine interface control, information collection and control, file management and peripheral device correspondences, remote access and control. Tasks priority dispatches according to time allocation in the main program.

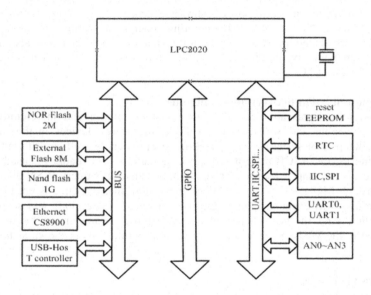

Fig. 1. Interface of ARM

4.1 The Realization of GUI

Touch screen in the design supports plenty of pictures save and graphic operation. Touch screen holds its own instruction set. All operations of the touch screen were implemented by sending a sequence of instructions in a certain format. In the project, we make the instruction set to sub-function for the convenience of calling. The sub-function format such as data source, operation set source, operator and operarand, the length of sub-function.

   Func_Type Func_Name([*set], [*source], [*operands], [length]);

   Through calling the sub-function we can carry on the user interface operation such as text demonstrated, drawing and database operation. The main program call sub-function in GUI program realizes the system display control.

4.2 System Software

The main program realizes the overall temperament of the system. After system powered on, program begin to work and the system realizes sensor auto-search in the area through the unique sensor ID. Compensation of sensor set parameters for polynomials, which standardizes the gathering data. Touch screen can display gathering information dynamically and may also review all history data. The information can also be displayed graphically. Watch-dog in software prevents system from halting. Back light of touch screen will be set to reduce the system's power loss. The primary tasks are shown in the Fig.2.

4.2.1 The Human-Machine Interaction Interface
The favorable man-machine interface and touch screen made the system operation simple. The controller has relative action according to different instructions. The system's human-machine interaction flow is shown as task 1 in Fig.2. The controller send commands to the LCD, then the LCD operating depends on commands parsing. And the controller distills the touch screen return information. The system can also display in multiple languages by download different word libraries. The touch screen can work as keyboard of PC.

4.2.2 Communication between the Controller and WSN
MODBUS protocol was adopted to enhance system communication reliability between the controller and WSN. It makes different control devices in the industry network realization of system's centralized monitoring. There are two different transmission patterns: the ASCII pattern and the RTU pattern. RTU pattern is used in this design. The system maps to sensor through unique sensor ID, query WSN data transmission through MODBUS protocol. When the system measure time is up, the communication task was executed, the data in the buffer of collect node was sending to the controller. And in the system sensor set, all the parameter set by user was downloaded to the WSN. System communication program flow is shown as task 3 in Fig.2.

4.2.3 Data Processing
In the software designing all sensors are defined as class. Relative operation of sensor depending on sensor's ID in program. Sensors attributes can be set by the user through the touch screen. Double-linked list is applied to carry on the dynamic node assignment for enhancing memory allocated. The operation of create or delete sensor node easily, and traversal all sensors which exists in the system very conveniently. Sensor information includes sensor's ID, sampling period, name and value. Compensation mathematical formula realized sensor intelligent compensation. After collecting data, system will transform it according to the offset parameter of the compensation formula. After the sensor loaded once, system will record it automatically, which will help sensor auto-search in the next time. All data acquired by the sensor is appended to the data saved document according sensors' ID. The corresponding time and the sensor value will be saved in file. All information will be saved in the board flash. It has 1GB capacity and able to export its information to U disk easily. All documents

are saved in a certain format and exported to generate document of EXCEL format, which help data processing easily. The flow diagram is shown as task 2 in Fig.2.

4.2.4 Remote Monitoring and Control

Server and system are connected by the B/S pattern. The communication is based on the TCP/IP network protocol. The controller listening in the port assign by server realize the server and the control system long-distance communication. All information frames transmit in network socket character. The system receive and action as the orders from the server. The flow diagram is shown as task 4 in Fig.2.

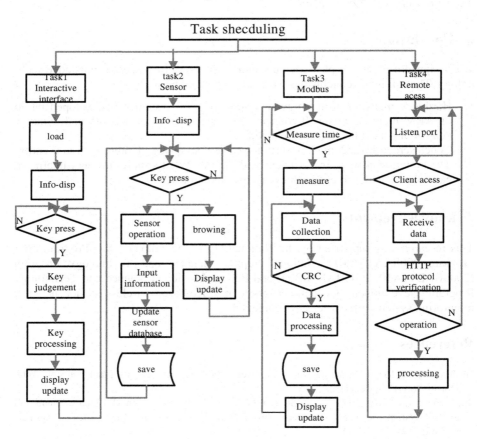

Fig. 2. Main program scheme

5 Discussion

System meets the requirements of analysis and monitoring plant eco-physiology information, which will alarm automatically when the eco-physiology information becomes abnormal. The system has the following features, such as flexibility, reliability, simple operation and scalability. System can be used as an accessorial tool for greenhouse control, which can support decision for greenhouse intelligent management.

The next step of developing, "knowledge-base management system" will be added in this system. According to different flowers and crops ' growth rule, revise the system set points to achieve environment optimization for plant. Working out optimization environment modeling for plant based on monitor and analysis of the plant eco-physiological parameters. As a result, system also has functions of alarm plant stress of water deficit, pest or environmental stress. Forecast and tutor in water-saving irrigation, accurate control and so on. It helps people to manage plant scientifically and use the resources more reasonably.

6 Conclusions

We designed a plant eco-physiological monitoring system based on embedded technology. The following conclusions were inferred from the study:

1. Hardware platform of autonomous data collection system was developed based on ARM microprocessor. It connects with wireless sensor network and dynamic control of data collection was realized.
2. Touch screen was adopted, operation was very simply, and interface was friendly.
3. The system can work independent without PC. What's more, through internet it can connect with server to achieve remote monitoring.

Acknowledgements

This research was supported by the National High Technology R&D Program (863 Program 2008AA10Z201), National Support Science and Technology Program (2006BAD11A10, 2006BAD30B03) and Beijing Municipal Science and Technology Program: Development and demonstration of voice-based wireless ad hoc network monitoring system in facilities environmental.

References

Hashimoto, Y.: Recent strategies of optimal regulation by the speaking plant concept. Acta. Hortic. 260, 115–121 (1989)

Speetjens, S.L., Janssen, H.J.J., van Straten, G., et al.: Methodic design of a measurement and control system for climate control in horticulture. Computers and Electronics in Agriculture 64, 162–172 (2008)

Helmer, T., Ehret, D.L., Bittman, S.: CropAssist, an automated system for direct measurement of greenhouse tomato growth and water use. Computers and Electronics in Agriculture 48, 198–215 (2005)

Zhaojiang, M., Aiwang, D., Zugui, L., et al.: Advances on diagnosis of crop moisture content from changes in stem diameters of plants. Transactions of the CSAE 21(2), 30–33 (2005) (in Chinese)

Weizbong, Y., Yiming, W., Haijian, L.: Distributed greenhouse intelligent control system based on the field bus concept. Transactions of the CSAE 22(9), 163–167 (2006) (in Chinese)

Guanghui, T., Changying, L.: DNCS—A New Scheme for the Automation of Greenhouse Environment Control. Transactions of the CSAE 18(5), 118–122 (2002) (in Chinese)

Motion Law Analysis and Structural Optimization of the Ejection Device of Tray Seeder

Xin Luo[1,*], Bin Hu[1], Chunwang Dong[1,2], and Lili Huang[1]

[1] Machinery and Electricity Engineering College, Shihezi University, Shihezi 832003,China
Tel.:13031336692
[2] Insititute of Cotton; The Chinese Academy of Agricultural Sciences; Anyang;
Henan 455000 China
hb_mac@shzu.edu.cn

Abstract. An ejection mechanism consisting four reset springs, an electromagnet and a seed disk was designed for tray seeder. The motion conditions of seeds in the seed disk were theoretical analyzed and intensity and height of seed ejection were calculated. The motions of the seeds and seed disk were multi-body dynamic simulated using Cosmos modules plug-in SolidWorks software package. The simulation results showed the consistence with the theoretical analysis.

Keywords: Tray seeders; ejection mechanism; numerical simulation; motion analysis.

1 Introduction

The Tray seeder is an essential component for modern cultivation techniques. Whether seeds are vibrated uniformly to suspension state by vibration device can affect the seeding capability. Now continual excited vibration by electromagnetic vibration device has been extensively employed to make the seeds "bouncing" and enable them being sucked. However, the uneven vibration intensity of seeds in different positions causes the seeds suspend in different height, so seeder has the poor air-suction effect for seeds. In recent years, the motion law of seeds based on continual excited vibration has been studied (cheng jin, 2002; Li Yaoming et al, 2009; Zhao Lixin et al, 2003). They reported that the seeds should obtain enough inertia force to get in appropriate suspension station for satisfaction of the air-suction requirements. But there is lack of information on how to make all seeds in different positions acquire same vibrating intensity. In this paper, an ejection mechanism consisting four reset springs, a electromagnet and a seed disk was designed. The motion conditions of seeders in the seed disk were theoretical analyzed and then ejection intensity and height were calculated. The motions of the seeds and seed disk were multi-body dynamic simulated using Cosmos modules plug-in SolidWorks software package. The simulation results showed the consistence with the theoretical analysis.

---

* Corresponding author.

D. Li and C. Zhao (Eds.): CCTA 2009, IFIP AICT 317, pp. 191–197, 2010.

2 Working Principle of Ejection Mechanism

The ejection mechanism schematic diagram is shown in Fig.1. It mainly consists of four reset springs, a electromagnet and a seed disk. The disk and the armature of electromagnet are connected together by pin bolts. The reset springs are installed in the lug boss between the worktable and the seed disk. The armature overcomes friction, pulls the disk down when the electromagnet is charged, and conversely, the disk is ejected by the reset springs and the seeds are tossed into suspension. The hopping seeds can overcome the force among seeds, effectively improving the adsorption rate of all seeds.

Structural Design Parameters :

The weight of disk is 3kg, the stiffness of spring is 4N/mm, the voltage of electromagnet is 220v, the maximum traction force is 80N, and the maximum stroke is 25mm.

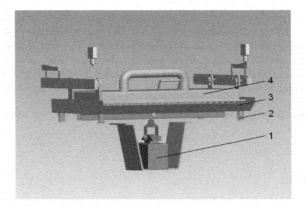

1. electromagnet 2. reset spring 3.disk 4.sucker

Fig. 1. schematic diagram of ejection mechanism

3 Analysis of Motion Regularity of Ejection Mechanism

3.1 Moving Condition and Calculation for Hopping of Seeds

The movements of monolayer seeds on the disk are studied. Gravity and force of inertia of the seeds are mainly considered (J. Theuerkauf P. Witt, 2006), air resistance is neglected. The critical condition for hopping of seeds is:

$$P \geq G \tag{1}$$

Where: P is the inertia force of seeds (N), G is weight of seeds (N)

The periodical harmonic excitation equation of disk is:

$$x = Ae^{-\varepsilon t} \sin(\omega_d t + \varphi) \tag{2}$$

Without considering the system damping, the motion is:

$$x = A\sin(\omega t + \varphi) \tag{3}$$

Where: t is vibration time(s), φ is the initial phase of vibration(rad), φ is natural frequency(Hz).

The initial acceleration of seeds is equal to disk's:

$$\alpha_s = \alpha_d = A\omega^2 \tag{4}$$

According to Newton Second Law:

$$P = m\alpha_s = \frac{G}{g}A\omega^2 \tag{5}$$

Puts (5) into (1):

$$P = \frac{G}{g}A\omega^2 \geq G \tag{6}$$

The ejection strength coefficient K is:

$$K = \frac{A\omega^2}{g} \geq 1 \tag{7}$$

When k is greater than 1, seeds are tossed.

As shows in Fig.2,

$$P + (M + m)g = k(\Delta x + \delta) \tag{8}$$

$$\delta = \frac{(M + m)g}{k} \tag{9}$$

Where: k is the coefficient of stiffness of spring Δx is the amount of compression, δ is the static Deformation between seed and disk.

Puts (9) into (8):

$$k = P / \Delta x \tag{10}$$

According to the structural characteristics of the ejection mechanism:

P=80 N, Δx =20mm, $k = 4N / mm$.

The natural frequency of the ejection system is:

$$\omega = \sqrt{\frac{\kappa}{m_s}} = \sqrt{\frac{4 \times 1000}{3}} = 36.5 rad / s$$

Where: m_s is the total mass of seed and disk

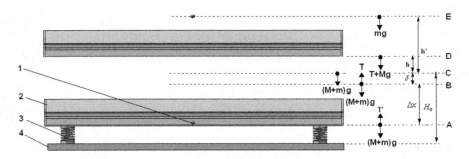

1. cotton seed 2. disk 3. reset spring

Fig. 2. Motion State of ejection mechanism

The amplitudes of disk are:

$$A = \sqrt{x_{t1}^2 + (\frac{\dot{x}_{t1}}{\omega_n})^2} = \sqrt{20^2 + 0} = 20mm$$

Put it into (7)

$$K = \frac{A\omega^2}{g} = \frac{0.02 \times 36.5^2}{9.8} = 2.718 \geq 1$$

The results meet the requirement.

3.2 Analysis on Ejecting Height of Seed

As shown in Fig.2, at the position of B, According to the conservation of mechanical energy:

$$\frac{1}{2}(M+m)v^2 + (M+m)g\Delta x + \frac{1}{2}k\delta^2 = Es \qquad (11)$$

Elastic potential energy is

$$Es = \frac{1}{2}k(\Delta x + \delta)^2 \qquad (12)$$

Puts (12) into (11)

$$v = \sqrt{\frac{k\Delta x^2(\Delta x + \delta)}{(M+m)} - 2g\Delta x} \qquad (13)$$

Where M is the weight of the disk, m is the weight of seeds, v is the speed before separation.

Throwing height of seed is:

$$h' = \frac{v^2}{2g} = \frac{k\Delta x^2(\Delta x + \delta)}{2g(M+m)} - \Delta x \qquad (14)$$

According to the conservation of mechanical energy:

$$kh^2 + 2Mgh - Mv^2 + 2Mg\delta - k\delta^2 = 0 \tag{15}$$

$$h = \sqrt{\frac{Mv^2 + k\delta^2}{k} - \frac{2Mg\delta}{k} - \frac{Mg}{k}} \tag{16}$$

Puts M =33kg,m=0.096g P=80N into (11)、(13)、(14)
v=735mm/s，h=12.59mm，h'=26.67mm，h'- h= 14.08 mm.

4 Multi-Body Dynamics Simulation of the Ejection Vibration Mechanism

4.1 Model Establishment of Ejection Mechanism

The ejection mechanism model consisting of seed disk, seeds and working table was established using SolidWorks software. In the seamless simulation environment of CosmosMotion, the distance between seed disk and working table was set 42.5mm, namely, electromagnet can draw spring 27.5mm from original position.

4.2 Definition of Constraint

The constraints of parts in ejection mechanism model can be automatically created in CosmosMotion environment according the mating counterpart. Seed disk and seeds were defined as motion body and working table was defined as stationary body. The simulation parameters were set as follows: material of disk is alloy steel; density of cotton seed is 1.06×103kg/m3, weight is 0.96g, restitution coefficient is 0.5, friction coefficient is 0.3; Stiffness of each reset spring (adding four spring between seed disk and spring seat) is 1N/mm, length is 70mm, diameter is 16mm, wire diameter is 1.6mm, cycle number is 11.3d. 3d elastic collision between disk and seeds is defined (friction is set dynamic).The parameters of equation solver was defined as follows: ADAMS GSTIFF was set 0.5s; frame numbers was 100 and other parameters are defaulted.

4.3 Simulation Results

Mesh was automatically calculated by equation solver. The curve of velocity and centroid displacement of seed and disk are as shown in fig 4 (red represents seeds, blue represents disk). The Seed moved to a static balance position while the disk reached the maximal rate, and then they began to separate. Subsequently the seed was ejected and the disk started to decrease in velocity, which is consistent with the theoretical analysis. The highest ejection height of seed above disk was 13.5mm, which is close to the theoretical calculated value of 14.08mm. The difference between values comes from the energy loss caused by collision friction.

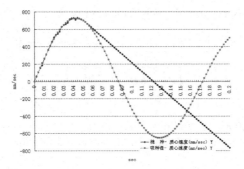

Fig. 3. Motion trajectory of speed

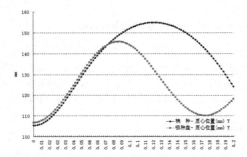

Fig. 4. Motion trajectory of centroid position

5 Conclusion

1. The results of theoretical analysis on ejection mechanism motion showed that seed could be ejected completely when the coefficient of stiffness of spring $k \geq 1$. The ejection height was determined by the static deformation between seed and disk (δ), ejection velocity (v), the coefficient of stiffness of spring (k).

2. Motion simulation of disk and seed showed the consistence with the results of theoretical analysis. The difference between values comes from the energy loss caused by collision friction. When the seed moved to the static balance position together with the seed disk, they began to separate for their velocities decrease at different rate.

3. The physical prototype is simple, which has a reliable performance and the rising height of the seeds are even. The tests have confirmed that the ratio of single-seed was over 97%, The ratio of non-seed was below 2%, completely to satisfy the requirements of precision seeder.

References

[1] Jin, C., Yaoming, L.: Study on Seeds Movement Law in Sowing Test Stand with Suction and Vibration. Transactions of The Chinese Society of Agricultural Machinery 33(1), 47–50 (2002)

[2] Yaoming, L., Zhan, Z., Jin, C., Lizhang, X.: Discrete Element Method Simulation of Seeds Motion in Vibrated Bed of Precision Vacuum Seeder. Transactions of The Chinese Society of Agricultural Machinery (03), 55–59 (2009)

[3] Yaoming, L., Zhan, Z., Jin, C., Lizhang, X.: Numerical Simulation and Experiment on the Seeds Pickup Performance of Precision Air-suction Seeder. Transactions of The Chinese Society of Agricultural Machinery 39(10), 95–100 (2008)

[4] Lixin, Z., Liyun, Z.: Seed suction performance of vibrational air-suction tray seeder. Transactions of the Chinese Society of Agricultural Engineering 19(04), 122–125 (2003)

[5] Singh, R.C., Singh, G., Saraswat, D.C.: Optimisation of Design and Operational Parameters of a Pneumatic Seed Metering Device for Planting Cottonseeds. Biosystems Engineering 92(4), 429–438 (2005)

[6] Theuerkauf, J., Witt, P., Schwesig, D.: Analysis of particle porosity distribution in fixed beds using the discrete element method. Powder Technology 165(2), 92–99 (2006)

[7] Shangping, L., Yanmei, M., Fanglen, M.: Research on the working mechanism and virtual design for a brush shape cleaning element of a sugarcane harvester. Journal of Materials Processing Technology 129(1), 418–422 (2002)

[8] Bangchun, W.: Vibration Utilization Engineering, pp. 53–54. Science Press, Beijing (2005)

[9] Edited by SolidWorks Corporation. COSMOS Advanced Course: CosmosMotion. Machinery Industry publishing house, Beijing (July 2008)

[10] Kurowski, P.M.: Ph.D, P.Eng. Engineering Analysis with Cosmos Works Professional (2006)

Nozzle Fuzzy Controller of Agricultural Spraying Robot Aiming Toward Crop Rows

Jianqiang Ren[*]

Department of Computer, Langfang Teachers College, Langfang, Hebei Province,
P.R. China 065000,
Tel.:+86-316-6817802
renjianqiang@163.com

Abstract. A novel nozzle controller of spraying robot aiming toward crop-rows based on fuzzy control theory was studied in this paper to solve the shortcomings of existing nozzle control system, such as the long regulation time, the higher overshoot and so on. The new fuzzy controller mainly consists of fuzzification interface, defuzzification interface, rule-base and inference mechanism. Considering the actual application, the fuzzy controller was designed as a 2-inputs&1-output closed-loop system. The inputs are the distance from nozzle to crop row and its change rate, the output is the control signal to the execution unit. Based on the design project, we selected the FMC chip NLX230, the EMCU chip AT89S52 and the EEPROM chip AT93C57 to make the fuzzy controller. Experimental results show that the project is workable and efficient, it can solve the shortcomings of existing controller perfectly and the control efficiency can be improved greatly.

Keywords: nozzle, fuzzy controller, spraying robot, crop-rows.

1 Introduction

The automation and intellectualization of the farmland working assignment is the development trend of today agriculture, the application of agricultural robot will reduce the agricultural labor intensity and raise the work efficiency greatly. The spraying robot is a type of intelligent agricultural implement. The conventional spraying robot adopts the overlapped spraying technology(Cui Jun et at., 2006), which can effectively solve the problem of leak spray but can lead to the serious consequences easily, such as the waste of pesticide resources, the unattainment of pesticide residues in agricultural products, the environmental pollution and so on. The precision pesticide-application technology has become one of the research directions of precision agriculture (Ma Hui, 2002, Fu Zetian et at., 2007). A pesticide system spraying with changeable quantity based on fuzzy control was studied by Anhui Agricultural University (Shao Lushou et at., 2005). The automatic target detecting electrostatic air assisted orchard sprayer was designed by China Agricultural University and Chinese Academy of Agricultural Mechnization Sciences (He Xiongkui et al., 2003). The

---

[*] Corresponding author.

D. Li and C. Zhao (Eds.): CCTA 2009, IFIP AICT 317, pp. 198–206, 2010.

precision sprayer for site-specific weed management was studied by the Davis Branch, California University of United States (Tian L. et al., 1999). Precision band spraying system with machine-vision guidance and adjust able yaw nozzles was studied by Giles D.K. and Slaughter D.C. (Giles D.K. et al., 1997). The spray control system to aim toward crops rows based on machine vision was studied by Nanjing Agricultural University (Rao Honghui et at., 2007). In these works, the spray control to aim toward crops rows is an important technology in the farmland working and the nozzle was moved to aim toward crop rows accurately by a step motor. But to the existing nozzle control system has shortcomings, such as the long regulation time, the higher overshoot and so on. To solve these problems, a new nozzle controller based on fuzzy control theory was proposed in this paper and the control efficiency was greatly raised.

2 Structure of Spraying Robot

The Structure of the spraying robot is shown in Fig.1. It mainly consists of the right camera, the left camera, the spray lance, the ultrasonic ranging device, the nozzles, the lifting gear, the robot body and the pedrails.

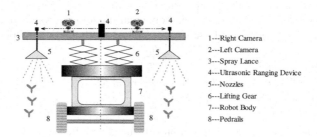

1---Right Camera
2---Left Camera
3---Spray Lance
4---Ultrasonic Ranging Device
5---Nozzles
6---Lifting Gear
7---Robot Body
8---Pedrails

Fig. 1. Structure of the spraying robot

The left camera and the right camera are used to make images of the crops. Based on these images, the crop rows can be detected and the distance between the crop rows and the central axis of robot can be calculated using stereoscopic vision algorithm (Rao Honghui et at., 2007). The ultrasonic ranging device is used to measure the distance between the nozzle poles and it is consisted of one ultrasonic transmitting device and two receiving devices, the transmitting device sits on the central axis and the receiving devices sit on the top of the nozzle poles. The spray lance (1meter long) is used to sustain the nozzles and nozzle control systems. Each nozzle is equipped with a fuzzy control system which is in the spray lance and the nozzle control system is consisted of difference calculator, fuzzy controller (based on FMC (Fuzzy Micro Controller) chip), execution unit, drive circuit, stepper motor and transmission gear. The structure of the nozzle control system is shown in Fig.2. The difference calculator can receive the distance (d_0) from a crops row to the central axis of robot and the distance (d_1) from the corresponding nozzle to the central axis, then calculate their difference value (ed). The fuzzy controller can receive ed as input and produce the control signal c to the execution unit based on fuzzy control theory. The execution unit and the drive circuit are used to produce the specific control signal to stepper motor according to the value of c, the absolute value of c represents the rotation speed of stepper motor and the sign of c represents the rotation direction. The stepper motor and the transmission gear are used to adjust the position of the nozzle in order to aim toward the crop row.

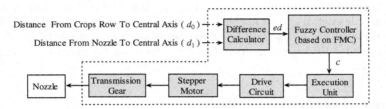

Fig. 2. Structure of the nozzle control system

In the course of the control, the stepper motor is the final controlled object, so the mathematical model of the stepper motor is very important. Duan and Yang studied the mathematical model of the stepper motors in their paper (Duan Yinghong et at., 2006). To be on the distinct side, the mathematical model is introduced simply as follows. The differential equation of the stepper-motors mathematical model is as follow

$$\begin{bmatrix} u_a \\ u_b \\ u_c \end{bmatrix} = \begin{bmatrix} R_a & 0 & 0 \\ 0 & R_b & 0 \\ 0 & 0 & R_c \end{bmatrix} \cdot \begin{bmatrix} i_a \\ i_b \\ i_c \end{bmatrix} = \begin{bmatrix} L_{aa} & L_{ab} & L_{ac} \\ L_{ba} & L_{bb} & L_{bc} \\ L_{ca} & L_{cb} & L_{cc} \end{bmatrix} \cdot \begin{bmatrix} \dfrac{di_a}{dt} \\ \dfrac{di_b}{dt} \\ \dfrac{di_c}{dt} \end{bmatrix} + \dfrac{\partial}{\partial \theta} \begin{bmatrix} L_{aa} & L_{ab} & L_{ac} \\ L_{ba} & L_{bb} & L_{bc} \\ L_{ca} & L_{cb} & L_{cc} \end{bmatrix} \cdot \begin{bmatrix} i_a \\ i_b \\ i_c \end{bmatrix} \cdot \dfrac{d\theta}{dt} \tag{1}$$

The transfer function of the stepper motor deduced as follows

$$G(s) = \frac{\theta_2(s)}{\theta_1(s)} \tag{2}$$

To simplify, we can ignore the self-inductances and the mutual-inductances. Under the assumption of $T_1 = 0$ and a-b-c-a phase sequence, the motion equation of stepper motor is as follows

$$J \frac{d^2\theta}{d^2 t} + D \frac{d\theta}{dt} - \frac{Z_r L_1 i_a^2}{2} \sin Z_r \theta = 0 \tag{3}$$

If the rotor arrives at the equilibrium position when $t = 0$, the transfer function of the stepper motor can be further deduced based on the conditions stated above and it is as follows

$$G(s) = \frac{\theta_2(s)}{\theta_1(s)} = \frac{Z_r^2 L_1 i_a^2 / 2J}{S^2 + \dfrac{D}{J} S + Z_r^2 L_1 i_a^2 / 2J} \tag{4}$$

Where: Z_r is the teeth number of stepper motor. L_1 is the inductance. i_a is the phase electric current. J is the rotational inertia. D is the coefficient of viscosity.

3 Design of Fuzzy Controller

3.1 Overall Design of the Fuzzy Controller

Considering the actual application, the fuzzy controller is designed as a 2-inputs&1-output closed-loop system. The first input is ed (in centimeters), the second one is the change rate of ed (recorded as ed') (in centimeters/second) and the output is c. As mentioned above, the absolute value of c represents the rotation speed of stepper motor and the sign of c represents the rotation direction. The structure of the nozzle controller is shown in Fig.3. In the controller, the original inputs (ed and ed') are tuned to $\widetilde{ed}$ and $\widetilde{ed}'$ firstly,and K_{ed} , $K_{ed'}$ are the proportional gains for tuning via scaling universes of discourse of inputs. Then $\widetilde{ed}$ and $\widetilde{ed}'$ are quantified to the fuzzy linguistic variables (eD and eD') by the fuzzification interface. The inference mechanism can receive $\widetilde{ed}$ and $\widetilde{ed}'$, determine the extent to which each rule is relevant to the current situation as characterized by $\widetilde{ed}$ and $\widetilde{ed}'$, and draw conclusion (C) using the previous inputs and the information in the rule-base (Kevin M. P., 2001, P60-66). The defuzzification interface and the output scaling gain K_c are used to convert the fuzzy linguistic variable C into the actual output c.

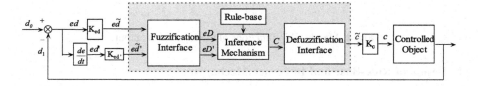

Fig. 3. Structure of the fuzzy controller

3.2 Tuning via Scaling Universes of Discourse

In order to enhance the efficiency of the fuzzy controller, we can tune the universe of discourse for inputs and output in the course of design. In this papae, the spraying robot has three nozzles and the first input is ed , so the effective universe of discourse for ed is set to [-33.3,+33.3] centimeters. The tuned universe of discourse for $\widetilde{ed}$ is selected as $\widetilde{Ed} = [-3, +3]$ and the input scaling gain $K_{ed} = 0.09$.

The second input is ed' and its value depends on the actual rotation speed of the stepper motor, the parameters of the transmission gear and some other specific factors. In this paper, taking universality into account, its effective universe of discourse is set to $[\alpha, \beta]$ centimeters/second, its tuned universe of discourse for $\widetilde{ed}'$ is selected as $\widetilde{Ed}'=[-3, +3]$ and the input scaling gain $K_{ed'} = \dfrac{6}{\beta - \alpha}$.

The output c is the control signal to the execution unit and its value depends on the actual factors too, its effective universe of discourse is set to $[\mu,\nu]$ in this paper. The tuned universe of discourse for $\tilde{c}$ is also selected as $[-3,+3]$ and the input scaling gain $K_c = \dfrac{6}{\nu - \mu}$.

3.3 Fuzzy Quantification of Inputs and Output

In the fuzzy controller, the fuzzification interface is used to convert the inputs $\tilde{ed}$ and $\tilde{ed}'$ into their linguistic variables so that they can be interpreted and compared to the rules in the rule-base and the defuzzification interface is used to convert the conclusions reached by the inference mechanism into the output $\tilde{c}$. eD, eD' and C are the linguistic variables corresponding to the inputs and output and their linguistic values set are all the set of {NB, NM, NS, NZ, PS, PM, PB}. Their membership functions are shown in Fig.4.

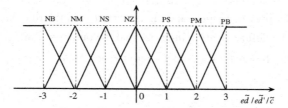

Fig. 4. Membership functions of the inputs and output

3.4 Design of Rule-Base and Inference Mechanism

The rule-base is the foundation of fuzzy inference and it contains a set of rules, which are the knowledge of how best to control the nozzle system and how to control the system without outside help. In this paper, taking into account the information from a human decision maker who performs the control task, the rule-base array that we use for the nozzle controller is designed as Table 1.

Table 1. Rule-Base for fuzzy nozzle controller

eD	eD'						
	NB	NM	NS	NZ	PS	PM	PB
				C			
NB	PB	PB	PB	PB	PM	PS	PS
NM	PB	PB	PS	PS	PS	PS	NZ
NS	PB	PM	PS	PS	NS	NM	NB
NZ	PB	PM	PS	NZ	NS	NM	NB
PS	PM	PS	PS	NS	NS	NM	NB
PM	PS	PS	PM	PB	PM	PS	NZ
PB	PM	PB	PB	PB	NM	NB	NB

The inference mechanism is used to emulate the expert's decision making in interpreting and applying knowledge about how best to control the nozzle system. In this paper, the "Mamdani" inference method is used to realize the fuzzy inference and the fuzzy implicative relation as follows

$$
\begin{aligned}
\mu_{C^*}(z) &= \underset{\substack{x\in X \\ y\in Y}}{V}[\mu_{ED^*}(x))\wedge\mu_{ED^*}(y)]\wedge[\mu_{ED}(x)\wedge\mu_{ED'}(y)\wedge\mu_C(z)] \\
&= \underset{\substack{x\in X \\ y\in Y}}{V}\{[\mu_{ED^*}(x))\wedge\mu_{ED^*}(y)]\wedge[\mu_{ED}(x)\wedge\mu_{ED'}(y)]\}\wedge\mu_C(z) \\
&= \{\underset{x\in X}{V}[\mu_{ED^*}(x))\wedge\mu_{ED}(x)]\wedge\underset{y\in Y}{V}[\mu_{ED^*}(y)\wedge\mu_{ED'}(y)]\}\wedge\mu_C(z) \\
&\overset{\Delta}{=}(\omega_{ED}\wedge\omega_{ED'})\wedge\mu_C(z)
\end{aligned}
\tag{5}
$$

Where: ω_{ED} is the adaptive degrees of ED^* to ED. $\omega_{ED'}$ is the adaptive degrees of ED^* to ED'.

3.5 Selection of Defuzzification Method

The defuzzification interface is used to convert the conclusions of the inference mechanism into the actual output through a defuzzification method. In this paper, we use the center-average method and the process of defuzzification is shown in Fig.5.

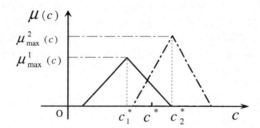

Fig. 5. The process of defuzzification

Where: c^* is the actual output and it is given by

$$
c^* = \frac{c_1^*\mu_{max}^1(c)+c_2^*\mu_{max}^2(c)}{\mu_{max}^1(c)+\mu_{max}^2(c)}
\tag{6}
$$

4   Implementation of Fuzzy Controller

The fuzzy controller can be implemented based on FMC (Fuzzy Micro Controller), and NLX230 chip is selected as FMC in this paper. NLX230 is a typical FMC chip produced by NeuraLogix, it has 40pins in DIP package, and its input and output are all 8-bits of digital data. The built-in 24-bit rule register can store 64 rules at most and the maximum inference speed is 30M rules/second (Huang Xiaolin et at., 2006). The internal structure of NLX230 is shown in Fig.6.

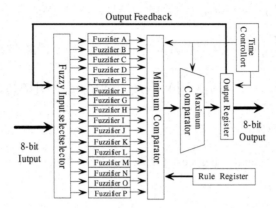

Fig. 6. Internal structure of NLX230

In this paper, the connection of fuzzy controller circuit is shown in Fig.7. The MCU AT89S52 is used to control all the NLX230 chips. In the initialization of system, the pins of P1.0-P1.2 of AT89S52 provide signals to control and synchronize all the NLX230 chips. When the RST signal become effective, NLX230 will generate the signal of SK and CS to AT93C57 and receive the rules data from AT93C57. AT93C57 is a 2Kb EEPROM (Wang xingzhi, 2004, P360-373) used to store the rules data. In work process, NLX230 receive the input ed through DI0-DI7 and produce the output c through DO0-DO7.

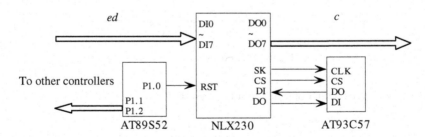

Fig. 7. Connection diagram of fuzzy controller circuit

5   Experiment and Analysis

In order to testify the performance of the proposed scheme, the experiment was carried out and the 57BYG4503 stepper moter was selected in it. The driving voltage of the 57BYG4503 is 36V, the phase current is 1.5A, the phase inductance is 2.4mH, the rotational Inertia is 0.3 Kg·cm^2 , the teeth number is 50 and the the stepping angle is 1.8°. In the experiment, the proposed control project was compared with the traditional control methods. The experimental results show that the proposed project has excellent control effect and ability of anti-jamming.

6 Conclusion

A novel nozzle controller of spraying robot aiming toward crop-rows based on fuzzy control theory was studied in this paper and the aim of this work was to solve the shortcomings of existing nozzle control system, such as the long regulation time, the higher overshoot and so on. In our design project, the fuzzy controller mainly consists of the fuzzification interface, the defuzzification interface, the rule-base and the inference mechanism. The proportional gains were used to tune the universe of discourse for inputs and output. The fuzzification interface is used to convert the inputs into their linguistic variables so that they can be interpreted and compared to the rules in the rule-base and the defuzzification interface is used to convert the conclusions reached by the inference mechanism into the output. The inference mechanism is used to emulate the expert's decision making in interpreting and applying knowledge about how best to control the nozzle system. In our Implementation project, we selected the FMC chip NLX230, the EMCU chip AT89S52 and the EEPROM chip AT93C57. The experimental results show that the project is workable and efficient, and the control efficiency can be raised greatly.

Acknowledgements

Funding for this research was provided by Hebei Provincial Department of Education (P. R. China) and the project number is 2009332.

References

Jun, C., Baijing, Q., Jun, Z.: Research on Overlap and Leak Spray of Spraying Robot. Journal of Agricultural Mechanization Research (12), 63–65 (2006)
Hui, M.: China should vigorously develop the precision of pesticide application technology 41(6), 47 (2002)
Zetian, F., Lijun, Q., Junhong, W.: Developmental Tendency and Strategies of Precision Pesticide Application Techniques. Transactions of the Chinese Society for Agricultural Machinery 38(1), 189–192 (2007)
Lushou, S., Zhixiang, D., Huailei, C., et al.: Study on a Pesticide System Spraying with Changeable Quantity Based on Fuzzy Control. Transactions of the Chinese Society for Agricultural Machinery 36(11), 110–112 (2005)
Xiongkui, H., Kerong, Y., Jingyu, C., et al.: Design and testing of the automatic target detecting, electrostatic, air assisted, orchard sprayer. Transactions of the CSAE 19(6), 78–80 (2003)

Tian, L., Reid, J.F., Hummel, J.W.: Development of precision sprayer for site-specific weed management. Transactions of the ASAE 42(4), 893–900 (1999)

Giles, D.K., Slaughter, D.C.: Precision band spraying with machine-vision guidance and adjustable yaw nozzles. Transaction of the ASAE 40(1), 29–36 (1997)

Honghui, R., Changying, J.: Study on spray control to aim toward machine vision. Journal of Nanjing Agricultural University 30(1), 120–123 (2007)

Yinghong, D., Shuo, Y.: Fuzzy-PID control of stepping motor. Computer Emulation 23(2), 290–293 (2006)

Passino, K.M., Yurkovich, S.: Fuzzy control [Photoprint edition]. Tsinghua University Press, Beijing (2001)

Xiaolin, H.: Research on NLX230 fuzzy controller and its embedded application. Foreign electronic components (3), 16–21 (2006)

Xingzhi, W., Aiqin, Z., Lei, W., et al.: Principle and Interface Technology of AT89 series Single Chip Processor. Press of Beijing University of Aeronautics and Astronautics, Beijing (2004)

Design of Intelligent Hydraulic Excavator Control System Based on PID Method

Jun Zhang[1,2], Shengjie Jiao[1,*], Xiaoming Liao[2], Penglong Yin[2], Yulin Wang[2], Kuimao Si[1], Yi Zhang[1], and Hairong Gu[1]

[1] Key Laboratory for Highway Construction Technology and Equipment of Ministry of Education, Chang'an University, Xi'an, Shangxi Province, P.R. China 710064
jsj@mail.chd.edu.cn
[2] Changlin Company, Changzhou, Jiangsu Province, P.R. China 213002

Abstract. Most of the domestic designed hydraulic excavators adopt the constant power design method and set 85%~90% of engine power as the hydraulic system adoption power, it causes high energy loss due to mismatching of power between the engine and the pump. While the variation of the rotational speed of engine could sense the power shift of the load, it provides a new method to adjust the power matching between engine and pump through engine speed. Based on negative flux hydraulic system, an intelligent hydraulic excavator control system was designed based on rotational speed sensing method to improve energy efficiency. The control system was consisted of engine control module, pump power adjusted module, engine idle module and system fault diagnosis module. Special PLC with CAN bus was used to acquired the sensors and adjusts the pump absorption power according to load variation. Four energy saving control strategies with constant power method were employed to improve the fuel utilization. Three power modes (H, S and L mode) were designed to meet different working status; Auto idle function was employed to save energy through two work status detected pressure switches, 1300rpm was setting as the idle speed according to the engine consumption fuel curve. Transient overload function was designed for deep digging within short time without spending extra fuel. An increasing PID method was employed to realize power matching between engine and pump, the rotational speed's variation was taken as the PID algorithm's input; the current of proportional valve of variable displacement pump was the PID's output. The result indicated that the auto idle could decrease fuel consumption by 33.33% compared to work in maximum speed of H mode, the PID control method could take full use of maximum engine power at each power mode and keep the engine speed at stable range. Application of rotational speed sensing method provides a reliable method to improve the excavator's energy efficiency and realize power match between pump and engine.

Keywords: Hydraulic excavator, energy saving, PID, CAN, negative flux system.

---

[*] Corresponding author.

D. Li and C. Zhao (Eds.): CCTA 2009, IFIP AICT 317, pp. 207–215, 2010.
© IFIP International Federation for Information Processing 2010

1 Introduction

Hydraulic excavator, which is widely used in mechanized construction such as in industrial and civil construction, farmland transformation, transportation, is multi-function earthmoving machinery. It consists of four parts namely power system, hydraulic system, working mechanism and electronic system. The energy flow in the excavator was like this: Mechanical energy was produced by the engine, then the mechanical energy was transferred to pressure energy by the hydraulic system to drive oil cylinder or hydraulic motor, finally, the pressure energy was change to mechanical energy to drive working mechanism. Its working state was a periodic operation, the periodic operation order is excavating, slewing, unloading and back to excavate.

Hydraulic excavator is a high power machine; the energy crisis attracted the energy saving research on excavator. The working time of the excavator is too long; most of them are working at day and night. Its operation load is of great variation, its average load is almost 50~60% of maximum load(Guo Xiang'En, 2004).It was investigated that the energy utilization of excavator is about 30%, most of power was wasted in hydraulic system and mechanical transmission(Xie GuoJing ,2008), so energy saving is a hot topic in excavator research.

Aiming to reduce the energy loss in hydraulic system, most of famous excavator companies united with hydraulic component companies to develop better hydraulic component and electronic technology, they search energy saving method in power system, hydraulic system and so on; from another method, they search energy saving method from power chain and focus on matching of the power between engine and hydraulic pump, power match between hydraulic system and load to optimize the power distribution of excavator. These researches gain great success. The hydraulic system was improved by the two methods, fix displacement pump were replaced by variable displacement pump; full power and sub power were replaced by cross power method. But it still couldn't meet the requirement of high performance, high reliability and automatic operation of excavator.

Along with the development of the sensor technology, control technology and servo-hydraulic technology, the excavator providers concentrated on developing electronic control system to save energy and improve the fuel utilization. Caterpillar company was the first to introduce the electronic control technology in excavator, it developed the first electronic monitor system to detect the operational fault in 1987(Zhang Dong, 2005). Then more famous companies took part in the development of electronic control system for excavator for energy saving, these system proved obvious energy saving, such as PMS system of O&K company, AEC system of CAT, EC and PVC system of Hitachi, ITCS system of KOBELCO, CAPO system of HYUNDAI, ACS system of VOLVO, OLSS system of KOMATSU and so on(Guo Yong et al., 2006; Cheng DuWang, 2007), of which the electronic control system of Hitachi and KOMATSU was more perfect.

From 1960s, China began to research on designing hydraulic excavator, and imported technology from German in1980s. Up to date, the market share of the domestic companies was only 10~15%, most of the market share was dominated by the foreign company such as DOOSAN, KOMATSU, HYUNDAI and Hitachi. The main reason was that the foreign companies own the core component of excavator, such as engine, hydraulic system and electronic control system, which were sealed for China. China adopt two methods to improve the domestic excavator performance, one way is creating joint venture Company with foreign famous company, but this is still not solving the technology sealed problem. Another is to develop excavator through research with colleges and universities. Guo Xiaofang(Guo Xiaofang et al., 1999) conducted an experiment on engine and pump united control system; later, Gao Feng(Gao Feng et al., 2001) research on valve control technology and give the power match between pump and engine in theory; Ji Yunfeng(Ji Yunfeng et al., 2003) research on the excavator control based on rotational speed of engine, Wu Xiaojian(Wu Xiaojian, 2005) analyze the performance of the engine and the pump and applied PID method to control the excavator. Even hybrid power was investigated on excavator for energy saving(Wang Dongyun et al., 2009).

Based on research with colleges and universities, domestic company has gain the core technology on fault monitor system in excavator, while electronic control system has a certain gap between domestic and foreign company. So, there is still scientific and economic value for researching on electronic control system and energy saving for excavator. In this paper, based on rotational speed of engine, an intelligent electronic control system was designed with CAN-BUS controller, and PID method was constructed to gain matching of power between the engine and the pump.

2   Power Matching between Engine and Pump

Inconsideration of the mechanical transfer efficiency between engine and pump, their power should be equal:

$$P_p(t) = P_e \tag{1}$$

Where $P_p(t)$ represents the power of hydraulic pump (kW), P_e represents the engine power (kW).

$$\text{Engine power: } P_e = \frac{T.n_e}{9549} \tag{2}$$

$$\text{Pump power: } P_p(t) = P_p \frac{.Q_p}{60} = P_p.q_p.\frac{n_p}{60000} \tag{3}$$

Where T is torque of engine(N.m), P_p represents for the pressure of pump outlet (bar), Q_p is pump flow (L/min), q_p is displacement of the pump at each cycle (mL/r); n_p is rotational speed of pump (r/min).

Because the pump and the engine is of rigid connection, when excavator is at working, their torque and rotational speed should be equal and the torque should be constant, and at this work state, the power of pump is matching to engine. So, it could be conclude from equation (1)~(3) that the pump torque M_p could be calculated like equation (4). In the hydraulic system, the torque of the pump (T) was decided by load pressure and pump flow rate.

$$T = M_p = P_p.q_p.2\pi \tag{4}$$

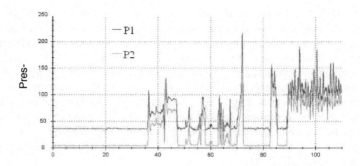

Fig. 1. Pump pressure changed with time during working

The electronic excavator was controlled by the constant power method. Fig.1 recorded the two pump pressure changed with time; it indicated that the pump pressure varied a lot during working. The pressure (P_p) was decided by load, and pump outlet (q_p) was controlled by a proportional valve, when the pressure changed, then $P_p \cdot q_p$ was varied with load, it would change the torque of the engine and caused the variation of rotational speed, the change of rotational speed would cause high fuel consumption. So, the drop of the rotational speed could reflect the variation of the load. The constant power controlled method expects the rotational speed be stable and the hydraulic system to be suitable with operator. In order to solve for adjusting the relation among the engine, the pump and the load, it need electronic controller to control the pump outlet based on the variation of rotational speed. The excavator electronic system was the key to constant power control method.

3   Electronic Control System and Ennergy Saving Strategy

3.1   Intelligent Electronic Control System

Fig.2 shows the schematic diagram of the electronic control system, the energy saving system includes Cummins 6BTA5.9 engine, T5V displacement pump, negative flux hydraulic control system, linear throttle motor, throttle position sensor, two pump outlet pressure sensor, working state detecting sensor (Px and Py pressure sensor), engine speed setting knob and power mode knob, pump flow rate adjusting valve, three working solenoid valve (pressure up, travel high speed and pilot valve).

The electronic energy saving control system was constructed with CAN bus PLC and LCD monitor and was coded with CODESYS software. The control system was consisted of engine control module, pump power adjusted module, engine idle module and system fault diagnosis module. The PLC detects and processes the sensors' input signal, and then sends the signal to control the motor or value. The hydraulic system

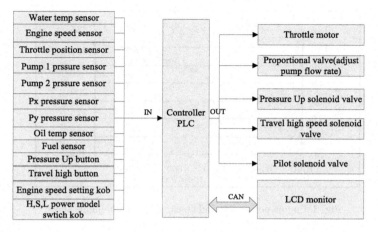

Fig. 2. Electronic control schematic diagram of excavator based on CAN bus controller

adopts two variable displacement pumps to transport power and drive load, electro-hydraulic proportional technology was employed to control the flux of pump; these fluxes were controlled by adjusting the current through the electro-proportional valve by PLC. Fault diagnosis module was used to inspected engine and hydraulic fault through water temperature sensor of the engine, oil temperature sensor of the hydraulic system.

3.2 Energy Saving Strategy

The working state of excavator is very complex, not only heavy load excavating, but also light load leveling. Even at excavation, there has loosen and hard soil excavation. So excavator is always working at different status, it should set different output engine power to meet the complex working status, and adjust the throttle motor and displacement pump according to the requirement of operation. Thus, power mode control, auto idle control, transient overload control and PID power matching control method were used to design the energy saving electronic control system.

3.2.1 Three Power Mode Control Method

In order to meet different working status, the system designed three power modes—H, S and L power mode, their maximum power rang is 100%, 85% and 70% of the maximum power of engine separately. Fig.3 is the engine's power and fuel consumption changed with engine speed. According to fig.3's data, the engine speed range of each power mode was defined in Table 1. H mode was designed for heavy load or rapid work status, S mode was designed for fuel economic utilization, and L mode was designed for finishing or leveling operation. At each power mode, operator could adjust the engine speed through engine setting knob, then the PLC drive the throttle motor to the setting speed.

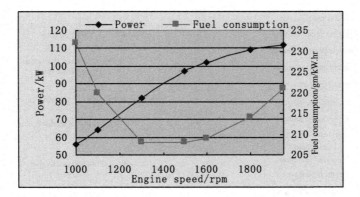

Fig. 3. Engine power and fuel consumption with engine speed

Table 1. Engine speed range of three power mode (unit: rpm)

L mode	S mode	H mode
1000~1600	1600~1800	1800~1950

3.2.2 Auto Idle Control Method

When the excavator is at waiting status, it should decrease the throttle opening position and let the engine run at idle status to save energy. The control system detects the idle status through two pressure sensors—Px and Py (see Fig.2). The pilot hydraulic system pressure is beyond 2Mpa during working, so when these two sensors' pressure is lower than 1Mpa, it indicated that the excavator is at idle status. Fig.4 is the control strategy for auto idle module. The auto idle speed is setting at 1300 rpm according to Fig.3; at this speed, the fuel consumption is minimum (208gm/kW.hr). When engine is working at 1950rpm (H mode) and now excavator is at idle status, the PLC will pull down the engine speed to 1300rpm, it could decrease 33.33% of fuel consumption when considering the engine power is proportional to the engine speed.

$$Fuel\ consumption\ drop = \frac{n_1 - n_e}{n_1} = \frac{1950 - 1300}{1950} = 33.33\%$$

3.2.3 Transient Overload Control Method

The hydraulic system of the excavator remains pressure allowance, such as its normal high pressure is 33Mpa, but when the pressure up solenoid is on, the system maximum pressure could be up to 35Mpa, which couldn't cause damage to the hydraulic system within short time. The maximum pressure can't last long; otherwise it could short the serve life the hydraulic components. In order to improve the excavator's working efficiency, the electronic control system allow the load pressure is bigger than normal maximum pressure (33Mpa) for 8s, then the control system would decrease the engine speed to 1300rpm (idle status) and set the alarm speaker on to notice the operator, the control strategy was shown in Fig.5. It could be used in deep digging operation without consuming more fuel.

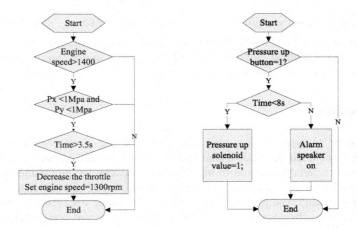

Fig. 4. Engine auto idle control module   **Fig. 5.** Transient overload control module

3.2.4 Power Matching between Pump and Engine with PID Control Method

In the three power control mode, the power of the displacement pump was adjusted by electronic proportional valve, and the maximum power of pump was different at each power mode, only when the power of the pump is equal to the setting engine power, it is the constant power control strategy. When excavator is at working, the power of engine would decrease. In order to avoid engine speed drop, usually, the pump power was set at 85~90% of engine power, so, it caused a lot of fuel wastage.

The rotational speed of engine variation could reflect the load variation; it could use rotational speed sensing technology to adjust the absorption power of pump through a proper current to the proportional valve. Consideration of the pump and the engine system belongs to large time-varying and nonlinear system; it couldn't get ideal control effect through engine speed drop. PID (proportion integral differential) method was a useful method to duel this nonlinear system; it was employed to adjust power matching between the pump and the engine during work. Fig.6 shows the PID control model for excavator control system. The PID control system was applied through increasing PID method in Codesys software. The PLC detects the throttle position timely and set the maximum engine speed n_0 at the chosen power mode. Then the PLC adjust the pump power through engine speed variation, if the detected speed $n_e(t) < n_0$, PLC decrease the proportional current to decrease the absorption

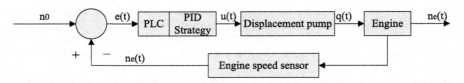

Fig. 6. Power matching adjustment though PID control method based on rotational speed sensing

of pump power; when the engine power increase and $n_e(t) > n_0$, then PLC increase the pump power to let the absorption power equal to the maximum power of engine. So the PID control method could adjust the pump power around the maximum power of engine and improve the fuel utilization.

4 Conclusion

On the basis of analysis of the power matching principle among engine, pump and load, it was concluded that the engine speed could sense the variation of the load, which provides a reliable method to match power between the pump and the engine. On this basis, based on negative flux hydraulic control system, rotational speed control method was employed to design an intelligent hydraulic excavator control system to improve energy efficiency. An intelligent control system was constructed with special CAN bus PLC. The control system was consisted of engine control module, pump power adjusted module, engine idle module and system fault diagnosis module. Engine speed sensor and throttle position sensor were employed to acquire load information and control the engine power, Px and Py pressure sensor were employed to judge whether the excavator is at working or idle status. Two pressure sensor of pump outlet could provide load information for PLC controller. The proportional valve was employed to control the absorption power of pump. Three solenoid valves were employed to control the excavator movement. Water temperature sensor and oil temperature sensor was used to inspected engine and hydraulic fault.

Four energy saving control strategies were employed to control the excavator. Three power mode (H, S, L mode) control method were designed by operator's requirement. When united with PID control method, the engine's maximum power could be taken full use of and the engine speed could be kept stable, which would result in less fuel consumption. Auto idle control method could decrease fuel consumption automatically; it could decrease 33.33% of fuel compared when the excavator is at high speed without working. Transient overload control method could be used to be deep digging without consuming more fuel.

References

Xiangen, G.: Reserach on energy-saving fuzzy control system of hydraulic excavator [PH.D]. Jilin University, Chang chun (2004) (in Chinese)

GuoJing, X.: The research on electronic energy-saving control system of hydraulic excavator based on arm processor [Master]. GuangXi University, Nanling (2008) (in Chinese)

Dong, Z.: Study on energy-saving technology of hydraulic excavator based on power match [Ph.D]. Jilin University, Changchun (2005) (in Chinese)

Yong, G., Yong, C., Qinghua, H., et al.: View the development of electronic control system of excavator from INTERMAT 2006. Construction Machinery and Equipment 37(11), 40–43 (2006) (in Chinese)

DuWang, C.: Research on power match on multi function excavator [Master]. Central South University, ChangSha (2007) (in Chinese)

Xiaofang, G., Weizhe, L., Zongyi, H.: Research on the combined control system of engine and oil pump in hydraulic excavator. Journal of TongJi University 27(3), 333–336 (1999) (in Chinese)

Feng, G., Yu, G., Pei-en, F.: Method of load matching control of hydraulic excavator's energy saving. Journal of TongJi University 29(9), 1036–1040 (2001) (in Chinese)

Yunfeng, J., Qiangen, C., Qinghua, H.: Analysis of control system of electromechanical internalization for hydraulic excavator of KOMATSU. Modern Machineary 2003(5), 4–6 (2003) (in Chinese)

Xiaojian, W.: Research on energy-saving control technology of hydraulic excavator [Master]. Central South University, Changsha (2005) (in Chinese)

Dongyun, W., Shuangxia, P., Xiao, L., et al.: Energy saving scheme of hydraulic excavators based on hybrid power technology. Computer Integrated Manufacturing systems 15(1), 188–196 (2009) (in Chinese)

A Control System for Tobacco Shred Production Line Based on Industrial Ethernet

Li Zhang[1], Guang Zheng[1], Xinfeng Zhang[2], Lei Liu[2], and Lei Xi[1,3,*]

[1] College of Information and Management Science, Henan Agricultural University,
Zhengzhou, He Nan, China, 450002
[2] Nanyang Cigarette Factory of China Tobacco Henan Industrial Corporation,
Nanyang, He Nan, China, 473000
[3] College of Information and Management Science, Henan Agricultural University,
63 Agricultural Road, Zhengzhou, 450002, P. R. China,
Tel.:+86-13803866921; Fax:+86-371-63558090
hnaustu@126.com

Abstract. The Industrial Ethernet based on IP realizes interconnection of industrial network and information network, and it is the most potential technology in the new industrial net products. In this paper, the defects of the original control system for tobacco shred production line are analyzed, and the new design plan of control system based on EtherNet/IP is presented. The control net adopts redundant 1000M fiber optic ring network that consists of six managed Industrial Ethernet Switches, and they are distributed to the central control room, leaf processing line, shred processing line, mixed stem shred processing line, online mositure regain processing line and cut tobacco dryer control cabinet. The switch in the central control room works in the pattern of redundancy management, which can switch the link in the event of the failure in link of ring net, the recovery time of link line is less than 500ms, and each main PLC of control section has dual Network Adapters. The plan has been applied for reform of 5000kg/h Tobacco Primary Processing Line in Nanyang Cigarette Factory of China Tobacco Henan Industrial Corporation, and the configurable software and Industry Ethernet network which has been used promots the capability of automatic control system fundamentally, showing much better transmission efficiency and reliability, realizing the goal of high cost performance and making equipment's ability of handling grow fast.

Keywords: Industrial Ethernet, tobacco shred prodution line, PLC.

1 Introduction

Along with the development of modern industry control technology, Industrial Ethernet network, which is easy to connect with Internet, emerging as a fast-growing unified network control technology. It has been gradually applied to the process control system (Hansen,k., 2003). The traditional industrial automatic control system

---

* Corresponding author.

D. Li and C. Zhao (Eds.): CCTA 2009, IFIP AICT 317, pp. 216–224, 2010.

generally divided into three layers: information layer, control layer and device layer(Xia Liang, 2007). The Ethernet technology based on the TCP/IP protocol has been widely used in the modern factory information layer. Because of the higher demanding for real-time data, the application of the technology is relatively small in the control layer and device layer, especially in the device layer.

The tobacco shred production line is the first working procedure of the cigarette production, which including leaf shred production line and stem shred production line. The aims of the shred production line is to resurge, charge, cut into shred and dry the tobacco leaf and stem which has been processed in the stem-dried factory. Then the processed tobacco shreds will be blended and added fragrance in accordance with the technology requirements and finally the shredded tobacco will be produced which is suitable for future cigarette production. The stability of the shred thread production control system and its reliable operation is the key point of the tobacco shred quality insurance (Sun Xin, 2007).

The Rockwell Automation Company introduced a new industrial Ethernet technology which is based on the normative TCP/IP protocol. It integrates the normative TCP/IP protocol and CIP information protocol together. The application of this new technology into the transformation of the tobacco shred production line allows the EtherNet/IP technology applying into the control layer and the device layer. It effectively resolves the network-based control and information management of various process parameters, such as temperature and pressure of the industrial field (Yang Benyu et al., 2007). It provides a theoretical and practical basis for the information exchange and integrated automation realization among the settlement from the industrial field device layer to the control layer, management and decision layer (Zhang sheng et al., 2007).

2   Overview of the Tobacco Shred Production Line Control System (Before Reconstruction)

The 5000KG/H tobacco shred production line in Nanyang Cigarette Factory of China Tobacco Henan Industrial Corporation, was put into operation in year 2002. It adopts the network structure of ControlNet and DeviceNet which is exploited by Rockwell Automation Company. This production line includes six control technology operating processing lines which are online moisture regain processing line, Leaf processing line, shred processing line, stem shred processing line, mixed stem shred processing line and flavoring processing line. Every operating stage includes a group of main control cabinet, one to two field operator stations, a distributed control box and an inverter which is distributed installed. The host controller uses ControlLogix processor and stores in the main control cabinet. The centralized control communicates though the ControlNet (fieldbus) by using RSView32 (configuration software). EtherNet standard-based C/S structure is used between monitor computer and server to finally realize the monitor of the whole production line. Network system plan before the transformation is shown in Fig.1:

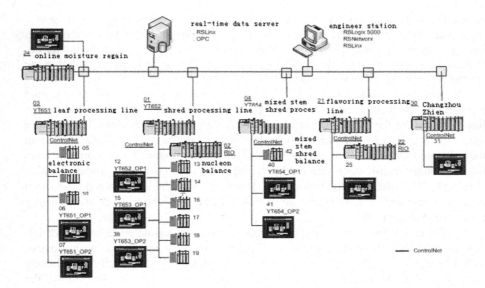

Fig. 1. The network structural plan (before reconstruction)

The problems of the system:

The network maintenance is quite difficult because it uses ControlNet network to connect all the PLC (Programmable Logic Controller) and its expansion rack together as a network. When one part of the network on PLC needs for changes (e.g. increasing templates, moving templates or increasing the expansion rack), all PLC have to stop and reorganize the network and unnecessary downtime will be happened.

Currently there are too many templates in the ControlNet Network. The scan cycle is too long which will reduce the instantaneity of the system. The communication delay and network congestion also happens sometimes.

In the central control room, all the monitors have to read PLC information though data acquisition server. When the server fails, the whole central control system cannot be used.

In the current system, the shred processing stage and stem shred processing stage together use only one CPU. As a result, the CPU will be overloaded. Moreover, in the product process, if any problems happen in one stage, the other one will also be affected.

The root cause of the above problems is using the fieldbus-based control technique. To overcome these drawbacks, the new industrial network technique-industrial Ethernet has to be used.

3   Anylazing of the Technical Issues in the Alteration

3.1 Network Architecture Design

In the early demonstration of equipment, single NIC and VLAN (Virtual LAN) technology was planning to adopt in order to improve network security and avoid broadcast storms. However, several problems were detected in the pilot network formation process:

VLAN settings brought about the suspension of communication across the whole network when the modules or sub-station added in the system.

Communal data transmission media made the common data channels congested.

The unclear division of authority brought about the busyness of the data source in the long term.

The system is inefficient because of two reasons: (1) there was no data sending and receiving priority; (2) different needs and different priorities could not be taken into account at the same time.

The broadcast storm, which was hoped to avoid, still came in time. It would lead to the collapse of switches easily.

Therefore, the following methods are prepared to use in the reconstruction process: 1) abolishing VLAN and using Dual NIC; 2) setting up a relatively high performance modular switches as regional automatic circuit exchange to facilitate the future expansion of facilities. Moreover, fiber link will be arranged to connect with branch switches; 3) the number of sub-stations and 1756-ENBTs (Ethernet card) should be calculated as follows: every 15 sub-stations treated as a unit, the corresponding network cards at main controller shall be increased by multiples to shunt the data streaming and improve the stability and reliability of the networks.

3.2 Switch Selection

Ethernet switch is one of the core equipment which plays a key role in connecting. If the switch is selected unreasonably or does not have a plenty of management features, the entire technical program would be greatly impacted. In the practical application of the process, in order to ensure industrial control, the industrial network node equipments normally issue a large number of real-time packets and a mass of network time will be used up. This phenomenon is called frequent packet in the industrial control which required lower network bandwidth and higher handling capacity of the switches. Generally, the processing capacity of the network node switches is 255 broadcast wave team groups. To meet the requirements of the network, the switches shall have management capabilities and also can discriminate every type multicast data and classify the priority of them. Otherwise, data obstruction or broadcast storms probably will happen in the network. In the alteration, fiber-optic port industrial Ethernet switches with extensive management functions will be used in the modified networks. The switch has rapid spanning tree function. Moreover, the switch in the central control room works in the redundancy management mode. Therefore, the working conditions of the ring network can be detected in the real-time. In the event of the ring link's failure, link switch can be implemented and the recovery time of the ring link would be much faster than before.

3.3 Network Cabling

Network cabling was another key problem in the entire engineering design. The following aspects should be fully taken into account in the initial design stage of an automated control production line: the distribution density of electrical equipment; the network structure and scale of the data transmission medium; the limit for data transmission; the distance and power, electromagnetic interference between the physical media; as well as the network optimization, distribution strategy and so on. Therefore, network cabling should pay attention to the following aspects: 1) the fiber-optic with good anti-interference ability should be selected between the regional switch and various branch switches. 2) Routing and connection mode which could effectively reduce the harmonic interference impact of harmonic interference must be taken for the frequency converter with larger Electro-Magnetic Interference. 3) Strong and weak electrical signal lines must be distributed separately and the professional network media need to be installed using professional equipments and tools. Professional testing tools also were required to detect the quality indicators. 4) The problem of the power interference and device node grounding, especially the problem of network grounding, must be taken into account. It is necessary to be one-point grounding for the network grounding. Multi-point grounding shall not be used in the network cabling. Otherwise, the wiring quality will be affected directly and many other problems, especially the problem of the network stability, will be found in the later stages. 5) Lightning and electrostatic protection also need be considered in the design of the machinery room environment.

4   Modified Control System of the Tobacco Shred Production Line

4.1 System Structure

The control network adopts the EtherNet/IP industrial Ethernet which developed by Rockwell Automation Company. The backbone network uses the 1000M fiber-optic ring network. The fiber-optic ring network is mainly composed by optical cables, optical connection device and industrial Ethernet switch which has network management function with extensive fiber-optic ports.

All the backbone switches are linked through multi-mode fiber and constitute a 1000M redundant ring network. As a communication link among monitoring layer, control layer and control device, the servers of the centralized monitoring layer and the monitors are connected to the industrial Ethernet through industrial Ethernet Switch for realizing communication, achieving data and information sharing.

Using the long-distance transmission, high anti-interference ability of fiber-optic and the rapid spanning tree function of the switches, redundant ring structure can be constituted by six managed industrial Ethernet switches. Switches are arranged in the central control room, Leaf and shred processing line, stem and shred processing line, mixed stem shred processing line, online moisture regain processing line and control cabinet of cut tobacco dryer. Among them, the switches in the central control room works in the redundancy management mode to real-time detect the working conditions of the ring network. In the event of the ring link's failure, link switch can be implemented and the recovery time of the ring network link will be less than 500 milliseconds.

The main PLC in each control stage is using dual NIC mode. One of NIC connects to the backbone network for the data communication of information layer and another NIC communicate with the underlying distributed I/O sub-station by a 100M network through star connection scheme. The inverter, soft starter, the smart devices instrumentation such as moisture meter and flow meter are connected by DeviceNet. The signal of the implementing agencies at the scene (e.g. motors, control valves etc.) and the detection devices (e.g. photoelectric tube, proximity switches, temperature transmitter and pressure transmitter etc.) are switching in site distributed I/O boxes. The site operator terminal and the independent control system (e.g. shred machines, electronic scales) are linked into the star network of each control stages.

Topology of the network has a good open, it links up the implementation of devices (e.g. converter, electronic scale, nuclear scale etc.), testing equipments (e.g. moisture tester of cut tobacco, infrared thermometer etc.) and automatic control equipments (tobacco cutter) from different manufacturers with Profibus. It reduced the number of modules of analog, raised speed and accuracy of data transmission, saved the Installation time, is convenient to adjusting and failure diagnostics, providing a guarantee for Precision of flow control, moisture control and temperature control. In the whole tobacco shred production line, Precision of moisture control is 1%, Precision of temperature control is $2°C$-$3°C$, and Precision of matching is 1%.

The network topology structure is shown in Fig.2.:

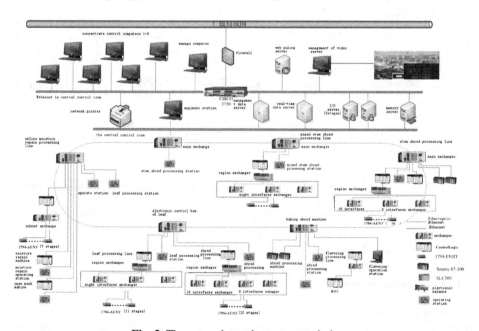

Fig. 2. The network topology structural plan.

The backbone network adopts fiber-optic ring network to connect six Hieschmann MS4128-L2PHC series of modular switches to link ControlLogix controller, two I/O server, engineer stations, real-time data server, management data server and the backup card of domain control server. The benefit of doing in this way is to reduce the number of PLC connections, lower the CPU load, and take into account for the accident which the two I/O server broken at the same time.

In addition, there is a Cisco switch connecting the two I/O server, management data server (fault tolerance), historical database server (fault tolerance), management of application server, Web publishing server, network printer and monitors together and links to the enterprise LAN through firewall at the same time.

The branch network adapts series of modular switch and integrative switch two types switches. The ControlLogix controller with dual-card which located between the backbone network and connected to two switches separately. When one NIC or one switch is at fault, another switch or NIC can switch to the normal network in a short time to avoid data loss and inoperable condition. Upper and lower layer network both can be connected with the controller and isolated unnecessary data. As the Leaf processing line an example, the network structure is shown in Fig.3.:

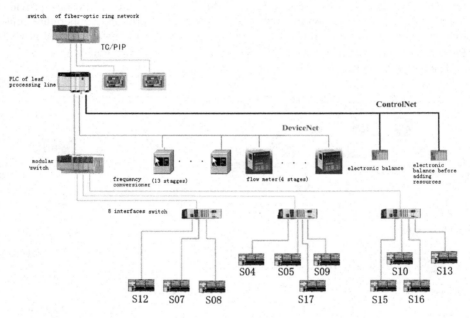

Fig. 3. The network structural plan of the leaf processing line

4.2 Monitoring System

The central monitor system comprises two sets of I/O server, one Oracle data server (fault tolerance), real-time database server (fault tolerance), management of application server, Web publication server and 6 monitors. It uses C/S structure to achieve monitor, data acquisition and information management functions. It sets up an engineer station to complete the programmed maintenance, system development and network maintenance. It also can communicate with other network or system though external switch and routing equipment. The field operator station uses the IFIX4.5 uniprocessor version to realize equipment monitoring. It also can solve the problem of different I/O servers (in the central control room) failure at the same time which would lead production suspension.

The operating system adopts Windows 2003 Server, Windows XP Professional. The monitoring software is IFIX 4.5. The communication software is RSLinx Gateway. The network software is Hirschman Hivision and the database software is Oracle 10G.

The redundant fiber optic ring network structure realizes the network redundancy and also reduces the network single point failure though a fiber -optic Ethernet ring structure. It strengthens the viability of emergency of the entire communication networks.

5 Conclusion

The application of the Ethernet technology in to the 5000KG/H tobacco shred production line transformation has achieved a good effect in the Nanyang cigarette factory which affiliated to Henan tobacco industrial corporation. The production proving: the entire control system works reliably, runs in good condition, has a perfect production management function. Through the upgrading, some of the unreasonable of the original technology and design have been improved, Economic and technical efficiency are significant. The new system is also very well reviewed by users.

The application of the system configuration software and the Ethernet technology fundamentally improved the performance of the automation control system, and also show a better transmission efficiency and reliability. It realized the cost effective and achieved the goal of rapidly upgrade the equipment management and control ability. At the same time, this new network solution provides great significance for the industrial field. It provides a whole set of seamless integration system for the device layer to the layer of enterprise information system. Its technique advantage can satisfy a wide range of automation field and also can be used as an universal technique expanded to other industries, such as the electronic power, petrochemical and steel industry.

References

Crockett, N.: Industrial ethernet is ready to revolutionise the factory - Connecting the factory floor. Manufacturing Engineer, 41 (2003)

Hansen, K.: Redundancy Ethernet in industrial automation, Emerging Technologies and Factory Automation. ETFA, 943–947 (2005)

Krommenacker, N., Rondeau, E., Divoux, T.: Genetic algorithms for industrial ethernet network design. Factory Communication Systems, 149–156 (2002)

Georgoudakis, M., Kapsalis, V., Koubias, S.: Advancements, trends and real-time considerations in industrial ethernet protocols. Industrial Informatics, 112–117 (2003)

Xin, S.: Application of PLC Control in New Technology Tobacco Primary Processing Line. Electric Transmission 37(12), 56–60 (2007) (in Chinese)

Liang, X.: Implementation of comprehensive automation network system in Dafosi Mine. Coal Science and Technology 35(11), 39–42 (2007) (in Chinese)

Benyu, Y., Weixing, Z., Donghong, W.: The design of a control system in waterworks based on configurable software and industry ethernet network. Measurement Control Technology and Instruments (9), 75–77 (2007) (in Chinese)

Xiwei, Z., Weiguo, L., Jun, K.: Study on Motor Network Control Using Industrial Ethernet. Micromotors 40(12), 59–62 (2007) (in Chinese)

Sheng, Z., Fengchun, X., Hui, C.: Supervisory system for filters in waterworks based on ControlNet. Journal of University of Shanghai For Science and Technology 29(5), 481–484 (2007) (in Chinese)

Intelligent Controlling System of Aquiculture Environment

Deshen Zhao[1,*] and Xuemei Hu[2]

[1] Department of Electric Engineering, Henan Polytechnic Institute, Nanyang,
Henan Province, P.R. China 473009,
[2] Department of Electrical and Mechanical Engineering, Henan Polytechnic Institute, Nanyang,
Henan Province, P.R. China 473009,
Tel.: +86-0377-632769922; Fax: +860-377-63270216
zhaodesh@163.com

Abstract. The paper has analyzed present aquiculture conditions and controlling problems of water environment factors of aquiculture, and constructed effective security aquiculture breeding intelligence controlling system suitable to Chinese situation, and presented the control strategy of neural network realizing dynamic decoupling for the factory aquiculture, and specially solved the water environment control and so on the key questions. The long term practice has shown that the system operation is simple and effective safe by applying some breeding bases in Zhenjiang, the system has met the requirements of culturists and enhanced international market competition for aquiculture.

Keywords: Aquiculture environment factors, Fuzzy control, Neural network, Decoupling control.

1 Introduction

Aquiculture environment is a complex system engineering which affects growth process of aquatic animals and plants, there are many questions of technology, information, management and aquaculture environmental control and other issues, the application of high technology is relatively weak. These won't make culturists take timely measures when they encounter the problems, which will delay the best growth time of plants and animals, and cause serious losses. It is the key for efficient application of water resources and technology integration to achieve efficient aquatic product, China's current aquaculture management has changed from mainly relying on the experiences to decisions of the scientific decisions, these can make rapid response and timely decision according to the status changes of aquatic animals and plants resources and guarantee the prevalence of efficient using resources model. The study will associate aquaculture expert technical knowledge, scientific research and long term practical experience of culturists with computer technology and fieldbus technology and establish a comprehensive, intelligent environment controlling system which will stead for the growth of aquatic plants and animals.

---

* Corresponding author.

D. Li and C. Zhao (Eds.): CCTA 2009, IFIP AICT 317, pp. 225–231, 2010.

2 System Hardware Structure

Controlling system of aquiculture environment regulates environmental temperature, dissolved oxygen, PH, and salinity such as breeding environmental factors by controlling furnishments and artificially creates their growth environment according to their different growth phases, the system detects real-time parameters of breeding environmental and automatically adjusts states of the various controlling equipment after comparing the measured parameters and the standard parameters of system configuration in order to meet aquiculture growth requirements to the various environmental factors.

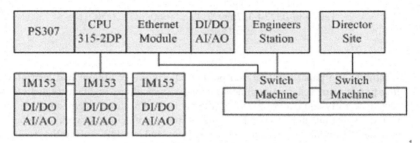

Fig. 1. System hardware structure

System hardware (Ma Congguo, et al, 2005) includes three components: 1) signal acquisition: including temperature, dissolved oxygen, PH, as well as salinity detection environmental factors. 2) Signal conversion and processing: Real-time data acquisition signals will be transformed to the digital and analog. 3) Output and controlling parts: speed of the machine of increasing oxygen and fan and opening of windows and electromagnetism valve, The system hardware is composed of monitoring computer and PROFIBUS-DP field controlling units, which detects and controls the ecological environment factors through numeric and analogy input and output modules of remote I/O controlling units, the PLC ,engineers stations and director Room can monitor the states of controlling furnishments and the ecological breeding environment. The system consists of three controlling modules, the overall framework as shown in Fig.1.

3 System Software Structure

The whole system software includes the PLC monitoring units software and computer monitoring software, the development software of field monitoring Unit PLC uses STEP7-Micro/WIN32; monitoring computer software adopts Siemens Wincc6.0; production management software applies Borland C + + Builder6.0, the system software structure includes: (1) PLC software: PLC program can be divided into three modules: ① Data acquisition modules sample the pretreatment sensor data of aquiculture breeding environment, ② Environmental parameters controlling modules control the states of controlling furnishments according to the error between system configuration and the actual value of sensor detection, ③Communication modules are mainly responsible for PLC and the I / O modules, monitoring computer communication.

(2) Monitoring software layer: monitoring computer software is operating platform of controlling breeding environment process, program includes four modules: ① Man-machine interface displays real-time data, regular store data, generating controlling commands of breeding process, ② Data management module defines four databases of the system: real-time database, technology database, task database, operating log database, ③ The system management module mainly answers for the system equipment failure , prediction ,diagnosis, guidance etc, ④ Data processing module mainly inquires about historical data and processes data. (3) Production management software: production management computer software provides user interface of different information management, which includes the following modules: ①User management ② human-machine interface ③ Data query ④ system status.

4 Controlling Algorithm

The aquatic product environment has many factors, such as temperature, dissolved oxygen, PH value, water level and so on, because the temperature and dissolution is the most important 2 environment factors for the aquatic product benefit, the paper has taken the temperature and dissolved oxygen as example, and regulated controlling equipment states to meet system request, the input of fuzzy controller (Hao jiu Yu,et al, 2004)is erroneous and erroneous change rate, the output of fuzzy controller is neural network input, the output of neural network controls equipment states. The change process of the dissolved oxygen and temperature of aquiculture environment is big inertia, time variable, nonlinear, strong coupling multi-inputs and outputs system (Li Xu Ming, 2001), the intelligent controller based on neural network decoupling has controlled aquatic product environment, and combined breeding expert experience with fuzzy control and neural network technology. It utilizes humanity experience knowledge, fuzzy logic inference and neural network study to solve the control strategy which adapted cultivation environment, and applies online study to seek its decoupling relations by using BP algorithm, this algorithm can realize the static and dynamic decoupling, fuzzy neural network decoupling control system is shown in Fig. 2.

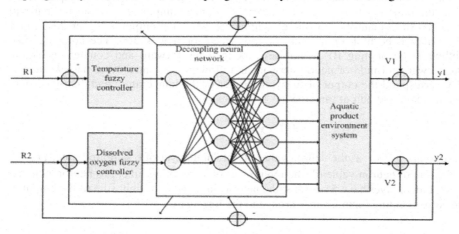

Fig. 2. Decoupling controller of aquiculture environment

Among: R1 and y1 as the temperature setting value and actual value, R2 and y2 as the dissolved oxygen setting value and actual value. The decoupling algorithm is composed of the fuzzy controller and neural network, the neural network is trained by off line, it will gradually adapt environment change through the study and undertakes the controlling duty; when the cultivation environment is disturbed or the parameters change suddenly, the parameter performance can't satisfy the request, the fuzzy controller and neural network will adjust proportionality factors of fuzzy controller and neural network weight to enable the control system to adapt the aquatic product growth environment.

4.1 Fuzzy Controller Design

A fuzzy regulator using self-adjustment factor consists of two parallel component parts of fuzzy control and integral role, its fuzzy control rules:

$$Uf = K0 * f(e,e') \tag{1}$$

Among formula: Uf as fuzzy regulator output, $K0$ as output coefficient, $f(e,e') = \alpha * e + (1-\alpha) * e'$. α for adaptive correction factor, $0 \le \alpha \le 1$; α reflects the size of the error e and error rate e' on impact of the regulator output, at the initial stage, the main tasks of the control eliminates error, Therefore, α value should be greater than 0.5; when the system is close to stable, and should avoid oscillation, α value should be less than 0.5, this paper takes to 0.7 and 0.3.

4.2 Neural Network Design

Neural network (Li Ming, 2000) is 3 layers BP networks of 2 input nodes, 5 latent nodes and 7 output nodes, this first layer of this networks is the output of the fuzzy controller, the output of the third layer is correspondence controlling equipment states, this layer uses sigmoid activation function, the output controlling quantity of decoupling neural network is limited between 0-1, and is multiplied by suitable output factor to control external equipment. To switch quantity controlling equipments, when the output is bigger than 0.5, correspondence equipment will open, or it will close; To shift quantity controlling equipments, when the output is bigger than 3 or 2 or 1 , it will be high or medium or low speed ,or it will stop. The computational method of decoupling BP neural network includes forward and counter algorithm. The forward formula of neural network is following: the input layer has two nodes, they correspond the output of fuzzy controller of the temperature and dissolved oxygen, its input and output function is:

$$o_i'(k) = x_i'(k) \tag{2}$$

Among: $x_i'(k)$ as the neuron value of the input layer of k sampling time, $o_i'(k)$ as the output neuron value of the input layer of k sampling time, i as neuron number, i=1,2. Latent layer have 5 neurons, the neuron value is the weight sum of the output of the corresponding input layer neuron, and the formula is:

$$x_j''(k) = \sum_{i=1}^{2} w_{ij}''(k)o_i'(k) \tag{3}$$

The activation function of latent layer takes the positive and negative symmetrical Sigmoid function, among formula: $w_{ij}''(k)$ as connection weight of input layer to latent layer, $x_i''(k)$ as the output value of latent layer neuron, $o_i''(k)$ as the input value of latent layer neuron, I=1,2, j=1,2,3,4,5. The output layer has 7 nodes, they correspond 7 controlling equipments, the input of each neuron is the weight sum of the output of latent layer neuron, and the formula is:

$$x_o''(k) = \sum_{i=1}^{5} w_{jo}'' \cdot o_j''(k) \tag{4}$$

Because the latent layer completely connects the output layer, during the latent layer mapping the output layer process, neural network has realized decoupling of overall system, and simultaneously sent the most superior decoupling results for the external controlling equipments, the nodes of the output layer select nonnegative sigmoid activation function.

4.3 Counter Algorithm of Decoupling Network

Learning rule of BP neural network (Sun Xin ,2005) is one kind of Delta rule, and adjusts connection weight by using error antigradient,and makes the outlet error reduce monotonously, the counter algorithm regards the decoupling network and fuzzy regulator as well as the multivariable controlled object as multi-layer generalizing network, the multivariable controlled object is the final layer of this network, the learning algorithm of erroneous reversion adjusts error, the system has considered various variables coupling function, and taken performance target function [4] is:

$$J = E(k) = \frac{1}{2}\sum_{s=1}^{2}(r_s(k) - y_s(k))^2 \tag{5}$$

According to the gradient descent law modifying coefficient of network weights, antigradient direction of E (k) to weighting coefficient searches adjustment, and attaches inertia item to make the rapid convergence to overall minimum, and obtains adjustment formula (Wang Da Yi,1999) of the output layer network weight:

$$w_{jo}'''(k) = w_{jo}'''(k-1) + \Delta w_{jo}'''(k) \tag{6}$$

According to similar principle, connection weight iterative formula of input layer to latent layer is:

$$w_{ij}''(k) = w_{ij}''(k-1) + \Delta w_{ij}''(k-1) = w_{ij}''(k-1) - \eta_2 \cdot \frac{\partial E(k)}{\partial w_{ij}} + \alpha_2 \cdot w_{ij}''(k-1) \tag{7}$$

Controlled object doesn't need mathematical model and the parameter according to the neural network algorithm, the system trains BP network by off line, it may be put into operation of the systems after training, the connection weight adjustment formula reference (6), (7).

4.4 Controlling Results

The decoupling system of fuzzy neural network has controlled aquatic product environment factors, this control system of neural network decoupling has solved both temperature and dissolve coupling function, and considered controlling quality low consequence of controlling equipment frequent start-up, and improved the controlling quality. The long term application has indicated neural network decoupling controller has obtained good controlling effect., Year trial and observation have shown that the system can maintain appropriate growth environment ,when it is applied to 20 m × 20m the breeding ponds. We may see change situation of breeding pond temperature and dissolved oxygen from the next table, the starting value of temperature and dissolved oxygen is 30℃ and 7.5mg/L, the actual value and given value are compared, the errors exist in 2%, and meets the system growth requirement, the controlled process doesn't basically have over modulation, the adjustment time is short, the stable errors doesn't basically have.

For example: the original temperature to 32°C, dissolved oxygen to 7.15mg/L, controlling equipment are all closed, they begin to control according to the detection of the errors and decoupling, the system sets temperature at 25°C and dissolved oxygen at 7.5 mg/L, The sampling period is 20s and the equipments have the same controlling state within 6 minutes, The state changes of controlling equipments are shown in table 1, among table: 0-3 representatives of the stall controlling, they represent off or stop, small or low-speed, medium or medium-speed, open or high-speed.

We can see the temperature will drop to 29.0°C and dissolved oxygen will rise to 7.25 in 6 minutes from table 1, under the new deviation, this opening of controlling equipments are changed (0,1,1,1,0,2,2), in another 6 minutes, the temperature will drop to 26.5°C and dissolved oxygen will rise to7.38, and the opening is (0,0,0,1,0,1,1) etc. The system will become small deviation in 20 minutes. At different initial value of temperature and dissolved oxygen, Experiments have been applied and the results are very good.

Table 1. Aquiculture environment control results

t (M)	T (°C)	D (m g / L)	U	S	N	C	H	F	M
1	31.5	7.15	1	2	2	2	0	2	3
2	30.0	7.20	1	2	2	2	0	2	3
4	29.0	7.25	1	2	2	2	0	2	3
6	28.0	7.30	1	2	2	2	0	2	3
8	27.0	7.35	0	1	1	1	0	2	2
10	26.5	7.38	0	1	1	1	0	2	2
12	26.0	7.40	0	1	1	1	1	2	2
14	25.7	7.45	0	0	0	1	0	1	1
16	25.3	7.48	0	0	0	1	0	1	1
18	25.1	7.50	0	0	0	1	0	1	1
20	24.9	7.49	0	0	0	1	0	1	0

Among table: t- controlling time, T-temperature, D-Dissolved oxygen, U-sunroof, S-south window, N-north window, C- cold water valve, H-hot water valve, F-fan,M- increasing oxygen machine.

5 Conclusion

The method of fuzzy neural decoupling has improved controlling level of aquiculture breeding environment and taken on the larger theoretical and practical significance, real-time testing results have demonstrated its feasibility, and it has greatly practical value. With the existing pattern compared to this controlling scheme, the system is open and achieves data sharing, thus has enhanced the level of controlling breeding environment. Intelligent decoupling control system of the aquatic product environment has strong comprehensive, usable, scientific and compatibility to breeding environment, the system needs aquaculture expert and engineering technology intelligence expert mutual coordination, and closely unifies the domain knowledge experience and project technology, the practice has shown the system has enhanced cultivation benefit, and improved aquatic growth environment.

References

Congguo, M., Wei, N.: Factorization aquaculture monitoring system design based on PLC. Industry appliance and automation device 12(2), 51–53 (2005)

Yu, H.j., Wei, C.: Decoupling for multivariable fuzzy Control System. Tianjin University Journal 37(5), 396–399 (2004)

Ming, L.X.: Fuzzy neural network control of multivariable system. Control and Decision 16(1), 107–110 (2001)

Ming, L., Yongjun, L.: Auto-adapted neuron non-model much variable system decoupling control. Computer simulation (3), 68–71 (2000)

Xin, S.: Quota moisture decoupling control analyzes of ma paper product process. Control theory and application 20(18), 121–124 (2005)

Yi, W.D., Rui, M.X.: Neuro-Optimal Guidance Law for Lunar Soft Landing. Engineering and Electronic Technology 21(12), 31–34 (1999)

Development of a Semi-controller for a Variable Rate Fertilizer Applicator[*]

Jianbin Ji[1], Xiu Wang[1,*], Yijin Mao[2], Liping Chen[1], and Lingyan Hu[3]

[1] National Engineering Research Center for Information Technology in Agriculture,
Beijing 100097, China,
Voice: +00-86-51503696
wangx@nercita.org.cn
[2] China Agricultural University, Beijing 100083, China
[3] TaiYuan University of TechnologyShanXi 030024, China

Abstract. In this paper, aiming at the current development of domestic agricultural production, introduced a variable fertilizer controller which suits for a domestic food-producing areas to promote output. The variable rate fertilization controller combined with the current Chinese made fertilizer equipment; mechanism of metering system for the applicator uses a flute wheel. The amount of fertilizer is determined by the rotation speed of flute wheel. The use of the control system can be inconsistent in soil fertility, pre-division of plots with different fertilizer way through the process of moving the driver automatically adjusts the amount of fertilizer direction buttons to precision variable fertilization. Papers on the electronic structure of the controller and software design in detail. In addition, we have examined the performance of this new product in the field by using different fertilizer and requirement. The results reveal that semi-automatic controller work well by controlling the quantity of fertilizer precisely. Specifically, coefficient of variation of fertilizing is controlled to be less than 5%. That means this new type semi-automatic controller worth developing in further.

Keywords: speed adjustment, speed signal collection, real-time control, variable rate fertilizer.

1 Introduction

Recently, farmers around the world generally use chemical fertilizers as the main means of increasing food production, farmers in most areas are evenly spread the fertilizer in the surface, and then make the appropriate agronomic operations. To take such operating procedures, soil nutrients in the high area and in the low area still use the same amount of fertilizers resulted in a high position in the nutrient over-use, and the location of low-nutrient may occur fertilizer shortage. In the over-use area, part of chemical

[*] Supported by National High-tech R&D Program of China (863 Program) (Grant No. 2006AA10A308); Supported by Beijing Rural Affair Committee (Grant No. 20070110).
[*] Corresponding author.

fertilizers in the soil break down or be absorbed, some of others by leaching in the soil, these fertilizer undergo eluviations absorbed by groundwater easily to pollute the underground water resources, and affect people's health. Thus, as the voice of sustainable development of agricultural production increasing, the environment pollution caused by chemical fertilizers be concerned more and more in the countries, most developed countries have established relevant laws to restrict the use of chemical fertilizers.

With popularization of the technology of the global positioning system (GPS), when the tractor carries on the field work, the tractor can obtain the real-time work position accurately under the GPS system help, Therefore, manage the soil by fixed position become a riper technology. But this technology is comparatively complicated, the short term, the nationwide promotion of access to a large area will be subject to funding, staff skills and the specific restrictions related to the environment. Therefore in order to promote the technology quickly, based on soil nutrient status of the field, we can divide the same block into different application management unit, through setting up different operating signs in each unit, using semi-automatic variable rate fertilization realize manual control of variable rate fertilization, drivers do not need to leave the driver's seat, by changing the button on the controller to achieve the purpose of accurate variable rate fertilization. Semi-automatic controller of variable rate fertilizer applicator can receive the speed of tractors and automatic adjust the fertilizer volume to achieve accurate variable rate fertilization. This paper aimed at the current equipment characteristics of agriculture of China; carried out semi-automatic variable fertilization technology research, the developed semi-automatic controller of variable rate fertilizer applicator can automatic calculate the fertilizer volume according to tractors speed. Drivers can also facilitate change the fertilizer volume in fertilization; overcome the traditional deficiencies of parking change fertilizer volume.

2 Controller Structure

Control functions by the two co-ordinations 51 single-chip realization, system architecture as shown in Figure 1. System has five major components: PWM pulse signal output module, operation module, LCD module, power supply module and drive module.

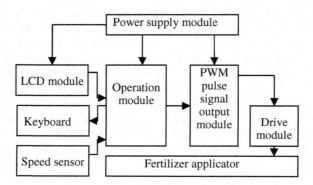

Fig. 1. The structure of semi-automatic variable rate

2.1 PWM Pulse Signal Output Module

PWM control chip generated and output PWM pulse signal, the PWM control chip adopted AT89S52 single-chip. This module is simple in design, the main reason is: if these are too many functions in design, chip need much time to run these functions, this may impact the output PWM signal, and it will affect the accuracy of the fertilization. The module structure is in Figure 2.

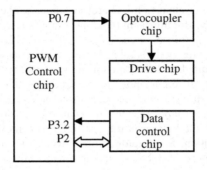

Fig. 2. PWM output module

In this module, as output PWM pulse signal pin, the P0.7 pin of PWM controlling chip, which outputs signal to amplifier drive module isolated by optocoupler chip, the output drives hydraulic Flow Valve. P2 interface of PWM controlling chip is connected with P2 interface of data controlling module, PWM controlling chip will change the output pulse width according to duty cycle computed and sent by data controlling chip. P3.2 pin is external interrupt pin, which can enable PWM controlling chip to respond immediately to adjust PWM pulse width in time, when data controlling chip computes duty circle needing varying immediately.

2.2 Operation Module

Operation module mainly operates data and harmonizes parts function, including input and control of keyboard, LCD module, tractor rate collection, and output of PWM duty circle signal and so on. Fig3 is module connecting block diagram. P0 interface of data controlling chip is connected with 4*4 keyboards, for the convenience of controlling data during the process of fertilization, one column keys are connected with interrupt interface P3.3 pin, when one is pressed, it will touch off interrupt of data controlling chip. P1 interface is used as data output interface of LCD module, which input the needing data to LCD module. The pins of P2.1 P2.2, P2.3 are multiplexing pins with different functions controlled by code; here they are used as LCD controlling pins. P3.2 pin is external interrupt pin, which is used for signal input pin to collect pulse signal of speed sensor. Data controlling chip adopts timer T1 to time the system.

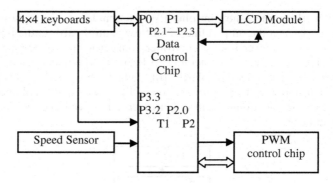

Fig. 3. Operation module

2.3 LCD Module

LCD module is convenient for human-computer interaction; it could display data and menu. The system designed four menus: pre-set fertilizer volume input, per-axis fertilizer volume set, tractor speed calibration, and fertilization. When pre-set fertilizer volume input menu, the value can be set manually according to the anticipant fertilizer quantity, per-axis fertilizer volume set menu adjust to different fertilizer mechanism and different kinds of fertilizer, its set value is the number of how many grams of fertilizer output when the axis makes one revolution, the system can not test fertilizer output when the axis makes one revolution automatically for the sake of decreasing cost. Tractor speed calibration menu is to overcome skid factor, tractor speed calibration menu responds tractor rate at different task conditions correctly to fertilize exactly. field operation, first measure one hundred meters, press enter key on controller when tractor starts at starting point, then press it again when the tractor reaches end point, the system can adjusts automatically and achieve accurate speed value. Fertilization menu watches fertilization state, here preset value, fertilizer output and tractor speed etc. can be seen.

2.4 Software Design Process

The controller software design is based on the fertilization process, figure 4 flow chart for software design. According to the current international fertilizer applicator, the axis of fertilizer volume parameter is set to check volume. Therefore, for different varieties of fertilizer, farmers must be carried out the fertilizer volume in each row of axis before fertilization, and the checked fertilizer volume is only suitable for such fertilizer, if change fertilizer, check it again. In addition, farmers need to pre-set the fertilizer volume before fertilization and check the speed of tractor. This concept, software designed menu displayed on the LCD screen, are as follows: pre-set fertilizer volume, per-axis fertilizer volume, tractor speed and fertilization.

3 Experiment and Testing

In order to achieve the purpose of accurate fertilization must pre-measure the per-axis fertilizer volume, as well as the relationship between the PWM pulse duty cycle, the revolution speed of axis and the tractor speed.

Preparatory work: connect the controller PWM output line to the input port of oil flow control valve, connect the oil pipeline of oil flow control valve to the input and output ports of tractor, the power line of controller connect to the battery, adding fertilizer into fertilizer hopper, start the tractor. Fertilization experiments using a total of 24 fertilizer exhaust port, select 24 plastic 10L capacity box marked No. 1-24 for each fertilizer exhaust port select fertilizer, to facilitate weigh the volume of fertilizer. Weighing use the electronic scales. Fertilizer-axis speed measurement using the contact tachometer.

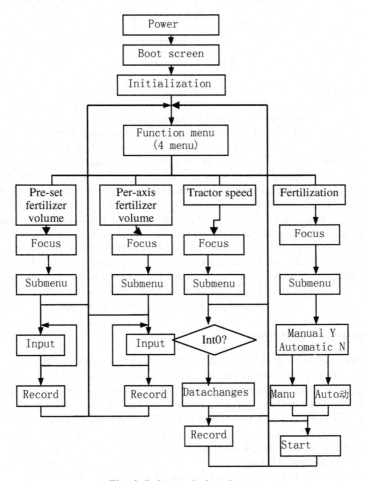

Fig. 4. Software design chart

3.1   The Per-axis Fertilizer Volume Measurement

Methods: The controller use the manual mode, put the PWM pulse width to a random value, at the same time record the number of fertilizer-axis turning circle, stop and weigh the total output fertilizer and divided by the number of turning circle. Change the PWM pulse width, re-measured and calculated. The average results of several measurements are the per-axis fertilizer volume.

3.2 The Relationship between the PWM Output Signal Width and the Revolution Speed of Axis

The controller of variable rate fertilizer applicator is manual mode, set PWM pulse width to a smaller value, measuring and recording the revolution speed of axis. Increase PWM pulse width, measure again. Repeatedly measured and calculated come to the relationship between the revolution speed of axis and the PWM signal pulse width, shown in Figure 5.

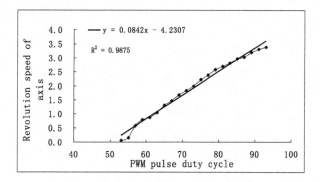

Fig. 5. The relation of PWM pulse width and axes rotates

When the width of the PWM output signal increases (the PWM duty cycle that is upgrading), revolution speed of axis also enhance, and is obviously linear increase. According to experiments, the linear relationship is y = 0.0842x-4.2307, y is the revolution speed of axis and x is the width of the PWM.

3.3 The Relationship between Pre-set Fertilizer Volume, PWM Output Signal Width and Tractor Speed

When the pre-set fertilizer volume is set, the relation between PWM pulse signal width and tractor speed is linear, if the speed go up, the rotate speed will go up, namely PWM pulse width will increase, and vice versa. The rotation axis with twenty teeth on fixed gear drives fertilizer axis with 26 teeth on gear through chain. The fertilizer output port is total 3.6 meter, formula 1 is deduced through integrating all parameters, put axis rotation speed and PWM pulse signal width into formula 1, relation between PWM pulse width and tractor speed could be achieved.

Use code to build the relation among these varies, writes the code in chips, Adjusting fertilizer automatically according to tractor rate is implemented.

4 Field Trials and Testing

At present, fertilization mainly use granular fertilizer, Therefore the trial use urea. Test Method for fertilizer in the fertilizer hopper width of 3.6 meters and the fertilizer exhaust port width of 18mm, driving a tractor traveling 100 meters, test the output fertilizer volume. such as Table 1. In the actual measurement, the permissible error in the 5% error range, so it meets the purpose of variable rate fertilization and its standards.

Table 1. Fact test

NO.	Pre-set fertilizer volume	Actual fertilizer volume	Actual error	Variation coefficient
1	14.4kg	15.19kg	0.79kg	0.05
2	10.8kg	11.17kg	0.37kg	0.03
3	7.2kg	7.51kg	0.31kg	0.04
4	50kg	54.24kg	4.24kg	0.80
5	80kg	83.52kg	3.52kg	0.03
6	100kg	102.86kg	2.86kg	0.02
7	150kg	154.53kg	4.53kg	0.03
8	200kg	196.20kg	3.80kg	0.03
9	250kg	245.74kg	3.26kg	0.02
10	300kg	297.56kg	2.44kg	0.01

As can be seen through Table 1, experimental error shows irregular changes. The main reason is that the errors are due to the physical properties of devices and operating factors, and even generate an error. Therefore, the operator can adjust the use of agricultural machinery and methods of operation to achieve the purpose of smallest error in fertilization.

5 Conclusion

The controller designs simpler, can not connect to GPS / GIS systems, and did not adding the feedback control functions, which greatly reduces the design cost. In addition, the controller can colloct the speed signal of tractor in real-time, realize the purposes of the fertilizer automatic adjustment according to the tractor speed.

The status of China's agricultural production is that Chinese farmers have less agricultural machinery and equipments, but the domestic product 37.8KW tractor which can conjunction with fertilizer application system for agricultural production is in more.

Most of the variable rate fertilization system which developed by the current domestic or international are complicated to use and control, and the prices are expensive, difficult to purchase for farmers. The controller described in papers has a simple structure, easy to operate, and at a great price, suitable for the vast rural areas to promote. Controller of the field-fertilization experiment can achieve described function; greatly enhance the efficiency of fertilization. In this paper, the controller of variable rate fertilizer applicator is based on traditional variable rate fertilizer applicator and innovation, operation more convenient, more obvious effect of fertilization.

Acknowledgements

Supported by National High-Tech R&D Program of China (863 Program)(Grant No. 2006AA10A308); Supported by Beijing Rural Affair Committee (Grant No. 20070110).

References

Sawyer, J.E.: Concepts of variable rate technology with considerations for fertilizer application. Prod. Agric. 7, 195–206 (1994)

Han, S.F.: Nitrogen sensing and site-specific application technology. In: Information Technology of Agriculture, Proc. of the International Conference on Agricultural and Technology, Beijing, China, June 2001, pp. 324–330 (2001)

Raun, W.R.: Improving Nitrogen Use Efficiency in Cereal Grain Production with Optical Sensing and Variable Rate Application. American Society of Agronomy Journal 94, 815–820 (2002)

Sawyer, J.E.: Concepts of variable rate technology with considerations for fertilizer application. Prod. Agric. 7, 195–206 (1994)

Le Maitre, R.W.: Numerical Petrology. Statistical Interpretation of Geochemical Data, 210 (1982)

Design of Intelligent Conductivity Meter Based on MSP430F149

Yaoguang Wei, Jianqing Wang, Daoliang Li[*], and Qisheng Ding

College of Information and Electrical Engineering, China Agriculture University,
Beijing, P.R. China 100083
dliangl@cau.edu.cn

Abstract. An intelligent conductivity meter based on MSP430F149 microcontroller was proposed in this paper. The intelligent conductive meter was composed by MSP430F149 microcontroller and its peripheral circuits, bipolar pulse excitation circuit and waveform transform circuit. The MSP430F149 was chosen to control the generation of bipolar pulse excitation signal and take temperature compensation. It adopted bipolar pulse excitation signal to avoid the electrode polarization effects. The measurement error generated by the solution temperature fluctuation could be amended by the temperature compensation. It could self- compensation and self-tuning to fit variance solution.

Keywords: Bipolar Pulse, Electrical Conduction, MSP430, Temperature Compensation.

1 Introduction

In recent years, water quality had become a hot issue in the fields of clean drinking water, distilled water, medicinal and biological water, power boilers, aquaculture and other areas [1,2]. Aquaculture was a fast-expanding mode of food production in the world. Global production of farmed fish and shellfish has more than doubled in the last 10 years. Ninety percent of the world's aquacultures undertaken in Asia, with China producing two thirds of the world total while Europe, North America and Japan, which produce only 10 %, consume the bulk of the seafood traded internationally. Fish will be the main farmed aquaculture species, but production of more extensive species like stichopus japonicas, bivalves and seaweeds would increase. In the aquaculture, water quality was the key influencing factors. The water quality problem was associated with both physical and chemical factors such as high or low dissolved oxygen, high concentration of nitrogenous compounds (ammonia-N and nitrate-N) and high levels of Electrical conductivity (EC). The water quality in aquaculture fish-ponds was controlled by a complex interplay of many factors. Air-water Environmental problems have resulted from the conversion of wetland habitats to aquaculture ponds. These include nutrient, sediment and organic waste accumulation leading to deterioration of water quality, one of the important factors that determine the viability of fish farming. The water quality problem was associated with both physical and

---

[*] Corresponding author.
The main research fields are agriculture information technology.

D. Li and C. Zhao (Eds.): CCTA 2009, IFIP AICT 317, pp. 240–247, 2010.

chemical problems such as too high or too low Electrical conductivity (EC). Electrical conductivity was the measure of total concentration of dissolved salts in water. When salts dissolved in water, they give off electrically charged ions that conduct electricity. The more ions in the water, the greater the electrical conductivity it had. Because there were almost no ions in distilled water, it had almost no electrical conductivity. Hard water contained more salts, and therefore more ions, had a high electrical conductivity. Electrical conductivity was the criterion to measure the conductivity of the solution[3]. It was inherit physical and chemical properties of the solution, and it was the main influencing factors to the water quality.

The researchers had taken great attention on the electrical properties of the solution and lots of methods to measure EC of the solution had been proposed in the last years. The commonly used measurements of EC were electrode conductivity, electromagnetic conductivity and ultrasonic measurements[4,5]. Restricted by the measurement mechanism, the last two methods were usually been used to measure the high-conductivity solution. The electrode conductivity measurement method was based on the principle of electrolytic conductivity. It had the features of simple electrode structure and wide range of measurement which promoted it to be widely used in the measurement of solution EC.

2 Measurement Principle of Conductivity Electrode

The conductance pool could be equivalent to the parallel resistor and capacitor circuit which has been shown at figure 1 (A). There, R_{L1}, R_{L2} was the electrode lead resistance; R_1, R_2 was the polarization resistance resulted in chemical polarization and concentration polarization, also known as Faraday resistance. C_P was the lead capacitor. The simple circuit of the conductance pool was shown at figure 1 (B). There R_x was the resistance of the conductance pool, C_x was the capacitance of the conductance pool.

Many conductivity electrode measurement methods had been proposed at home and abroad. But there were many shortcomings about the methods to measurement the EC through conductivity electrode. Firstly, the measurement results were inevitable influenced by the polarization resistance R_1, R_2 , especially in the measurement of high concentrations of solution. Secondly, as shown in figure 1(A), the value of C_x was also influenced the measurement result. Thirdly, the temperature of the solution

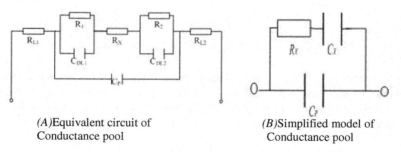

(A)Equivalent circuit of   (B)Simplified model of
Conductance pool   Conductance pool

Fig. 1. Conductance pool and its Equivalent circuit

could influence the ionization, solubility, ion migration rate, solution viscosity and expansion of the solution, which could also take effects on the accuracy of the EC measurements[6]. When the temperature increased, the solution viscosity become lower, the ion movement accelerated under the electric field, the conductivity changed. It could be concluded that the EC measurement based on conductivity electrode should avoid the influence of polarization effect, capacitance effect and temperature effect.

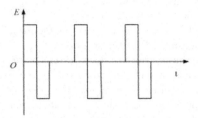

Fig. 2. Bipolar Pulse Signal

To solve these problems, we designed the voltage bipolar pulse method which pulse width could be easily adjusted to measure EC. As shown in figure 2, the excitation signal was bipolar pulse voltage source, which has same amplitude, opposite polarity. The pulse excitation and sampling diagram was shown in figure3. In one measurement circle, the system sampled the output voltage twice at T moment. It selected the average of the twice sample value as the measured value to eliminate the low-frequency noise which could be generated by the DC excitation system. T moment was at the moment of positive or negative half-cycle which output the 80% of output voltage. It could be considered that the output voltage has reach to stable when the average of the two sampling was less than some value.

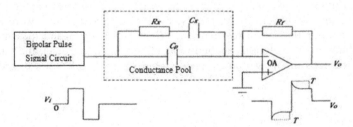

Fig. 3. EC Measurement based on Bipolar Pulse Excitation

If the sampling time T was far less than product of Rx and C_x, then the voltage of the C_x should be small enough. At that time, there was no current flowing through C_p because the charging of C_p has finished. At T sampling moment, the relation between the measured transient voltage and solution resistance was shown following:

$$R_x = -\frac{V_t \times R_f}{V_0} \qquad (1)$$

It could be concluded from equation (1) that the R_x has no relation with C_x, C_p. That was to say, the measurement method had eliminated the effects of the capacitance. The effects of polarization resistance was related with the voltage between the two electrodes, voltage duration time and composition of the solution. Therefore, polarization resistance calculation was very difficult, even impossible. Research had found that there would not appear polarization when the voltage was less than a certain value.

3 Design of Measurement Circuit

As shown in figure 4, the structure of EC measurement circuit was composed by MSP430F149 control circuit and its peripheral circuits, bipolar pulse excitation circuit, range switching and amplifier circuit, waveform transformation circuit and temperature measurement circuit. The function of the circuit was to generate bipolar pulse signal, range switching, converted the measurement signal into digital signal, temperature compensation, data storage and transmission. We adopted the MSP430F149 to capture analog signals of conductivity electrode, convert them to digital values, process measurement data and transmit the data to a host system. The MSP430F149 was microcontroller configurations with two built-in 16-bittimers, a fast 12-bit A/D converter, two universal serial synchronous/asynchronous communication interfaces (USART), and 48 I/O pins. The timers make the configurations ideal for industrial control applications such as ripple counters, digital motor control, EE-meters, handheld meters, etc. The hardware multiplier enhances the performance and offers a broad code and hardware-compatible family solution.

The traditional electrical conductivity measurement system adopted sinusoidal excitation signal to avoid electrode polarization. It should to add the complex filter circuit for signal processing. The system has been proposed in this paper adopted the

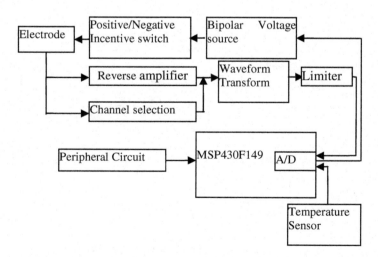

Fig. 4. *Structure* of the Intelligence Electrode Conductivity Meter

bipolar pulse voltage source as the excitation source. It was equivalent to add a DC power source to the electrode when electrode received excitation. There were no filter circuit and capacitance compensation section which could simplify the structure.

For certain solution, there was no polarization effect when the excitation voltage value less than some special value. The polarization could be negligible when the voltage time was very small. The pulse width was not only restricted to the capacitor charging and discharging time, but also restricted to the polarization time. So the pulse width should be as small as possible to eliminate the polarization. Meanwhile, it should enhance the signal strength in order to separate the signal from noise. The circuit to generate bipolar pulse excitation signal was shown as figure 5.It adopted 78L05, 79L05 series of integrated three terminal voltage regulators as the incentive signal generator. 78L05 was the positive voltage regulator and 79L05 was the negative voltage regulator. The devices had the features of overheating and over-current protection circuit, few external components, small size and low cost. The switch of the bipolar pulse source was controlled by the SPDT Analog Switches MAX303. The switch closure time was less than 150ns, its disconnect time was less than ns, the resistance less than 22Ω.

As shown in figure6, the pulse control signal was produced by the I/O port of MSP430F169. To ensure the drive capability of the signal, the global weak pull-ups was set to inhibit. The output of the port P1.0, P1.1 was set to push-pull mode. The frequency of Control signal was controlled by the internal Timer_A of MSP430F169. It chose appropriate feedback resistor to ensure the measurement accuracy at different measurement range. The range switch was divided into four stalls and adopted feedback resistor of 390Ω, 3900Ω, 39000Ω, 390000Ω separately. It adopted the internal voltage reference of MSP430F139 as its A/D converter voltage reference. The conversion range was 0V~2.5V. It adopted MAX313 as the gating control device to

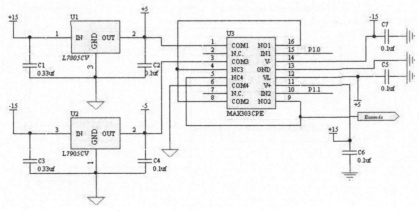

Fig. 5. Bipolar pulse excitation circuit

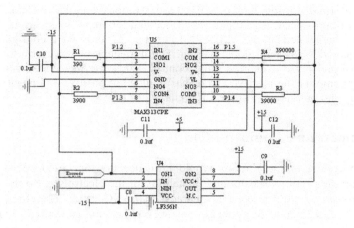

Fig. 6. Range Switching and Amplifier Circuit

choose the feedback resistor. Maxim's MAX313 analog switches feature low on-resistance (10Ωmax) and 1.5Ω on-resistance matching between channels. These switches conduct equally well in either direction. They offer low leakage over temperature (2.5nA at +85°C). The MAX313 was quad, single pole/single-throw (SPST) analog switches. The control signal was generated by the MSP430F149. The I/O ports were configured as push-pull mode and weak pull-global ban on.

The waveform transformation circuit was shown at figure7. Because the input signal of A/D in MSP430F149 was the positive polarity signal, so it should transform the AC signal into DC signal. It adopted two operational amplifiers and one CD4053 analog switch to construct the waveform conversion circuit. In order to ensure

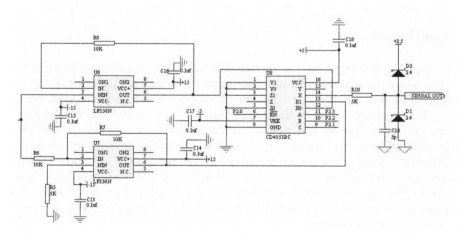

Fig. 7. Waveform Transformation Circuit

non-waveform distortion, it chose LF356 as operational amplifier. The LF356 had the features of high-speed, high input impedance and low input bias current. The control signal was generated by the MSP430F149. The I/O ports were configured as push-pull mode and Weak pull-global ban on. The frequency of waveform transform control signal and pulse excitation pulse control signal were synchronization. The two amplifiers amplify the positive and negative signal at positive and negative phase respectively and output the positive DC signal.

4 Temperature Compensation

The temperature of the solution would influence the ionization degree, solubility, ionic migration ratio, viscosity and dilatability of the electrolyte. So the changing of the temperature would affect the accuracy of the EC value. Therefore, temperature compensation of conductivity measurements had become particularly important, and took temperature compensation could reduce the error and increase the accuracy of the measurement. Research has found that when the temperature was deviate from 25℃, the error of the EC was become larger. To reduce the error and increase the accuracy, it divided the temperature between 1℃ to 30℃ into five intervals. Every temperature interval was set different correction factor by the experiments results. The correction equation was shown as follows:

$$K_s = \begin{cases} K_t/(0.01t + 1.21), 1^\circ C \leq t \leq 10^\circ C \\ K_t/(0.04t + 1.17), 10^\circ C < t < 20^\circ C \\ K_t/(0.05t + 1.45), 20^\circ C \leq t \leq 30^\circ C \end{cases} \tag{2}$$

Where, K_s was the Electrode Conductivity value of the solution at 25℃, Kt was the Electrode Conductivity value of the solution at t℃, t was the solution temperature.

5 Conclusion

A new seawater electrical conductivity measurement system has been proposed in this paper. It adopted conductivity electrode as the sensor to measure the resistance of the solution. It adopted bipolar pulse excitation signal to avoid effects of electrode polarization. When being powered by the bipolar pulse which excited by the MSP430F149 microcontroller, the circuit generated voltage signal which corresponded to the resistance of the solution. The analog output signal was amplified by the range switching and amplifier circuit. Then it changed the waveform through the waveform transformation circuit and sent the signal to the A/D of MSP430F149. The temperature signal was also being measured to take temperature compensation. The system which has been proposed in this paper had the characteristics of easy to realization, self-compensation and self-tuning.

Acknowledgements

This work was supported by the Key Projects in the National Science & Technology Pillar Program during the Twelve FIVE-YEAR Plan Period (2006BAD10A02-05)

References

[1] Randall, E.W., Wikinson, A.J., Cilliers, J.J., et al.: Current Pulse Technique for Electrical Resistance Tomography Measurements. In: Proceedings of the World Congress on Industrial Process Tomography, Hannover, Germany, August 29-31, pp. 493–501 (2001)

[2] Cilliers, J.J., Xie, W., Neethling, S.J., et al.: Electrical Resistance Tomography Using a Bi-Directional Current Pulse Technique. Meas. Sci. Technol. 12(8), 997–1001 (2001)

[3] Teijun, L., Zhiyao, H., Baoliang, W., et al.: High Speed Electrical Resistance Tomography System Based on Bi-directional Pulse Current Technique. In: Proceedings of the 4th International Symposium on Measurement Techniques of Multiphase Flows, Hangzhou, China, September 10-12, pp. 192–198 (2004)

[4] Haibo, H., Halfeng, J., Zhiyao, H., Haiqing, L.: Design of Electrical Resistance Tomography System based on bi-directional pulsed current technique. In: ICEMI 2003: Proceedings of the sixth international conference on electronic measurement & instruments, Taiyuan, China, August 18-21, vol. 1-3, pp. 2300–2303 (2003)

[5] Xiaoping, C., Hongxian, C.: Double Frequency Method to Measure Water Conductivity. Chinese Journal of Scientific Instrument 27(5), 520–522 (2006)

[6] Jianping, Z., Baoliang, W., Zhiyao, H., Haiqing, L.: A New Electrical Conductance Measurement Instrument Based on Bi-directional Pulsed Voltage Technique. Chinese Journal of Scientific Instrument 26(8), 57–58 (2005)

Intelligent Control Technology for Natural Ventilation Used in Greenhouse

Yifei Chen and Wenguang Dai

College of Information and Electrical Engineering, China Agricultural University,
Haidian, Beijing, P.R. China 100083

Abstract. It is very important that the comfortable indoor climate is built for crop growing in greenhouse. To adjust the temperature and humidity, the natural ventilation operation and indoor airflow operation is necessary to be controlled. Depending on many researches and analysis for indoor climate of greenhouse, it is proved that the two main parameters, temperature and humidity in greenhouse, correlate with ventilation, and the relation equation is indicated tunambiguous. On the other hand, there is other research which indicates that the growth result of crop is affected by the ways of airflow in greenhouse as well as natural ventilation. For all of these factors, with advance control theory and intelligent control technology, the control system of natural ventilation is presented in this paper. The main point in the paper is that depending on replacement ventilation principle the natural ventilation will be controlled by driving the windows open. First, the ventilation model using in greenhouse is constructed specially using actuators for driving windows one by one, and control model is analyzed on the principle of natural ventilation too. It is present firstly that using fuzzy control method depending on the temperature indoor to drive windows to work is feasible. Then the system designing of driving windows is given, and the intelligentized controlling system for natural ventilation in greenhouse is constructed finally. According to the smart fuzzy calculating and crop growth need condition, the windows open angle can be controlled accurately with detecting indoor parameters such as temperature, humidity, CO_2 density as well as outdoor temperature, sun light, wind and rain. Also, the characteristic and designing factors used in this control system, with one example, are shown in this paper.

Keywords: greenhouse indoor climate, natural ventilation, fuzzy control, driving window.

1 Introduction

It is fact that the natural ventilation in greenhouse has many advantages for crop growth. General speaking, the ventilation in greenhouse is forced to work by the machine like blower, so, there are many disadvantages clearly to see such as the wind force and direction cannot be controlled. On the other hand, not only the ventilation effect is affected, but the electrical energy cost is caused to increase. So, how about natural ventilation is valid used is the researching direction for greenhouse ventilation.

D. Li and C. Zhao (Eds.): CCTA 2009, IFIP AICT 317, pp. 248–253, 2010.
© IFIP International Federation for Information Processing 2010

The basic principles of natural ventilation are discussed in reference (T.Boulard et al.,1977; Roy J C et al.,2002; Boulard T et al.,2002), and the calculating model and simulation usind CFD are presented based on the indoor thermal pressure and outdoor wind pressure. Moreover, the general characteristic of parameters in greenhouse are stated too in reference (Xu Fang et al.,2005; Li Wiyi et al.,2005) as well as calculating model for ventilation. But how about making the control model for natural ventilation is not given as well as how to realize the control system. The ventilation control is discussed in reference (Gu Jinan et al.,2001). The author thinks that the indoor climate parameters like temperature, humidity, day lighting is assumptive inertial controlled factors, and the control model is given by time difference concept. But the control method is BANG-BANG with ON-OFF function, so control precision is lower and the veracity of dynamic adjusting for above parameters has been not proved.

As well as know, the dynamic adjusting for the indoor climate of greenhouse is very difficult for the fact that many parameters are strong coupling each other, and the control models are non-linear and complex. So, basing on this situation, it is hard to build the control system. According to the statement in reference (Roy J C et al.,2002), there is the basic change process that indoor air is mutual exchanged without air of greenhouse. We think this exchanging is taking place in parts of greenhouse, such as the enclosure, soil and windows. It is shown as Fig.1.

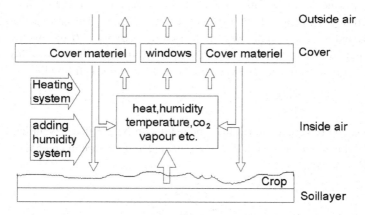

Fig. 1. Schematic representation of air exchange inside and outside of greenhouse

As Fig.1 showing, the exchanging is composed by two kinds: one is passive exchanging from greenhouse roof and wall etc. Another one is active exchanging from roof windows in greenhouse. That the roof windows are exactly controlled can resentful affect the indoor climate of greenhouse (Cecilia et al., 1995).

In this paper, for the windows controlling, the fuzzy control technology is used firstly, and the intelligent control system for greenhouse natural ventilation is built as following.

2 The Control System Model Designing for Natural Ventilation in Greenhouse

2.1 Building of Relation Model on Natural Ventilation in Greenhouse

The temperature and relative humidity in greenhouse associate strongly with ventilation (Cecilia et al., 1995), and under the condition of natural ventilation, the relation equation of relative humidity and ventilation flux is as following:

$$V = g_{vent}(\chi_a - \chi_o) \qquad (\text{kg m}^{-2}\text{ s}^{-1}) \qquad (1)$$

Here, χ_a means indoor relative humidity, χ_0 means outdoor absolute humidity, g_{vent} means ratio of ventilation, V means bentilation flux density. It is presented that the relation is linear between ventilation flux and indoor humidity. On the other hand, if the roof window is as Fig.2, the relation equation between the temperature and ventilation flux, based on the vapor buoyancy effects, is as following:

$$\Phi_{b,front} = C_f(g \beta)^{1/2} \frac{L_o}{3} H^{3/2}\left(\bar{T}_a - \bar{T}_o\right)^{1/2} \qquad (\text{m}^3\text{ s}^{-1}) \qquad (2)$$

Here, $\Phi_{b,front}$ means ventilalation flux, T_a and T_0 mean indoor and outdoor temperature respectively, C_f means discharge coefficient, β means thermal expansion coefficient of air, and g means conductance to heat or mass transfer. H means the roof window open height as Fig.2, and H_0 and L_0 stands for the size of roof window. So, the relation equation between the open angle and open height is as following:

$$H = H_o\left[\sin \psi - \sin(\psi - \alpha)\right] \qquad (\text{m}) \qquad (3)$$

According to above equations, the conclusion can be got that if ventilation flux is controlled right, and than, the relative humidity and temperature in greenhouse also can be adjusted accordingly.

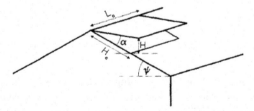

Fig. 2. Schematic representation of a roof greenhouse ventilator

2.2 Control Model Structure

Fig.3 is shown the structure of intelligent control system for natural ventilation on greenhouse. There are some the windows driven by electrical actuators one by one designed in the tilted roof. The detecting parameters are included outdoor temperature

To, outdoor lighting *L*, outdoor wind *W* and rain *R*,whole which are input into MCU, main control unit. With the indoor detecting factors, and by the program calculating of computer in MCU, the electrical actuator is controlled to work, and ventilation flux of greenhouse is adjusted too.

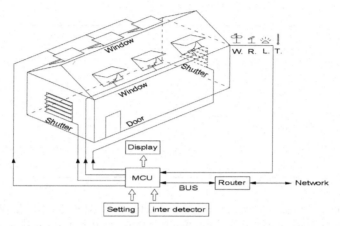

Fig. 3. Control model of nature ventilation used in greenhouse

3 Fuzzy Control Technology for Nature Ventilation

Certainly, the ventilation flux has been related with open angle of windows directly, and the angle has also a direct proportion with the stroke of actuators. For the control need of veracity and speediness, it is valid that fuzzy control technology is used for control window operation (Chen yifei et al.,2009). The fuzzy control system for driving windows in greenhouse is shown as Fig.4.

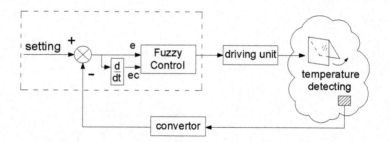

Fig. 4. Fuzzy control system model for natural ventilation

Where, setting value is the temperature that is known and at indoor pressure middle point, the output from driving unit is used to control stroke of actuators for window, such as rod length.

Depending on the control window open area and type of rod actuators used, the total length of rod is thought as six segments to operate. By the testing, the relation of rod work delay time and window open angle is as Table 1.

Table 1. Rod segments setting opreation

segment	Window open angle(°)	Actuator work delay time(s)
1	5	3
2	10	6
3	15	9
4	20	12
5	25	15
6	30	18

Where, the parameters of the electrical rod actuator used is: 24VDC,25W, type S1, by Mingardi srl. Italy, max.stroke 750mm, operation speed 20mm/s with load. The size of roof window is 1200mm*750mm.

All of parameters of E, EC, U in fuzzy system are defined as error of temperature 、 the change ratio of the error、 the run delay time of actuators, and their language values are definite as:{PB, PM, PS, PO, NO, NS, NM, NB}.Therefore, the fuzzy control regular table is built (Chen yifei et al.,2009).

The following Fig.5 and Fig.6 are respective shown that the test situation and the electrical rod actuators are used to drive the windows of greenhouse for natural ventilation.

Fig. 5. Test of driving window with fuzzy control

Fig. 6. The electrical rod actuators used in greenhouse window driving

4 Conclusion

For the need of natural ventilation in greenhouse, exact ventilation control will be regarded. According to the electrical rod actuator using to drive the window of greenhouse, the relation both open angle of window and rod length is established in this paper. Finally, the fuzzy control technology is used in the control system, and establishing the control strategy that the veracity and speediness of ventilation control can be achieved by control window open area with some limited conditions. After testing, it is proved that the ventilation control effect is improved with fuzzy control system being built, and the intelligent control is developed obviously. Moreover, the control model is simply and the intelligent control network is reached too. There are some advantages for application promotion, such as long distance monitoring and controlling for greenhouse etc. Of course, in next step, some points in this control will be researched like the smart control for indoor air flow organizing, and energy-saving potential control as well as the effect testing for different crops.

References

Boulard, T., Papadakis, G., et al.: Air Flow and Associated Sensible Heat Exchanges in a Naturally vVentilated Greenhouse. Agricultural and Forest Meteorology 88, 111–119 (1977)

Roy, J.C., Boulard, T., Kittas, C., et al.: Convective and Ventilation Transfers in Greenhouses, Part 1:the greenhouse considered as a perfectly stirred tank. Biosystems Engineering 83(1), 1–20 (2002)

Boulard, T., Kittas, C., Roy, J.C., et al.: Convective and Ventilation Transfers in Greenhouses Part 2: determination of the distributed greenhouse climate. Biosystems Engineering 83(2), 129–147 (2002)

Fang, X., Linbin, Z., et al.: Modeling and Simulation of Subtropical Greenhouse Microclimate in China. Transactions of the Chinese Society for Agricultural Machinery 36(11), 102–105 (2005) (in Chinese)

Wiyi, L., Zhaoli, l., et al.: Research on Agricultural Greenhouse Micoclimate and Analysis of Theoretical Model. Transactions of the Chinese Society for Agricultural Machinery 36(5), 137–140 (2005) (in Chinese)

Jinan, G., Hanping, M.: A Mathematical Model on Intellingent Control of Greenhouse Environment. Transactions of the Chinese Society for Agricultural Machinery 32(6), 63–65 (2001) (in Chinese)

Stanghellini, C., de Jong, T.: A Model of Humidity and Its Application in a Greenhouse. Agricultural and Forest Meteorology 76, 129–148 (1995)

Yifei, C., Ku, W., Wenguang, D.: Natural Ventilation Control System by Fuzzy Control Technology. Accept by International Conference on Intelligent Networks and Intelligent Systems (ICINIS 2009), Tianjin, China (November 2009)

An Application of RFID in Monitoring Agricultural Material Products

Jianhui Du[1,2], Peipei Li[1], Wanlin Gao[1,3,*], Dezhong Wang[1], Qing Wang[1], and Yilong Zhu[1]

[1] College of Information and Electrical Engineering, China Agricultural University, Beijing, P.R. China 100083
[2] Yunnan Agriculture Department, Yunnan Province, P.R. China 650224
[3] College of Information and Electrical Engineering, China Agricultural University, No. 17, Qinghua Dong Lu, Haidian, Beijing 100083, P.R. China,
Tel.: +86-010-62736755; Fax: +86-010-62736755
gaowlin@cau.edu.cn

Abstract. With the development of modern agriculture, more and more agricultural material products are used in it. While how to keep these things safe is a big problem at present, which needs to be paid more attention. This article develops an agricultural material products monitor system based on RFID which gives alarm as soon as possible if there is anything unmoral. Every warehouse exit is equipped with a RFID reader, while each agricultural material product has a tag on them. When passing though, the reader identifies the tag's information and transfer it to the PC, The PC inquiries the database storing all tags' information, and tells which one is not taken out legally by alarming aloud.

Keywords: agricultural material products, RFID, monitor, reader, tag, alarm.

1 Introduction

Nowadays with socio-economic development, based on abundant agricultural resources, the agriculture of our country has been developing a lot, more and more agricultural material products are used in it, such as fertilizers, pesticides and so on. However this has brought much pressure to daily management and many problems occur at the same time. Products theft is one of these problems, which does cause properties loss and hurt the owners in psychology. It is necessary for us to find a way to monitor agricultural material products from being taken out of the warehouse by thefts.

Usually we take products supervision with cameras, which can't tell whether the things are taken out of the warehouse legally or not in real time. When the truth is found, they are already in miles away. Or the thieves may hide the things when pass by the camera to avoid exposure. This paper aims to develop a real time monitoring system based on RFID (radio frequency identification) which is widely used in Warehouse custody, access control, highway inspection, rail inspection, retail stores, libraries and many other areas.

---

* Corresponding author.

D. Li and C. Zhao (Eds.): CCTA 2009, IFIP AICT 317, pp. 254–259, 2010.

2   RFID System

RFID (Radio Frequency Identification) is a non-touch automatic identification technology rising in 1990s. In the early information period, manual data-input and identification brought a large number of problems such as labor-intensive, high bit error rate and lack of real-time, etc. In order to solve these problems, people developed many kinds of identification technologies such as bar code, magnetic card, IC card, biometrics, voice recognition and wireless radio frequency identification, which are very common in our present life. RFID is one of the most promising means which makes use of electromagnetic or inductive coupling between the reader and the tag adhesive to objects for data communications, automatic target recognition and retrieve relevant data without human intervention(Dingyi Dai et al., 2006; Automatic Identification Manufacture Association of China., 2003).

A complete RFID system consists of three parts: the transponder (tag), reader, as well as back-end database. The radio frequency identification includes tag and reader (Zhanqing You et al., 2005).

Tag: consists of coupling components and chips. Each electronic tag is attached to the target object and has a global unique identifier (ID) preserving electronic data in prescribe format, which can not be changed or fabricate. In practical applications, tags are attached to the surface of the objects to be identified. According to the power supply, tags can be classified into active tags and passive ones. As the name suggest, active tags means the tag has power (i.e. batteries) inside, while passive tags do not have.

Reader: an equipment for reading (or writing) RFID information, which can be hand-held or fixed. Reader can identify the preserved electronic data given by RFID tags so as to achieve the purpose of automatic identification of objects. Reader is usually connected to the computer and sends the recognized information to the computer for the next process.

Electronic tags and readers complete spatial couple of RF signals (non-touch) through the coupling components (antenna) for successful energy transfer and data exchange on the coupling channel according to temporal relations. There are two types of radio frequency signals between reader and tag: inductive coupling and electromagnetic backscatter coupling. The former achieves coupling through the spatial high-frequency alternating magnetic field, based on the law of electromagnetic induction; the latter model is based on the principle of radar: when electromagnetic wave is launched out, touched the target and reflected, it will bring back the target information.

Inductive coupling manner is suitable for the intermediate frequency and low-frequency radio frequency identification system at close range. Typical operating frequency include: 125kHz, 225kHz and 13.56MHz. Distance identified is less than 1m, typically 10cm ~ 20cm. 13.56MHz frequency band is widely used in public traffic card.

Electromagnetic backscatter coupling is generally suitable for high frequency, ultra-high frequency, as well as microwave radio frequency identification systems for long distance. Ultra-high frequency (UHF) band radio frequency and microwave tags are referred as microwave radio frequency tags with the typical operating frequency is: 433.92MHz, 862 (902) ~ 928MHz, 2.45GHz, 5.8GHz. The RF tag reader antenna is located within the far zone radiation field of the reader antenna. Coupling between tags and readers is electromagnetic. Reader antenna radiation field provides energy for passive radio frequency tags and wakes up the active tags. The corresponding distance is generally greater than 1m, typically 4 ~ 6m, up to longer than 10m (Xue-zong Tao et al., 2006; Xiaoguang Zhou et al., 2006).

Based on current technological level, the relatively successful passive microwave RF products work mainly on 902 ~ 928MHz frequency bands. 2.45GHz and 5.8GHz radio frequency identification system are more available with the semi-passive microwave RFID products. Semi-passive tags are generally button battery powered, with a relatively long reading distance. Typical characteristics of a microwave RF tag are mainly concentrated on passive or active, wireless reading-writing distance, whether to support the multi-tag reading and writing, whether to suitable for the high-speed applications identification, the reader's transmission power tolerance, prices of tags and reader, etc. Typical applications of microwave radio frequency tag include: mobile vehicle identification, electronic ID cards, storage and logistics applications, electronic anti-theft locking (electronic controller for remote control door) and so on.

In general, RFID works as follows: tags enter the magnetic field; receive signals emitted by the reader; send the product information stored in the chip through the induced currents obtained by the energy (Passive Tag), or send a frequency signal actively (Active Tag); readers read the information and decoded; send the information to the central system to be processed. The basic model of Radio frequency identification system is shown as in Fig.1.

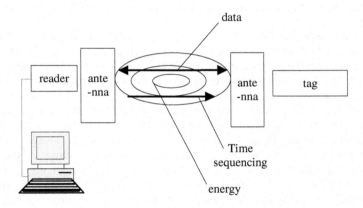

Fig. 1. The basic model of radio frequency identification system

3 Warehouse Agricultural Material Products Monitoring System

In the warehouse products monitoring system, the owner and their agricultural material products are equipped with a unique ID card each. The radio frequency readers which are connected with the PC are equipped at each warehouse exit. The PC has a backstage database for the monitoring system keeping tag information of all the owners and products. If the valuable is taken out by someone who is not the owner, the system will alarm, thus rises the owners' attention and achieves timely inspection.

The working process of system is explained as follows: When someone leaves the warehouse, the reader reads the cards' information. If there are label information of human and products identified simultaneously, check   if the human information is legal(correspond to the database information) and gives warning or not; If only products information, but nobody's information is identified, that means a non-owner carries over the products, the system alarms.

Supposes all the owners' card number is A1/A2/A3…, the products cards are numbered as B1, B2… respectively. If the system identifies A1/A2… and B1/B2… (one or several) at the same time (possible have difference of some amount of time, like several seconds), it does not alarm; If A1/A2…but not B1/B2/… (one or several) is identified, it gives no alarm; If B1/B2/… but without A1/A2… (one or several), it alarms. Fig.2. shows the hardware configuration of monitoring system.

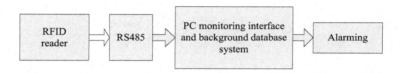

Fig. 2. Hardware configuration of monitoring system

This system is based on the prerequisite of that all owners and their products have been equipped with a tag and their information are stored in the database, Fig.3. shows the flow chart of this system.

Multi-tag reading technology involved in this system is a key point. It is different from the access control system or traffic card system we usually see, which can only identify one card each time. This monitoring system involves issues of legality of people and products, and when products are brought out and people walk out, their information must match. The system can identify a number of tags at same time or within a short period of time (seconds). Moreover, system should be passive to reduce energy loss and save the cost of tags.

According to the introduction to the RFID systems, as well as the actual needs, the system uses 920-925MHz ultra-high frequency section for the identification system and passive UHF RFID tags with corresponding standard as the system tags. The system uses U-CODE GEN2 card, which operates on 860-960MHz, following ISO 18000-6C standards and EPC C1 Gen2 standard, with 512 bit memory, reading-writing distance of up to 3-10m (with the distribution of the antenna). It is a reading-writing card and can work at the speed of up to 60km / h, suitable for the rapid identification. It has anti-collision mechanism, suitable for multi-tag reading.

S1871 is a short-range multi-purpose mini-UHF reader module. It is based on the Intel R1000 micro UHF reader module and can be embedded in fixed or mobile devices. Power range is in line with the mainstream standards such as European (ETSI EN 302 208) and U.S. (FCC part 15). It's adjustable with maximum power of 1300mW, supporting EPC Class1 Gen2 protocol and using dipole antenna(RFID World Forum, 2007).

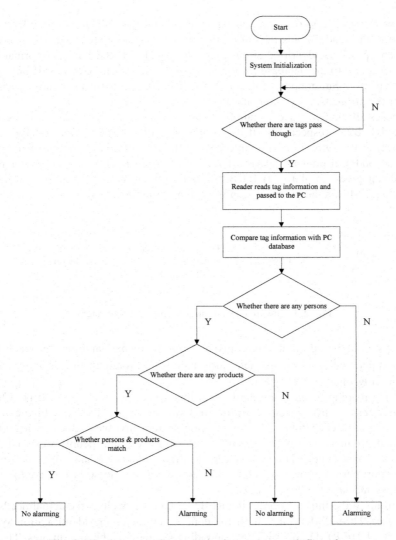

Fig. 3. Warehouse products monitoring system work flow chart

The system uses 920-925MHz UHF frequency band, following the latest ISO / IEC, ISO 18000-6C and EPC global organization, EPC C1 Gen2. The usage of these standards makes preparation for the idea of the Global objects of great networking in future. Moreover, since it is passive system, and can identify objects within a long distance, it is widely used in many areas, such as high-speed train system which are familiar to us(Embedded online., 2007).

Communications between PC and reader are achieved through the serial port, generally RS232 and RS485(Chaoqing Li et al., 2002). The system adopts RS485 for it working relatively within a long distance. PC software is developed with Visual Studio2005, using SerialPortComm to achieve serial communication. Access database in PC has all information of owners and their products. The program can receive and store tags' information at real-time, query database and display the corresponding information through graphic interface. (Weichen Lv et al., 2006; Qiming studio., 2004).

When tag information is identified, the system will test if it is owners' or products' or both. Then:

Only owners', no alarming;
Only products', alarming;
If information of owners and products are both identified, the system queries the owners' tag information in the database to tell who is passing; then queries the products' tag information in to know whose products they are, According to the result , the system alarms or not.

4 Conclusion

1. The valuables monitoring system based on RFID technology, using UHF frequency band, supports accuracy identification and multi-passive-tags to meet the monitoring requirements completely, and therefore achieves the real-time and accuracy.

2. We have applied this system to monitor the warehouse agricultural material products in the real life, the correct recognition rate could be up to 99.5%, which meet our demand primely.

3. The system follows the most advanced international standards and can be widely applied in future.

References

Automatic Identification Manufacture Association of China. Barcodes and RFID application note. Mechanical Industry Press, Beijing (2003)

Li, C.: PC machine and single-chip data communications technology. Beijing University of Aeronautics and Astronautics Press, Beijing (2002)

Dai, D.: RFID technology in China Development. RFID technology and application (2006)

Embedded online. Abroad RFID: expand the functions of the application to broaden the growing popularity [R/OL] (2007), http://www.mcuol.com/News/147/8105.htm

Qiming studio. Visual C++ + SQL Server database application system development and examples. Posts & Telecom Press, Beijing (2004)

Lv, W., Huo, Y., Lv, B., et al.: Primary and increased Visual C #. Tsinghua University Press, Beijing (2006)

Zhou, X.: Radio Frequency Identification (RFID) technology and application. Posts & Telecom Press, Beijing (2006)

Tao, X.: China's RFID technology analysis. RFID technology and application (2006)

RFID World Forum. Sense UHF RFID reader module for micro embedded [R/OL] (2007), http://www.rfidworld.com.cn/bbs/disptopic.asp?boardid=56&topicid=8497

You, Z., Li, S.: Radio Frequency Identification Technology Theory and Application. Publishing House of Electronics Industry, Beijing (2005)

The Development Model Electronic Commerce of Regional Agriculture

Jun Kang[*], Lecai Cai, and Hongchan Li

Institute of Computer Application, Sichuan University of Science and Engineering,
Zigong 643000, Sichuan Province, P.R. China,
Tel.: +86-135990066775; Fax: +86-813-5505966
kj_sky@126.com

Abstract. With the developing of the agricultural information, it is inevitable trend of the development of agricultural electronic commercial affairs. On the basis of existing study on the development application model of e-commerce, combined with the character of the agricultural information, compared with the developing model from the theory and reality, a new development model electronic commerce of regional agriculture base on the government is put up, and such key issues as problems of the security applications, payment mode, sharing mechanisms, and legal protection are analyzed, etc. The among coordination mechanism of the region is discussed on, it is significance for regulating the development of agricultural e-commerce and promoting the regional economical development.

Keywords: regional, agricultural information, agricultural e-commerce, development model.

1 Introduction

At present, the majority of the group focuses on higher education of the population from the Internet knowledge and use in China's, although it has distributed in agricultural industry, but overall, the proportion is very low. However, the really agricultural users are limited by objective conditions, the real demand of users have been playing a discount. In recent years, the developments of the e-commerce in agriculture have gradually become the transformation of agricultural marketization, and realize the power of the agricultural modernization.

A new development model electronic commerce of regional agriculture base on the government is advanced; it is significance for regulating the development of agricultural e-commerce and promoting the regional economical development.

2 The Present Situation of the Development of Agricultural e-Commerce

In recent years, the Chinese government attaches great importance to the development of e-commerce in the country, On issuing the Summaries of Middle/Long Term

---

[*] Corresponding author.

D. Li and C. Zhao (Eds.): CCTA 2009, IFIP AICT 317, pp. 260–267, 2010.

Science and Technology Development Plans of China, and the Development of Information Industry Plans in the Eleventh Five-year and Long Plan in 2020 Years in The Ministry of Information Industry, "the application of e-commerce platform technology" and "the agricultural informatization technology" have been listed as the key points.

China is a large agricultural nation, agricultural informatization has started. At present, the national rural e-commerce sites have more than 2000, agriculture website have more than 10,000, a lot of vegetables and fruits, sapling, livestock, cultivation of supply and demand information and related agricultural economy, the investment information are issued by these agriculture websites, play an important role in enlivening the circulation, realizing the agricultural efficiency and increasing farmers' income. There are the influence of Chinese agricultural science and technology information network, the Chinese agricultural information network, the Chinese seed group company, the seed information center, gold dragon network etc [1] .This shows that the Chinese agricultural electronic commerce is moving into a fast development period.

But, according to statistics, the Internet users of engaging in agriculture in proportion in less than 1% in the nearly 78 million Internet users, and most of users are the agricultural management and technical personnel. And they are highly focused on the economy developed cities like Beijing, Shanghai, Guangdong, Zhejiang, Jiangsu and other regions, the real Internet farmers are almost negligible.

3   The Factor Analysis of Restrict Regional Agricultural e-Commerce Development

To sum up, there are some factors about restricting agricultural development of e-commerce as following [2-3]:

(1) The agricultural informatization level is low. There are lots of phenomenon in China, for example, the agricultural company informatization level is lower, the information of the network is not popular; the number of the agricultural e-commerce websites are limited, and online trading function is not complete; Internal management informatization is serious insufficient.

(2) Enterprise electronic commerce ideas are lagging. Most of our agricultural enterprises of the network to be understood, acceptant and application of ability are not high by itself; the concept of the utilizable network business are unclear, modern management ideas, method and technology of consciousness are lack, Electronic commerce constructions have large investment, long period, the large amount of maintenance.

(3) Consciousness of e-commerce is decreasing. Some enterprises in the "blind leading" process of implementing e-commerce pursue advanced hardware equipments and mature technology, and neglect integration and optimization of the enterprise internal and external information resources, and hard to obtain high quality, high efficiency and benefit of the business enterprise.

(4) e-commerce environment is imperfect. At present, China's e-commerce involved banking, information industry, taxation, customs, finance, law and other related

standards, norms are still not perfect, agricultural market standardization and organization are low, the trust of e-commerce have an effect on farmers friends.
(5) Agricultural products are particular. Many products have seasonal, uneasy storage,
   so it is difficult to process in the late of preservation and transportation, it is
   more difficult to make its logistics link relative to the industrial products, and
   many people have more difficult for agriculture in developing electronic commerce scruples.
(6) The availability of network resource is bad. Due to scarce online resources, high
   online fair, high rent charge on private data lines, it is not high in expensive period
   of e-commerce high operating costs.
(7) Personnel qualities and skills. At present, overall culture quality and application of
   information technology skills of our country agricultural workers are lower; the
   development of agricultural electronic commerce is limited.

4   The Development Model Electronic Commerce of Regional Agriculture

(1) Establish (G2G) 2B mode
This model is the innovation mode of the agricultural industry of e-commerce,
which is based on G2G and emphasized on G2B. Cohesion in the two flows forms a
new concept of agricultural e-commerce mode. Provincial agricultural electronic
commerce center is external agricultural information gateway of unity which agricultural management departments are leading, including agricultural information of the
area around the city state. Agriculture information resources of agricultural industry
uploading of this area are integrated by the background platform of every city and
state agricultural electronic commerce center, be the public platform of the regional
agricultural information. The information can be shared and unified integrated between platform, and the agricultural service information which is comprehensive,
detailed and accurate can be provided to the agricultural e-commerce center, consumers can search information by the unified platform. These regional platforms of the
government have strong operability, enhance participation of the agricultural business
enterprise, the agricultural business enterprise of agriculture's cognition and trust are
improved.
(2) Establish (B2B) 2C mode
This mode bases on the two operation mode of traditional B2B and C2C
e-commerce, the regional agriculture enterprises are combined the front and background closely by through the information integration and consumer. The information of agricultural enterprises can be shared between regional areas, the system to
solve the enterprise internal and external E of disconnection issues is provided
capital and information flow mutual coordination and two-way communication, information query across the region becomes more convenient. In this mode, consumer
could realized by the unified platform, namely could query information of
omni-directional comprehensive agricultural and self-help customized service of
agricultural projects.
(3) Establish regional agriculture information database of e-commerce invested
based on government

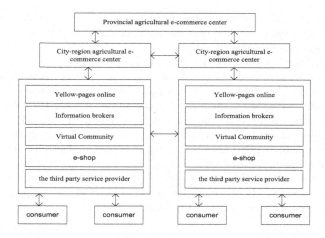

Fig. 1. The development model e-commerce systematic structure of regional agriculture

In the new mode, the government as the macroeconomic manager of agriculture e-commerce has the effect of norm and support. The investment and maintenance costs of agricultural electronic commerce are very large; it is difficult to construct and maintain by the enterprise alone. The region unified enterprise information database is established by the government, many costs of maintenance, personnel training, functional development, etc, are be reduced when the government build their own information platform. Internal management information network of agricultural enterprises are built, information mode transfer to horizontal, agricultural research, forecasting, planning and management of digital, procedure and standardization are realized by inside information management platform. The advantage of regional resource sharing makes database information, accuracy and timeliness strong, the platform has the strict standard model of information, information quality, strict management mechanism and standardization of the management for funds and service platform, there are positive function for after-sales service and guarantee.

(4) Establish many technical integrated application modes

(a) ASP mode: ASP by hiring equipment, software system and communications lines from server vendors, network communication equipment suppliers, operating system developer, database system developers, network security system developers, network management system developers and lines of communication, to construct the application system operation platform[4-5].

(b) Safe management mode: It strengthens the safety of network through the establishment of personnel management, system maintenance and data backup, emergency measures and virus prevention system, etc.

(c) Resource sharing mode: Unstructured information the information may lead to split information, disagree information. Based on the new generation of information cooperation, establish regional platform realizing information resources sharing, and improve the system of regional platform of internal and external information sharing.

(d) Mobile information mode: Developing of 3G and credit card, maturing of internet technology, and state policies, making the use of mobile business trend more widely. Using standardized data as sharing platform, basing on modern information, using modern information technology makes agriculture E-commerce, as the most convenient way to use by public [6-7].

(5) Establish comprehensive service function mode

In the building of new mode, information resources regional sharing is realized, the most personal service to consumers is provided. For instance, professional consulting could be provided by network guide, and according to the needs of different users, comprehensive information and combination are dynamically provided, easy operation interface allows users to design agricultural needs database by them, the information required standard is provided, According to these standards the information for users can be the most qualified provided by the database [8-10]. Various forms of services form as internet, telephone, mobile digital terminals, etc. make users enjoy comprehensive services. These overall images of agriculture can be promoted vividly by the openly agricultural information distribution system.

5   Study of the Regional Agricultural Electronic Commerce Development Mode

Through a new development model electronic commerce of regional agriculture base on the government as above, and analyzed such key issues as problems of the security applications, payment mode, sharing mechanisms, and legal protection [11-13], etc. The discussion on the among coordination mechanism of the region, it is significance for regulating the development of agricultural e-commerce and promoting the regional economical development.

(1) Security mechanism

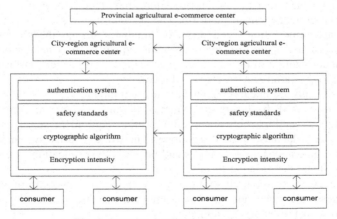

Fig. 2. Security Application mechanism

Agricultural e-commerce security mechanism, such as authentication system, safety standard, encryption algorithm and encryption intensity, they are mainly applied in trading, the share of information, network, etc, G2C G2B G2G in, and B2C and B2B, C2C mode application security mechanisms have been very good.

(2) Payment mechanism

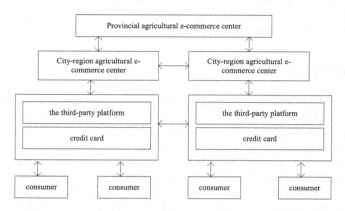

Fig. 3. Payment mechanism

In numerous agreements and method, the credit card is the most convenient and fast, it provides accurate data encryption, electronic signature and identity authentication, it has been mainly agricultural electronic commerce in developed countries. But in China, agricultural electronic commerce is mainly to provide the third's platform for the trade, such as the trades between regional and regional, between customers and consumers, between enterprise and enterprise, between consumers and business, can be accomplished by using the third's platform.

(3) Shared mechanism

Fig. 4. Shared mechanism

The sharing mechanisms of establishment need of enterprise and consumer, which can analyze, investigate, forecast and plan. The data obtained are datamation, programmed and standardization, then this information are stored to database, some enterprises or governments can share them, but some between the enterprise and government which are limited can not share with resources.

(4) Legal assurance mechanism

Legal guarantee mechanism suitable G2C, G2G, G2B, C2C, B2C and B2B model, at the framework of the legal system, we should pay attention to coordination between the construction of legal system and special form. In the value orientation of the legal system, the related legal should be paid to consumers, especially the individual consumer's protection. In the specific contents of the legal system, they need to set up necessary access conditions, the rights between the bank and the customer, Signature and certification, Trading evidence, the responsibility when accidents caused loss, the problem between International law appliance and the jurisdictional.

6 Conclusion

A new development model electronic commerce of regional agriculture base on the government of (G2G) 2B is constructed; the environment of e-commerce is regulated. At the same time, the (B2B) 2C resource sharing public support platform of the participation and cooperation is built, the guide and resource integration of enterprise website, e-commerce site agriculture business is strengthened, the overall level of the agricultural informatization is improved. In this, the regional government of agricultural information database is built, and some related mechanisms of managements or other aspects are also built, standardize trade standards. The integrated applications mode of various techniques are established, it provides the Omni-directional support for the development of agriculture. At last, the integrated service mode is established, humanization, intelligent, personalized service is provided, self-help customized service become the possibility.

References

[1] Lee, X.: Agricultural E-commerce Spring. Farmers' Daily (September 28, 2006)
[2] Zhao, Y., Lee, L., Lv, J.: Analysis of Agricultural E-commerce Development. Countryside of economy and technology (5), 24–25 (2006)
[3] Zou, J.: The Analysis of Agricultural the Third-Party E-commerce Mode in China. Trade Time (9), 75–79 (2007)
[4] Christiaanse, E., Huigen, J.: Institutional Dimensions in Information Technology Implementation in Complex Network System. European Journal of Information System 6, 77–85 (1997)
[5] Henderson, J., Dooley, F., Akridge, J.: Internet and E-Commerce Adoption by Agricultural Input firms. Review of Agricultural Economics 26(4), 505–520 (2004)
[6] Kuan, K.K.Y., Chau, P.Y.K.: A Perception-Based Model for EDI Adoption in Small Business Using a Technology Organization Environment Framework. Information& Management 38, 507–521 (2001)
[7] Kraemer, K.L., Dedrick, J.: Strategic Use of Internet and E-commerce: Cisco System. Journal of Strategic Information System 11, 5–29 (2002)
[8] Walsha, S., Linton, J.D.: The measurement of technical competencies. Journal of High Technology Management Research 13, 63–86 (2002)
[9] Christophe, R.D.: Building core competencies in crisis management through organizational learning.Technological Forecasting and Social Change 19(2), 113–127 (1999)
[10] Barney, J.B.: Firm resource and sustained competitive advantage. Journal of Management 17(1), 99–120 (1991)

[11] DeCarolis, D.M.: Competencies and Imitability in the Pharmaceutical Industry: An Analysis of their Relationship with Firm Performance. Journal of Management 29(1), 27–50 (2003)

[12] Lado, A.A., Wilson, M.C.: Human resource systems and sustained competitive advantage: a competency-based perspective. The Academy of Management Review 19(4), 699–727 (1994)

[13] Johannessen, J.-A., Olsen, B.: Knowledge management and sustainable competitive advantages: The impact of dynamic contextual training. International Journal of Informational Management 23(23), 277–289 (2003)

Food Traceability System Tending to Maturation in China

Fengyun Wang[1,*], Jianhua Zhu[1], Minghua Shang[1], Yimin Zhao[2], and Shuyun Liu[1]

[1] S&T Information Engineering Research Center,
Shandong Academy of Agricultural Sciences, Jinan 250100,
Shandong Province, P.R. China,
Tel.: +86-531-83179076; Fax: +86-531-83179821
wfylily@163.com
[2] Quality Assure Department, Shandong Shengli CO., Ltd., Jinan 250101 China

Abstract. The study on food safety traceability system in China started from 2002. During the study and implementation, some related standards and guides have been gradually made, many traceable foods and traceable enterprises were created and a series of traceability subsystems were developed. This paper introduces the importance of food safety traceability system construction, the present construction situation in China and the food safety traceability platform by bar code, deeply analyzes the existing problems and puts forward some advices. It shows that a sound food traceability system has been built in China and tends to maturate. It provides the reference for food safety traceability system construction in developing countries.

Keywords: Food safety, Traceability, China.

1 Introduction

In recent years, there have been many problems in food safety from mad cow disease and foot-and-mouth disease (FMD) in foreign country to water injected meat, inferior milk powder and Sudan red event in China which has been attracted the attention of the world (Li, G. et at., 2007). Food safety has been the concerned issues by the consumers and the businessmen together and become the important factor that affected the international competitiveness of Chinese agriculture and food industry.

As the biggest developing country and WTO membership country, China actively coped with various food problems, carried out the preliminary study on food safety traceability, made some related standards and guides, preliminarily created some food traceability institutions and issued some regulations in some local governments and enterprises. Article Numbering of China (ANCC) cooperated with China National Food Industry Association (CNFIA) to build the food safety traceability platform by bar code, establish a great deal of traceable foods and enterprises and develop a series of traceability subsystem (Chen, H. et at., 2007).

---

* Corresponding author.

D. Li and C. Zhao (Eds.): CCTA 2009, IFIP AICT 317, pp. 268–274, 2010.

Food safety traceability system (FSTS) is referred to an information management system that can connect the production, inspection, supervision and consumption etc processes to let the consumers know about the sanitary and safe production and circulation process and improve the safety trust of consumers on the food (MchKean, J.D. 2001). FSTS provides the traceability mode "from farm to table", selects some common traceable factors concerned by the consumers from production, processing, circulation and consumption etc supply chains, creates food safety information database, once there is food safety problem, it can effectively control and call back the food according to the traceability so as to ensure the legal rights and interests of consumers from the food source (Huang, W. et at., 2006). Fig.1 is the diagram of FSTS

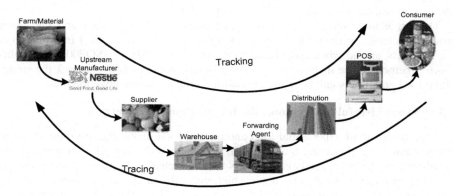

Fig. 1. Food Safety Traceability System

2   Importance

2.1   FSTS Is Necessary for Ensuring the Food Safety

With the improvement of living standard of consumers and the arising of food safety problems, food safety is increasingly concerned by every country in the world.

From domestic environment, facing the frequently happened food safety accidents, the consumers hope to know about the whole process of food production and circulation. In order to meet the consumers' requirements, enhance the consumers' trust on the food, food enterprises and food industries, build a well-off society in all around way, improve the living quality of the people and reduce the economic and social efficiency loss caused by food safety accidents, it is necessary to build the FSTS.

From international environment, European Union (EU) and America etc developed countries and areas require that some imported foods must have the traceability requirement. The EU administration statute No. 178/2002 requires that all meat sold within EU areas must be tracking and traceable since Jan., 1st, 2005, or else can't be sold in the market (Guan, E. et at., 2006). Japan decided to build good agricultural products certification system before 2005 to certificate the identity of agricultural products into the Japanese market (Ozawa Y et al., 2001). The FSTS built by the developed country plays increasingly obvious trade barrier role, except that it can effectively ensure the food safety and food traceability.

Therefore the FSTS not only provides the safe food with high quality for the people, but also breaks the trade barrier caused by the food safety traceability which plays important role for improving the international competitiveness of Chinese agricultural products (Qiao, J. 2007).

2.2 FSTS Is an Effective Means to Control the Food Quality

FSTS has been one of the key factors to study and make the food safety policy. ISO 9000, GMP, SSOP and HACCP etc many effective management measures have been introduced to control the food safety and achieved a certain effect in the practice (Lu, C. et at., 2006). But the above mentioned measures are mainly for processing chain and lack the means to connect the whole supply chain.

The traceability system emphasizes the unique mark of product and the whole process tracking, for traceable food, it can track and trace the product information of each process on the whole supply chain by HACCP, GMP or ISO9001 etc quality control methods, once there is food safety problem, it can effectively track the source of the food and timely call back the unqualified product to minimize the loss.

2.3 FSTS Is Helpful to Overcome the Information Asymmetry

The characteristic of experiential food and trustful food makes the information asymmetry among the producer, businessman, consumer and even the government, the details are: (1) the food safety character is the inner character and difficult to identify from the appearance for the consumer which make the safety information asymmetry among the producer, businessman and consumer. (2) Though the government etc supervisors can measure the food safety level, under small-scale peasant economic regime, the food has distributed production site, wide area and large quantity and lacks the mark, the food safety responsibility traceability is bad and the measure cost is high and speed is low which greatly improve the supervising cost. In addition, for a long time, China partly pursues the amount and commercial character of food and lacks the sound and effective food safety supervising regulation and means which lead to the safety information asymmetry among the governmental supervisor, producer and businessman. (3) The food safety information from government etc supervisors can't be quickly and effectively sent to the consumers which lead to the safety information asymmetry between the government and consumer so as that the consumer lacks the information for selection. (4) The food safety information between the top supervisor and the bottom supervisor is different and can't be completely shared and communicated which lead to the safety information asymmetry between the top supervisor and bottom supervisor (Qiao, J. 2007).

3   Present Construction Situation of FSTS in China

3.1   Established Some Related Regulation and Standard

The study on food traceability system in China started from 2002 and gradually made some related standards and guides. For example, in order to respond to implement the aquatic products trade traceability from 2005 in EU, the General Administration of Quality Supervision, Inspection and Quarantine of the People's Republic China drew up «Traceability regulations for emigrated aquatic products (try out) », ANCC made «Traceability guide for beef products» on the base of the experience of EU. Shaanxi Standardization Academe made «Applicable scheme of quality tracking and traceability system for beef products» (Chen, H. et al., 2007).

Some local governments and enterprises built the traceability regime of some food and issued some rules. In Jul., 2001, Shanghai city government issued «Temporary methods of safety supervision for edible agricultural products in Shanghai» and put forward that created "market archives traceability system" on circulation chain. In 2002, Commerce Committee of Beijing City established the food information traceability system to clearly require that the food businessman should have subsidiary ledger for purchased and sold foods i.e. the purchased food should be created the archives of origin, supplier, purchased date and batch. On Sept., 20th, 2005, Shunyi District firstly launched the graded package and quality traceability regime for vegetable in Beijing City. In order to ensure the citizen purchase the reliable non-pollution vegetable, Tianjin City carried out non-pollution vegetable traceability regime and launched non-pollution vegetable order service on the internet.

3.2 Built the Food Safety Traceability Platform by Bar Code

ANCC cooperated with CNFIA to build the food safety traceability platform by bar code, through the platform, established a great deal of traceable foods and enterprises and develop a series of traceability subsystem. The platform framework is shown in Fig.2 (Yang X. et al., 2006).

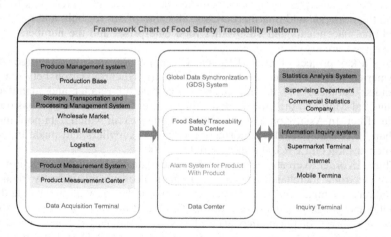

Fig. 2. Framework Chart of Food Safety Traceability Platform

3.2.1 Established a Great Deal of Traceable Foods

As of present time (Feb., 26th, 2008), through the platform, there have established 10362 meet and poultry products, 12323 fruits, vegetables, nuts and seeds, 2686 sea products, 8433 dairy products and eggs, 5028 edible oil and grease, 7322 chocolate, sugar, sweet products and candies, 20427 flavorings and preservatives, 12150 bread and baking foods, 27769 prepare instruments and cans, 40028 beverages, 349 tobacco and its products, 6185 grain and legume products, 4348 others and the numbers of traceable foods are increasing.

The basic information of these traceable foods includes: global trade item number (GTIN), main picture, global location number (GLN), Chinese name of product, English name of product, Chinese name of trademark, English name of trademark, specification, classification, aimed market, package mode code, height, width, depth, shelf life, origin, marketing time, keyword, short description of product, package material etc, in addition, also includes additional information, extended information and multimedia information. The manufacturer information of the product can be traced by the system, including company introduction, main product or service, main business place, management system certification, business trademark, enterprise name (Chinese), enterprise name (English), registered address (Chinese), registered address (English), registered postcode, business place (Chinese), business place (English), business postcode, contact, telephone, fax, email and enterprise website etc.

3.2.2 Established a Great Deal of Traceable Enterprises

As of present time (Feb., 26th 2008), based on the global data synchronization (GDS), there have established 250 enterprises for planting and cultivation, 6194 enterprises for manufacturing the agricultural and sideline products, 2984 enterprises for manufacturing the baking food, 2075 enterprises for manufacturing the prepare food and can, 519 enterprises for manufacturing the dairy product, 1519 enterprises for manufacturing the health food, 179 enterprises for manufacturing the additive, 2643 enterprises for manufacturing the alcoholic liquor, 2169 enterprises for manufacturing the beverage, 32 enterprises for manufacturing the tobacco, 937 enterprises for manufacturing the pesticide and pharmaceuticals, 883 enterprises for bulk and retail selling the food pharmaceuticals and 1835 other food enterprises and the numbers of traceable enterprises are also increasing.

GDS is a key project of ECR (Efficient Consumer Response) committee. At present, America and European are implementing the synchronization between GDS-merchandise database and the merchandise database of retailer and manufacturer, including Wal-Mart and the large scaled manufacturer in America, and it has obtained a certain effect. In Asia, Korea also established this system which data pool included more than 10,000 merchandise information over 200 suppliers and made the speech and demonstration on ECR Expo in Asia in Oct., 2003.

In China, in order to create the content database (data pool) and data synchronization system of electronic merchandises in quick circulation consumable industry (retailer and manufacturer) in China, ANCC is actively carrying out the study of global data dictionary (GDD) and global product classification (GPC) and internationally has signed with UCC to take charge of the maintenance of UNSPSC standard (Chinese version) and signed with EAN to take charge of the Chinesization of GDD and the registration, management, maintenance and standard establishment of domestic EPC system.

3.2.3 Developed a Series of Traceability Subsystem

China has made preliminary experiments on food traceability system and developed a series of traceability subsystem. In 2004, the vegetable quality safety traceability system is tried out on Shouguang Tianyuan Vegetable Base and Luocheng Vegetable Base under the cooperation of General Administration of Quality Supervision, Inspection and Quarantine of the People's Republic China, Weifang City and Shouguang

City Bureau of Quality and Technical Supervision in Shandong Province together. ANCC promotes the application of bar code technology in food traceability in China through the China bar code promotion project and successively carried out the food traceability technology study and try-out in Shaanxi, Beijing, Shanghai and Shandong etc places, e.g. "Agricultural products inquiring system In Shanghai City", application and demonstration system of tracking and traceability automatic identification technology for beef product in Beijing Jinweifuren Hala Food Co., Ltd and fruit traceability information system in Jiangxi, Olympic FSTS etc.

In July, 2005, Beijing Municipal Government began to launch the action plan for the Olympic green food engineering program and supervised, aiming for every section involving with production, processing, transportation, storing, packaging, testing and hygiene. The solution of Olympic FSTS mentioned by Aerospace Golden Card Company has been adopted. Olympic FSTS will be used in "Lucky Beijing" sports competition. It includes fruit & vegetable, animal, pre-packaged food and Olympic foods four subsystems, covering the main food varieties and planting, cultivation, production and manufacturing and logistics etc processes of Olympic foods. The related material producer and supplier of food and beverage and food and beverage service unit will be recorded in the system. No matter what process, from the farm and athlete table, and no matter what food have problem, from aquatic products, livestock, poultry to fruit and vegetable products, it can be found out.

4 Main Problems

On Apr., 13th, 2007, China Chain-Store & Franchise Association (CCFA) released «Study report about food safety traceability» in Beijing. It indicated that China faced six great obstacles to establish FSTS as following:

(1) In China, the production of food, especially agricultural product, is quite distributed, the intensive degree of production is not high, the technology and standardization level is low; (2) The circulation mode of food is backward, the traditional circulation channels, e.g. wholesale market and bazaar, occupy great proportion, the modern circulation channel, e.g. chain-supermarket, isn't popular; (3) Food safety law system and standard system aren't sound, related regulations and standards lack and lag behind the actual development and there are many cases that can't meet the international standard; (4) The systematicness and unity of food safety supervising regime isn't still enough; (5) The whole social cognition about FSTS isn't sufficient; (6) The cost for creating the traceability system is higher and the enterprise lacks the prior devoted drive (http://www.tech-food.com/news/2007-4-16/n0105260.htm).

5 Advice

According to the construction situation in China and the main problems, the FSTS construction should be perfected and improved from the following aspects:

(1) The laws, regulations and policies related to the food safety should be perfected to provide the system guarantee for implementing the food safety traceability; (2) The supervising regime of food safety traceability should be harmonious and a relatively

centralized and uniform professional supervising regime should be established to provide the regime guarantee for implementing the traceability; (3) The related sound standard system should be constructed to provide the technology basis for implementing the traceability; (4) Promote the construction of agricultural products etc food production base, advance the product quality level and optimized the food supply chain; (5) The role of modern circulation modes represented by the chain supermarket in implementing the traceability should be enough exerted; (6) Use the chance of food safety cared by all societies to stress the propaganda of related traceability knowledge and create the social foundation of implementing the traceability; (7) Stress the traceability technology study and provide more convenient and cheaper technology; (8) Construct demonstration projects of food safety traceability to provide the reference and promote the application (Zou, Y. et at., 2005).

References

[1] Li, G., Huang, L., Zhan, J., et al.: RFID Application in Food Safety Traceability. Logistics & Material Handing 12(3), 85–87 (2007)
[2] Chen, H., Tian, Z.: Comparative Study of Traceability System on Domestic and Foreign Agricultural Products. Market Modernization (7X), 5–6 (2007)
[3] MchKean, J.D.: The Importance of Traceability for Public Health and Consumer Protection. Scientific and Technical Review 20(2), 363–371 (2001)
[4] Huang, W., Wang, M., Zheng, Z., et al.: Create Modern Animal and Animal Products Identification and Traceability System. China Journal of Animal Quarantine 23(11), 1–4 (2006)
[5] Guan, E., Zhang, Y.: Studies on Implementation of Food Traceability Management. Chinese Journal of Food Hygiene 18(5), 449–452 (2006)
[6] Xu, S., Li, Z., Li, Z.: Study on China Food Safety Informaiton Sharing and Public Management System. China Agriculture Press, Beijing (2006)
[7] Ozawa, Y., Ong, B.L., An, S.H.: Traceback Systems Used During Recent Epizootics in Asia. Scientific and Technical Review 20(2), 605–613 (2001)
[8] Qiao, J., Han, Y., Li, D.: Importance and Restricting Factor Analysis on Implementing Food Safety Traceability in China. Guide To Chinese Poultry 43(6), 10–12 (2007)
[9] Lu, C., Xie, J., Wang, L., et al.: Completion of Digital Tracing System for the Safety of Factory Pork Production. Jiangsu Journal of Agricultural Sciences 22(1), 51–54 (2006)
[10] Wang, L., Lu, C., Xie, J., et al.: Review of traceability system for domestic animals and livestock products. Transactions of the CSAE 21(7), 168–174 (2005)
[11] Yang, X., Qian, J., Sun, C., et al.: Implement of Farm Product Archives Management System Based on Traceability System. Chinese Agricultural Science Bulletin 6(6), 441–444 (2006)
[12] Bai, Y., Lu, C., Li, B.: Design of Traceability System for Broiler Secure Production Monitoring. Jiangsu Journal of Agricultural Sciences 21(4), 326–330 (2005)
[13] Vitiello, D.J., Thaler, A.M.: Animal Identification: Links to Food Safety. Scientific and Technical Review 20(2), 598–604 (2001)
[14] Zou, Y., Jin, X.: Problems and Countermeasures Analysis in Food Safety Management. Chinese Health Quality Management 12(5), 57–59 (2005)
[15] Tech-food.com (2007)-Study Report about Food Safety Traceability website, http://www.tech-food.com/news/2007-4-16/n0105260.htm (document accessed on April 16, 2007)

Study on Intelligent Multi-concentrates Feeding System for Dairy Cow

Yinfa Yan, Ranran Wang, Zhanhua Song, Shitao Yan, and Fa-De Li*

College of Mechnical and Electronic Engineering, Shandong Agricultural University,
DaiZong Street NO.61, Tai'an, Shandong Province, P.R. China 271000
Tel.: +86-538-8246106
li_fade@yahoo.com.cn

Abstract. To implement precision feeding for dairy cow, an intelligent multi-concentrates feeding system was developed. The system consists of two parts, one is precision ingredients control subsystem, the other is multi-concentrates discharge subsystem. The former controls the latter with 4 stepper motors. The precision ingredients control subsystem was designed based on Samsung S3C2440 ARM9 microprocessor and WinCE5.0 embedded operating system. The feeding system identifies the dairy cow with passive transponder using RFID (Radio frequency identification) reader. According to the differences of based diet intake and individual dairy cow milk yield, the system can automatically and quantificationally discharge 4 kinds of different concentrates on the basis of the cow identification ID. The intelligent multi-concentrates feeding system for dairy cow has been designed and implemented. According to the experiment results, the concentrate feeding error is less than 5%, the cow inditification delay time is less than 0.5s and the cow inditification error rate is less than 0.01%.

Keywords: Precision feeding, intelligent concentrate feeding system, Dairy cow, RFID.

1 Introducton

With the development of dairy cow breeding industry, more milk yield, higher milk qulity and top-priority feeding cost for each dairy cow are the most important aspects for dairy cow farmer (Halachami et al., 1998; Hua et al., 2006). In recent years, concentrates automatic feeding system for cow is used in few of large dariy farms (Hua et al., 2006). However, in the absence of precision feeding system in the common scale dairy farm, food intake of each cow can not be calculated, feeding cost can not be estimated, milk yield and milk qulity are difficult to be increased (Halachami et al., 1998). In other words, both milk yield and quality depend on food intake, especially on the concentrates intake. In fact, the food intake varies

---

* Corresponding author.

D. Li and C. Zhao (Eds.): CCTA 2009, IFIP AICT 317, pp. 275–282, 2010.
© IFIP International Federation for Information Processing 2010

significantly with cows, thus the computer controlled concentrates, self-feeders, feeding robots are designed to solve the automatic feeding problem of individual cow in the small scale cow farm (Halachami et al., 1998; Fang, 2005; Kuang, 1999).

Automatic TMR (Total Mixed Ration) feeding system is applied in the UK and the US firstly since 1960's. TMR feeding technique well mixes the concentrates and the coarse fodders, and realizes nutritional balance of cow, but TMR can not agree with the demand of the nutrition of the individual, cow, especially the cow with the high capability of the milk yield. Feeding robots can identify the cow ID with RFID system and can deliver the different amount of the concentrates and coarse fodders for individual cow, but the discharging mode of the robot is to discharge concentrates at first, and then to discharge the coarse fodders in sequence, or to deliver the coarse fodders at first, and then the concentrates, which results in that the concentrates and the coarse fodders can not be mixed well (Tan et al., 2007; Halachami et al., 1998; Fang, 2005).

In order to deliver automatically and accurately 4 kinds of concentrates, simultaneously mixing well during discharging, on the basis of the individual cow identification, the intelligent multi-concentrates feeding system was developed based on WinCE, 32bit high performance microprocesser, 4 precision stepper motors and a flute-wheel concentrate feeding device. Furthermore, the system includes a cow feeding database which consists of individual cow feeding information, such as the weight of the concentrates agreeing with the recipe, the feeding times and the feeding interval. The feeding database can record every individual cow's each intake concentrate weight, the intake times and the intake time per day.

2   System Design

2.1 System Overview

The system consists of two parts, one is a precision ingredients control subsystem, and the other is a multi-concentrates discharge subsystem which includes 4 concentrate dischargers driverd by 4 stepper motors. The former controls the latter with 4 stepper motors. The precision ingredients control subsystem was designed based on Samsung S3C2440 ARM9 microprocessor and WinCE5.0 embedded operating system. The flute-wheel concentrate feeding device was used on the discharge subsystem.

Fig. 1 shows the one concentrate discharger which mainly consists of a base frame, a feeding trough, a discharge funnel, a flute-wheel discharger, a hopper, a stepper motor, a coupling and a stepper motor driver. When the cow with passive transponder steps into the feeding area, the cow RFID reader gets the transponder data and transimates the data to the control subsystem using RS485. Then the control subsystem decodes the data, identifies the cow's ID and searches the cow's feeding information. As the weight of the concentrates agreeing with the recipe, the feeding times and feeding interval, from the feeding database. If the cow is permitted to take

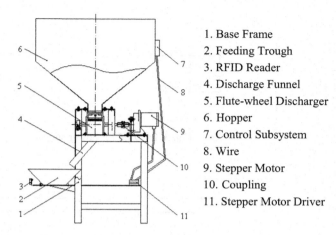

1. Base Frame
2. Feeding Trough
3. RFID Reader
4. Discharge Funnel
5. Flute-wheel Discharger
6. Hopper
7. Control Subsystem
8. Wire
9. Stepper Motor
10. Coupling
11. Stepper Motor Driver

Fig. 1. The concentrate discharger diagram

the concentrates, the control subsystem will calculate the respective rotation steps of each stepper motor. The steps are calculated on the basis of the discharge weight per step and the weight of the concentrate needed to be discharged. The 4 concentrate dischargers will start and complete their own task simultaneously.

2.2 Control Subsystem Design

The system hardware is designed based on Samsung S3C2440 ARM920T microprocessor, the S3C2440 provides a comprehensive set of common system peripherals and minimizes overall system costs and eliminates the need to configure additional components, the S3C2440 is perfect for general applications with cost-effective, low-power consumption, and high performance microcontroller solution in a small form-factor (Samsung, 2004).

2.2.1 Display and Touch Panel System

The user control interface is achieved using Sharp LQ080V3DG01 640×480 TFT LCD. The LQ080V3DG01 LCD connects with S3C2440 LCD controller as shown in Fig. 2. VCLK, VLINE, VFRAME, VW are the LCD control signals generated by S3C2440 LCD controller, VD23:18, VD15:10, VD7:2 are the data ports for video data. LCD dedicated DMA of LCD controller can transfer the video data in frame memory to LCD driver Automatically without CPU intervention, sharp LQ080V3DG01 need AC power supply for backlight, the LQ080V3DG01 backlight driver circuit is designed using CXA_L10A that delivers sine wave output for CCFL (Cold Cathode Fluorescent Lamps) as shown in Fig. 3. In order to improve friendly HMI (Human Machine Interface), The LQ08V3DG01 is equipped with a 4-wire resistance touch panel. The driver circuit of the touch panel shown in Fig. 4 is designed using dual N&P channel specific IC FDC6321, and the driver circuit connects the interface unit of S3C2440 internal touch panel (Samsung, 2004; Sharp, 2003).

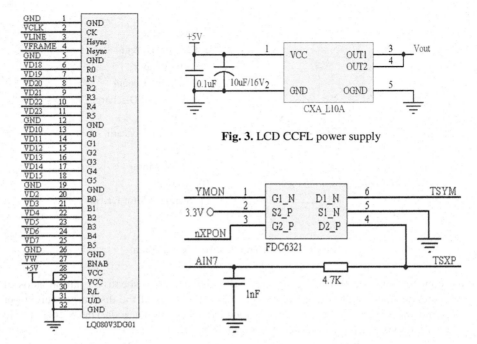

Fig. 2. LCD connection diagram   **Fig. 4.** Touch panel driver circuit

2.2.2 Identification System

The individual cow identification is achieved using Beijing WMTech WM-18 series RFID system. WM-18 consists of a WM-181 reader and EM4100 compatible ID cards, the work frequency is 125 kHz, and the maximal reading distance is 100 cm. The EM4100 compatible ID card is a passive battery-free transponder weared on the cow neck. When the cow with the transponder steps into the feeding area, the WM-181 reader gets the transponder data and sends the data to S3C2440 for processing through RS485 bus. It takes the system less than 150ms to transmit the data to WM-181 reader and send the data to S3C2440. The RFID system working current is less than 250 mA. The cow inditification error rate is less than 0.01%.

2.3 Discharge Subsystem Design

The discharge subsystem includes 4 dischargers. Each discharger mainly consists of a flute-wheel discharger, a stepper motor, a stepper motor driver, a coupling, a hopper and a discharge funnel. The output shaft of the stepper motor connects the input shaft of the flute-wheel discharger with the coupling.

2.3.1 Flute-Wheel Discharger

Due to that the configuration and manufacture technology are simple, that it does not make the pellet broken, as well as the higher discharge rate with high accuracy, the flute-wheel discharger is selected as the delivering device. The discharge capability of the flute-wheel discharger is influenced by the effective length of the flute-wheel, the flute number, and the cross-section area of each flute. According to the result of experiment, the suitable range of the rotation speed of the flute-wheel is from 9r/min to 60r/min, if the rotation speed is out of the suitable range, the flute can not be filled with pellet because of the higher liner velocity. Fig. 5 shows the flute-wheel designed as taper arc flute (Tai, et al., 2003).

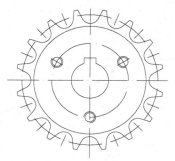

Fig. 5. Flute-wheel diagram

2.3.2 Stepper Motor System Design

According to the recipe for the individual cow, the control subsystem generates 4 stepper driver pulse signals and drives the flute-wheel concentrate feeding devices to discharge the 4 kinds of the concentrates simultaneously. In order to improve the concentrates discharge accuracy, a 3-phase stepper motor 110BYG350 equipped with a 3-phase stepper motor driver 3H110MS is used. The DIP switchpack is provided on the driver to select the basic step angle of the stepper motor from 1 to 16 divisions. The stepper motor rotates smoothly at the basic step angle 0.3o/pulse and it is low vibration even at low speed.

2.4 Software Design and User Operation Interface

The system software is designed based on WinCE 5.0 embedded operationg system, and is programmed in C/C++ language under embedded VC development environment. The user operation interface is a very concise and friendly graphical interface, by operating LCD touch panel, the user can calibrate the concentrates discharge weight, input the individual cow intake multi-concentrates recipe, set up the cow management database, manage the concentrate feeding database, query and report the individual cow feeding data. On the other hand, the system can communicate with the host management PC through RS485 bus. The system receives the cow database data, the individual cow recipe, the individual cow concentrates intake times and the concentrates intake interval, in addition, the system transmits the intake times and the weight of individual cow concentrates to the host manage PC. The flow chart of the intelligent multi-concentrates feed for the individual cow is shown in Fig. 6.

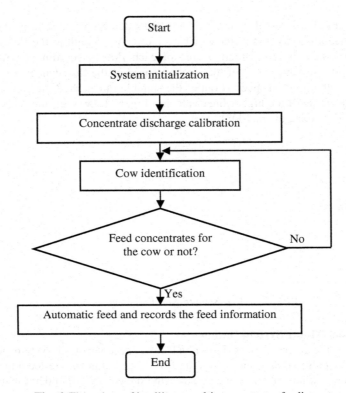

Fig. 6. Flow chart of intelligent multi-concentrates feeding

3 Result

The intelligent multi-concentrates feeding system for dairy cow has been developed and used in the experimental dairy cow farm of Shandong Agriculture University. The data presented below were collected in the experimental farm on the trial period.

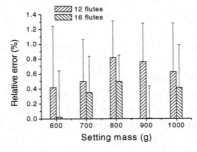

Fig. 7. Large pellet and high displacement

Fig. 7 shows the influence of the flute number on the discharge accuracy of the concentrate with great discharge capacity when the large pellet feed was delivered. The results indicated that the discharge accuracy and the stability of the flute-wheel with 16 flutes is higher than that of the flute-wheel with 12 flutes, even thouth the number of flute does not have significant difference on the measurement accuracy ($p>0.05$). The discharge error of the flute-wheel with 16 flutes is less than 1.5% in this case.

Fig. 8 shows the relationship between the different flute number and the discharge accuracy under the conditions of discharge large or small pellet and low discharge capacity. The results indicated that the discharge accuracy of the flute-wheel with 16 flutes is nearly the same as the flute-wheel with 12 flutes, but the discharge stability is better than the flute-wheel with 12 flutes. The discharge error of the flute-wheel with 16 flutes is less than 1.5%.

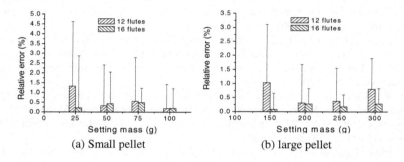

(a) Small pellet   (b) large pellet

Fig. 8. Low displacement

According to the experiment results, the total error of the concentrate discharging is less than 5%.

4   Conclusion

The intelligent multi-concentrates feeding system for dairy cow has been designed and implemented. According to the individual cow food intake recipe and the feeding times as well as the feeding interval, the feeding system can discharge 4 kinds of concentrates by the cow indentification with RFID technique. Moreover, the system can online record cow's feeding information such as each concentrate weight, the intake times and the intake time. Because of the friendly graph user interface and easy update in system, the feeding system developed in our research is different from the previous feeding systems based on PLC and the single chip microprocessor. In addition, the system is cheaper than the similar system, and has easy maintenance, reliability. The system is suitable for the small cow farm short of funds.

The flute-wheel with taper arc flutes was used on the discharger, and the results indicated that the discharger has well pellet discharge accurancy and discharge stability. The mechanism has the characteristic of simple structure and manufacturing technology. The discharge accuracy can be satisfied with this mechanism.

Acknowledgements

The authors would like to acknowledge The Ministry of Science and Technology of the People's Republic of China for their financial support (2006BAD11A0904). The authors also would like to acknowledge professor Wang Zhonghua and Guo Yuwei from College of Animal Science and Technology of Shandong Agriculture University for their intensive support.

References

Fang, J.J.: Study and development of feeding robot. Journal of Agriculture Machanization Research 1, 158–160 (2005)

Halachmi, I., Edan, Y., Maltzet, E., Peiper, U.M., Moallem, U., Brukental, I.: A realtime control system for individual dairy cow food intake. Computers and Electronics in Agriculture 20, 131–144 (1998)

Hua, J.G., Zhou, Y.L., Hua, J.Z., Yu, Y.C., Zhang, W.F.: Development of the auto-feeding equipments for dairy cattles. Transaetions of the CSAE 22(2), 79–83 (2006)

Kuang, P.S., Liu, G., Kuang, J.S.: On the precision-agriculture technological system. Transaetions of the CSAE 15(3), 1–4 (1999)

Meng, H.W., Kan, Z., Li, Y.P.: Design and 3D modeling way of screw conveyor in the fodder feeds device for cow. Journal of Agriculture Machanization Research 10, 61–63 (2008)

Samsung. S3C2440 32-bit RISC microprocessor user's manual prelimanry revision 0.14. Samsung electronics, Korea (2004)

Sharp. Technical literature for LQ080V3DG01 TFT-LCD module. Sharp Corporation, Japan (2003)

Tai, M.Y., Yuan, J.E., Liu, X.Z.: Design of Incline outer-fluted feed with upper distribution. Chinese Agriculture Machanization 34, 34–35 (2003)

Tan, C.L., Kan, Z., Zeng, M.J., Li, J.B.: Analysis of the situation about cow-feeding technique and equipment. Journal of Agriculture Machanization Research 12, 240–245 (2007)

Tan, C.L.: Design and study of cow feeding equiment. MS thesis. Shihezi Xinjiang P.R. China: Shihezi University, College of Mechnical and Electronic Engineering (2007)

Tan, C.L., Kan, Z., Zeng, M.J., Li, J.B.: RFID technology used in cow–feeding robots. Journal of Agriculture Machanization Research 2, 169–171 (2007)

Yu, Y.C., Li, W.Y., Hu, F.S., Li, C.L., Zhang, W.F.: The research and development of the dairy cow precision feeding system based on SCM. Journal of Agriculture Machanization Research 9, 108–111 (2008)

Construction of Traceability System for Quality Safety of Cereal and Oil Products

Huoguo Zheng[1,2,*], Shihong Liu[1,2], Hong Meng[1,2], and Haiyan Hu[1,2]

[1] Agricultural Information Institute, Chinese Academy of Agricultural Sciences,
Beijing, P.R. China 100081,
Tel.: +86-10-82106263; Fax: +86-10-82106263
huoguos@caas.net.cn
[2] Key Laboratory of Digital Agricultural Early-warning Technology, Ministry of Agriculture,
The People's Republic of China, Beijing, P.R. China 100081

Abstract. After several significant food safety incident, global food industry and governments in many countries are putting increasing emphasis on establishment of food traceability systems. Food traceability has become an effective way in food quality and safety management. The traceability system for quality safety of cereal and oil products was designed and implemented with HACCP and FMECA method, encoding, information processing, and hardware R&D technology etc, according to the whole supply chain of cereal and oil products. Results indicated that the system provide not only the management in origin, processing, circulating and consuming for enterprise, but also tracing service for customers and supervisor by means of telephone, internet, SMS, touch machine and mobile terminal.

Keywords: traceability system, quality and safety, cereal and oil products, traceability encoding.

1 Introduction

Food traceability system, also called food tracking and tracing system, has become an effective way in food safety management. After several food safety related issues, particularly several food sandals, the global food industry and governments in many countries have paid increasing attention to systems along the food chain (Liu Yin. 2003). The implementation of traceability systems is important: one is to give consumers the right to know, and the second is to strengthen the responsibility for the enterprises which produce food products, the third is to find the root causes when the food safety issue emerged (Pu Yinyan. 2008).

Traceability system was initially promoted by the EU to control the risks of mad cow disease in 1997. From then on, most agriculture developed countries are active to implement food.safety traceability systems, but they have different focus.In June 2002, the Canadian federal government established an ambitious goal that, before 2008, the

---

* Corresponding author.

D. Li and C. Zhao (Eds.): CCTA 2009, IFIP AICT 317, pp. 283–290, 2010.

country would achieve tracing back 80 percent of agricultural products to its source, supporting the "Brand Canada strategy"(Lu Changhua et al.. 2007), of which a mandatory identification system for cattle and beef on July 1, 2002 came into operation. In December 2003, the United States developed the statutes of tracking food safety, which required all enterprises involved in food transportation, distribution and import recording their trade information for tracking and tracing back (Cheng Hao. 2007). In July 2003, The EU published a White Paper on Food Safety, proposing a new framework for food safety system. Most countries in the EU have implemented mandatory livestock and meat products traceability system (Zhu Haipeng. 2007). The Japanese Government has passed new legislations on cattle and beef requiring a mandatory traceability system from farm to retail (Wang Lifang et al. 2005). The system allows consumers via the Internet to enter identification numbers of beef on the packaging box getting access to production information of the beef. Australia has plans for general mandatory traceability. NLIS is currently carrying out, which enables trace backward and forward from farm-of-origin to abattoir (Schroeder et al. 2005).

China has also achieved important progress in food tracing. In April 2004, the State Food and Drug Administration and 7 other departments chose meat industry as a pilot industry, started meat and meat products traceability institution construction and system implementation (General Administration of Quality Supervision. 2002). In June 2004, Administration of national barcode management promoting investigated on vegetable products traceability and started an application project on two vegetable production bases located in Shouguang and Luocheng respectively in Shandong province (Zhou Yingheng et al., 2002). Shanghai Livestock Bureau legislated to build digital archives for pigs, cattle, sheep and other critters, and the residents can now get access to the egg production information through internet (Shanghai agricultural committee. 2001). In August 2008, Beijing enforced a food traceability system along the full supply chain for the food supplied for Olympic games to secure food quality and safety (Chang Xiang. 2007).

Cereal and oil products are the common agricultural products in our country, therefore, construct the traceability system for cereal and oil products of great significance. Food traceability system can trace the quality and safety information from "farm to table "(Liu Shihong. 2008). To construct the traceability system, we should study and analyses the critical quality safety point basic the whole supply chain of cereal and oil products, then carry out the research of key technologies and build the traceability system for cereal and oil products.

2 Traceability System Design

2.1 System Architecture Design

Aiming at the major quality problems of cereal and oil products existing in the production (cultivation), inspection, storage, processing, circulation and other links, this paper starts with systematical analysis of the source, channel and the kind of contamination, which forms the traceability system framework. In this paper, integrated with the information flow analysis, FMECA (Failure Mode Effect and Criticality Analysis) (Kang Ri. 2006) approach is used as a tool to detect the possible critical control points. HACCP (Bao Dayue. 2007) (Hazard Analysis and Critical Control Point) quality system and some related national standards on cereal and oil products are referred to determine the key quality indicators in the supply chain. Then, the traceability framework and an optimal set of quality indicators come into appearance. At the same time, study the key technology and design the coding principle. Finally, construct the traceability system for cereal and oil products.

The traceability system for cereal and oil products, which will not only meet the day-to-day management and internal tracing needs, but also provide services for consumer and regulatory by means of SMS, telephone, network and mobile portable-terminal., has achieved origin, processing, detecting, quality and safety management and other functions. From another perspective, the system can be broken down into three layers: information collection, information processing, and information services. The main function of the level of information collection include filtering the key factors in the quality and safety indexes, acquisition the information about origin, processing, producing, quality inspection, logistics and sales, supporting the ultimate tracing. Information processing level's role is building the tracing platform through information encoding technique, information collection, information exchange and hardware R&D technology. We provide different kinds of services for consumers and regulators through SMS, telephone, network platform and mobile terminals. The diagram of system architecture is shown in Figure 1.

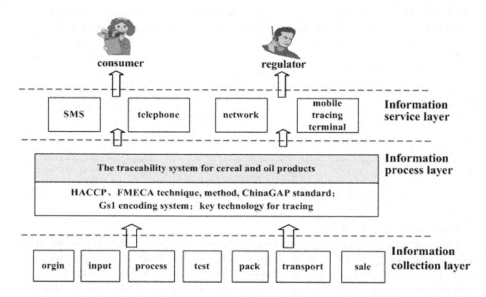

Fig. 1. System architecture diagram for cereal and oil products tracing

2.2 Function Design for the System

The traceability system for cereal and oil products consists of four sub-systems: origin management system, production control system, indexes management system and tracing management. The origin system achieves the information management of natural environment of origin, input materials, producing process based on the China GAP standard. Production control system management is the foundation of the whole system, which cover the information from raw material purchase to final product sale. The indexes management system's main function is to determine the critical quality and safety factors of the cereal and oil products, after that, management them. The tracing management system is the display layer of the system, which provides services for users by a wide range of interface.

3   Key Technologies

3.1   Traceability Encoding

Unified encode system is the premise of cereal and oil products traceability, is also the basis of information exchanging and processing. The encoding system should comply with the following principles: uniqueness, stability, commonality, expandability and applicability.

1 origin code design

Origin code is the key to get the quality and safety information of the raw materials planting stage of the cereal and oil products, is also the core code for the tracing. Considering the commonality and the standardization of the origin code, we adopt the 《Rules for Coding of Agricultural Land》 standard, which was published in Dec 1, 2007. The origin code is composed of 5 parts. The code structure is shown in Fig 2.

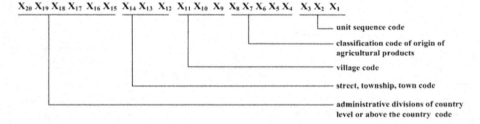

Fig. 2. Origin coding rules for cereal and oil products

2 Traceability code design

The traceability code is the unique identification of cereal and oil products. All of the quality and safety information of the products can be obtained through this code, combined with the production control system. This code is from commodity code, but contains more information than commodity code. Compared to the animal products, fruits products, vegetables products and other agricultural products, cereal and oil products have long production chain and involved many factors, which can't identification individual as animal products, can't identification producer of raw materials as fruits and vegetables products (Yang Xinitin et al., 2007).

The traceability code follows the UCC/EAN-128 standard (Lin Ling et al., 2004), which is a worldwide standard for exchanging data between different companies. We put forward a unique identification number for encode the data concerning cereal and oil products. The identification number consists of five parts, including enterprise identification number, commodity identification number, batch number and two kinds of verifying number. The first three items in fact form the Global Trade Item Number (short for GTIN). The code structure and example for cereal and oil products is shown in Fig.3.

Fig. 3. The code structure and barcode example

3.2 Multi-platform Tracing Technology

The tracing management system can provide different services for three kinds of user such as consumers, enterprise managers and supervisors, through five patterns including SMS, telephone, computer, fixed terminals in the supermarket and mobile terminal. The system locate at the data center, which preserve the whole information of supply chain. The diagram of multi-platform is shown in Fig 4.

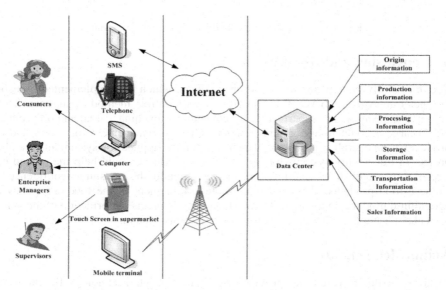

Fig. 4. The diagram of multi-platform tracing

3.3 Mobile Terminal R&D Based on GPRS Network

Since the wireless application widely spread and for the consumers convenience, we developed a kind of mobile electronic terminal based on GPRS data transmission, which was consisted of three important parts: the motherboard, GPRS communication module and a micro Liquid-crystal display (short for LCD). The main board is used for processing information, while the GPRS communication module used for receive the information and send the results back. The micro LCD is simply for displaying the results in a friendly way. The mobile terminal is shown in Fig.5.

The mobile terminal can trace the cereal and oil products under the condition of absence of cable networks, to facilitate regulatory agencies at any time, anywhere to monitor the quality of agricultural products. It is also applicable to other agricultural products tracing.

Fig. 5. The diagram of the hardware of the mobile terminal

4   Conclusion and Discussions

The traceability system for cereal and oil products, which was implemented by using the .net framework, RBAC (Role-Based Access Control) method, component technology, has been used in supermarket for comsumers. In china, cereal and oil products take a large account in agriculture products. Quality management for cereal and oil product has an overall significance. The methodology, application system and the pilot project of traceability systems for cereal and oil products would help the related enterprises to establish traceability systems and rebuild the consumer confidence of Chinese food products, also can promote the enforcement of the food safety law of agricultural products in a practical way. However, the precise information of the whole food chain is another critical factor for the traceability system.

Acknowledgements

Funding for this research was provided by National High Technology Research and Development Program of China called as "863" (2006AA10Z268).

References

Dayue, B.: Guide to implementation of HACCP. Chemical Industry Press (2007)

Schroeder, C.T., Tonsorb, T.G.: Australia'S Live—stock Identification Systems: Implications for United States Programs [EB/OL], http://www.agmanager.info/events/risk_profit/2004/Schroeder (April 27, 2005)

Xiang, C.: The food quality and safety traceability systems for Beijing Olympic games will be put into use in August, http://www.chinanews.com.cn/ (July 09, 2007)

Hao, C.: Animal product safety control and traceability technologies. Modern Agriculture science and technology (13), 169–170 (2007)

Hongping, F., Zhongze, F., Ling, Y., Aisheng, R.: Appliance and Discussion of Traceability System in Food Chain. Ecological Economy 17(4), 30–33 (2007)

Schwagele, F.: Traceability from a European perspective. Meat Science (71), 164–173 (2005)

General Administration of Quality Supervision, Inspection and Quarantine of the People's Republic of China. GB/T 15425-2002 (2002)

Ri, K.: FMECA approach and its application. National Defense Industry Press (2006)

Ling, L., Deyi, Z.: On the Construction of Food Quality and Safety Traceability System. Commercial research (21) (2005)

Opara, L.U.: Traceability in agriculture and food supply chain: A review of basic concepts, technological implications and future prospects. Food, Agriculture & Environment 1(1), 101–106 (2003)

Shihong, L., Huoguo, Z., Meng, et al.: Study of full-supply-chain quality and safety traceability systems for cereal and oil products. IFIP. Springer, USA (2008)

Yin, L., Licheng, C.: Traceability Production System of Beef in EU and USA. Food science (8), 182–185 (2003)

Changhua, L., Changjiang, W., Sinong, H., et al.: Identification and Traceability System for Animals and Animal Products, pp. 35–36. Chinese Agricultural Science and Technology Press (2007)

Massimo, B., Maurizio, B., Roberto, M.: FMECA approach to product traceability in the food industry. Food Control 17(9), 137–145 (2004)

Yinyan, P.: Construction of traceability system for quality safety of apple and apple juice. Transactions of the Chinese society of agricultural engineering 24(2), 289–292 (2008)

Shanghai agricultural committee. Notification in carrying out animal identification in Shanghai. Shanghai agricultural committee (126), August 24 (2001)

Lifang, W., Changhua, L., Jufang, X., Yinong, H.: Review of traceability system for domestic animals and livestock products. Transactions of the Chinese Society of Agricultural Engineering (07) (2005)

Wenying, X.: The Policy of Traceability in Quality and Safety of Agricultural Products in USA. World Agriculture (04) (2006)

Jin, X.: Agricultural product supply chain——Guarantee food safety. China Logistics & Purchasing (07) (2005)

Xintin, Y., Chuanheng, S., Jianpin, Q., et al.: Application of UCC/EAN-128 bar code technology in agricultural product safety traceability system. Computer engineering and applications 43(1) (2007)

Hui, Y., Yufa, A.: Theoretical discussion of implementation Traceability System in food supply chain. Agricultural Quality and Standards (03) (2005)

Gumin, Z., Gongyu, C.: Food safety and Traceability System. China Logistics & Purchasing (14) (2005)

Yingheng, Z., Xianhui, G.: Application of Traceability in Food Safety. Research of Agricultural Modernization (06) (2002)

Haipeng, Z.: Research and implementation on traceability system of key cereal enterprise. Chinese Academy of Agriculture Science, 10–11 (2007)

The Study and Implementation of Text-to-Speech System for Agricultural Information

Huoguo Zheng[1,2,*], Haiyan Hu[1,2], Shihong Liu[1,2], and Hong Meng[1,2]

[1] Agricultural Information Institute, Chinese Academy of Agricultural Sciences,
Beijing, P.R. China 100081,
Tel.: +86-10-82106263; Fax: +86-10-82106263
huoguos@caas.net.cn
[2] Key Laboratory of Digital Agricultural Early-warning Technology,
Ministry of Agriculture, The People's Republic of China, Beijing, P.R. China 100081

Abstract. The Broadcast and Television coverage has increased to more than 98% in china. Information services by radio have wide coverage, low cost, easy-to-grass-roots farmers to accept etc. characteristics. In order to play the better role of broadcast information service, as well as aim at the problem of lack of information resource in rural, we R & D the text-to-speech system. The system includes two parts, software and hardware device, both of them can translate text into audio file. The software subsystem was implemented basic on third-part middleware, and the hardware subsystem was realized with microelectronics technology. Results indicate that the hardware is better than software. The system has been applied in huailai city hebei province, which has conversed more than 8000 audio files as programming materials for the local radio station.

Keywords: text-to-speech, software, hardware device, microelectronics.

1 Introduction

After year's development, China has made the amazing progress in the development face the farmer's information service. Agricultural information service played a more positive role in solving the problem of "Farmers, Rural Areas and Agriculture Production" (Guo Hongmin et al, 2007). In the recent two years, there had emerged a number of new agriculture information service patterns, which providing timely and accurate information for farmers by the media of TV, phone and network, such as integrated service pattern in Hebei province (Gao Jikui et al, 2005), farmer's mail in Zhejiang province (Wu Yitian et al, 2007), given number "12316" in Jilin province (Jilin farmer website, 2007), 110 in science and technology in Hainan province (Wu Yuanbin et al, 2007). However, we should see that the radio and TV are still playing the main role of information service infrastructure in rural. According to the latest statistics, China's radio and television coverage has increased to more than 98%.

---

* Corresponding author.

D. Li and C. Zhao (Eds.): CCTA 2009, IFIP AICT 317, pp. 291–296, 2010.
© IFIP International Federation for Information Processing 2010

Compared with other services, broadcasting as a carrier of information services has a wide coverage and low cost, easy-to-grass-roots farmers to accept and so on.

In the investigation and study we discovered that broadcasts the information service to emerge quietly in many places and the prospects for development are huge. To develop the broadcast information service well, we have researched and developed the rural broadcast information service system, including the network information acquisition, text-to-speech, the program arrangement; provide the information resource for the rural Broadcasting station. This article introduced the realization of text-to-speech system, which can transform the text information into the audio document, and provide program resource.

2 Design of Text-to-Speech System

Text-to-speech system, also known as text speech synthesis system, concludes two sub-systems, TTS software and TTS hardware. The system's functions are shown in Figure 1. Both of them can construct audio database by means of translating the text user defined or export from database into audio.

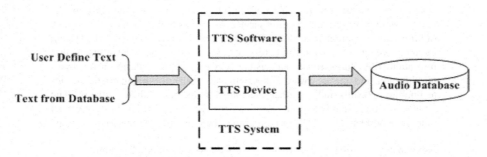

Fig. 1. Text-to-speech system's functions

2.1 Design of TTS Software

TTS software, which was implemented base on the third party middleware, can translate the user-defined text, text export from database into audio files. Its core function includes parameter setting, text-input, text-pretreatment and Text-to-Speech.

2.2 Design of TTS Hardware

TTS hardware sub-system includes the text speech synthesis hardware and the text-editing software. The synthesis hardware is overall system's core, which function is transforms the text information into the pronunciation, the text-editing software is the systematic input interface, which sent the text message to the hardware. Hardware must be associated with the computer through the USB interface to get power and adoption of the text-editing software to achieve text waiting to be converted between PC and hardware.

1 Design of hardware

The hardware is the text-to-speech sub-system's foundation; its functions must be careful, comprehensive low-level hardware support. The hardware's connection principle is shown in Figure 2. The text speech synthesis hardware's main component include 8 microprocessor ATmega128 (Li Hua et al, 2008), the USB main line general connection chip, the XF-S4240 module, the low voltage audio power amplifier.

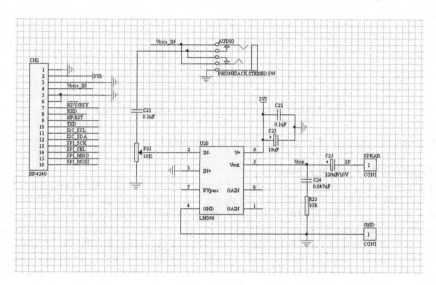

Fig. 2. The principle figure of text speech synthesis hardware

2 Design of text-editing software

The text-editing software is the controller of text data transmission between PC machine and the text speech synthesis hardware. After the process of edition, operation, pretreatment, the text can be accurately loaded, recognized, read by the outside loudspeaker. The test-editing software's major function includes: imports the text document (the .txt form), set special tags, preprocess the document, save the file and play voice. The text adopted by the special tags settings, you can effectively control the speed of play voice, tone, volume and so on, then have a clear, natural, accurate text-to-speech effect.

3   Implementation of Text-to-Speech System

3.1   Implementation of Text-to-Speech Software

The text-to-speech software has implemented the following functions, text-edit, audition, audition pause, audition stop, save the text as file and translate the text into audio file, base on the object-oriented software design thought, as well as the consideration of the interface of the software. The work flow of the software is shown in Figure 3.

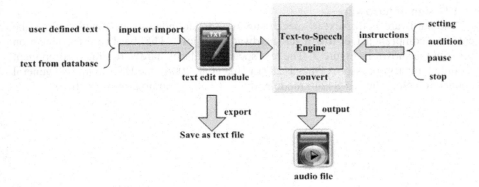

Fig. 3. The work flow of the text-to-speech software

3.2 Implementation of Text-to-Speech Hardware

1 Implementation of software

The hardware platform is composed of four components: the microprocessor, USB interface, the text speech synthesis module, as well as audio player. The hardware gets power and the text data by the USB interface from the computer, undergoes an 8-bit microprocessor system programming processing, to the information which receives carries on the correct judgment, the recognition again through the XF-S4240 text speech synthesis module, finally play the voice through the audio amplifier module. The audio can be recorded, preserved. The text speech synthesis hardware's work flow is shown in figure4.

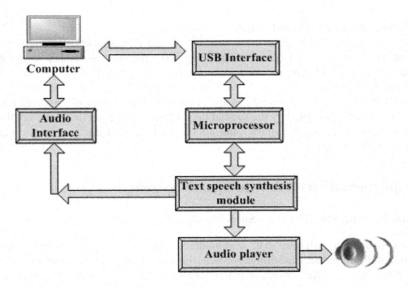

Fig. 4. The work flow of text speech synthesis module

The single-chip embedded program realized by ICCAVR (Ma Chao et al, 2007) programming in the course of hardware programming (Qin Changjiang et al, 2007), and its function is getting text data through USB interface by means of interrupts, then sent the text data to XF-S4240 module to synthesis pronunciation, in addition gains the transition status in interrupt service through the serial port, back to the PC machine in order to control the continued transmission of data.

2 Implementation of text-editing software

Basic on the detailed analysis of whole text-to-speech system, we carry the work of design and realization on the text-editing software. According to the characteristic of the function, divides it into four modules: import text file, edit the tags, text speech synthesis and save the audio file. The text-editing software's work flow is shown in figure 5.

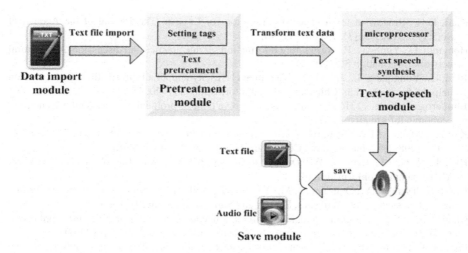

Fig. 5. The work flow of text-editing software

While the text-to-speech system running, we input or import the text which will be deal into the pretreatment module of the text-editing software, set format tags and pre-standardized, and then sent the processed text to the hardware, to judge the accuracy, identification. Finally, the text will be converted to voice, then play.

4 Discussions

Results indicate that the hardware works better than software. The audio file transformed by hardware sub-system is smoother than software sub-system. The text-to-speech system has translated more than 8000 piece of agricultural science and technology information into audio file in huailai radio station hebei province. Those audio files provide for the local farmers as information service by broadcast. Good effect achieved, according to the local people.

China's rural reform has entered a new period of development. The effect of agriculture information service in the promotion of rural development, agricultural efficiency and increase farmers in rural appear already. Agriculture information will play a more and more important role in rural area. All sectors of society participate in the growing movement. Under this kind of advantageous environment, how to display the function of traditional media fully, innovate rural information service pattern, is the question which the scientific research worker and the department of information service must consider.

References

AIC Monaghan. Intonation in a text-to-speech conversion system. University of Edinburgh dissertation (1991)

Klatt, D.H.: Review of text-to-speech conversion for English. The Journal of the Acoustical Society of America (3), 737–743 (1990)

Hongmin, G., Yunhong, P.: Agriculture information service practice and discussion. Agricultural network Information 8(3) (2007)

Jikui, G., Yongqiang, L., Jianyu, C.: The three electricity and a hall information service system help the farmer be rich. China agricultural information (11) (2005)

Jilin farmer website. 12316 new rural hotline resolve famer's problem. Jilin agricultural and rural economic information (7) (2007)

Kain, A., Macon, M.W.: Spectral voice conversion for text-to-speech synthesis. In: Proceedings of the 1998 IEEE International Conference, vol. 1, pp. 285–288 (1998)

Hua, L., Wei, M., Baoyou, W.: Design of CAN-GPRS Gateway Based on ATMEGA128. Micro-computer information (3) (2008)

Lee, L.S., Tseng, C.Y., Ouh-Young, M.: The synthesis rules in a Chinese text-to-speech system. IEEE Transactions on Acoustics, Speech and Signal processing (9) (1989)

Chao, M.: AVR Monolithic integrated circuit embedded system principle and application practice. The Press of Beijing University of Aeronautics and Astronautics (2007)

Page, J.H., Breen, A.P.: The Laureate text-to-speech system: architecture and applications. BT technology journal (14), 57–67 (1996)

Changjiang, Q., Ziquan, Y., Yuquan, L.: Design and Implement of the Communication Bus Interface between PLD and AVR in VHDL. Micro-computer information, 6–2 (2008)

Yitian, W.: "farmer mailbox"– practice "new countryside" by means of informationization. The construction of informationization (3) (2007)

Yuanbin, W., Jinhua, W.: Agricultural science and technology 110 is the effective way to service the "Farmers, Rural Areas, Agriculture Production". Agricultural science and technology communication (1) (2007)

General Framework for Animal Food Safety Traceability Using GS1 and RFID

Weizhu Cao, Limin Zheng, Hong Zhu*, and Ping Wu

College of Information and Electrical Engineering, China Agricultural University,
Beijing, P.R. China, 100083
Tel.: +86-13693562232
zhuhongxie@cau.edu.cn

Abstract. GS1 is global traceability standard, which is composed by the encoding system (EAN/UCC, EPC), the data carriers identified automatically (bar codes, RFID), electronic data interchange standards (EDI, XML). RFID is a non-contact, multi-objective automatic identification technique. Tracing of source food, standardization of RFID tags, sharing of dynamic data are problems to solve urgently for recent traceability systems. The paper designed general framework for animal food safety traceability using GS1 and RFID. This framework uses RFID tags encoding with EPCglobal tag data standards. Each information server has access tier, business tier and resource tier. These servers are heterogeneous and distributed, providing user access interfaces by SOAP or HTTP protocols. For sharing dynamic data, discovery service and object name service are used to locate dynamic distributed information servers.

Keywords: animal food safety traceability, GS1, RFID, general framework.

1 Introduction

Many countries had suffered disasters and economic loss because of the diseases occurred in recent years such as Bovine Spongiform Encephalopathy(BSE), Dioxin Toxicosis, Clenbnterol Hydrochloride, foot and mouth disease (FMD), and these diseases make the whole world focus on the issue of food safety(Zhao Jinyan et al., 2008). The European Union, the United States, Australia and other developed countries have introduced laws and regulations to make a mandatory requirement on animal food safety traceability (Zan Linsen et al., 2006). The strengthening of the supervision of animal food safety management is in favor of improving product value and market competitiveness.

GS1 represents a global standard (GS1 standard), a global system (the global identification system, GS1 System), and a global organization. It is responsible for establishing the standards of bar code and data exchange for the world's goods, and determine RFID standard by EPCglobal. GS1 is a global food standard for traceability, with components of coding systems (EAN/UCC, EPC), data carriers (bar code,

---

* Corresponding author.

D. Li and C. Zhao (Eds.): CCTA 2009, IFIP AICT 317, pp. 297–304, 2010.
© Springer-Verlag Berlin Heidelberg 2010

RFID) and data exchange (EDI, XML). "GS1 traceability standard" launched by GS1 retroactively supplied standard operating procedures for the design and implementation of food safety programs, which used successfully in beef, fish, vegetables and other foods, and such track and trace systems of food industry chains "from farm to fork" had established successfully (GS1, 2009).

Existing animal food traceability systems are starting from the slaughter, and cannot be achieved from "the source" (Kong Hongliang et al., 2004; C. Shanahan, 2009). Databases of food traceability systems can be centralized or hierarchical. Centralized databases can only save the limited information, without particular information of products, and are not flexible (C. Shanahan, 2009). While hierarchical database need to resolve the data sharing between the databases at all levels, and data replication will lead to redundancy (Zhao Jinyan et al., 2008). Compared with the bar code, RFID possesses of advantages such as non-contact, multi-objective automatic identification, which make it used in animal food safety monitoring more and more, but the standardization of RFID application needs to be improved (Ci Xinxin et al., 2007).

2   General Framework of Traceability

Animal food industry chains (supply chain, sales chain) are long, mainly including breeding, slaughtering, processing, logistics and sales. Applications of animal food safety traceability systems in monitoring animal food safety, includes not only the records and monitoring during the whole growth process of animals from birth to entering the slaughterhouse (feeding management, vaccine injection, disease treatment, feed used), but also the query for information of every stage, which is from consumers by the unique identifier of each animal, when animal products are send to the consumer market (supermarkets, etc.).

Realization of the whole process trace of animal food safety mainly depends on the effective documentation, efficient management and seamless transfer of the key information of each stage. RFID tag, the data carrier links the information flow to real object flow, and implements data exchange of all steps. Each step, products are not

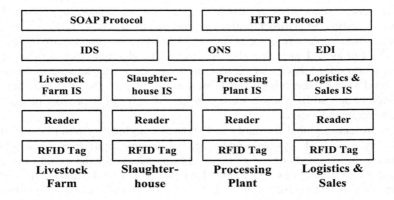

Fig. 1. General framework for animal food safety traceability

the same form, and we need to identify these different forms of products. Synchronously, we use the unique identifier of each animal to associates each pattern of the products.

Taking GS1 standards and RFID identification technology and the characteristics of animal food industry chains into account, fully considering of the traceability systems at home and abroad, taking their public properties, we have designed the general framework of the animal food traceability, shown in Fig.1.

RFID is used to identify cattle individual, and we label different form product with different identification, such as using ear tags on farm to get the cattle information. RFID reader is used to achieve non-contact and multi-objective automatic identification, and improves collection efficiency of data. Because of geographical location, functions and other factors, breeding, slaughtering, processing, logistics and marketing, each basic stage builds their own information server (IS) to maintain the static and dynamic product information of the whole industry chain. These servers are heterogeneous and distributed, providing user access interfaces by SOAP or HTTP protocols. For tracing and tracking of each individual, each livestock is corresponding to a unique ID, and the information servers save the individual animal husbandry management, vaccine injection, disease treatment, feed used and other essential information. For sharing dynamic data, discovery service and object name service are used to locate dynamic distributed information servers.

3 RFIDTAG

Product identification is one fundamental aspect for food safety traceability. RFID is a non-contact, multi-objective automatic identification technique, widely used in traceability from farm to supermarket. RFID tags are low-capability devices, because cost and size requirements mean that power consumption, processing time, storage and logical gates are extremely limited.

EPCglobal Inc. has published the EPCglobal Tag Data Standard 1.4 for RFID tag contents (EPCglobal Inc. 2008). Its coding schemes include SGTIN, SSCC, SGLN, GRAI, GIAI, and so on.

In order to trace from farm to supermarket, a SGTIN (Serialized Global Trade Item Number) is used to identify individual logistical units (beef, feedstuff, medicine), a GLN (Global Location Number) is used to identify physical locations (farms, abattoirs, marts). In addition to a Header, the SGTIN-96 is composed of five fields: the Filter Value, Partition, Company Prefix, Item Reference, and Serial Number.

We propose that the EPC tag for goods identification contain the same identity number as the current ear tags, and a standard format for the content of the RFID tags should be compliant with the EPCglobal standard.

4 Design and Implementation of Information Server

4.1 Design of Information Server

Subject to capacity, RFID tags can only identify products, without the information of specific producing and the circulation. But consumers and suppliers want to query for various aspects of the industry chain and product information at all levels, which needs information servers to support it. Information Servers are used to record the process of producing and circulation, and provide users with such information service interfaces as capture and query. All sectors of the traceability system set up their own servers in accordance with the characteristics of the animal food chain. These servers are distributed, and even heterogeneous, different from centralized and hierarchical servers.

The information server saves the static and dynamic information of products throughout the industry chain. It provides static product-related data, when it belongs to the manufacturer's information system, such as manufacture date, validity period, raw materials and other related information. Participants provide dynamic databases for the management of the measured data and tracking data of mobile products, to provide such tracking content as delivery times, locations, transport ways.

Information server of each link is structured on TCP/IP network, distributed, even heterogeneous, with HTTP/SOAP to provide user with interfaces, and using XML to transfer message. Information servers of breeding, slaughtering, processing and other stages use C/S structure, which logistics and marketing stages B/S structure, supporting queries through the supermarket terminal, the Internet, and messaging platform.

4.2 Implementation of Information Server

Each information server has three-tier architecture (access tier, business tier and resource tier), in Fig.2. Access tier provides user access interface to information server, with the form of Web services, including the capture interface and query interface. Business tier is the core of information servers, implements specific capture processing, query processing, query callback interface, authentication, and pursue a variety of business rules and logics. Resource tier stores product management information to provide data to the business tier, using relational database. JAVA language is adopted for implementing information servers, MySQL 6.0 for databases, and ADO technology for data access.

In the business tier, authentication module completes authentication by the user name and password. Capture handling module bind Java Servlet with HTTP. When capture interface receives capture request of HTTP POST from the user, it binds the XML format for the capture message, and hands to the capture handling module. Capture handling module maps message captured with SQL statements, and stores in the database.

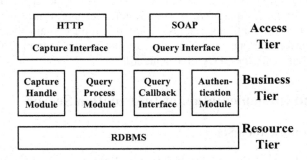

Fig. 2. Architecture of information server

Query Processing uses Simple Object Access Protocol (SOAP) to package the message as Web service. SOAP makes different applications exchange information to with each other through the HTTP protocol, and by XML format. When query interface receives the SOAP request of XML format from the user, it binds XML to query parameters, and send to the query processing module. The query processing module generates SQL statements dynamically according to query parameters, query databases, and returns the results with the package of SOAP response of XML format. Users can reserve asynchronous query from callback interface. The asynchronous query is scheduled at a certain time or triggered by an event. And the returned query results call HTTP/HTTPS POST requests from the designated URL.

5 Realization Mechanisms of the Dynamic Information Sharing

5.1 Dynamic Information Sharing

Animal food chain involves several participants to work together, and each participant has its own Information Server. When recalling, or protecting the privacy, the producer might hope to track all information about the product. Because of geographical location, functions and other factors, each link of the animal food industry chain can configure for its own Information Server. These servers are geographically distributed, heterogeneous and cross-platform. In the industry chain, what specific supply links will the product pass is not known in advance, so the information server is dynamic. Data sharing between distributed and heterogeneous databases is the key problem to be solved in animal food safety.

Data transfer and data sharing are very important in animal food safety traceability. For example, in Ireland beef GS1 retrospective case (J. Brackenm et al., 2005), slaughterhouses and processing plants are in one company, share the same database system, while the sale is taken charge by another company, which has its own traceability system. When the cattle is carried to the slaughterhouse, the ID of cattle ear tag and some of the historical information are copied to the database of the slaughterhouse. Data synchronization and data redundancy are required to consider. Into the marketing chain, the bar code information of body division will be transmitted into the company's system of safety traceability. This case can only trace back to the farm name and location, without specific information of the cattle individual, such as the information of growth, feed, disease treatment and vaccine injection.

The distributed retrospective database query can be achieved while introducing the query engine (A. Cheung et al., 2007). Query engine queries the local database, and analyze the results. New query submitted to other databases on the network can be reconstructed, depend on whether the local search results are complete or not. Results returned from other databases are added to the local search results for post-processing, and returned to the user. In EPCglobal Network system, EPC Information Service provides a query interface to support simple local inquiries, but do not support cross-organization, distributed and complex queries of the database (EPCglobal Inc., 2009). Object name service (ONS) can support distributed database query, but the ONS is a static query engine, can only provide static information resources of product EPC coding. ONS uses the domain name system of Internet, has a single service model, updates slow, hasn't strong resource description, and security cannot be guaranteed (B. Fabian et al., 2005; S Beier, 2006).

5.2 Implementation Mechanism

In the animal food safety traceability general framework, Information Discovery Service and ONS are used to share data. The discovery service of distributed system responds to requests of inquiry for resources, locates them or find a number of locations of resources for cross-system. Information discovery service is quite a dynamic discovery engine to provide dynamic trace information.

Information services are provided by the database and a series of Web services interface. Database records include the following attributes:

1)   Identification Code.
2)   URL of the information server submitting records.
3)   Certificate: the certificate of the company which owns the information server and has submitted the record.
4)   Security Marker: specify whether the records can be read for all, or only for partners on the industrial chain
5)   Timestamp: recording time.

Web service interface in the information discovery service is the interface for users to access. When the interface is connected to the Internet, it can be called by any computer on the Internet, but only those authorized computers can read and write the specific data elements. These Web service interfaces allow authorized companies to register a record when they produce or receive a new product. In addition, these authorized companies use identification code to enquire the URL of the Information Server saving the product information.

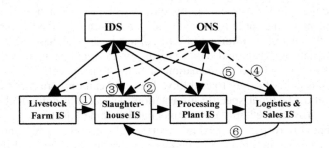

Fig. 3. Work principle of IDS and ONS

Before enterprises in animal food industry chain use discovery services, they need to apply for authorization to the authority to obtain a certificate. The information server configured register from the ONS static, the working principle in Fig.3. Registration process of dynamic information server to the IDS:

1)   Animal food mobile in the industry chain, RFID readers in slaughterhouses identifying need for registration to the IDS;

2) Send the request to the ONS, and return to IDS's URL;
3) The abattoir information server register registers to the IDS to indicate that it has the product information of traceability.

Process of product traceability:

4) Logistics and sales information server inquiries ONS by the product identifies, and returns with IDS's URL;
5) Query the IDS, and it returns with the URL of product information server of ownership, including the slaughterhouse information server URL.
6) Query the Information Server of slaughterhouse, and we can get the retrospective information of the product.

6 Conclusion

This paper takes GS1 standards and RFID identification technology and the characteristics of animal food industry chains into account, fully considers of the traceability systems at home and abroad, and takes their public properties, then proposes a general framework of the animal food traceability. In the framework, we used RFID tags, distributed information servers, Information Discovery Services and ONS, to complete the whole process of animal food tracing and tracking from farm to fork. And we have technically solved some urgent problems of animal food safety traceability systems, such as the Source Tracing, standardized, dynamic data sharing.

Data and management are key elements in food safety traceability, which calls for national legislation to change the management concept of enterprises. At present, some enterprises have taken action to products traceability management, but mostly limited to security, without using internationalGS1 standards in encoding. For tracking and tracing in the distribution center, supply chain, logistics and other links, it is necessary to promote international standards.

Acknowledgements

This work was supported by the National High-Tech Research and Development Plan of China under Grant No D07060501720702, 2006BAD22B06, 2006g35(3).We thank professor Zheng Limin, Zhu Hong, and Wu Ping for their helpful suggestions.

References

Cheung, A., Kailing, K., Schonauer, S.: Theseos: A Query Engine for Traceability across Sovereign, Distributed RFID Databases. IEEE Comm., 1495–1496 (2007)
Fabian, B., Günther, O., Spiekermann, S.: Security analysis of the Object Name Service for RFID. In: Workshop on Security, Privacy and Trust in Pervasive and Ubiquitous Computing (2005)
Xinxin, C., Subin, W., Shuo, W.: Radio Frequency Identification (RFID) technology and application system. Posts & Telecom Press (2007)
Shanahan, C., Kernan, B., Ayalew, G., et al.: A framework for beef traceability from farm to slaughter using global standards: An Irish perspective. Computers and Electronics in Agriculture 66(1), 62–69 (2009)

EPCglobal Inc. EPCglobal tag data standards v. 1.4. (June 2008),
http://www.epcglobalinc.org/standards/tds/
tds_1_4-standard-20080611.pdf

EPCglobal Inc. The EPCglobal architecture framework v.1.3. (March 2009),
http://www.epcglobalinc.org/standards/architecture/
architecture_1_3-framework-20090319.pdf

GS1. GS1 Global Traceability Standard 1.1.0. (February 2009),
http://www.gs1.org/docs/gsmp/traceability/
Global_Traceability_Standard.pdf

Brackenm, J., Mahtews, G.: GS1 Ireland Beef Traceability Case Study. GS1 Ireland (February 2005)

Hongliang, K., Jianhui, L.: Application Review of the Global Identification System (EAN·UCC System) in the Traceability of the Food Safety Supply Chain. Food Science 25(6), 188–194 (2004)

Jinyan, Z., Linli, T., Shizheng, G., et al.: Studies on Animal Food Safety Traceability System Using RFID Technology. Journal of Yunnan Agricultural University 23(4), 528–531 (2008)

Beier, S., Grandison, T., Kailing, K., et al.: Discovery Services—Enabling RFID Traceability in EPCglobal Networks. In: International Conference on Management of Data 2006, Delhi, India, December 14-16 (2006)

Linsen, Z., et al.: Design and Development of Quality Traceability Information Management System and Safety of the Beef Production's Entire Processes. Scientia Agricultura Sinica 39(10), 2083–2088 (2006)

The Function Analysis of Informationization in New Rural Cooperatives Medical Service Management

Yuefeng Zhou and Min Liu

College of Economics and Management, Dalian University

Abstract. The establishment of new rural cooperative medical system is an important action for comprehensive affluent society. It is an important measure for Central Party Committee and State Council to solve "three rural" issue effectively and to overall urban and rural, regional, coordinated economic and social development, building a well-off society in the new situation. It has important role to alleviate farmers to see a doctor expensively, see a doctor difficultly, reduce the burden on farmers and improve their level of health protection and quality of life, solve the problem of poor because of illness and the problem of returning poor due to illness, promote the production and rural economic development and stability in the rural areas. This article will analyze the function of informationization in new rural cooperative medical service management selectively.

Keywords: the new rural cooperative medical service, informationization, function analysis.

1 Introduction

The question of "Three agriculture" has been the most important question for developing our country and building socialism. The key for constructing the harmonious society, achieving the comprehensive better-off society in our country is countryside, the difficulty is also to solve rural question. Especially farmer's healthcare question immediately has influence to our country's economic development and social stability. "The Central Party Committee and the State Council's Decision about how to further strengthen rural sanitation Work" proposed that we should establish new rural cooperative medical service system to reduce farmers' disease economic burden. Therefore, along with the rapid development of our country comprehensive well-off construction, various areas gradually are also carrying out the new rural cooperative medical service system; in order to solve farmers' difficulties in seeing a doctor, seeing a doctor expensively, solve the problem of poor because of sickness and the problem of returning poor due to sickness". The new rural cooperative medical service system is organized and supported by the Official organization; the farmers could participate in voluntarily. The fund of system is collected by individuals, the collective and the government in every way. It is a system that farmers help each other and it is for treating serious sickness primarily. In the implementation of this system, farmers participate in the

D. Li and C. Zhao (Eds.): CCTA 2009, IFIP AICT 317, pp. 305–311, 2010.
© IFIP International Federation for Information Processing 2010

cooperative medical system with the family as the unit and pay certain fee, village collective (if the village has collective property) and All levels of government subsidize certain fund proportionally, all the Fund compose family account and the overall plan account. First farmers have to pay the medical expense personally, and then go to the office of cooperative medical management to obtain the compensation. In brief, the implementation of new rural cooperative medical service system is a popular sentiment project and policy which benefits to everyone in countryside.

2   The Necessity of New Rural Cooperative Medical Informationization Construction

First, in general countryside area, because of the disorder management of farmers' medical Fund and the low capabilities of the fund management, it causes farmers not trust to the policy, all of that cause new rural cooperative medical service system decline in the 80s,which is once largely succeed and has an effect internationally. Second, to the traditional medical management, not only the work load is big and tedious, the working efficiency is low. The applicant to reimbursement for farmers to see a doctor is difficult, the procedural is complex. It does not favor the prompt compensation for farmers who go to see doctors. Once more, the management of new rural cooperative medical service system involves new rural cooperative medical service Control section, the organization, the fixed-point Medical organization, farmers and so on. Its coverage is broad; the personnel are complex; the fixed-point Medical organization disperses, and the related information is large, and each aspect has close connection each other, depending on traditional management tool, it is difficult to combine these resources validly. Finally, the development and application of informationization technology has become the prime motor to promote the national economy and social development. Moreover, in order to achieve the goal that the new rural cooperative medical service system covers the entire countryside resident basically until 2010, With the guidance of scientific development concept, and based on the standard management, The National Medical department also explicitly pointed out that we must establish new rural cooperative medical service information system adapting to the national health information system comparatively effectively in "medical department's instruction opinions about how to construct new rural cooperative Medical service information system", in order to improve work efficiency and do convenience to farmers. In whole, the new rural cooperative medical service management must realize informationization management.

3   New Rural Cooperative Medical Informationization Construction

New rural cooperative medical informationization construction takes the scientific development concept as instruction. using present information technology, we should establish the computer network among new rural cooperative medical service Control section, the fixed-point Medical institutions and other related departments, all the work should be done on-line by computer, realize standard management. The construction of new rural cooperative medical service informationization is to develop many systems in three stratification planets, take the farmer electron healthy database as a center, realize all the work be done by internet, the construction content mainly can be summarized as "a database center, three planet, many systems".

4   The Function Analysis of Informationization Management in New Rural Cooperative Medical Service System

The implementation of the new rural cooperative medical service system is a complex system; the service involves many departments' cooperation and coordination. The function of informationization will be presented on the chain link's main body in medical service system, the main body is named benefit counterparts, which mainly include Medical service Control sections, various fixed-points Medical institutions, farmers, village-level doctors, governments, sanitation and ant-epidemics departments and enterprises of drugs production and Circulation. Following, we will analyze the function of informationization aiming at each benefit counterparts separately.

4.1   To Farmers

First: the informationization construction facilitates the reimbursement of farmer medicine expense. Before infprmationization construction, when farmers saw a doctor, they must take receipts to the office of new rural cooperative medical management and then convert them into cash waiting a long time, the process is quite complex. However, after informationization, when farmers are going out of hospital, they can attain reimbursement directly in the fixed-point Medical institution. All of procedurals are operated by computer directly. The method of Informationization management facilitates farmers to reimburse greatly, and also reduces the unreasonable expense.

Second: facilitates farmers to handle the procedure of extension examine and reduce unnecessary examine expense. There is an electronic health file which record farmers' basic health and personal history, cure history in the new rural cooperative medical informationization management system. When farmers go to see a doctor, doctor can assign out the patients' health records during the diagnosis. Especially when patients want to go to anther hospital, the records about their health history can be transferred by computer. Therefore, patients do not do repeated inspection and repeat tests which cost much money and time. Finally to farmers, they can see a doctor at one place, but get reimbursement at another place. So it brings great convenience to farmers.

Third: benefits farmers to understand their own disease condition and diseases can be treated better. With the construction of long-distance consultation and the medical service advisory service system, Some diseases that can not be treated by village-level doctor can be discussed by outside areas medical experts through long-distance consultation system, they may provide better Therapeutic schedule for patients, then farmers' diseases can be treated well. Thus it avoids the situation which the farmer went see a doctor randomly and also benefits farmers to understand their own disease condition.

4.2 To Fixed-Point Medical Institutions

First: The application of informationization raises the level of hospital internal management service enormously. The internal management of the hospital includes patient management, hospital management, nurse station management, financial management, patient records management and so on; there is a large amount of information which should process. However, after informationization management, each work becomes very simple, only clicks the button, we can complete the work. Informationization management greatly improves the quality of management, work efficiency and service levels, thereby raising serious hospital out-patient and hospitalization rates.

Second: it is advantageous in raising the medical officers' medical service level. Online studying has become the mainstream for learning; the construction of distance learning system brings great convenience to medical officers to learn related medical technology. The medical officers may watch the medical video frequency at home for studying medical technology, when they meet difficult problems, they can exchange opinions with online experts, they can discuss together, and through discussion they can raise their own medical level. Then they will provide better medical service to general farmers.

Third: solves the drugs prompt supply problem and reduces the drugs cost. This system sounds warning automatically with the construction of drug control allocation management system. When the drugs stock reduces to the safety stock, this system will send the drugs table of contents demand to the hospital purchase center and remind the hospital purchase center to buy related drugs promptly. Through the construction of this system , on the one hand, it will safeguard the prompt supply of hospital drugs, on the other hand ,the hospital may know the drugs used medication situation of other various Medical institutions through internet, with the information of that, many medical institutions can buy drugs together, thus reduces the medicine price.

Fourth: Raises the outpatient rate of hospital and increases the hospital income. Through the construction of automatic extension examine system, many patients with no-treated diseases in the Village-level hospital can be sent to the related fixed-point Medical institution automatically. So it will increase the fixed-point Medical institution's sick person enormously, enhance Hospitalization rates, and also enhance the hospital income correspondingly.

4.3 To Medical Service Control Section

First: Regulates the internal management and raises working efficiency. Regarding a county, there is large population, in addition, the implementation of new rural cooperation medical system is operated by the family; the work load to manage Materials is too big, thus the efficiency is low, and the work quality cannot obtain guarantee. Now, with information management system, the administrative officers can be released from the tedious work. The procedure of reimbursement and calculation of expense can be completed with the computer; information management system regulates department's internal management, raises working efficiency obviously.

Second: facilitates the medical service Control section to supervise the fixed- point Medical institutions. It records each expense list of farmers detailed in the new rural cooperative medical informationization system, so officers could supervise the condition of drugs use and expense condition on-line directly through internet. In order to safeguard the Fund security and make the Fund coming from farmers use to farmers, to maximize the effectiveness of the Fund, to solve farmers' anxiety that Fund may be misused and embezzled. With informationization management, it also can guarantee safety of farmers who use drugs, at the same time, increase the transparency of the Fund service condition.

4.4 To Village-Level Doctor

First: Facilitates Village-level doctor to settle accounts. Since the information management of new rural medical service system, all of procedurals are completed on the platform of new rural cooperative medical service information management, such as swiping cards, keeping account and transferring accounts. So long as the Village-level doctor takes the receipt to the office of new rural cooperative medical service system management, the receipt can be converted into cash directly.

Second: Enhances the outpatient rate of Village-level hospital, information management of system solves the question that many farmers go to see a doctor in big hospital frequently but not the public health center, community hospital and village medicine room, increase the income of village medicine room outpatient service.

4.5 To the Government and Disease Guard Department

Information management provides scientific basis for the government and disease guard department to make hygienic decision. Through the construction and the application of new rural cooperative medical information management system, we can establish the life-long effective computerization health files for general farmers. Based on the computerization health files, we can form the region farmer disease spectrum, and disease spectrum provides most reliable materials to make public health decision for government. With this system Simultaneously, we may also establish informationization mechanism to prevent epidemic disease early and the public health thunderbolt's rapid reaction and the stitch in time, we may take the anti-epidemic measures in view of the high disease incidence rate disease, we may prevent the dissemination of infectious diseases and reduce diseases' occurrence, from the early time, we can pay great attention to safeguard the farmer health.

4.6 To Enterprises for Pharmaceutical Production and Circulation

The computerization health files show the market orientation to the business for pharmaceutical production and circulation. Research and development of any kinds of new drugs is based on market-oriented, it not only takes cost but much time to research the market demand, the results of survey will directly affect accesses to the future development of production. So market research for pharmaceutical manufacturers is a major challenge. However, farmers' computerization health files indicate the direction for them, manufacturers only need to carry on analysis to the file, and then they may understand the market direction basically. Going with what is desired, it will bring the huge benefit finally to manufacturers. Likewise, flowing of drugs is also a big difficult problem, enterprises for circulation must collect massive materials and analyze, only then, and they can have a clear goal to transferred drugs to. However, collecting these materials is not easy, the data is scattered, and the work load is so big. The establishment of farmers' medical service computerization health files has brought gospel for it, enterprises for Circulation also only carry on analysis to files, and they may know the market direction at ease. In brief, the establishment of computerization health files has brought enormously convenient to manufacturers and enterprises for drugs Circulation.

5 Summary

In short, through the construction of new rural cooperative medical information, we can: (1) realize Information and network management of farmers' electronic health material and the sharing of the farmers' information to each fixed-point medical institution. It's convenient for farmers to see a doctor and transfer another hospital, re-alize" treating in one place, but applying for reimbursement in another place", improve the capacity of social security of farmers, reduce the gap between farmers and urban residents, and it's favorable to promote the balance urban and rural development. (2) The establishment of all life effective electronic health information of farmers is helpful to enhance farmers' awareness of health care, realize disease rely on preven-tion-oriented, solve the Pyramid structure problem: overcrowding in major hospitals, but less people in institutions of Health and Community Hospital (80% farmers en-joy20% health care resources, the status quo of Dr. See a cold), so as to enhance the level of people's health.(3) norm business management of the fixed-point medical institutions, improve working efficiency and service level. It's favorable for medical management to supervise and manage the fixed-point medical institutions, ensure safety of fund, and drugs manufacturers and suppliers have a clear market-targeted, so they can shoot to targets directly. (4) it's useful for government to make decision about public health policy. At the same time, it also can establish information mechanisms for epidemic early warning , rapid response and timely processing about public health emergencies(for example: through connecting the phone of villagers to village medical room, sudden illness can be alarmed on village doctors' computer or mobile phone by pressing a key, and the location of the patient, the most recent line are showed),and then the first defense line will be established to resist and control of communicable diseases. The new rural cooperative medical information technology brings significant benefits to the entire business chain, so, In order to promote the sustainable development of the new rural cooperative medical system, countries and regions should strengthen the informationzation construction of the New-type rural cooperative, and make this policy carry out.

References

Jiu, S.Q., Yong, S., et al.: Discussion about new rural cooperative medical service informationization construction. China health economy (3), 25–26 (March 2004)

Runlong, X., Xinle, Y., Jianying, Y.: The report of new rural cooperative medical service informationization construction in Xiao Shan area, Hangzhou, Zhejiang Province. China rural sanitation enterprise management (12), 29–30 (Decmber 2004)

Shoujue, W.: Research on informationization management of the new rural cooperative medical system. Shanghai Transport University (June 2007)

Feng, Y.Y.: The research on strengthening the new cooperative medical system funds management. Shanghai Transport University (May 2008)

Current Situation and Countermeasures of Agricultural Information Construction in Jiamusi Area

Dongwei Shao[*], Junfa Wang, Donghua Jiang, and Qisheng Liu

College of Mechanical Engineering, Jiamusi University, Jiamusi,
Heilongjiang Province, P.R. China 154007,
Tel.: 0454-6553306; Fax: 0454-8550757
sdwshao@126.com

Abstract. Beginning with the current situation of agricultural information construction in Jiamusi area, the achievements obtained from agricultural information construction are clearly known by discussing current work conditions. On the basis of thorough investigation, the existing problems of agricultural information construction are found. The reasons of these problems are analyzed in detail, and then the general idea of agricultural information construction is laid out in this paper. At last, aiming at how to develop agricultural information construction, the feasible measures are proposed from the aspects of strengthening infrastructure construction, enhancing the quality of talent teams and integrating information resources.

Keywords: Jiamusi, agricultural information construction, current situation, countermeasures and suggestions.

1 Introduction

Agriculture is the foundation of the national economy, and agricultural information is an important aspect of the national information. However, China is still at the primary stage of socialism, and the degree of agricultural information is not very high, especially for the relatively backward region, Jiamusi. Significant development has been made in agriculture and the problem of food and clothing is basically solved after 30 years development of China's reform and opening up. However, to Jiamusi region, because of more agricultural population, relatively backward economic development level, geographical remoteness and larger proportion of agricultural production, the production concept of peasant is relatively backward; information exchange is poor; the quality of the farmers is generally low;agriculture production remains at the traditional stage and so on. To solve these problems, it is necessary to reform the traditional mode of agricultural development and to use agricultural information to promote agricultural modernization.

---

[*] Corresponding author.

D. Li and C. Zhao (Eds.): CCTA 2009, IFIP AICT 317, pp. 312–317, 2010.

2 The Main Problems of Agricultural Information and Running Status of Jiamusi Area

2.1 The Operational Status of Agricultural Information of Jiamusi Area

Agricultural information construction in Jiamusi starts late, but in recent years some progress has been made in the construction of infrastructure, information services system and the areas of application of information technology under the guidance of the national and provincial ministries and commissions.

2.1.1 Infrastructure Construction and Network Service Functions Have Been Increasingly Improved

All the administrative villages of Jiamusi have telephones, the rate reaching 96%, and the telephone network has reached 804 villages;the number of having phones is 136,806. Currently more than 800 villages have realized broadband Internet access in the city; the rate of the villages having broadband network is more than 80%.Towns and counties throughout the city are built in cable transmission networks, and cable networks have been built more than 3000km. Up to now, the total of radio and television agencies in the city is 70; the number of township and village having radio and television is the 70 and 948. Radio coverage of the city's population is 94.6 percent; the coverage of television population is 94.5 percent, and cable television subscribers reached 70%, basically formating a CCTV, Heilongjiang TV programs as the main body, radio and television, wireless and wired, and FM, satellite and network, such as the combination of a variety of technical means, differently covered by TV networks and radio transmission coverage.

2.1.2 The Channels of Information Dissemination Continue to Be Broaden

Agricultural Information Service Station of the city opens the "Sanjiang Agriculture" web site. "Three-River Agriculture" plate is the Jiamusi city government Web site's main section. The plate has opened up superior crops and leading products, and farmers are guided by new products, new projects and the supply and demand of information, market price information to adjust the industrial structure, so as to sell grain; by taking full advantage of online information in agriculture, selecting information and opening up the page, column, feature, etc., releasing all the information to the grass roots and the majority of farmers through traditional media (television, radio, newspapers, magazines, etc.).

2.1.3 The Market of Information Showed a Prototype and Online Transactions Achieved Initial Success

With the increasing agricultural restructuring, regional characteristics, industry characteristics, product characteristics are gradually emerging. Everywhere the Internet tools of the modern information transmission are in the full use, and agricultural production is actively done, in particular the work of characteristic online products, on-line market of agricultural products has been taking shape now, online sales become a larger scale. So far, the city's internet agricultural products have reached more than 300 , playing an active role in broadening the market of agricultural products and increasing the promotion of farmers.

2.2 The Main Problems of Agricultural Information of Jiamusi

Jiamusi agricultural information construction is characterized by late start, but the pace of development is relatively fast. The problem is still very prominent, and the concrete embodiment is in the following areas.

2.2.1 The Construction of Agricultural Information Is Understood Lackly

From the investigation of the county and township government, the situation is prevailing, which is understanding not in place, not enough attention. The diversity of the information services systems is not recognized, forming one-sided understanding of the single information service system, not conducive to the effective transmission of information; the network functions generally stay in using networks to send and receive information, do not play the role of using network information to improve operating and developing markets; from the survey of the leading enterprises, because of the size of the business, product sales object, as well as the different qualities of business leaders, and its different understanding of the network. Some enterprises of large-scale and a large proportion exports in city outskirts attach great importance to the network; but other strong traditional concept enterprises of small scale and low overall quality, rarely keep in touch with domestic and international market, also its awareness of using the network is not strong, not paying enough attention. The role of agricultural associations and the rural economy organizations has not fully developed. From the investigation of farmers, a small number of farmers pay more attention to the construction of agricultural information, so they benefit a lot from the network in actual production. Most farmers are still not aware of the role of networks informatizing, needing to step up publicity.

2.2.2 Agricultural Information Technology Infrastructure Is Relatively Backward

Jiamusi region is a less developed region, with limited revenue, the state investing less, which leads to the fact that the construction of agricultural information of Jiamusi region is poured less and less. So now the lack of personnel training, hardware and software equipment changing and the follow-up input of technology up grades have a direct impact on the daily operation after the completion of the site, resulting in low levels of information technology work. At the same time, a comprehensive information database and system are not healthy enough, low information exchange and a low degree of resource sharing. Network market of agricultural products having realized online trading has not yet formed, and the development and application agricultural geographic information systems, satellite remote sensing information system are still at the infancy stage.

2.2.3 Lack of Personnel in Agricultural Information

The requirements of the quality of personnel on the construction of agricultural information and other information industry have obvious differences. The construction of agricultural information needs the "compound" talents who are skilled in agriculture using a variety of information tools, also understanding internet information technology. There are many agricultural experts among agricultural population of the county and township levels in our city, and there are some technicians understanding computer, but the "compound" talents who know computers and agriculture are not enough, which does not suit the development of agricultural information needs in the new period.

3   The Countermeatures and Suggestions of Agricultural Information Construction in Jiamusi Area

3.1   Strengthen the Leadership and Advocacy Efforts of Agricultural Information Construction

Information technology in rural areas is an inter-departmental, cross-sectional, regional and multi-technology integration of business systems engineering, related to all aspects. All levels of government must strengthen the unified leadership of the construction of agricultural information work, proceeding from actual conditions, and firmly establishing the concept of information, recognizing that information is the resources, that is, the idea of wealth. Government must attach great importance to establishing a strong leadership system, and strengthen the organization of agricultural information management.Set up a special organization to further raise awareness, clear responsibility to implement the task. All departments should coordinate with each other to form a work force, take effective measures to ensure that capital investment, technical guidance, supervision and management are "three in place".

3.2   Establish the Construction of System Including Government, Science and Education Units, Associations, Enterprises (Information Enterprises, Agriculture-Related Enterprises) and Individuals, Households

Counties (cities) and districts of the city unify development and management of the rural information resources system, unify the training and management of agricultural information in accordance with the development of information construction in rural areas and rural implementation of integrated information service system in the overall program requirements and plans from actual situation to solve the organizational issues, personnel issues and financial implications of the agricultural information questions. Enhance the development and utilization of agricultural information resources and accelerate the pace of construction of agricultural information and close around the optimization of agricultural structure, improving the rural ecological environment, the dissemination of agricultural science and technology knowledge, the publishment of information about agricultural products and the increase of agricultural income of the farmers, paying close attention to implementation, and paying close attention to efforts. At the same time, we should also make full use of news media, such as radio, television, publicity boards, network and so on, to promote important role on the construction of agricultural information further and extensivly in agricultural growth, farmers' income, and further promote the popularization of information technology knowledge, continuously improve their understanding of information technology so that farmers benefit from them, happy to accept. The role of the media should be fully played; the extent of participation of the farmers' agricultural information should be improved, carry out a variety of publicity and education, accelerate the popularization of information knowledge. At the initial stage of information of the rural areas, government departments should actively guide, increase investment, establish a diversified investment and financing mechanism, improve management supporting system of the construction of agricultural information, assume the main role of agricultural information, promote the building of agricultural information, and improve the face of relatively backward of the rural areas, create the conditions for farmers to abandon the old production modes, to accept new concepts, to find new ideas.

4   The General Idea of Scientific Planning on the Construction of Agricultural Information Work in Jiamusi

In the next few years, agricultural information construction in Jiamusi region must take accomplishment of the agricultural modernization as the goal, take the market as the guidance, take science and technology as support, take the revitalization of northeast old industrial base as an opportunity, stand on a high degree of development of the future agricultural globalization, knowledge and commercialization, further establish and improve the agricultural information network system, strengthen the information services function, fully provide the trend of agricultural products and markets, rural economic information and technical information on agricultural production; strengthen the agricultural database building, information analysis and forecast release; take Jiamusi city as the core, take Huanan City and Fujin County as the wings , make full use of advantages of modern agricultural demonstration base construction on Jiansanjiang and Hongxinglong farms Authority, guided by the principle of the "one plan, the implementation step by step, constructs while uses, and high-performance running ", strive for a high planning starting point, novel methods, and bright features, and drive the development of agricultural information work around the cities and counties. At the same time, agricultural information construction in Jiamusi region must be guided by the principle " in the light of local conditions and classified instruction", further establish and improve the agricultural information system, get a current foothold, focus on long-term benefits, and achieve a modest investment, convenient operation, prompt and effective goal. In the network construction, the professional production base, agricultural products market, leading enterprises, the certain scale cooperative economy organization, the farmer managers and the agricultural wealthy and powerful family networking work should be together integrated to the plan as key. In the resources construction, various aspects strength should be conformed, and the city county agriculture information resource unit exploitation pattern should be established, and the information resource on-line exchange, on-line issue and on-line propaganda should be overall considered. In the information issue, the concrete time, the way and the content should be thoroughly considered and arranged, and thus the whole city agriculture information network's integrated function will be truly displayed.

Acknowledgements

The first author is grateful to the Jiamusi University for providing her with pursuing a Master's degree at the Jiamusi University.

Fuzzy Comprehensive Evaluation of Rural Information Poverty in China — Case Study of Hebei Province

Guizhen Sun[1], Shuanjun Wang[2], Yaqing Li[1], and Huijun Wang[3,*]

[1] College of Humanities and Social Sciences, Hebei Agricultural University,
Baoding, Hebei Province, P.R. China 071001
[2] Economic Management Department, Software and Technological Institute of Hebei,
Baoding, Hebei Province, P.R. China 071000
[3] Hebei Academy of Agricultural and Forestry Sciences, Shi jiazhuang,
Hebei Province, P.R. China 050051,
Tel.: +86-311-87652003; Fax: +86-311-87066140,
nkywanghj@yahoo.com.cn

Abstract. Information poverty is a new form of poverty in information society. With the growing information-gap between urban and rural areas, information poverty is prevailing in the vast rural areas in China. It is largely restricted the new rural construction and the social harmonious development of villages and towns and must be resolved. The evaluation of rural information poverty is the premise to resolve it. In order to estimate the problem, index system of rural informatization evaluation of Hebei province was designed by means of Delphi. Then, according to the survey of farmers' information demand, AHP and FCE were used to estimate rural information poverty of Hebei province. The purpose of this study is to provide a new operational approach in evaluating or solving rural information poverty and constructing rural informatization in China.

Keywords: rural areas, information poverty, fuzzy comprehensive evaluation (FCE).

1 Introduction

Information poverty, a new form of poverty, has become a common problem around the world. With the development of information and communication technology, it is more prominent and must be resolved. Abroad, the first concern of it is from the "digital gap". Lloyd Morrisett (1995) pointed out that the "digital divide" was awareness of a divide between the information haves and the information have-nots. It is real concerned by the whole world derived from the series of reports: Falling Through the Net published by United States long-distance communications and information administration (NTIA). Hu Angang (2005) pointed out that the Governments of developed countries provided universal services as a principal method to resolve it. In United States, Government has adopted a series of positive measures such as legislation, social participation and Government's encouragement and so on. In Japan, Government has promoted the computer's popularization in rural

---

* Corresponding author.

D. Li and C. Zhao (Eds.): CCTA 2009, IFIP AICT 317, pp. 318–325, 2010.
© IFIP International Federation for Information Processing 2010

areas through agricultural networks, financial support and computer training (Tian Ye, 2001). In Korea, Government started the "Information Network Village" project in 2001 to eliminate the "digital divide" (Ren Guisheng, 2006).

In China, information poverty is becoming more prominent in rural areas. Digital divide is huge between villages and towns (Mei Fangquan, 2007), which has seriously hindered agricultural development, rural progress and farmers' income. In order to solve rural information poverty, Government has promulgated a series of policies to promote the rural informatization. The majority of researches have been in the macro-field such as the significance, influencing factors, policies and measures and so on (Wen Jianlong, 2005; Liu bin, 2006; Wei Gang et al., 2006; Ding Kuili, 2007). At present, researches on the evaluation of rural information poverty are almost blank. In this study, index system of rural informatization evaluation of Hebei province was designed. Based on this system, Analytic Hierarchy Process and Fussy Comprehensive Evaluation were used to estimate rural information poverty of 6 different regions in Hebei province. The purpose of it was to provide an operational approach and a reference for evaluating or revolving rural information poverty in Hebei province or other areas.

2 Designing Index System of Rural Informatization Evaluation of Hebei Province

Informatization is not only an important cause of information poverty but also an effective way to solve it. The level of informatization is inversely proportional relationship to the extent of information poverty. In this study, index system of rural informatization evaluation of Hebei province was the basis of the evaluation of rural information poverty of Hebei province. According to index system of national informatization of China, index system of rural informatization of China and the characteristics of rural informatization of Hebei province, index system of rural informatization evaluation of Hebei province was designed through the Delphi method to evaluate rural information poverty of Hebei province. It has 3 stair indexes and 8 second grade indexes, shown in Table 1.

Table 1. Index system of rural informatization evaluation of Hebei province

stair index	second grade index
subjective environment of rural informatization	1. per capita annual net income of farmers (yuan)
	2. proportion of high school or above of farmers (%)
infrastructure of rural informatization	3. popularization rate of telephone (include mobile telephone) (sets/100 households)
	4. number of computer owned per 100 rural households (set)
	5. number of TV set owned per 100 rural households (unit)
effect of rural informatization	6. popularization rate of information service station at village (%)
	7. ration of internet users in rural areas (%)
	8. ration of information from telecommunication networks, radio and television networks and the internet (%)

3 Counting the Index Weight of Rural Informatization Evaluation System of Hebei Province by AHP

Rural informatization evaluation of Hebei province was a multi-index comprehensive evaluation, the core algorithm of it focused on the calculation method of the index weight. In this study, Analytic Hierarchy Process (AHP) was used to calculate the weight.

3.1 Steps of AHP

AHP was put forward by professor T.L.Saaty (Lu Taihong, 1998), including the following four steps.

(1) Establishing stepped hierarchy model. The stepped hierarchy model of this study shown in Figure 1 was established based on Table 1, with 3 layers and 8 principal evaluation indexes.

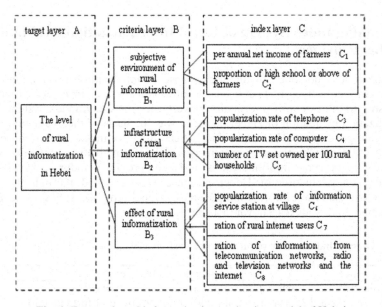

Fig. 1. Frame of rural informatization evaluation model of Hebei

(2) Constructing comparison judgement matrix. By pairwise comparing elements of the same layer in accordance with certain upper factor, each element of the judgement matrix is able to be defined. The relative importance of each element follows 1-9 scale of comparison. On the basis of the scores that experts provided, several judgement matrixes can be established.

(3) Monolayer weights order and its consistency test. Monolayer weights order is defined as the importance that each element of No.(k+1) layer relative to No.k layer is ordered according to the judgement matrix. In order to ensure the effectiveness of monolayer weights order, the judgement matrix should be dealed with consistency test judged by the random consistency ratio CR. If $CR<0.1$, the result can meet the requirements and so the order weights accepted. Otherwise, it must be adjusted to meet the consistency test. CR is calculated in accordance with the following formula.

$CR=CI/RI$

Where: CI is the consistency index, RI is the average random consistency index, and $CI=(\lambda max - n)/(n - 1)$ (λmax is the largest eigenvalue, n is the number of the order of judgement matrix)

(4) Total weights order of hierarchy and its consistency test. Total weights order of hierarchy refers to the relative importance that each element of No.(k+1) layer relative to the certain element of No.k layer is ordered according to the monolayer weights order. Similarly, its consistency test should be done. If $CR <0.1$, it can be satisfied and accepted. Otherwise, it is necessary to be adjusted.

Where: $$CR = \frac{CI}{RI} = \sum_{j}^{m} a_j CI_j / \sum_{j}^{m} a_j RI_j$$

3.2   Result of AHP

The monolayer weights order of this study passed the consistency test. And the result of total weights order of hierarchy was shown in Table 2.

Table 2. Estimation matrix and weight of rural informatization index system of Hebei

criteria layer B index layer C	B_1 0.101	B_2 0.226	B_3 0.674	total level ranking weight	ranking result
C_1	0.750	0	0	0.075	4
C_2	0.250	0	0	0.025	7
C_3	0	0.236	0	0.053	6
C_4	0	0.682	0	0.154	2
C_5	0	0.082	0	0.018	8
C_6	0	0	0.091	0.062	5
C_7	0	0	0.218	0.147	3
C_8	0	0	0.691	0.466	1
CR	0	0.057	0.047	0.057	$CR<0.1$
RI	0	0.580	0.580	0.580	meet the test of consistency

Table 2 shows that the main indexes to influence the level of Hebei rural informatization are index C_8 (information ration from telecommunication networks, radio and television networks and internet), C_4 (popularization rate of computer) and C_7 (ration of rural internet users) shown in Figure 1. The total weight of which is up to 0.767. It points out that the main task of rural informatization of Hebei in the future is to strengthen the development of informational technology infrastructure and to improve the popularization rate of computer and network. And it can also provide a reference in evaluating the level of rural informatization.

4 Fussy Comprehensive Evaluation of Rural Information Poverty of Hebei

Hebei province, located in the eastern part of China, has diverse types of terrain divided into 6 geo-economic regions—suburbs, piedmont areas, mountainous areas, Bashang plateau areas, littoral areas and low plain areas. According to index system of rural informatization evaluation of Hebei province, the level of rural informatization of 6 different regions was evaluated by Fussy Comprehensive Evaluation (FCE) to evaluate rural information poverty of Hebei province. FCE, a fuzzy decision-making method, was put forward by L.A.Zadeh and R.E.Bellman in 1965. By constructing a hierarchy of fuzzy subset, the membership degree of fuzzy indicator reflected the evaluated object can be identified and then the integration of each indicator can be obtained by fuzzy principles.

4.1 Steps of Fussy Comprehensive Evaluation

4.1.1 Determination of Factor Set of Evaluated Object
Based on Figure 1, the evaluated object of this study was divided into 1 target set (U) and 3 criteria subsets (U_1, U_2, U_3).

$U=(U_1, U_2, U_3)$=(subjective environment of rural informatization, infrastructure of rural informatization, effect of rural informatization)

$U_1=(U_{11}, U_{12})$=(per annual net income of farmers, proportion of high school or above of farmers)

$U_2=(U_{21}, U_{22}, U_{23})$=(popularization rate of telephone, popularization rate of computer, number of TV set owned per 100 rural households)

$U_3=(U_{31}, U_{32}, U_{33})$=(popularization rate of information service station at village, ration of rural internet users, ration of information from tele- communication networks, radio and television networks and internet)

4.1.2 Definition of Comment Set
This study took 5 comments—(better, good, average, poor, poorer), and their corresponding scores were shown in Table 3.

Table 3. Comment set and score comparison chart

comment set	better	good	average	poor	poorer
score	90-100	70-90	50-70	30-50	10-30

4.1.3 Single-Factor Evaluation

It is to determine the membership grade of each hierarchy subset and then to establish the fuzzy relationship matrix R.

4.1.4 Comprehensive Evaluation

It is to compute the fussy comprehensive evaluation vector B based on the weight vector W and the fuzzy relationship matrix R. The formula is $B=W \cdot R$.

Where: the weight vector W can be checked in Table 2.

4.1.5 Computing the Final Value of Comprehensive Evaluation

Fussy comprehensive evaluation vector B multiplied comment set score vector (100, 90, 70, 50, 30)'equaled comprehensive evaluation value T. That is, $T = B \cdot (100, 90, 70, 50, 30)'$.

4.2 Data Source

First of all, the data came from 446 rural households located in 6 different regions in Hebei province. Secondly, statistical results of the survey were counted by K-means clustering analysis for 5 categories with SPSS 12.0 statistical software. Finally, the clustering result of each index was ranked in descending order, then the fuzzy subset of membership grade was obtained.

4.3 Results and Discussion

The followings were comprehensive evaluation results of rural informatization of 6 regions in Hebei province by FCE.

$$B_{Suburbs} = \begin{pmatrix} 0.1252 & 0.3907 & 0 & 0 & 0 \end{pmatrix}$$

$$B_{Piedmont} = \begin{pmatrix} 0.0348 & 0.1252 & 0 & 0.3141 & 0.0418 \end{pmatrix}$$

$$B_{Mountainous} = \begin{pmatrix} 0 & 0.0120 & 0.0424 & 0.1475 & 0.3141 \end{pmatrix}$$

$$B_{Bashang} = \begin{pmatrix} 0 & 0 & 0.3559 & 0 & 0.1600 \end{pmatrix}$$

$$B_{Littoral} = \begin{pmatrix} 0.3634 & 0.0025 & 0.0991 & 0.0468 & 0.0041 \end{pmatrix}$$

$$B_{Low-plain} = \begin{pmatrix} 0 & 0.4132 & 0.0534 & 0.0076 & 0.0418 \end{pmatrix}$$

The result of FCE is a fuzzy vector. If compared on a number of objects, it is often difficult to obtain a clear comparison of the conclusions. In order to resolve this problem, the value T of a comprehensive evaluation is calculated according to the formula: $T = B \begin{pmatrix} 100 & 90 & 70 & 50 & 30 \end{pmatrix}'$

Where: vector B is the result of FCE, (100 90 70 50 30) is the greatest score of the grading comment sets shown in Table 3.

According to the formula, the comprehensive evaluation value T of 6 regions in Hebei province were obtained and shown in Table 4. Then, based on the inversely proportional relationship between the level of informatization and information poverty, the evaluation of rural information poverty of 6 regions in Hebei province was obtained and shown in Table 4. The extent of information haves was prescribed (richer, rich, average, poor, poorer) corresponding with 5 comment set shown in Table 3.

Table 4. Evaluation of rural informatization and extent of information haves of 6 regions in Hebei province

area	comprehensive evaluation result		level of rural informatization	extent of information haves
	scorce T	ranking		
suburbs	47.6832	1	poor	poor
littoral	45.9686	2	poor	poor
low plain	42.5532	3	poor	poor
piedmont	34.4380	4	poor	poor
Bashang plateau	29.7112	5	poorer	poorer
mountainous areas	20.8400	6	poorer	poorer

Table 4 shows that the level of rural informatization of 6 different regions is poor and information poverty is prevalent in Hebei province. And with the value T gradually decreasing, the extent of information poverty is gradually increasing. The value T of mountainous areas is the minimum. Accordingly, the extent of its information poverty is the most serious. Then, followed by Bashang plateau areas. Therefore, more attention should be paid to the two regions during rural informatization in Hebei province.

The level of informatization directly reflects the extent of information poverty. The higher level of informatization in a certain area, the more achievements of informational civilization shared by people and the more convenient way to get information, and the lower level of information poverty. Instead, the opposite is true. Therefore, strengthening rural informatization is an effective way to resolve information poverty in rural areas in China.

5 Conclusion

In this study, the evaluation system of rural informatization of Hebei province was designed by Delphi to estimate the rural information poverty of Hebei province. The weight coefficients of indexes at all levels of this evaluation system were fixed by Analytic Hierarchy Process. Then, based on the results, Fussy Comprehensive Evaluation was used to estimate the rural information poverty of 6 different regions in Hebei province. Result of this study indicated that information poverty was prevailing in rural areas in Hebei province, especially more prominent in mountainous areas and Bashang plateau areas where more attention and more investment should be given to solve rural information poverty and to promote rural informatization and the coordinated development between urban and rural areas in Hebei province.

The evaluation of rural information poverty is the premise to resolve rural information poverty. Rural information poverty involves many aspects, so the method to evaluate it is very important. The method AHP and the method FCE are a combination of qualitative and quantitative analysis. It is a new way to evaluate or measure the extent of rural information poverty by AHP and FCE. This study provided a new operational approach in evaluating and resolving information poverty in rural areas. Due to the different level of rural informatization in different provinces or regions, the index system of rural informatization evaluation should be in the light of the local conditions to design and to conduct an objective evaluation of rural information poverty.

Acknowledgements

Funding for this research was provided by Hebei Provincial Science and Technology Department (P. R. China) and Agricultural University of Hebei. The authors are grateful to Hebei Agricultural University, Software and Technological Institute of Hebei and Hebei Academy of Agricultural and Forestry Sciences for providing them with help in the research.

References

Kuili, D.: Analysis of rural public information products under the perspective of information poverty. Journal of Hubei Administration Institute 33(3), 85–87 (2007) (in Chinese)

Angang, H.: Accelerate the common telecom services in rural areas to narrow the information-gap between villages and towns. Journal of Zhongguancun 28(8), 62–66 (2005) (in Chinese)

Bin, L.: On China's rural information poverty and its causes and responses. Journal of Jinggangshan University 27(9), 76–79 (2006) (in Chinese)

Taihong, L.: Information Analysis, pp. 293–299. Sun Yat-sen University Press (1998) (in Chinese)

Fangquan, M.: Choice of development model of low-cost rural informationization in China. Journal of China Information Times (9), 12–13 (2007) (in Chinese)

Guisheng, R.: Enlightenment to narrow the digital divide in Korea. Journal of Management World (7), 157–158 (2006) (in Chinese)

Ye, T.: Agricultural informationization and its inspiration in Japan. Journal of Global Technology Economy Lookout 181(1), 47–48 (2001) (in Chinese)

Gang, W., Qin, G., Weihong, Z.: Bridging the digital divide and building a harmonious society. Journal of Gansu Social Science (1), 184–185 (2006) (in Chinese)

Jianlong, W.: The issue of information divide and vulnerable groups. Journal of Socialist Theory Guide (8), 21–23 (2005) (in Chinese)

Realization of the Regional Advantageous Agricultural Industries Analysis System

Kaimeng Sun

Institute of Agricultural Information, Chinese Academy of Agricultural Sciences, Beijing, P.R. China 100081

Abstract. In this paper, a system for analyzing the strategic adjustment of regional agricultural industries in China is briefly introduced, particularly on several system functions of database, region selection, statistics, comparative analysis and result presentation in terms of their technical realizations. The introduction is focused on the realized comparative advantageous agricultural analysis computation methods and assessment recommendations. From this paper, we can see an application of a knowledge based system for regional advantageous agriculture analysis. Technical process of system analysis and system design for developing such a knowledge-based system is presented.

Keywords: GIS, Agricultural regions, Computer, System.

1 Introduction of the System

1.1 Objectives

The Regional Advantageous Agricultural Industries Analysis System (or the System hereafter) is designed to analyze country-based advantageous agro industries and their indicators based on a region's natural resources, current agricultural production, and agro-economic performances in China. Through analyzing cross-regional (neighboring regions and within the province) comparative advantages, current regional development will be assessed. Meanwhile, by using temporal data analysis, development of the regional advantages can be obtained so as to identify the long-term trends of the regional economy. Therefore, the System is designed by focusing on regional advantages to be taken. However, due to unavailable agricultural economic data sets, some regional industrial advantages may not be fully analyzed. Data in this study have included that of regional natural resources, labor force, major crop production, and local enterprises. By analyzing these datasets, the overall development of regional advantageous agricultural industries can be understood.

1.2 Major System Modules

By considering data flow, functional requirements and system characteristics, the System is designed with modules of data management, data processing, knowledge reasoning, result output, and man-machine interface. Working together, a complete system flow from data selection, model computation, inference, and analysis result

D. Li and C. Zhao (Eds.): CCTA 2009, IFIP AICT 317, pp. 326–332, 2010.

output is realized to obtain the overall analysis and assessment of regional agriculture economy.

Therefore, the System is composed of region selection, statistics analysis, advantage comparative analysis, assessment, and data management functional modules.

2   Realization of Database System

In order to precisely define the regional industries, large amount of basic data are necessary. The analysis results rely heavily on availability of accuracy of detailed information on resources, agricultural production, and local economic activities. For this purpose, relevant databases are developed containing spatial, temporal, economic, and knowledge. Databases in the System are classified into five areas:

2.1   Spatial Database

In the regional advantageous agricultural industrial analysis system, a 1:4,000,000 national country spatial database developed by Super Map is used; including the map referenced spatial attributes and county names and codes. Using this spatially relational database, county level spatial locations and related agricultural basic agro-economy data are utilized to extract datasets for functional modules for statistics, computation, analysis, and assessment.

2.2   Agricultural Economy Basic Database

According to the System design, a total of 64 datasets are selected on agricultural resources, crop plantation of both grain crops and cash crops, livestock farming, local enterprises, and economic indicators, and as well as planting areas, population, stable harvesting area, summer crop plantation area and summer crop yield. All the data are original and maintained in ACCESS data tables.

2.3   Interim and Final Computation Result Database

The database is developed by ACCESS, including:

Interim computation results: Storage of agricultural economic original data, such as wheat planting area, total yield, for ranking and comparative advantages computation.

Computation results: Storage of computation results such as statistics and comparative advantages computation results.

2.4   Knowledge Based Database

It is used to store knowledge for reasoning computation. This information set is basis for inference and rules for analysis and judgment. So it is like a combination of knowledge base and rule base.

2.5   Code Reference Database (Data Dictionary)

It contains data dictionary to be used in the system, including names of provinces, counties, project title, menu, and database field references.

3 Realization of Data (Region) Selection Module

3.1 Region Selection on Map

The GIS Supermap platform is used to select a region. After a region is selected on map, the program will create a dataset for the selected region including region name and code. The two data attributes are accessed to query matched agricultural data. According to the design, simple statistics will be computed such as unit crop yield and computation results will be exported into the result database for future use. Once a region is selected by user, the region will be marked red. Relevant adjacent regions will be selected by user.

3.2 Region Selection by Name

User can choose a province name at first and the system will automatically display related counties in a popup. According to user selection in the popup, system will extract region code from county database and then access corresponding spatial database to the county and finally mark red of the selected region on the map. Meanwhile, the system will provide simple statistics such as crop yields and then store the computation result in the database for future function use.

3.3 Adjecient Region Selection

To select an adjacent region is to select border regions with the selection region, with common border on the map. The selection of adjacent region is based on user selection and extracted from spatial analysis. The selected province is based on the first two digits of region code (provincial code). Agricultural data are extracted from agro-economy database according to the first two code digits automatically by the system.

4 Statistics Module

Statistics computation includes ranking of the indices of the region by comparing with adjacent regions. The ranking and computation method is not presented in detail here.

For overall status assessment, method can be different for selected and adjacent regions. For six classes of data such as agricultural resources, grain crop plantation, cash crop plantation, animal farming, rural enterprises, and economic indices, items can be different for different regions. The overall assessment in terms of excellent, good, moderate, poor, or extremely poor will be resulted.

To achieve the overall assessment, according to selected and adjacent regions and their indices, rankings for the indices will be computed, including 6 items of agricultural resources, 8 items of grain crop indices, 7 items of cash crop yields, 10 items of animal farming indices, 5 items of rural enterprises, and 9 items of major economy indices.

5 Omparative Advantages Computation

Computation of comparative advantages of a region is firstly focused on agricultural crop indices such as for summer crops, wheat, corn, beans, oil crops, and sweet beet. Scale production advantages, productivity advantages, and efficiency advantages will be firstly computed and then the overall advantages based on their squares and square roots. For other non-crop indices, only scale advantages will be computed such as arable land area, rural population, paddy land area, rural enterprises, rural enterprise employment, total income, etc. Therefore different processing methods are used for the two different types of data.

The computation method for comparative advantages is detailed as follows:

5.1 Scale Advantages

An index of a selected region (county), such as crop plantation area, fresh water fishing, or enterprises, is compared with mean of adjacent regions to represent its scale advantage. The formula will be:

$$S_{ij} = \frac{A_{ij}}{\dfrac{1}{n}\displaystyle\sum_{j=1}^{n} A_{ij}}$$

Where S_{ij} is the scale advantage index for the j th region and the i th index; A_{ij} is the value of the j th region and the i th index; j is total number of regions (to compare with adjacent regions, it include the selected region and adjacent regions. If to compare with its province data, it include all regions in the province totaled n), $j = 1,2,3,\ldots, n$; i is indices to be compared such as rice, wheat, corn, and other grain crops.

5.2 Productivity Advantages

The productivity advantages will be presented based on natural conditions of major agricultural crops and unit yield indices, i.e.:

$$P_{ij} = \frac{Q_{ij}/A_{ij}}{\displaystyle\sum_{j=1}^{n} Q_{ij} \Bigg/ \sum_{j=1}^{n} A_{ij}}$$

where P_{ij} is productivity advantage index at j th region and for i th crop; Q_{ij} is total production of the i th crop at the j th region; Q_{ij}/A_{ij} represents the land productivity level at the j th region for the i th crop; the denominator is total number of regions (to compare with adjacent regions, it include the selected region and adjacent regions. If to compare with its province data, it includes all regions in the province totaled n), or the average land productivity for the i th crop; $j = 1,2,3,...,$ n ; i is the comparative index to participate in the assessment.

5.3 Efficiency Advantages

Per capita agricultural product production at selected region is compared with the average to present production efficiency (to compare with adjacent regions, it include the selected region and adjacent regions. If to compare with its province data, it includes all regions in the province), i.e.:

$$E_{ij} = \frac{Q_{ij}/L_{ij}}{\sum_{j=1}^{n} Q_{ij} / \sum_{j=1}^{n} L_{ij}}$$

where E_{ij} is efficiency advantage index at the j th region for the i th crop; L_{ij} is number of labors at the j th region for crop i ; therefore Q_{ij}/L_{ij} reflects the labor efficiency level at the j th region for crop i ; the denominator is the average production efficiency over i crops such as rice, wheat, corn, and other gain crops (to compare with adjacent regions, it include the selected region and adjacent regions. If to compare with its province data, it include all regions in the province totaled n).

5.4 Comparative Advantage Index

The comparative advantage is assessed by an integrated consideration of scale advantages, productivity advantages, and efficiency advantages in a region. Two methods can be considered. If to compare with adjacent regions, it includes the selected region and adjacent regions. This is called comparative advantage with adjacent regions. If to compare with its province data, it include all regions in the province. This is called comparative advantage in the province.

$$C_{ij} = \sqrt[3]{S_{ij} \bullet P_{ij} \bullet E_{ij}}$$

where C_{ij} is the comparative advantage index at the j th region for crop i , representing the overall advantage for the region j and crop i .

When the value of S_{ij} , P_{ij}, E_{ij} , or C_{ij} is greater than, it represents the region is advantageous, otherwise no advantage. If the value S_{ij} for the region is greater than 1, it demonstrates a scale advantage in the region compared over other regions. If the value C_{ij} for the region is greater than 1, it demonstrates an overall advantage in the region compared over other regions. The greater the value is, the bigger the overall advantage, vice versa (see Nie Fengying, 2006).

5.5 Different Comparative Indices for Different Crops and Economic Indices

In the agricultural basic database used in the study, some data items directly related to crop production such as yield of summer crops, rice, wheat, corn, beans, oil crops are used to compute S_{ij}, P_{ij}, E_{ij}, and then C_{ij} is computed to present the overall comparative advantage. If the value of C_{ij} >1, the index in the region is called comparative advantageous over adjacent or province.

For some indices not directly related crop production, such as arable land area, rural population, rural enterprises, and agricultural income, the system will only compute their scale advantage S_{ij} by comparing mean value over the adjacent regions. If the value is greater than 1, it is scale advantageous, vice versa.

The comparative advantage computation can be for adjacent or for province. Purpose of the two methods is to represent the region's advantageous agricultural industries in the overall region. In fact, an advantage over adjacent regions may not be so over the overall province, vice versa.

6 Output Display

6.1 Graphical Output Display

Two kinds of output graphical displays in the regional advantageous industry analysis system will be provided: comparing a single index with all regions and ranking comparison in terms of each of the six index types including agricultural resources, grain crop plantation, cash crop yields, animal farming, rural enterprises, and economic performance. The comparison result can be displayed by bar chart as the most illustrative way.

6.2 Text Output Display

For objectives of the system to process regional data statistics, computation, and inference and final assessment, text output will facilitate precise description and understanding. As the system can be used for all counties in China, common text output will be necessary.

6.2.1 Text Presentation of Statistics and Advantage Computation Results
In general, the statistics and comparative advantage result can be described by simple text of excellent, good, etc. The following subsection will in particular illustrate the output assessment recommendations.

6.2.2 Assessment Recommendation Creation and Text Presentation
According to computation result of some indices, for example, both wheat planting area and rural population at a region are advantageous. It can infer that the region's per capita wheat land area is also advantageous. In this case, its total wheat production is generally advantageous. However, through comparative advantage and ranking, it may not be the case since its total wheat production is less advantageous.

Under the similar conditions of soil, climate, and other natural resource compared with adjacent regions, there must be some problems probably in wheat species, plantation management, or others.

Based upon the above assumption, a knowledge base is developed in the system to store and compare the computation results, as well as judge rules, and relevant knowledge. The knowledge base also incorporated reasoning mechanisms to make the system have artificial intelligent capability.

For this further system function, based on the advantageous agricultural industry analysis and comparative advantage computation, the knowledge in the base is used to judge and reason the computation result in order to create assessment and action recommendations expressed in text. This can be referred to be knowledge reasoning. Instead of index values, text is used to represent the assessment with expert knowledge and reasoning result.

This is an innovative design and new experiment. So far, the knowledge base is still rich and for further development. However, the knowledge base can be easily maintained and upgraded under the assistance by regional planning, agro-economy, and agricultural experts. According to a partial investigation, knowledge based system or artificial intelligence technology is less used and seldom reported in regional analysis and regional planning applications. Further studies in this aspect are worth strengthening.

Acknowledgements

This research was supported by National Scientific and Technical Supporting Programs Funded by Ministry of Science and Technology of China (2006BAD10A06, 2006BAD10A12), Special Fund of Basic Scientific Research and Operation Foundation for Commonweal Scientific Research Institutes (2008J-1-06).

Reference

Fengying, N.: Grain security and its studies. China Agricultural Science and Technology Press, Beijing (2006) (in Chinese)

Research of Development of Agricultural Knowledge Service in China

Junfeng Zhang, Cuiping Tan*, Huaiguo Zheng, Sufen Sun, and Feng Yu

Institute of Agricultural Science and Technology Information, Beijing Academy of Agricultural and Forestry Sciences, Beijing, P.R. China,
Tel.: +86-01-51503318; Fax: +86-01-51503318
tcpspring@163.com

Abstract. With the global development of knowledge economy, the knowledge requirement of farmers is more personalized and solution-oriented, so there is pressing needs to develop agricultural knowledge service. The paper analyzes characteristics of agricultural knowledge service, and summarizes typical cases of agricultural knowledge service development in China.

Keywords: agricultural knowledge service, solution-oriented, personalized service.

1 Introduction

As Rural Informatization is more emphasized, a series of focus special projects are launched, such as the three electrics gather one, the integration of three networks, the rural remote education and so on, which have achieved very good results. But there are still some problems that the lacking of overall-planning and coordination-management, duplicate construction have resulted in wastage of funds; and urgently needed information for farmers is still inadequate. Agricultural experts' experience in the tacit knowledge does not be fully developed and utilized; and farmers' diverse information needs have not been deeply met.

What exactly was the need for farmers'information service? Farmers not only need the original information for retrieval, but also need comprehensive information, dynamic information, analysis and forecast information. They prefer one-to-one consulting service to general information service. In brief, they need the realistic solution throughout the whole agricultural production process, which is just the aim of agricultural knowledge service.

2 Origin of Agricultural Knowledge Service

The concept of agricultural knowledge service is rarely found in all kinds of literature. According to the paper of Design and Realization of Agricultural knowledge Service

---

* Corresponding author.

D. Li and C. Zhao (Eds.): CCTA 2009, IFIP AICT 317, pp. 333–337, 2010.
© IFIP International Federation for Information Processing 2010

System Based on Internet, the agricultural knowledge service is a course of shifting visible knowledge which store in computer and invisible knowledge which store in agriculture expert's brain to farmer. (Zhou Guomin et al., 2005). Another paper of Comparison of Agriculture Information Service and Agricultural Knowledge Service added that agricultural knowledge service is special service industry built on knowledge and service functionality of institutions and libraries on agricultural science and technology intelligence research. For the goals of resolving the users' problems, according to users' special needs, high value-added service of intelligence is provided for users based on agricultural or other field knowledge. (Yanxia et al., 2008) The authors think that agricultural knowledge service is high value-added service of intelligence support and intelligence services just for agriculture, to meet special needs of rural and farmers, whose final goal is to resolve their realistic problem, and to promote knowledge of integration, innovation and shared.

3   The Characteristics of Agricultural Knowledge Service

3.1   User-Centered and Aiming at Users' Satisfaction

Knowledge service of agricultural science and technology institutions is totally user-centered to cater for the user's knowledge. According to the users' actually requirement, they select and gather a variety of information resource, and provide new knowledge production after refining, processing and reorganization, just for ease of comprehension and absorption. It is emphasized that directly putting their own unique knowledge and capability into the agricultural production process, to help them resolve their own insurmountable problems, and provide users with the value of knowledge and creativity.

3.2   Solution-Oriented and Throughout the Whole Process of User Information Activities

Agricultural knowledge service is committed to users' requirements, helping users finding or forming a solution. It is affected by several factors that the process of user information requirement generation and utilizing information to solve problem. With the change of external factors and user problem-solving progresses, the user information needs or behavior will change. Staying outside the user's information activities, if you do not see those changes, you cannot really solve the user's problem. Therefore, agricultural knowledge service require that they provide forwardly knowledge match users' constantly-changing needs. (Du Yeli el al. 2005)

Agricultural knowledge service must provide knowledge forwardly that matches users' changing requirements, and help them forming solutions, in the whole process of participation, comprehension, analysis of the users' information activities.

3.3   Facing Knowledge Innovation

The most difference between agricultural knowledge services and traditional information services is that it provides value-added knowledge rather than simply transferring information. The process of value-added knowledge is innovation. They select all kinds of information resource, collating, filtering, and reorganizing in use of a variety of information technology, and eventually form new knowledge production for users. (Du Yeli el al. 2005)

3.4 Personalized Service

Agricultural knowledge service is always standing in the user's perspective, and measuring the knowledge and information to fit users, help users solve actual problems. Each user's scheme is targeted and specific. Exactly according to the user's information requirement characteristics and behavioral habits, agricultural knowledge service is realized by adopting the diversified and personalized service.

3.5 Comprehensive Integration, Collaboration and Sharing

Agricultural knowledge service is to integrate organically a variety of resources (including specialist resources, information resources and technical resources), play the overall advantages of the service sector in manpower and intelligence, information and technical resources. Based on open-service mode, through system integration, service integration, and team work, they joint, coordinate and utilize a variety of knowledge resources to provide knowledge-based services. (Du Yeli el al. 2005)

4   Typical Case Analysis of Agricultural Knowledge Service

4.1 Agricultural Experts and Decision Systems

The representative of the earlier agricultural knowledge service is agriculture expert systems. The Agriculture Expert Systems are designed to emulate the logic and reasoning processes that an Expert would use to solve a problem. The expert systems in agriculture are based on the integration of knowledge and experience of specialists from different fields, such as Agronomy, Breeding, Soil science, etc. These systems have the capability to answer relevant questions and explain its reasoning process and will be able to interact with farmers and end users in a way that can be understood by them. Expert System can be developed in any specific domain for some specific purpose. Supported by National 863 plans and the national important science and technology project During The 7th Five-Year and The 8th Five-Year etc, the expert systems about corp varieties, agricultural production management, and disease pest control are developed in China.

4.2 Pluralistic Agricultural Advisory Services

4.2.1 Agricultural Science and Technology Information Service 110

Agricultural science and technology information service model of 110 is a typical innovation model of rural technology services in the practice of the Starfire Enriching Farmers. On the mission of science and technology services for farmers, to information resources as the core, to service hotline as a link, based on data networks, it is devoted to promote information low-cost and efficient communication in the wider rural, and realizes the zero distance between science-technology and farmers.

(Zhangbo el al, 2007) The premier of China Wen Jiabao thinks high of the model as a creation of serving for farmer. The service model is mainly on information dissemination, but experts can directly solve the problems with telephone online, so it is one basic form of agricultural knowledge service.

4.2.2 Agricultural Experts Two-Way Video Advisory Diagnostic Service

Because the cultural level of farmers is generally ragged and not high, it is urgently needed that the methods of online farmer-expert interaction and consultation. It is possible with computer communication and streaming media technology. Since 2006, Institute of Agricultural Science and Technology Information of BAAFS carries out expert and farmer one-to-one video Advisory diagnostic services, implements real-time remote technology consultation, pests and diseases remote diagnosis etc. Farmers can transmit easily the sample pictures of field pests and diseases to experts by face to face through the video system, and experts will diagnosis and provide practical solutions according to the actual disease. Agricultural experts two-way video Advisory diagnostic service is orient-solution knowledge service, involved in agricultural production and management decision. It has received more recognition and welcome of farmers since it began.

4.2.3 Agricultural Engineering Consulting

Agricultural engineering consulting business is intelligence service industries, following independent, scientific and impartial principles, which provide advisory services of decision and implementation about agricultural economic construction and engineering for government departments and investors, to enhance the macro and micro economic benefits. Foundation and development of Agricultural engineering consulting is accompanied by rapid development of our agricultural investment management reform. In 1992, Ministry of Agriculture has established the Ministry of Agriculture Engineering and Construction Services Center and the Chinese Association of Engineering Consultants, as a sign of a formal engineering consulting industry in China. until 2006, our agriculture engineering consulting organization has reached 50, the number of jobholders is up to 3000, and the overall strength and technical level is more improving. Agricultural engineering consulting fully utilizes the advanced agricultural technologies, insisting on the independently advisory services, and provide decision-making evidence for government and enterprises, thereby improving the investment results. In integration of expert teams force, orienting knowledge innovation service, agricultural engineering consulting is designed to add knowledge value, and realize effectiveness and industrialization of knowledge service business. It is relatively successful agricultural knowledge service mode.

4.2.4 Agricultural Knowledge Integration and Sharing

(1) Chinese Agricultural science data sharing centre

Agricultural Information Research Institute of Chinese agricultural science Academy is in charge of national agricultural science data sharing centre. Based on agriculture sector, to meet state and society requirements on agricultural scientific data sharing service, taking the data source sectors as the mainstay, surrounding data sharing center, the national and international agricultural scientific data resource can be put together through integration, importing and exchange. The data should be standardized and classified, and eventually rapid sharing service network system will be constructed. With database construction, system application and providing service, they gradually expand the scope of building and sharing in China. At present, 13 categories of the huge database groups have been built.

(2) Provincial and municipal agricultural science and technology knowledge-sharing center

Since 2003, the provinces gradually established many regional centers of agricultural science and technology knowledge sharing, especially Beijing, Guangdong, Anhui. Beijing agriculture digital information resource center, for example, based on the services for urban style of modern agriculture in Beijing, under integration of agricultural experts' experience and knowledge, with a combination of centralized and distributed building patterns, realizes joint construction and sharing with agriculture-related institutions in Beijing, to meet the needs of local agricultural science and technology. Until now, More than 200 databases, millions of agricultural data has been completed, forming a dynamic agriculture knowledge warehouse. it becomes a comprehensive agricultural Center for digital information resource sharing in Beijing. In cooperation with county government, the communication channels of agricultural knowledge is widened. As a result, it has been the main source of the agricultural knowledge in Beijing.

5 Conclusion

Agricultural knowledge service is still in the initial stage in China. There are some meaningful attempts in the fields of the expert consultation, knowledge integration and expert systems. With agricultural knowledge economy growing, Only joint agricultural information research industries can complete the process of knowledge organization, communication, application and innovation. Meanwhile agricultural knowledge service is core competence power of information research industries. It is the all-dimensional and targeted solution that is deeply needed for farmer in fact, which require our long-term efforts for the topics.

Acknowledgements

Funding for this research was provided by Project of Upgrade Agricultural information services to Agricultural knowledge services Research and Project of Personalized Knowledge Service Platform Development and Application Based on Network of Beijing Municipal Science and Technology Commission (P. R. China).

References

Guomin, Z., et al.: Design and Realization of Agricultural knowledge Service System Based on Internet. Journal of Library and Information Sciences In Agriculture (2), 238–240 (2005)
Yanxia, et al.: Comparison of agriculture information service and agricultural knowledge service. Guangdong Agricultural Sciences (11), 135–138 (2008)
Yeli, D., et al.: Knowledge service pattern innovation. Beijing Library Press (2005)
Zhangbo, et al.: Agricultural Information Service Models Innovation in the Construction of Socialist New Village. Chinese Agricultural Science Bulletin (4), 430–434 (2007)

A Study on the Digital Integrated Platform of China Participatory Rural Community Informationization

Hong Zhou[1,*], Guangsheng Zhang[1], Xin Ning[2], and Jin Luo[1]

[1] Economic Management College of Shenyang Agricultural University,
Shenyang, Liaoning Province, P.R. China 110161,
Tel.: +86-24-88487153; Fax: +86-24-88487248
zhouhong003@163.com
[2] Information Centre of Shenyang Agricultural Economic Committee,
Shenyang, Liaoning Province, P.R. China 110024

Abstract. Agriculture informationization is the inevitable trend of the world modern agricultural development, at present it is the most main bottleneck that rural informationization progress is slow in the course of Chinese rural informationization development, the last kilometer becomes a difficult problem to solve urgently. Based on the theory of participatory rural community, this article lead to analyze the core reason why exits the endocardial power deficiency of the last kilometers problem in China rural community and the restraint obstruct to the exterior intervention impetus function, researching the original data gained by agricultural informationization investigations and studies in Liaoning Province, and moreover proposes the effective solution is that a Digital integrated platform should be build in China rural community, which play two aspect of the function meanwhile the rural community management and information service, marking a feature of extreme interactivity, integrativity and comprehensive consistent, furthermore it should choose a specific informationization construction plan according to the different community characteristic together with the farmer and the rural community in view of establishing a reasonable effective development mechanism, realizing the fair and reasonable disposition and management to the information resource, ultimately realizing the sustainable development of the rural informationization.

Keywords: participatory rural community informationization, digital integrated platform.

1 A Last Mile Problem in China's Agricultural Informationization: Based on the Survey of Liaoning Province

The research team has carried on a investigation in August 2008 about rural informationization condition in different village area of Liaoning Province, Investigated more than 800 households in rural households, which drew a following conclusion:

---

* Corresponding author.

D. Li and C. Zhao (Eds.): CCTA 2009, IFIP AICT 317, pp. 338–344, 2010.
© IFIP International Federation for Information Processing 2010

1.1   Farmers Gain the Information Still Based in the Traditional Method

The peasant household generally gains the information by television, villager dissemination, and the village technical talent, SMS, newspaper, magazine as well as the Internet. 96.46% of peasant households through TV gain information, 60.17% of peasant households gain information by depending on villager dissemination.

1.2   Farmers Cannot Timely and Full Make Access to the Technology That Agricultural Production Needs

Peasant household gains agricultural production technology mainly from TV, Broadcast and Newspaper, secondly from farm technology station or the village model household. It stands a few proportions through network to gain agricultural technology, which depends on the lower volume of computer in rural households.

The peasant household like to listen and watch such TV & Radio program as CCTV news, CCTV Agricultural program, Black Earth, Law Online, Weather Forecast and so on. Farmer pays more attention to changes in weather and in national rural policy very much, because it relates to their production behavior.

Every year peasant can accept technology training that is organized by county and township farm technology station. The farmer is generally good response to the training, but the training time is short and pertinence is not enough strong.

Tab Survey how Farmer obtain production technology   unit: household, %

City	Item	1	2	3	4	5	6
Shenyang	sample size	54	85	17	59	57	61
	proportion	47.79	75.22	15.04	52.21	50.44	53.98
Anshan	sample size	8	37	20	31	19	22
	proportion	20.51	94.87	51.28	79.49	48.72	56.41
Chaoyang	sample size	25	18		8	8	10
	proportion	69.44	50.00		22.22	22.22	27.78
Tieling	sample size	25	23	1	9	13	11
	proportion	73.53	67.65	2.94	26.47	38.24	32.35
Dandong	sample size	11	25	4	18	11	12
	proportion	32.35	73.53	11.76	52.94	32.35	35.29

1. hand down one generation after one another 2.TV, Radio, Newspaper   3.Internet 4.agro-technical station 5.village model household   6.Rural Cadres 7.SMS

1.3   The Peasant Household is in an Information Inferiority Status in Agricultural Product Sales

An investigation found that the peasant household has a broad marketing   channel, and it's have no difficult to sale the agricultural product. Being not grasp the market information accurately, or being at the inferiority position in business, famer have to accept the purchase business in a low price, thus has affected enhancement income of peasant household. The peasant household gains selling price information mostly by the way of dissemination between neighbor villager, accounts for 38.05%; secondly by querying price in the market, accounts for 32.14%; Inquiring to the old customer and searching in the internet accounts for 18.58% separately and 4.42%. When carrying on the agricultural product sale, however the peasant household was at the weak trend position, they could not adjust price according to market quotation to sales situation.

1.4 Information Acquisition Is an Important Influence Factor to the Peasant Household Income

Result of study shows, peasant household thought the primary factor which hinder the improvement of their income include the education level, the information factor, the geography factor, the policy factor and market price. 52.21% of peasant households thought that they are not good able to increase income by grasping market information.

1.5 The Peasant Household Underuse Information Service Project of Government

The peasant household is not very satisfied about the village information officer who uploads and downloads information that they require indeed. Regarding whether village information officer could promptly upload or download the information for the peasant household, which they require indeed, 4.42% of answer of peasant thought very unprompted, 5.31% of peasant households thought not prompt, 16.81% of peasant households think neutral, 25.66% of peasant households thought quite promptly, 16.81% of peasant households think not clearly.

The peasant household is not satisfied with village information officer, 28.85% of peasant households thought strongly unsatisfied, 35.40% of peasant households think satisfied. The peasant household thought they would seek such information as seed, chemical fertilizer and other rural purpose commodity through the information officer, which stand 58.41% of all opinions. 46.90% and 41.59% of peasant household hope through the village information station they could order agricultural books or the agricultural insurance, but at present this kind of service not have done.

Since the national uniform agricultural service hotline 12316 has been established, a new channel is opened to the farmers to consult information, which enables the farmer to be possible obtaining consultation about agricultural production, management and so on from the expert; it is a very good kind of rural information service pattern. However, among 113 peasant households which accepted survey there were only 6 households who have used the agricultural information expert system hotline 12316, the above-mentioned survey indicated that rural information service hotline 12316 Needs to strengthen vigorous promotion.

2 Analyzes about Reason of the Last Mile Problem of Rural Informationization in China Village

2.1 Endogenous Deficient in Motivation

Participatory theory think that development is internal change in the social economy system, because it does not impose from exterior but occurs voluntarily from interior, development motivation supply from the subjective activity itself, outside assistant is only an incentive means, which couldn't substitute for the subjective activity itself of development. The participation essence lies in undertake, share, and altogether takes on risk.

At present, the main reason why agriculture informationization progress in Chinese village seems slow is that farmer plays main roles in agricultural production and management, however they lie in the passive position actually in the agricultural informationization practice, the farmer does not grasp the technology, the resources and the project, in additional they are in a lower education level, and then the administrative order and the expert is in the dominant position in the informationization project promotion, whereas the agricultural information provided for the peasant in the existing rural information service channel is poor targeted, and lack of integration of practical information resources, poor interactive information channel, in short it's in urgent need of change to this kind of situation in order to develop modern agriculture.

2.2   Exterior Intervention Impetus Function Occurs the Restraint Barrier

In participatory development pattern, the participant includes the villagers, as well as the governmental organization, technical department, non-governmental forces and so on. The core factor of participatory countryside development approach is how to develop human and talent resources, to transform role, to enhance communication, exchange and dialogue. It should devolve authority of voice, analyze, decision and management to the localities and the people, the concerned outside factor is the catalyst, which assist to develop countryside and transform process.

At present, the agricultural information dissemination system mainly is both at higher and lower levels, which generally terminates at county level or at the villages and towns government department, they usually carry on the hardware and software construction deferring to the request, however regarding the final practical information needed to peasant household, there have not been established an inspected KPI and clear request.

Moreover, the dispersive distribution characteristics of village brings certain difficulty to gather and update information, in some areas the concurrent-job-personnel manages the software and hardware system, even if there is the full time personnel to manage, regarding in practical information gathering and processing for peasant household, then the less time and cost be spend. Some information provided by department officer mutually reprints and duplicates that seems extremely high duplication and lower usability. Because the above-mentioned information resource is dispersible, nonsystematic, overlapping and duplicated, which does not carry authority, result of that sometimes peasant household thought themselves into a dilemma. Being characterized by poor practicability for application, lack of pertinence, less information that the farmer most cared about, the peasant household made lower satisfaction rate of comment on government information service degree, some farmers even more entirely do not catch existence of the government information department, also some has not obtained the corresponding information service.

Sometimes the government officer and the expert penetrate to the village to provide some training, in somewhat system reason the experts are not impossible to systematically and persistently carry on instruction in the village basic unit, therefore it was not enable to guarantee training to be practical and conform to the farmer season, in quantity cannot satisfy the farmer's need.

Being above reason external force intervention impetus functions such as government, experts and so on is extremely limited to agricultural informationization construction in the village basic unit in long period.

3 Effective Solution of Last Mile Problem: Built the Digital Integrated Platform of China Participatory Rural Community Informationization

It is extremely vital significance to solve the village last mile problem to Apply Participatory Rural development theory in agricultural informationization domain, to rebuilt participatory village agriculture informationization service pattern, to establish an unified comprehensive information service platform that extend truly to the township, the village and the peasant household, in turn enabling each peasant household to be possible to interact truly the participation, and to change the situation of top-down one-way information dissemination in village.

3.1 Project Plan of Participatory Rural Digital Integrated Platform

The peasant household starts to involve the project from the beginning of plan definition, and should pass through the whole process in the project cycle, recharging monitor and appraisal to involved shareholders in the project construction and implementation. The main important difference of the general monitor and an appraisal is that peasant household not only provides the information for the monitor appraisal, moreover they are also responsible to collect and analyze information, and therefore it is benefit in enhancing their participation and application consciousness.

3.2 Character of Participatory Rural Digital Integrated Platform

3.2.1 Be Great Strengthened Interactive and Information Is Comprehensive and Timely

Each villagers may registry as a member, who enjoys the corresponding member service and own mailbox, who can issue circulation of the land, the agricultural product sales, purchase and transfer of the agricultural material, life consumable product purchase and transfer through BBS (forum). Blogs: Every peasant register may open Blog space, where they introduce the agricultural product sale information in detail; information station may issue promptly some policies, the laws and regulations, the meteorological information, news and so on by sending SMS, enabling farmer to be possible to promptly obtain the news most cared about, as well as such information as health, entertainment, rural education and so on.

3.2.2 Be Good at Powerful Integration

This platform has powerful function to enable E-commerce, accomplishing the opportunity to the both sides of seller and buyer to sell and buy agricultural materials and product. The platform integrates ERP system of the agricultural enterprise, the agricultural materials manufacturer, the distributor and business agent, enables the farmer to be possible to learn such information as farming, crops grow, prevention of plant disease, the agricultural material and product business and market, preventing themselves be deceived and so on through the computer and the network. At the same time, the platform has definite advantage to rebuilt and improve the countryside basic database, to broaden the application space, to develop alerting monitor system for quality and safety of agricultural products, agricultural product market supply and demand system, management system of rural labor force migration, agricultural product safety traceability system, realizing issue and inquiry of agricultural product supply and demand, quality and safety of agricultural products and so on.

3.3 Content of Participatory Rural Digital Integrated Platform

The system mainly has two big main functions of countryside management and agricultural information service, it connects and shares with all levels of agricultural information network resources, taking adequate advantage of computer, television, telephone, broadcast, SMS and so on to rebuilt the information network platform The Digital Village That serves for three agriculture, countryside and peasant, in order to form the information network that can link the province, city, county, township, and village. In all counties (city), towns and village will setup the subnet. By subnet the information about the county, towns, and villages as well as the information of various levels of commission will be browsed. On the subnet the pages Agricultural Product Information will be set up, the related department will take charge of uploading such local information as agricultural product output, price and so on at the right moment, which put up the bridge to moves towards the agricultural product to the market. The system covers all countryside information network service system, realizing purpose "every township has the website, every village has the homepage". Through the homepage any information such as natural resource, the appearance, infrastructure, farmhouse, school and so on should be browsed, including all king of form demonstration such as picture, movie and text. All levels of enterprises, the association, the cooperative society and other specialty industries also should display the corresponding multimedia demonstration, at the same time such item as village economical situation, infrastructure, social undertaking and local folk custom should be displayed in multimedia form.

3.4 The Guarantee Mechanism for Digital Integrated Platform of Participatory Rural Community Informationization

3.4.1 Establishment of the Perfect Personnel Training for Participatory and Sustainable Rural Informationization Development

To establish agriculture informationization training mechanism, it should involves all levels of personnel in view of the basic unit agricultural informationization integrated digital platform to establish the suitable training plan, paying more attention to entirely receiving training personnel to participate.

3.4.2 Participatory Rural Informationization Management

The key function of integrated digital platform mainly includes countryside management and rural information service. Firstly it should promote countryside management informationization, and carry out an advanced management method in informationization fields, guaranteeing the smooth realization of informationization function in countryside information service.

3.4.3 Comprehensive Utilization of Many Kinds of Media Form

Facing to the various educational levels of farmers to provide service, it must unify organically the traditional media and many kinds of modernized media, it should play radiating and impetus function of main item enterprise, the planting and cultivating peasant, the professional association, the cooperative organization and the broker troop.

Multipath for Agricultural and Rural Information Services in China

Ningning Ge[1], Zhiyuan Zang[2], Lingwang Gao[1,*], Qiang Shi[2],
Jie Li[3], Chunlin Xing[2], and Zuorui Shen[1]

[1] IPMist Lab, College of Agriculture and Biotechnology,
China Agricultural University, Beijing, P.R. China, 100193,
Tel.: +86-10-62731884
lwgao@cau.edu.cn
[2] Beijing Candid soft Technology Co. Ltd. Beijing, P.R. China, 100083
[3] Plant Protection Institute, Shanxi Academy of Agricultural Sciences,
Taiyuan, Shanxi Province, P.R. China, 030031

Abstract. Internet cannot provide perfect information services for farmers in rural regions in China, because farmers in rural regions can hardly access the internet by now. But the wide coverage of mobile signal, telephone line, and television network, etc. gave us a chance to solve the problem. The integrated pest management platform of Northern fruit trees were developed based on the integrated technology, which can integrate the internet, mobile and fixed-line telephone network, and television network, to provide integrated pest management(IPM) information services for farmers in rural regions in E-mail, telephone-voice, short message, voice mail, videoconference or other format, to users' telephone, cell phone, personal computer, personal digital assistant(PDA), television, etc. alternatively. The architecture and the functions of the system were introduced in the paper. The system can manage the field monitoring data of agricultural pests, deal with enquiries to provide the necessary information to farmers accessing the interactive voice response(IVR) in the system with the experts on-line or off-line, and issue the early warnings about the fruit tree pests when it is necessary according to analysis on the monitoring data about the pests of fruit trees in variety of ways including SMS, fax, voice and intersystem e-mail.The system provides a platform and a new pattern for agricultural technology extension with a high coverage rate of agricultural technology in rural regions, and it can solve the problem of agriculture information service 'last kilometer' in China. The effectiveness of the system was certified.

Keywords: integrated technology, information services, agricultural and rural, multipathIntroduction.

1 Introduction

The use of ICT enables the production of goods in a short amount of time with the assistance of computerised systems. Services are also provided more efficiently and

---

* Corresponding author.

D. Li and C. Zhao (Eds.): CCTA 2009, IFIP AICT 317, pp. 345–351, 2010.
© IFIP International Federation for Information Processing 2010

rapidly (Bongo, P., 2005). Advances were made in promoting informatization of agriculture which is the key to speed up agricultural development in recent years in China. Agricultural and rural information service based on internet is developing rapidly, but the communication between the internet and farmers in rural regions is inefficient (Jiang Yongmei, et al., 2007; Michelle W, 2009) because most framers cannot afford the expenses for buying computers and accessing the Internet and they are not proficient in the skills of using the computer. It means that the farmers can hardly get the necessary information, include marketing information, extension advice, information about rural development programmes, and other information from government and private sources. So, the potential of information and communication technology (ICT) for the speedy dissemination of information to farmers need to be realised (Meera et al, 2004). Developing countries face challenges when harnessing ICTs potential for economic development(Michelle W. L. Fong, 2009). The method of integrating other information and telecommunication products such as television, mobile telephone, fixed-line telephone which have been popularized and have been brought into extensive application with the computer can improve the information services for farmers and provide interactive communication through the E-mail, telephone-voice(Lin Chuyou, et al., 2008), short message, voice mail, videoconference or other format anytime and anywhere.

Integrated technology refer to a new communication pattern that combined the computer technology and traditional communications technology or a platform which includes telecommunications networks, computer networks and cable television networks, achieveing the goal of offering the information service through multipath such as telephone, fax, data transmission, videoconference, call centers, real-time communication (http://www.skyoa.com). The user can connect the platform with many kinds of communication terminals and access any module in this platform to get the desired information, not limited to the only way of operating a computer.

Pests of fruit trees such as borers have caused great damage to the fruit growers (Myers, et al, 2006). Therefore it is necessary to investigate and research the biological characteristics, perniciousness feature, factors of influence, and prophylactic-therapeutic measures of the pest of fruit trees to reduce the loss. Moreover, it is more important to prvide information services to the farmers to control the pests damages.

The integrated pest management platform of fruit trees in Northern China were developed based on the integrated technology, which can integrate the internet, mobile and fixed-line telephone network, and television network, to provide integrated pest management(IPM) information services for farmers in rural regions in E-mail, telephone-voice, short message, voice mail, videoconference or other format, to users' telephone, cell phone, personal computer, personal digital assistant(PDA), television, etc. alternatively.

2   Integrated Pest Management System of Fruit Trees in Northern China

The integrated pest management platform of fruit trees in Northern China consists of four main parts by now: (1) Monitoring data acquisition and management System; (2) Expert response system based on the internet and telecommunications networks; (3) Information dissemination system; (4) Administration system of agricultural production.

2.1 Monitoring Data Acquisition and Management System

The system of monitoring data acquisition based on PDA and GPS is developed by visual studio with three functions: collecting data, processing data, and output. PDA receives the signals of GPS through internal or external GPS hardware devices. Binding the manual input data about the dynamic of the fruit tree, the environment and the pests with latitude and longitude information, PDA stores all data in XML document which can be transferred using wireless transmission into the remote database. This can significantly improve the efficiency and guarantee accuracy of the data comparing with the traditional methodes of data acquisition which write the data on the paper before input them into the computer(Wu Shou-zhong et al, 2005).

Environmental parameters, such as illumination, temperature, humidity can be recorded with the sensor devices which can be set the time interval and automatically send to the remote database through the GPRS network.

Beside the data come from PDA and sensor device, the user also can input the historical data into the database through the interface based on the computer and manage them.

2.2 Expert Response System

An expert database was set up in the system which contain related expert s' information, such as address, contact details, academic field on which response the

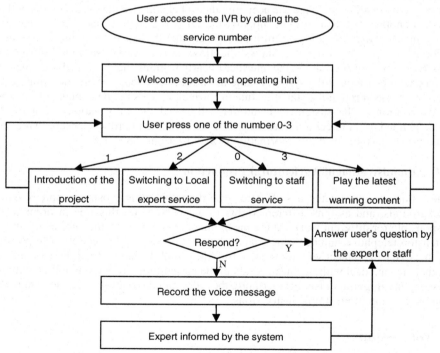

Fig. 1. *T*the flowchart IVR in the system

consultation service. The system automatically assigns the expert to the consultant who dials the service number and access the interactive voice response (IVR) in the system on the basis of dialing area code; the expert can answer the questions including diagnosis of pest, market information, prevention and control measures from the consultant. If the assigned expert have no time to answer the phone, the consultant can leave a voice message which would be recorded in the system and sent to the expert electronic mailbox in the system as a .wav format file. When it is possibel, the expert can call back or leave a voice message in the system to answer the consultant's questions. Beside making a call, the consultants also can send Short Messaging Service(SMS) and Multimedia Message Service (MMS) with picture and video about the pest or symptom damaged by the pest to the system using cell phone. The system can record the ask-answer process in a process tracking database. When the process ends, the expert collates the questions and answers, adding them into the database of questions and answers to enrich the information contents that can reduce the expert future work. In addition, entering some common knowledge the fruit grower need at earlier time helps the intelligent retrieval (Dong Xiaoxia, et al, 2009).

2.3 Information Dissemination System

Agency in charge of the project issues the early warnings about the fruit tree pests when it is necessary according to analysis on the monitoring data about the insect pest of fruit trees. The early warnings can be sent in variety of ways including SMS, fax, voice and intersystem e-mail. A message can accommodate 44 characters, the excess words will be send in another one. The system supports the .tif (for fax) and .wav (for phones and voice mail) format file comes from local-storage, and the voice can come from online recoding by phones also. *Fig2* shows the flowchart of early warning Information dissemination, and *Fig3* shows the interface of early warning information dissemination.

The administrator of the system is able to view the efficiency of broadcasting of the early warnings through the statistics function which reveals the percentage of respond to the warning. The system can automatically send the information again to the un-responded users in certain interval before the deadline in order to ensure that every fruit grower has received the information to avoid serious damages.

2.4 Administration of Agricultural Production

The administrator creates a staff tree according the titles, positions and research backgrounds, and assigns different rights to the persons on this tree in order to be responsible for different work. At the mean time the system also allocates every staff a extension telephone number which can be set feature service by oneself. The persons on this tree can connect with the system by MMS, mobile phone, fixed-line telephone, intenet etc. and deal with the business about the agricultural production in time without missing any important task. Furthermore, they also can query or browse the data in the database by inputting simple instructions.

3 Discussion

Integrated pest management system of fruit trees in Northern China, a platform based on the integrated technology, enables the fruit growers in rural region, experts, the agricultural officers to easily access to the system via the variety of device, such as PDA, cell phone, fixed-line telephone and the computer. This would change the situation that most information system involving agriculture cannot provide the efficiency information services to the farmer in rural region. The system provides a platform and a new pattern for agricultural technology extension with a high coverage rate of agricultural technology in rural regions, and it can solve the problem of agriculture information service 'last kilometer' in China.

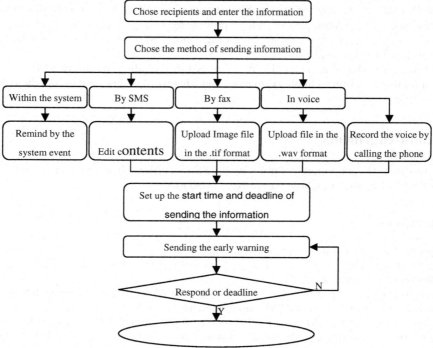

Fig. 2. The flowchart of early warning information dissemination

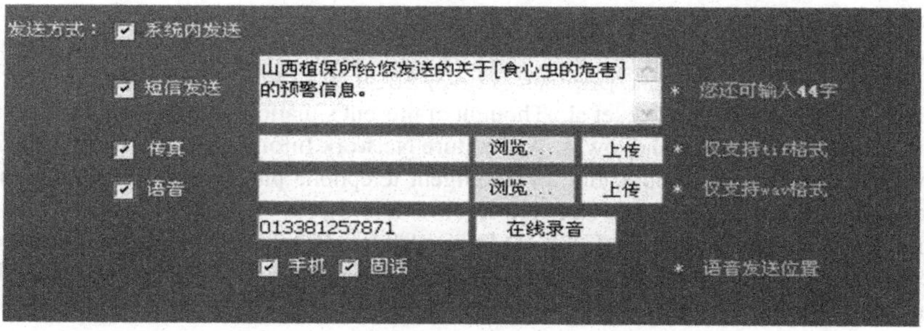

Fig. 3. The interface of early warning information dissemination system

The system is still in expanding and it would be a robust, reliable, and expandable integrated pest management information system in the future. Scalable architectural design facilitates the addition of a wide variety of new functions. Main current limitations of the system is the data analysis which depends on the data collection by investigator in the later practice, and it needs developing more functional sub-systems, such as an expert system for forecast and prediction of pests (Gao Lingwang, et al.,2006; Liu Minghui, et al, 2009), a pest spatial analysis information system based on WebGIS and remote sensing (Gong Yanping, et al., 2008), and a remote diagnosis system based on technologies of image processing (Yang Hongzhen, et al.,2008), expert system(Prasad, et al, 2006; Lingwang Gao, 2009), and video conferencing system (Liu Yuexian, et al., 2002).

Additionally, the system can also be expanded to providing the information of the agricultural production technology, market, and whatever farmers need except the IPM information.

Acknowledgements

This research was supported by Public Welfare Project from Ministry of Agriculture of the People's Republic of China (Grant No: 200803006).

References

Bongo, P.: The impact of ICT on economic growth (2005),
  http://129.3.20.41/eps/dev/papers/0501/0501008.pdf
  (Retrieved November 30, 2007)
Xiaoxia, D., Guifa, T., Fang, W., et al.: The Ontology - based Agricultural Information Service System. Journal of Agricultural Mechanization Research (6), 137–140 (2009)
Lingwang, G., Jiguang, C., Xinwen, Y., et al.: Research and development of the expert system platform for forecast and prediction of agricultural pests. Transactions of the CSAE 22(10), 154–158 (2006)
Gao, L., Yan, C., Shen, Z.: Designing and Algorithm Implementing of the Expert System Platform for Assistant Identification of Agricultural Pests Based on a Dendriform Hierarchical Structure. In: Proceedings of 2009 World Congress on Software Engineering, WCSE 2009, Xiamen, China, May 19-21, pp. 205–210 (2009)
Yanping, G., Wenjiang, H., Yuchun, P., et al.: Construction of a Web GIS-based forecast system of crop diseases and pests. Journal of Natural Diastase 17(6), 36–41 (2008)
Integrated technology,
  http://www.skyoa.com/phpcms/rhsy/2008/0723/article_12.html
Yongmei, J., Yaohui, X., Ping, G., et al.: Thought of present situation of agricultural information service system in villages and towns. Agriculture Network Information (1), 44–57 (2007)
Chuyou, L., Yanyong, X.: Application of intelligent telephone phonetic system in agriculture information service. Agriculture Network Information (3), 39–40 (2008)
Minghui, L., Zuorui, S., Lingwang, G., et al.: Expert System Based on WebGIS for Forecast and Prediction of Agricultural Pests. Transactions of the Chinese Society for Agricultural Machinery 40(07), 180–186 (2009)

Yuexian, L., Zuorul, S., Xinyan, C., et al.: Development of the computer-aided species identification and information service system for agricultural pests management. Computer and Agriculture (1), 9–11 (2002)

Meera, S.N., Jhamtani, Q., Rao, D.U.M.: Information and communication technology in agrudultural development: Acomparative analysis of three projects from india. ODI: Agrudultural Research & Extension Network, Network paper No. 135 (2004)

Fong, M.W.L.: Digital Divide: The Case of Developing Countries. Issues in Informing Science and Information Technology (6), 471–478 (2009)

Myers, C.T., Hull, L.A., Krawczyk, G.: Comparative survival rates of oriental fruit moth (Lepidoptera: Tortricidae) larvae on shoots and fruit of apple and peach. J. Econ. Entomol. 99(4), 1299–1309 (2006)

Prasad, R., Ranjan, K.R., Sinha, A.K.: AMRAPALIKA: An expert system for the diagnosis of pests, diseases, and disorders in Indian mango. Knowledge-Based Systems 19(1), 9–21 (2006)

Shou-zhong, W., Ling-wang, G., Da-zhao, S., et al.: Development Data Collection System of the Rodent Pest on the Grassland Based on the PDA with GPS. Acta Agrestia Sinica 15(6), 550–555 (2007)

Mahajan, S.: Impact of Digital Divide on Developing Countries with Special Reference to India. SERALS Journal of Information Management 40(4), 328–329 (2003)

Hongzhen, Y., Jianwei, Z., Xiangtao, L., et al.: Remote automatic identification system based on insect image. Transactions of the CSAE 24(1), 188–192 (2008)

The Design and Development of Test Platform for Wheat Precision Seeding Based on Image Processing Techniques

Qing Li[1], Haibo Lin[1], Yu-feng Xiu[1], Ruixue Wang[2], and Chuijie Yi[1,*]

[1] Department of Vehicle and Traffic, Qingdao Technological University,
Qingdao, Shandong Province, P.R. China 266033,
Tel.: 13963924026
chuijieyi@vip.163.com
[2] College of Engineering China Agricultural University, Beijing, 10083

Abstract. The test platform of wheat precision seeding based on image processing techniques is designed to develop the wheat precision seed metering device with high efficiency and precision. Using image processing techniques, this platform gathers images of seeds (wheat) on the conveyer belt which are falling from seed metering device. Then these data are processed and analyzed to calculate the qualified rate, reseeding rate and leakage sowing rate, etc. This paper introduces the whole structure, design parameters of the platform and hardware & software of the image acquisition system were introduced, as well as the method of seed identification and seed-space measurement using image's threshold and counting the seed's center. By analyzing the experimental result, the measurement error is less than $\pm$ 1mm.

Keywords: image processing, dropping distance, precision seeding.

Introduction

The precision seeding technique is the combination of technology about agricultural mechanics and agronomic, which proves to be a practical way to lower the cost and improve the efficiency. The seed metering device is the important part in the technology of precision seeding. And the seed metering device test platform is the main tool to test the seed metering device's performance. Working state of seed metering device in the field can be simulated by this test platform so as to perform repeated experiments on the seeding metering device. Using technology of computer detection, the test platform of wheat precision seeding based on image processing techniques is designed. Through the processing and analysis on the images, the distance between seeds can be measured by the method based on image-binary threshold. By processing the distances of the seeds, the qualified rate, reseeding rate and leakage sowing rate can be given.

---

[*] Corresponding author.

D. Li and C. Zhao (Eds.): CCTA 2009, IFIP AICT 317, pp. 352–358, 2010.
© IFIP International Federation for Information Processing 2010

1 Total Structure of Test Platform and Working Principle on the System of Measurement

1.1 Total Structure of Test Platform

The total structure is shown in Fig.1.It is composed of test-platform bench, species-bed conveyor, conveyor motor, seed metering device (air-suction seed metering device), drive motor, vacuum pump devices, computer vision detection system. The test-platform bench is 0.7 meters high; roll wheel's space of species-bed conveyor is 2.5 meters long. The bracket-wheel is resettled in upper and lower levels of conveyor belt. The conveyor is joined seamlessly so as to ensure a smooth running, small vibration, low noise to meet the standard laboratory vibration and noise requirements.

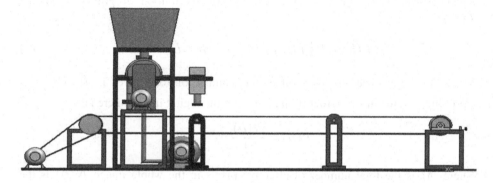

Fig. 1. Total structure of test platform

1.2 Principle of Measuring System

The seed metering device is installed on the test bench. In the driven of VVVF motor, species-bed conveyor and metering device are relatively uniform operation; seed metering device's inner chamber results a vacuum degree under the vacuum so that the seeds in the box are adsorbed to the seed plate in the negative pressure, then the seeds fall to conveyor belt as soon as the negative pressure disappears with the rotation of seed plate (coated with the lubricating oil, conveyor belt can paste seeds so that it does not move relative to the conveyor);the seeds' images are collected and transferred to a computer when conveyor pass through the camera, where after which are processed by software, at last realize the measurement of the distance of the seeds.

2 Constitution of the Image Processing System in Software and Hardware

2.1 Constitution of the Software

Software is NI Vision module in Labview 8.6 of National Instruments Company. Compile programs for the acquisition and processing using graphical programming language to achieve the purpose of image processing and analysis.

2.2 Constitution of the Hardware

A Pentium IV computer; IEEE 1394 image acquisition card; CCD monochrome industrial camera: DH-SV1410FM.

3 Image Analysis and Processing

3.1 Statistical Characteristics of Image

The overall statistical features of an image can be described by joint probability density function. For discrete digital images, histogram is obtained by statistics of the pixel gray value which can be used to estimate the probability distribution of the image. If we define the first order probability distribution of gray-scale image $f(i,j)$:

$$P(b) = P\{f(i, j) = b\}, 0 \le b < L \tag{1}$$

Where: b is quantization layer value, the number of total layer is L. $P(b)$ is approximate value of first-order probability. For the digital image, there is:

$$P(b) = \frac{N(b)}{M} \tag{2}$$

Where: M is the total number of effective pixels in the whole image. $N(b)$ is the total number of b gray value in the whole image. So we can draw the first-order gray-scale image histogram with gray value as the abscissa and the number of pixels or pixel frequency appearing as the longitudinal coordinates. Fig.2 shows a typical histogram when some light objects appear on a dark background.

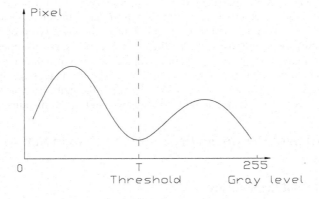

Fig. 2. Curve of typical gray histogram distribution

3.2 Binary Converting in Image Processing

There are no clear boundaries between the objects(wheat seed)and background (conveyor of species-bed) because they mix together so we must use an appropriate threshold method to partition the objects. Here, we use the method of setting maximum variance threshold based on discrimination and principle of least square method.The basic idea is: dividing the histogram into two groups in some gray-scale, the gray value shall be the threshold when the variance between the two groups is the largest.

If the gray threshold which is selected is t, the gray levels is divided into two groups: $C_0 = \{0, t-1\}, C_1 = \{t, S-1\}$, C_0 and C_1's probabilities respective are :

$$\omega_0 = \sum_{i=0}^{t-1} P(i) = \omega(t) \tag{3}$$

$$\omega_1 = \sum_{i=t}^{L-1} P(i) = 1 - \omega(t) \tag{4}$$

The averages of gray value about C_0 and C_1 respectively are:

$$q_0 = \sum_{i=0}^{t-1} \frac{iP(i)}{\omega_0} = \frac{q(t)}{\omega(t)} \tag{5}$$

$$q_1 = \sum_{i=t}^{S-1} \frac{iP(i)}{\omega_1} = \frac{q - q(t)}{1 - \omega(t)} \tag{6}$$

Where: $q = \sum_{i=0}^{S-1} iP(i)$ is the average gray value of the whole image. $q(t) = \sum_{i=0}^{t-1} iP(i)$ is the average gray when the threshold is t.So interclass variance we can obtain is as follows:

$$\sigma^2(t) = \omega_0(q_0 - q)^2 + \omega_1(q_1 - q)^2 = \frac{[q\omega(t) - q(t)]^2}{\omega(t)[1 - \omega(t)]} \tag{7}$$

In this way, we regard t as integer variable and take value from 0 to $S-1$ respectively the values, then substitute t into function of $\sigma^2(t)$, and the t shall be the threshold T we desire when $\sigma^2(t)$ is the largest.

3.3 Binary Converting in Image Partition

According to the threshold T calculated earlier, we can transform the gray image into binary image. Transformation is as follows: The gray value is 0 when the pixel gray value is less than threshold of T; When the pixel gray value is larger than or equal to T, the gray value is 1. Fig.3 shows the collection of the original image.Fig.4 shows the binary image with the threshold of T.

Fig. 3. Original image

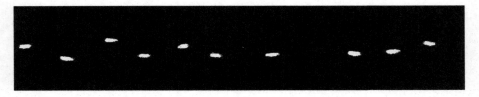

Fig. 4. Binary image with the threshold of T

3.4 Measurement of the Spaces of Seeds

To measure the distances of seeds, firstly we have to determine the poid of seeds, firstly we have to find the poid of seeds, and then regard this poid as the center of seeds. This article is only concerned about the projector distance of two seeds which are in the sowing centerline. So as long as we calculate coordinates of seeds in this direction, the distances of seeds will be got. Set upper-left corner of image as the origin coordinate, right is the positive X-axis, down is the positive Y-axis. Assuming a seed's coordinates of pixel location is $(x_i, y_i)(i = 0,1,...,n-1; j = 0,1,...,m-1)$, the center of each seed center's coordinate of $\overline{X}$-axis shall be the result. The equation is as follows:

$$\overline{X} = \frac{1}{k}\sum_{i=0}^{k} x_i \qquad (8)$$

Where: k is the quantity of pixel about seed's image.

4 Measurement Results and Error Analysis

4.1 Measurement Results

When the seed plate is in a speed of 40rmp, conveyor belt is in a speed of 0.72m/s, vacuum pump's pressure is -2kp, we intercept a portion of images about the seeds on the conveyor belt, as shown in table 1, lists a set of measurement results using this system and the results of manual measurement results. We carried out two experiments using this system so that to test the system's consistency on the results.

Table 1. Contrast table of test results

Measure-ment times	Results of manual measurem-ent (mm)	Results using this system 1 (mm)	Deviation 1 (mm)	Results using this system 2 (mm)	Deviation 2 (mm)
1	40.21	40.01	0.20	40.02	0.19
2	52.08	51.25	0.83	51.25	0.83
3	45.26	46.04	-0.78	46.04	-0.78
4	78.43	77.96	0.47	77.90	0.53
5	23.15	23.01	0.14	23.02	0.13
6	62.05	63.02	-0.97	63.01	-0.96
7	39.68	39.19	0.49	39.32	0.36
8	40.21	39.80	0.41	39.80	0.41
9	42.13	42.07	0.06	42.07	0.06
10	41.50	40.78	0.72	40.76	0.74

In this test qualified rate is 96.46%; leakage sowing rate is 1.63%; reseeding sowing rate is 1.91%. Table 1 shows measurement results got by this system have a good consistency with the manual measurement results, and the deviations are between ±1mm, which meets the test performance requirements. Further, the results have better consistency when we carry two compared tests using this system.

4.2 Error Analysis

The measurement results using this system are not exactly the same as the manual measurement results. The reasons for this inconsistency are:

1) Manual measurement has inherent error.
2) Speed quantization error.

5 Conclusions

In this paper, the application of the binary image threshold in the measurement of seeds' distance is studied. Aiming at features of seeds' space, we discuss the collection and processing for images. This system successfully realizes testing the parameters of seed metering device, shortening the test time, reducing the man-made error, increasing the system's ability of anti-interference, making the test data more accurate. At the same time this system simplifies the complexity of the operation and shortens the test's cycle.

Acknowledgements

Funding for this research was provided by Shandong major scientific and technological professional project(2007ZHZX10401). I'm grateful to my teachers for their help in my research.

References

Song-lie, Y.: China theory and practice of wheat cultivation. Shanghai science and Technology Press, Shanghai (2006)

Di, H., Li-dong, C., Yu-feng, X.: Experimental Study on the Factors of Seeding Quality about Air-suction Metering Device. Journal of Agricultural Mechanization Research 1, 175–176, 179 (2006)

Yu-jin, Z.: Image Partition. Science Press, Beijing (2001)

Sezan, M.I.: A Peak Detection Algorithm and its Application to Histogram Based Image Data Reduction. Graphical Model sand Image Processing, 47–59 (1985)

Yun-de, J.: Machine Vision. Science Press, Beijing (2000)

GB/T 6973-1986, Single (Precise)Test Method of Seeding Machine

Using L-M BP Algorithm Forecase the 305 Days Production of First-Breed Dairy

Xiaoli Wei, Guoqiang Qi*, Weizheng Shen, and Sun Jian

Agricultural engineer center, Northeast Agricultural University,
Harbin, Heilongjiang Province, China, 150030,
Tel.: +0451-55191146; Fax: +86-0451-55191146
liuhaiyang790214@163.com

Abstract. Aiming at the shortage of conventional BP algorithm, a BP neural net works improved by L-M algorithm is put forward. On the basis of the network, a Prediction model for 305 day's milk productions was set up. Traditional methods finish these data must spend at least 305 days, But this model can forecast first-breed dairy's 305 days milk production ahead of 215 days. The validity of the improved BP neural network predictive model was validated through the experiments.

Keywords: Dairy Milk production, prediction Neural Networks, BP algorithm-M algorithm.

1 Introduction

Milk yield is the main indexes to reflect and measure the cow's performance, and it is generally measured by a standard of 305 days milk yield in the world, But in the breeding work, we must calculate the milk production of 305 days, whether determining the descendants of bulls or assessing the quality of cows, which will take a long time and then extend the select cycle, thus how to use the cow's early productive performance to judge it good or not, and speed up the process of selection response has become the concerned and important topic(Qin zhirui,2005).

In this paper, A prediction model based on L-M BP neural network will be introduced to use in estimating the dairy 305 days milk quantities (Jiaoji Cheng, 1992), significantly, the model not only can forecast 305 days milk quantities production of the first-breed cow ahead of 215 days but also speed up the process of breeding dairy.

1.1 Improved BP Algorithm

Traditional BP algorithm uses the steepest gradient descent method to amend the weights, training course from starting point to reach the minimum point gradually along the slop of error function to reach iterate of each time, As for the complex network, error surface is in the multi -dimensional space, will therefore possibly fall

---

* Corresponding author.

D. Li and C. Zhao (Eds.): CCTA 2009, IFIP AICT 317, pp. 359–363, 2010.

into some- local smallest dot in the training process, so that will be unable to restrain, in addition, on the one hand, the stability of traditional BP stable request study rate is very small in the, therefore, the gradient descent law makes the training process very slow, on the other hand, the learning rate of momentum method usually be quicker than the pure gradient descent method, however, it is still slow in the practical application, all of above, two methods are only usually suitable for the increment training (Yin chaotian,2005).

This thesis will bring up the new method Levenberg-Marquardt training method which based on the characters of nonlinear system and considered to improve the traditional BP network, Levenberg-Marquardt training method can decrease sensitivity of the network regarding the erroneous surface local detail, and can limit the network to fall into the local minimum effectively, its basic idea is no longer along the sole anti-gradient direction to iterate of each time .but it permit error carry on the search along the worsened direction, greatly enhance the convergence rate and the pan-ability of network. This algorithm is much quicker than trainbp and train-bpx function's gradient descent law, but need more memories (Wang yu, 1992).

2 Prediction Model for First-Breeding Dairy

2.1 Sample Deal and Realize

Using the survey data which come from Heilongjiang province about the first breeding cow primary production and extracting 200 data recordings mainly include cow's maximum daily production, the 90 days and the 305 days production, Take 100 randomly selected records as the learning and training sample, other 100 cows records are used as forecast model examination sample, because the data range of variation is wide and the dimension is different, but the input and output based on neural network should be limited in certain scope, making the biggest input still to fall in the big place of neuron transforms function gradient, this may speed up the training speed of network, and can make the network training to be more effective, so it must be standardized for the network.

2.2 Network Design

In order to build up the neural network model, we firstly need to determine the neural network structure of forecast model, using three-layer neural network structure (Wen xin,Zhou lu, 2005):Input, Hidden-layer, Output .After building model, using this model can obtain cow's 305 days milk production on Output. The number of input nodes (N) is related with the input sample pattern. If the cow's highest milk production and 90 days milk production are took as the input pattern, then the number (n)=2, taking the cow's 305days milk production as the output of network, because the samples in this article are few, and only by an implicit strata, many strata nodes must be established, the established strata nodal point number is 6n, therefore,the network architecture of forecasting model is n×6n×1, namely 2×12×1. The transfer function of input node uses sigmoid hyperbolic tangent as formula (1)

$$f(x) = th(x) = \frac{e^x - e^{-x}}{e^x + e^{-x}} \qquad (1)$$

The transfer function of output node uses the linear function as formula (2).

$$f(x) = x \qquad (2)$$

2.3 Analysis

To forecast the cow's 305days milk production by improved BP network, under Matlab programming environment, network setting allow to make the max error is 0.01, the largest circulation is 10000 times, After training in 3904 iteration step-by-step, consuming one minute and three seconds, the network is restrain, then we use the traditional bp network to test samples(100 first-breeding cows), compare with the milk production between predicted production and actual production in 305 days milk yield, we can find the max error of network is about 77kg, the smallest error is about5 kg .

In order to reflect the superiority of improved BP(L-M BP) or traditional BP algorithm, fig.1 and fig.2 give the network's convergence situation of improved and traditional BP network in the same experimental data. As shown in fig 2 about the

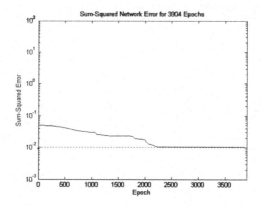

Fig. 1. Improved BP network

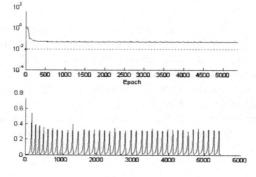

Fig. 2. Traditional BP network

traditional BP network, the SSE is no longer to renew when training achieves the certain extent, and the slackness of network appeared, causing the network not to go on training, but the improved BP network can overcome this question, it can jump out local minimum by regulating network parameter constantly, thus the network could be trained rapidly(Wei xiaoli,2006).

3 Experiments

L-M BP will not only improve the speed but also to improve the operation stability, we use L-M BP for the three months testing in several farms, the following example to illustrate this point, Through the forecast about twenty first-breeding cows' 305 days milk production from Mudanjiang Hailing dairy farm, 8511 cattle farm and XiangFang cattle farm, the result as shown in table1.

Table 1. The result of the forecast model

num	CowID	the maximum daily milk production	the 90 days production /kg	the 305 days production /kg	the forecasting production of 305 days /kg	error/kg
1	491007	20.2	1541	4597	4641	44
2	20004001	28.3	2213	6977	7047	70
3	20004002	15.2	1026	3841	3852	11
4	20004003	22.1	1713	5758	5826	68
5	20004005	32.3	2143	6780	6687	7
6	7977	20.6	1351	5020	4997	-23
7	7822	28.1	2286	6836	6798	-38
8	7945	28.6	1930	7045	6978	-67
9	7936	22.3	1717	5755	5832	77
10	7969	28	2115	6065	6070	5

4 Conclusions

To use the forecast model, we can forecast first-breeding dairy's 305 days milk production when the dairy has just begin to product for three months(90 days), in other word we will forecast the cow's production ahead of 215 days, by this method, you can choose a good milk production dairy in favor of breeding work.

It will supply us the condition for the advantaged of choosing excellent cows. It is known from the experiment, although this forecast model also has certain error, but it in the allowed scope, it is totally feasible to create a nonlinear system forecast model about cow's 305 days milk production with improved BP network in the production.

Acknowledgements

This paper is supported by the animal science center of northeast agricultural university, dairy experimental data, Experimental information and experimental methods are all coming from the topics "dairy science and technology major project" subject No is 2006BAD04A09, thanks of topic groups' helping, significantly, I want to express my thankful for Harbin Intelligent Agricultural Center of Northeast Agricultural University. Xiang-Fang Farm, Animal Medicine Center of Northeast Agricultural University.

References

Zhirui, Q.: Introduction of dairy cattle breeding guidance, pp. 1–5. Jindun Press, Beijing (2005)

Cheng, J.: Neural network introduction, vol. 32, pp. 22–28. Xi'an University of Electronic Science and Technology Publish, Xi'an (1992)

Yu, W.: The prediction model and using of the improved BP model. Compute Measurement and Control 13(1), 39 (2005)

Xin, W., Lu, Z.: Matlab neural network using and design. Science Publish, Beijing

Chaotian, Y.: Artificial intelligence and expert system, Beijing

Changhong, D.: Matlab network design and application, vol. 54, pp. 23–28. Science Publish, Beijing (2004)

Xiaoli, W., Yonggen, Z.: Using BP forecast the first-breeding dairy's 305 production. Beijing, China dairy (7), 24–25 (2006)

The Fuzzy Model for Diagnosis of Animal Disease

Xiao Jianhua, Shi Luyi, Zhang Yu, Gao Li, Fan Honggang,
Ma Haikun, and Wang Hongbin[*]

College of Veterinary Medicine, NorthEast Agricultural University,
Harbin, Heilongjiang Province, P.R. China
Voice: +86-451-55190470
66229894@qq.com

Abstract. The knowledge of animal disease diagnosis was fuzzy; the fuzzy model can imitate the character of clinical diagnosis for veterinary. The fuzzy model of disease, the methods for class the disease group of differential diagnosis and the fuzzy diagnosis model were discussed in this paper.

Keywords: disease, fuzzy, diagnosis, model.

1 Introduction

Diagnosis knowledge of animal disease is fuzzy, and the establishment of representation and reasoning model for animal diagnosis knowledge based on fuzzy theory can reflects the nature of medical diagnosis to some extent.

It was reported that the fuzzy inference model for diagnosis of bone tumors based on fuzzy set theory and medical statistics knowledge can not only simulate the diagnosis of bone tumors, but also can adjust the parameters to improve the accuracy of diagnosis with the accumulation of cases(Huikang Zhang, 2003).

In addition, the mathematical model for diagnosis of gastric cancer based on Fuzzy clustering and stepwise discriminate analysis method(Xiaoyan Lu,2003), the expert system for diagnosis of liver disease based on fuzzy reasoning(Carvalho V,2005) heart disease classification system (Shiomi S,1995) Dairy claudicating Forecast Expert System based on fuzzy logic(Pop HF,2001) etc.

2 Materials and Methods

2.1 Knowledge Is Key in Government and Governance

The symptoms, medical history, signs, laboratory data, pathology tests and other information of animal can be used as evidence of diagnosis. X is the universe composed with all evidence in diagnosis of animal diseases, x is one of evidence, different evidence with different effect which can be expressed as ambiguity when diagnosis of

---

[*] Corresponding author.

D. Li and C. Zhao (Eds.): CCTA 2009, IFIP AICT 317, pp. 364–368, 2010.
© IFIP International Federation for Information Processing 2010

diseases. Therefore the fuzzy mathematical model can be build for disease diagnosis, namely:

For any evidence of $x \in X$, the mappings: $x \mapsto \mu_A(x) \in [0,1]$ exist.

The above mappings can be regard as the relationship between x and one disease. The evidence collection of one disease can be established through this Mapping

$$\underset{\sim}{A} = \{(x \mid \mu_A(x))\}, \forall x \in X$$

$\underset{\sim}{A}$ is the collection of evidence for diagnosis of one disease, which may include the symptoms, history and epidemiological data, $\underset{\sim}{A}$ is the subset of X , F(X) is all collection in X . Then $\underset{\sim}{A} \subset F(X)$. $\mu_A(x)$ is the membership function of elements x in disease A, for a specific x , $\mu_A(x)$ is Degree of membership of x for $\underset{\sim}{A}$, is extent of the role in the course of the disease diagnosis in another words.

Supposing:

A= Rumen indigestion

$x_1 =$ The potable water is insufficient after crossing the food. $\mu_A(x_1) = 0.1$

$x_2 =$ Has eaten feed which easily to swell $\mu_A(x_2) = 0.25$

$x_3 =$ Change feed suddenly $\mu_A(x_3) = 0.2$

$x_4 =$ The rumen contents was hardly, assumes the pasta type $\mu_A(x_4) = 0.8$

$x_5 =$ Left side of girth of paunch expands obviously $\mu_A(x_5) = 0.6$

$x_6 =$ Rumen discharge gas when puncture upside rumen, and discharge digestion food when puncture lower part $\mu_A(x_6) = 0.7$

$x_7 =$ Indentation appeared in skin when palpate rumen $\mu_A(x_7) = 0.8$

The collection of element for diagnose rumen indigestion can be express as:

$$\underset{\sim}{A} = \frac{0.1}{x_1} + \frac{0.25}{x_2} + \frac{0.2}{x_3} + \frac{0.8}{x_4} + \frac{0.6}{x_5} + \frac{0.7}{x_6} + \frac{0.8}{x_7}$$

2.2 The Division of Differential Diagnosis Disease Group

Similarity or Relevance may exist between one disease and another disease. In order to diagnose, the relationships between diseases must be find and express in some forms. Diseases of one kind of animal often have as many as hundreds of species, it is impossible to express hundreds of the relationship between theses diseases, in fact, there is no need for such a gesture. Between each kind of disease has the relations by no means, it is easy to distinguish among diseases with few relations. The key of diagnosis is to differentiate diseases that exist relation. Therefore hundreds kind of diseases must be divide into many differential diagnosis group at first. Diseases that have big similarity must be division is one group.

The relations need to be established among theses diseases which belong to one differential diagnosis group. It is helpful to discover similar diseases that to established this kind of relations. The similarity among diseases is fuzzy, and can be express in degree of membership. And the relations among diseases may express for the following fuzzy relationship.

Supposes A one group of diseases that need to be differential diagnose, $A = \{x_1, x_2, \cdots, x_n\}$, x_i is the i_{th} disease.

$$R = \begin{pmatrix} r_{11} & \cdots & r_{1n} \\ \vdots & \ddots & \vdots \\ r_{n1} & \cdots & r_{nn} \end{pmatrix} \quad \text{or} \quad R = (r_{ij})_{n \times n}$$

And $r_{ij} = R(x_i, x_j)$. There was relationship between i_{th} disease and j_{th} disease.

By this model, diseases was divided into many groups, when diagnosis, the conclusion can be discover according to the similarity among diseases.

The differential diagnosis group may determined use the fuzzy distance law, Supposes D is a collection that need to be differential diagnose, D=$\{D_1, D_2, \cdots, D_m\}$, S is the collection composed of symptoms, $S = \{s_1, s_2, \cdots, s_n\}$. The distance between D_x and D_y can be express as:

$$d(D_x, D_y) = \frac{1}{n} \sum_{i=1}^{n} W(s_i) \mid D_x(s_i) - D_y(s_i) \mid$$

d is the distance between D_x and D_y, If one threshold value λ is determined, when the distance between two diseases $d < \lambda$, then those two diseases belong to one differential diagnosis group.

2.3 Animal Disease Fuzzy Diagnosis Mathematical Model

The fuzzy model was established between disease and symptoms at begin, supposed D is the collection that need to be differential diagnosed, $D = \{d_1, d_2, \cdots, d_m\}$, S is the collection of symptoms $S = \{s_1, s_2, \cdots, s_n\}$. The fuzzy relationship between diseases and symptoms can be express as fuzzy matrix of n row and m list:

$$R_{DS} = \begin{pmatrix} r_{11} & \cdots & r_{1m} \\ \vdots & \ddots & \vdots \\ r_{n1} & \cdots & r_{nm} \end{pmatrix} \quad \text{or} \quad R_{DS} = (r_{ij})_{n \times m}$$

R_{DS} is the fuzzy relationship between disease and symptom, $r_{ij} = R(x_i, y_j), 0 < r_{ij} < 1$。

The fuzzy distance model of diagnosis can be established based on the above model. Supposes D is the collection of diseases that need to be diagnosed, D=$\{D_1, D_2, \cdots, D_m\}$, S is the collection of symptoms, $S = \{s_1, s_2, \cdots, s_n\}$. Then the distance between disease D_x and D_r can be express as:

$$(d_1, d_2, \cdots, d_m) = (s_1, s_2, \cdots, s_n) \circ \begin{pmatrix} r_{11} & \cdots r_{1j} \cdots & r_{1m} \\ \vdots & & \vdots \\ r_{i1} & \cdots r_{ij} \cdots & r_{im} \\ \vdots & & \vdots \\ r_{n1} & \cdots r_{nj} \cdots & r_{nm} \end{pmatrix}$$

$$d_j(\underset{\sim}{D}_x, \underset{\sim}{D}_r) = \frac{1}{n} \sum_{i=1}^{n} W(s_i) \mid \underset{\sim}{D}_x(s_i) - \underset{\sim}{D}_r(s_i) \mid \quad \underset{\sim}{D}_r(s_i) = r_{ij}$$

$d = \{d_1, d_2, \cdots, d_m\}$, the most impossible disease is $\min\{d_1, d_2, \cdots, d_m\}$. the disease with most impossible can be determined by single factor fuzzy diagnosis model when there exist two diseases need to be diagnosed, however multi-factors fuzzy diagnosis model need to be established when many diseases need to be diagnosed, suppose D is one group disease need to be differential diagnosed, $D = \{\underset{\sim}{D}_1, \underset{\sim}{D}_2, \cdots, \underset{\sim}{D}_m\}$, suppose $\underset{\sim}{D}_x$ is the disease need to be diagnosed. σ is matching between treating disease and any disease in D, suppose:

$$\sigma(\underset{\sim}{D}_i, \underset{\sim}{D}_x) = \max\{\sigma(\underset{\sim}{D}_k, \underset{\sim}{D}_x) \mid k = 1, 2, \cdots, m\}$$

$$\sigma(\underset{\sim}{D}_i, \underset{\sim}{D}_x) = 1 - \frac{1}{n} \sum_{j=1}^{n} W(s_j) \mid \underset{\sim}{D}_x(s_j) - \underset{\sim}{D}_i(s_j) \mid$$

$$i = \{1, 2, \cdots, m\}, \quad j = \{1, 2, \cdots, n\}$$

It is most close that $\underset{\sim}{D}_x$ and $\underset{\sim}{D}_i$, then $\underset{\sim}{D}_i$ is determined the conclusion.

2.4 Conclusion

There are hundreds of diseases for animals, moreover the specialization degree is relatively low in veterinary, any veterinarian have to solve diseases of every kind of animal, and can solve internal medicine sickness, surgical sickness, the obstetrical disease, parasitic disease, the infectious disease and so on. Therefore, the expert system for animal disease must conform to the objective law of veterinarian clinical. The animal disease should be divides into some big group such as internal medicine sickness, surgical sickness, and then divide into many differential diagnosis groups. By different character of disease, the different knowledge expression and the inference method was developed. The fuzzy model in this paper has been used in some diseases, but it is necessary to apply in more diseases.

Acknowledgements

The authors thank the financial support from National Project of Scientific and Technical Supporting of China during the 11th Five-year Plan 2006BAD10A02-04.

References

Carvalho, V., Naas, I., Mollo, M., et al.: Prediction of the Occurrence of Lameness in Dairy Cows using a Fuzzy-Logic Based Expert System-Part I. Agricultural Engineering International: the CIGR Ejournal VII, Manuscript IT 05 002, June 1-12 (2005)

Zhang, H., Qian, Z., Qu, J., et al.: Fuzzy reasoning model for bone tumor auxiliary diagnoses expert system. Journal of Forth Military Medical University 24(2), 182–186 (2003)

Shiomi, S., Kuroki, T., Jomura, H., et al.: Diagnosis of chronic liver disease from liver scintiscans by fuzzy reasoning. J. Nucl. Med. 36(4), 593–598 (1995)

Pop, H.F., Pop, T.L., Sarbu, C.: Assessment of heart disease using fuzzy classification techniques. Scientific World Journal 17(1), 369–390 (2001)

Lu, X., Guo, J.: The auxiliary diagnostic model for medicine mathematics based on fuzzy and gradually analysis. Journal of shanxi medical college 34(6), 499–502 (2003)

Texture Detect on Rotary-Veneer Surface Based on Semi-Fuzzy Clustering Algorithm

Wei Cheng[*], Ping Liang, and Suqun Cao

Faculty of Mechanical Engineering, Huaiyin Institute of Technology,
Huaian, Jiangsu Province, P.R. China 223003,
Tel.: +86-517-83559196
hychw@sina.com

Abstract. The texture of rotary-veneer can interference in defects detection, this paper presented a modified semi-fuzzy clustering (SFC) algorithm. SFC algorithm incorporates Fisher discrimination method with fuzzy theory using fuzzy scatter matrix. By iteratively optimizing the fuzzy Fisher criterion function, the final clustering results are obtained. SFC algorithm exhibits its robustness and capability to obtain well separable clustering results. This algorithm can detect the texture and defects on rotary-veneer surface exactly.

Keywords: Rotary-veneer, Semi-fuzzy clustering (SFC), Defects detection, Texture.

1 Introduction

Plywood grade was decided by the surface quality of rotary-veneer. The defect of rotary-veneer surface affects the quality of the wood-board panels products such as plywood, the computer vision system that forms using linear CCD gathers the image of veneer tape can be used automatic control and optimization of veneer tape cut out offer tech support. A goal of the rotary-veneer defects detection is that presents according to the defects different with the veneer background characteristic, and separates it from the veneer background. The superficial backgrounds of rotary-veneer include mainly each kind of shape and the relative order texture pattern. Due to textures and the most detects of veneer surface are the nature formations, their quality are the same, and their characteristic are similar. Rotary-veneer surface texture regular session disturbance defects characteristic extraction. Therefore, the research of the rotary-veneer surface texture detection may enhance the rotary-veneer detection precision, realizes automatic grading based on the computer vision veneer.

In reference (Ping Liang, Wei Cheng 2008), the texture of rotary-veneer can interference in defect detects, this paper presented a modified Fuzzy C-Mean algorithm (FCM). The approach of this algorithm is sample density of inter-class and distances of intra-class as comprehensive parameters, thereby to obtain the validity initial cluster centers. This algorithm can detect the texture and defects on rotary-veneer surface. Most of these methods like FCM are essentially rooted at the within-cluster scatter

---

[*] Corresponding author.

D. Li and C. Zhao (Eds.): CCTA 2009, IFIP AICT 317, pp. 369–374, 2010.
© IFIP International Federation for Information Processing 2010

matrix as a compactness measure, which means that the assumption that the clusters are hyperspheroidal is taken. In fact, real data sets seldom accommodate such an assumption. Except for the compactness measure, the concept of the separation measure such as the between-cluster scatter matrix should also be involved in the design of clustering methods such that well separated cluster can be obtained.

In this paper, we extend Fisher linear discriminant to its fuzzified version and define the fuzzy Fisher criterion function. Then a novel semi-fuzzy clustering (SFC) method is presented. Compared with FCM, the proposed algorithm has the following characteristics: (1) It directly uses the fuzzy FLD as its objective function, therefore, it is an unsupervised fuzzy partition clustering method. Its objective function integrates the fuzzy between-cluster scatter matrix well with the fuzzy within-cluster scatter matrix. (2) It incorporates the discriminating vector into its update equations such that the obtained update equations do not take commonly-used FCM-like forms. (3) It is more robust to noise and outliers than FCM.

2 Fisher Criterion and Fisher Linear Discriminant

Given c pattern classes $X^{(i)} = [x_i^1, x_i^2, \ldots, x_i^{N_i}]$ in the pattern set which contains N d-dimensional patterns, where $i = 1, 2, \ldots c$, N_i is the number of all the patterns in the ith class, thus $N = N_1 + N_2 + \cdots + N_c$. The between-class scatter matrix S_b and the within-class scatter matrix S_w are determined by the following formulae:

$$S_b = \sum_{i=1}^{c} \frac{N_i}{N}(m_i - \overline{x})(m_i - \overline{x})^T \tag{1}$$

$$S_w = \frac{1}{N}\sum_{i=1}^{c}\sum_{j=1}^{N_I}(x_j^i - m_i)(x_j^i - m_i)^T \tag{2}$$

Where m_i denotes the mean of the ith class, $\overline{x}$ denotes the mean of all the patterns in the pattern set.

According to the scatter matrices, the Fisher criterion function can be defined as follows:

$$J_{FC}(\omega) = \frac{\omega^T S_b \omega}{\omega^T S_w \omega} \tag{3}$$

Where ω is an arbitrary vector in d-dimensional space. The Fisher optimal discriminate vector is ω^* corresponding to maximum of $J_{FC}(\omega)$, which is the eigenvector corresponding to maximum eigenvalue of the following eigensystem equation:

$$S_b \omega^* = \lambda S_w \omega^* \tag{4}$$

Where λ is diagonal and consists of the corresponding eigenvalues.

3 Fuzzy Fisher Criterion and the Proposed Algorithm

First, let us fuzzily the concept of the above Fisher linear discriminate (FLD).

Suppose that the membership function $u_{ij} \in [0,1]$ with $\sum_{i=1}^{c} u_{ij} = 1$ for all j and the index $m > 1$ is a given real value, where u_{ij} denotes the degree of the jth d-dimensional pattern belonging to the ith class, we can define the following fuzzy within-class scatter matrix S_{fw}:

$$S_{fw} = \sum_{i=1}^{c} \sum_{j=1}^{N_I} u_{ij}^{m} (x_j - m_i)(x_j - m_i)^T \tag{5}$$

And the following fuzzy between-class scatter matrix S_{fb}:

$$S_{fb} = \sum_{i=1}^{c} \sum_{j=1}^{N_I} u_{ij}^{m} (m_i - \bar{x})(m_i - \bar{x})^T \tag{6}$$

Thus, we can derive a novel fuzzy Fisher criterion called fuzzy FLD as follows:

$$J_{FFC}(\omega) = \frac{\omega^T S_{fb} \omega}{\omega^T S_{fw} \omega} \tag{7}$$

In terms of the fuzzy FLD as above, we will derive a novel semi-fuzzy clustering algorithm based on fuzzy Fisher criterion. Maximizing J_{FFC} directly in Eq.(7) is not a trivial mathematical derivation task due to the existence of its denominator. However, we can reasonably relax this problem by applying the following Lagrange multipliers λ and $\beta_j (j = 1,2,\cdots,n)$ together with the constraint $\sum_{i=1}^{c} u_{ij} = 1$ to formula (7):

$$F = \omega^T S_{fb} \omega - \lambda \omega^T S_{fw} \omega + \sum_{j=1}^{N} \beta_j (\sum_{i=1}^{c} u_{ij} - 1) \tag{8}$$

Setting $\dfrac{\partial F}{\partial \omega}, \dfrac{\partial F}{\partial m_i}, \dfrac{\partial F}{\partial u_{ij}}$ to be zero, we respectively have

$$S_{fb}\omega = \lambda S_{fw}\omega \tag{9}$$

Where λ may be taken as the largest eigenvalue.

$$m_i = \frac{\sum_{j=1}^{N} u_{ij}^m (x_j - \frac{\overline{x}}{\lambda})}{\sum_{j=1}^{N} u_{ij}^m (1 - \frac{1}{\lambda})} \qquad (10)$$

$$u_{ij} = \frac{F_1}{F_2} \qquad (11)$$

Where

$$F_1 = (\omega^T (x_j - m_i)(x_j - m_i)^T \omega - \frac{1}{\lambda} \omega^T (m_i - \overline{x})(m_i - \overline{x})^T \omega)^{\frac{1}{m-1}}$$

$$F_2 = \sum_{k=1}^{c} (\omega^T (x_j - m_k)(x_j - m_k)^T \omega - \frac{1}{\lambda} \omega^T (m_k - \overline{x})(m_k - \overline{x})^T \omega)^{\frac{1}{m-1}}$$

When Eq.(11) is used, as stated in the above, u_{ij} should satisfy $u_{ij} \in [0,1]$, hence, in order to satisfy this constraint, we let
$u_{ij} = 1$ and $u_{ij} = 0$ for all $i' \neq i$, if

$$\omega^T (x_j - m_i)(x_j - m_i)^T \omega \leq \frac{1}{\lambda} \omega^T (m_i - \overline{x})(m_i - \overline{x})^T \omega \qquad (12)$$

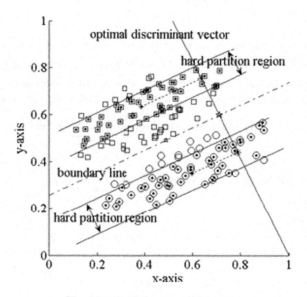

Fig. 1. FFC-SFC hard partition regions

That is to say, if Eq.(12) holds, we take a hard partition for the pattern. The rational can be intuitively explained from a geometric viewpoint as shown in Fig. 1. In Fig. 1, all the patterns with one class □ and the other class ○, the two clusters center with * and the total mean point (i.e. the average point of all samples) with ☆ are projected along dotted lines onto the optimal discriminating vector. Obviously, if the Euclidean distance between the projection of a pattern and the projection of certain cluster is equal to or less than multiplied by the Euclidean distance between the projection of this cluster and ☆, then we should take a hard partition for this pattern.

From the above analysis, we can obtain a novel semi-fuzzy algorithm based on fuzzy fisher criterion (FFC-SFC).

Algorithm FFC-SFC

Step1. Set the given threshold ε, initialize $U = [u_{ij}]_{c \times N}$ and $m = (m_1, m_2, \cdots, m_c)$ using K-mean;

step2. Compute S_{fw}, S_{fb} using Eq.(5), Eq.(6) respectively;

step3. Compute the largest eigenvalue λ and the corresponding ω using Eq.(9);

Step4. Update m_i and u_{ij} using Eq.(10), Eq.(11) and Eq.(12) respectively;

Step5. Compute J_{FFC} using Eq.(7);

Step6. If $J_{FFC} < \varepsilon$ or the number of iteration $\geq$ the given value, output the clustering result and then terminate, otherwise back to Step 2.

4 The Analyses of Rotary-Veneer Surface Texture Detection

In the research of rotary-veneer surface texture detection, a goal is to enhance the detection precision of rotary-veneer surface defects. Due to the textures show large expanse, continuously long linearity distribution. The textures occupy the big region in the rotary-veneer surface. But the defects of rotary-veneer surface occupy the small region, and they are regular, such as the shape of knots is an ellipse. Compares with the texture color, the defects color is also deep. Therefore, when we segment the rotary-veneer defect image, shape and color was chosen for this segmentation characteristic.

In fig.1, (a) image show rotary-veneer image; (b) image show used the FCM algorithm which this reference (Ping Liang, Wei Cheng 2008) proposed to segment this texture image; (c) image show used SFC to segment this texture image.

We can see that SFC has better segmentation accuracy, and is comparatively competitive to FCM.

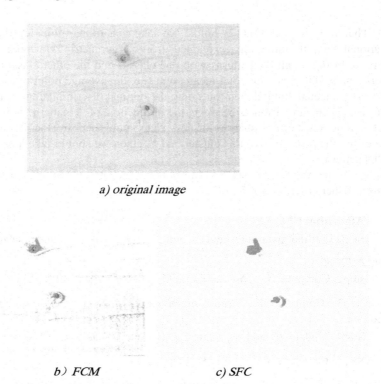

a) original image

b) FCM   *c) SFC*

Fig. 2. Rotary-veneer image detection result

Acknowledgements

Funding for this natural science research was provided by Jiangsu Provincial Department of Education (P.R. China) (No. 2005KJD520033).

References

Cao, S., Wang, S., et al.: Fuzzy Fisher Criterion Based Semi-Fuzzy Clustering Algorithm. Journal of Electronics & Information Technology 30(9), 2132–2165 (2008) (in Chinese)

Liang, P., Cheng, W.: Texture Detection on the Rotary-Veneer Surface Based on FCM. Techniques of Automation and Applications (11), 79–85 (2008) (in Chinese)

Three-Dimension Visualization for Primary Wheat Diseases Based on Simulation Model

Li Shijuan and Zhu Yeping

Agricultural Information Institution, CAAS, Beijing, P.R. China 100086

Abstract. Crop simulation model has been becoming the core of agricultural production management and resource optimization management. Displaying crop growth process makes user observe the crop growth and development intuitionisticly. On the basis of understanding and grasping the occurrence condition, popularity season, key impact factors for main wheat diseases of stripe rust, leaf rust, stem rust, head blight and powdery mildew from research material and literature, we designed 3D visualization model for wheat growth and diseases occurrence. The model system will help farmer, technician and decision-maker to use crop growth simulation model better and provide decision-making support. Now 3D visualization model for wheat growth on the basis of simulation model has been developed, and the visualization model for primary wheat diseases is in the process of development.

Keywords: 3-dimension, visualization, wheat diseases, simulation model.

1 Introduction

Growing with the development of computer technology, system analysis theory and agricultural scientific research, crop simulation model has been becoming the core of agricultural production management and resource optimization management, and the basis of precision agriculture which is implemented now in China. After 50 years evolvement crop simulation model and becomes more mature possesses of more mechanism. America, Holand, England and Australia developed many crop models, some of which had been used in agriculture successfully such as DSSAT, SUCROS series models, RZWQM and APSIM etc. Now crop simulation model mainly is used to forecast yield with typical application such as studying world food and agri-ecological belt, forecasting regional yield, evaluating the effects of environment and social economy changes on agriculture.

China started crop simulation model study since 1980s. After introducing, analyzing and improving foreign crop models, researchers developed a lot of application systems. Some units exploited their own crop simulation model. For example, Zhao chunjiang et al constructed wheat management expert system based model for Beijing (Zhao Chunjiang et al., 1997); Soil and Fertilizer Institute of Chinese Academy of Agricultural Sciences developed wheat and maize optimal fertilization expert system of Yucheng county on Huanghuai Plain; Nanjing Agricultural University integrated wheat growth model and expert system to wheat intelligent management and

D. Li and C. Zhao (Eds.): CCTA 2009, IFIP AICT 317, pp. 375–381, 2010.
© IFIP International Federation for Information Processing 2010

decision-making system (Zhu Yan et al., 2004); Jiangsu Academy of Agricultural Sciences constructed rice cultivation simulation-optimization-decision making system (Gao Liangzhi et al., 1992).

Displaying crop growth process makes user observe the crop growth and development intuitionisticly. To make the crop growth and development process visualized, many scientists have made great efforts. With respect to crop virtual technology, L system put forward by theoretical biologist Lindenmayer in 1968 became one of the main method for plant modeling (Hu Baogang et al., 2001). In order to effectively deploy short-rotation woody crop plantations for energy and fiber production at regional scales, Host et al (1996) described an object-oriented strategy for scaling ECOPHYS, an individual tree growth process model for hybrid poplar, to a plantation, which overcame the shortage of traditional individual tree growth process models, which were too complex to use at the plantation scale (G. Host et al., 1996). Deng et al (2005) developed a mathematic model for changes on morphology in corn leaf, aimed at visualizing the information for simulation model (Deng Xuyang et al., 2005). Chen et al (2005) discussed the modeling of leaf growth dynamics in winter wheat, and simulated the leaf growth dynamic using Logistic equation (Chen Guoqing et al., 2005). China Agricultural University and Beijing Academy of Agriculture and Forestry Sciences developed virtual model for maize growth (Ma Yuntao et al., 2006; Guo Xinyu et al., 2007).

Wheat is the dominating food crop in China, which possesses planting area and yield just less than rice. Wheat diseases restrict the grain quality and quantity, and imperil the food safety. From the ninety's of twenty-first century, more and more expert system for wheat diseases and insect pests prediction had been developed (Rossing WAH et al., 1994). In 1993 Denmark researched probabilistic model to predict and cure wheat mycosis, and the simulation system EPIDEMIC simulating the popular wheat diseases by analyzing the information concerning disease occurrence incidence, local climate, production and management etc. Winter wheat bactericide model WDM belongs to Decision Support System for Arable Crops (DESSAC) took weather, crop varieties, diseases occurrence incidence into account, can provide decision support for making use of wheat bactericide (D. Brooks, 1998). Gonzalez-Andujar et al developed aphid recognition expert system (J. L. Gonzalez-Andujar et al., 1993). Guo et al analysed field prevalence trend and factors of main diseases in Wheat in Luohe (Guo Chunqiang, 2007), predicted the prevalence of diseases based on systematic analysis of the impact elements of main diseases using weighted crosstabs, Fisher's two group discriminant analysis and stepwise regression analysis and built up the models for forecast. The forecast and management system for main wheat disease, wheat eyespot, wheat head scab, etc., was set up by means of Visual Basic 6.0 sharp and mufti-media technology. The system was made up of diagnosing module, forecast and control decision module, system explanation module.

From these existing studies, we can see that some only analyzed the topology structure; some model systems made plant growth visualized, but couldn't couple with the growth simulation model data effectively. Displaying growth dynamics of leaf, stem and stamen as well as the occurrence rule and symptom for primary wheat diseases is important for teaching, research and management, especially for these users such as decision-maker and generic agronomic technician. It is necessary for them to develop a visualization system to enhance enthusiasm. This study developed 3D visualization model for wheat growth and disease, and integrated it with wheat simulation model.

2 Wheat Simulation Model

Based on the past studies (Li Shijuan et al., 2007), we collected related literatures and agronomic expert information in a large scale, then designed wheat cooperative models in accordance with wheat growth and development discipline, and combined the models with corresponding database and repository, and constitute wheat simulation model system using technologies of system engineering theory, software engineering theory, computer and animation and image processing. The system consists of cooperative models, database and interface etc.

Wheat model system evolved from CERES-Wheat, which is a crop specific model aimed at dynamic simulation of wheat growth as affected by climatic, plant and soil properties along with certain farm management practices (J.T. Ritchie and S. Otter), computes LAI, light Interception, photosynthesis, and dry matter production and distribution in wheat, and calculates N uptake and distribution. Wheat development is divided into 8 stages in wheat development model: from sowing to germination, from germination to seeding emergence, from seeding emergence to juvenile stage, from juvenile stage to jointing stage, from jointing stage to silking stage, from silking stage to beginning of grain filling, from beginning of grain filling to physiological maturity. Temperature, water, photoperiod and genetic parameter restrict the replacement of development stages. Genetic parameter can be input by user or decided automatically by parameter determination program in system. Water balance model is built to simulate water leakage, runoff, soil evaporation, plant transpiration and root water absorption in each soil layer according to water movement rules, soil water status and wheat absorption characteristic adopting Priestly-Taylor equation and SCS Curve Number Method. N balance model mainly simulates N mineralization and fixation of organic matter in soil, N losses and uptake by crop. N deficit index calculated by the model affects directly daily accumulated value of wheat dry matter and LAI. This model considers the amount of nitrate leached out of wheat root zone (here define it as 2 meters) with water movement, and evaluated possible effects of the leached nitrate on groundwater. The effects on main wheat quality (protein, starch and fat) of variety trait, weather, cultivation management and nutrition are analyzed by grain quality model which deducted the algorithm with Logistic equation by drawing up the relation between quality and impact factors such as density, days after grain-filling, water and nutrition.

3 3D Visualization Model System for Wheat Growth

3D visualization model mainly simulates the wheat growth based on agronomic shape knowledge, image and 3D animation technology, and the simulation data in results database from above simulation model. After getting the wheat growth dynamic data from growth simulation model, 3D Studio MAX was used to develop basic crop model files such as seed, root, stem, leaf, leafstalk, corncob and stamen, which were in the format of MAX. Then these model files formatted as MAX were transferred to format as 3DS, and then these 3DS files were transferred to data files formatted as filename.h and filename.gl. In the program compilation, files formatted as .h were used, and files formatted as .gl were used in the program execution. With wheat growth data and files formatted as .h and .gl, the 3D visualization function would be implemented by analyzing wheat growth principle in a computer, with the operational platform Windows 2000 in Chinese version, and Microsoft® Visual C++ 6.0 was used as programming language.

3D visualization output of wheat growth includes 7 stages, i. e. seed, germination, leaf development, root growth, tiller dynamic, heading, grain-filling process. Fig.1 shows the 3D visualization results. The system supports many kinds of databases such as Access, Excel, SQLSERVER, Oracle, MySql, SqlServer etc. and shows the 3D wheat plant and population according to the setting of plant density. When simulating single plant, the effect of rain, water and nitrogen on wheat grown should be expressed. When simulating wheat population, system can show the effect of plant density. It creates animation file named as wheat.avi which supports network as display material. It makes it easy to observe and predict wheat growth, development and yield formation directly. The methods such as rigid body kinematics and flexible system were applied to model construction in order to realize the functions such as zoom, growing longer, growth, horizontal movement, rotation, color change and total shape change. User can observe whole wheat growth from different angle by means of horizontal moving, zoom in, zoom out, and so on. Horizontal axis and vertical axis are the circumgyration center.

4   Visualization Design for Primary Diseases

Consulting a great deal of research material and literature to in-depth understand and grasp the occurrence condition, popularity season, key impact factors for main wheat diseases such as stripe rust, leaf rust, stem rust, head blight, powdery mildew, we designed the flow chart for wheat diseases visualization model (Figure 1). Each disease has its key impact factor, among which air temperature is the most important one, and is related to each disease we study. In addition, the rainfall in April affects the occurrence rule of leaf rust, and the occurrence degree of head blight depends on the days of more than 0.1mm rainfall after tasseling.

The final occurrence degree lies on the key impact factors for each disease. For leaf rust, stem rust and powdery mildew varietal resistance determine their degree. The occurrence degree of stripe rust is influenced by varietal resistance and the rainfall in april. Varietal resistance and the days of more than 0.1mm rainfall after tasseling decide the final occurrence degree of head blight. As illustrated by the case of stripe rust, according to the disease-resistant ability we classify the wheat varieties to following 4 types: high-resistance, middle-resistance, middle-susceptivity and high-susceptivity. For certain kind of variety, the more rainy days or rainfall April has, the more serious the final occurrence degree is. Here the rainy days or rainfall in April is divided into 3 groups shown in table 1. Figures in table 1 express the final occurence degree: 1 indicates there's no further development and plant retains the earlier symptom; 2 means the ratio of the leaves catching the disease to the normal leaves reaches 20%; 3 shows above ratio reaches 40%; 4 expresses more than 60% leaves catch the stripe rust, and stem catches the disease too.

By means of 3D animation technology too and analysing the occurrence rule and fractal feature, We'll develop 3D visualization model for primary wheat diseases on

Table 1. Influence of rainfall in April on different kinds of resistant varieties

Varietal resistance	Rainfall in April		
	Rain days >15 or Rainfall>50mm	Rain days: 10-15 or Rainfall 15-40mm	Rain days<5 or Rainfall<15mm
High-resistance	2	1	1
Midding-resistance	3	2	1
Midding-susceptivity	3	3	2

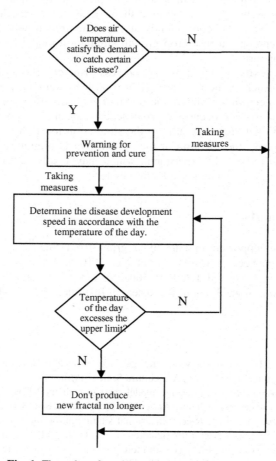

Fig. 1. Flow chart for wheat diseases visualization model

the basis of 3D visualization model for wheat growth and phase development, and express the occurrence and development process of main diseases realistically. Our main works consist of three steps listed below. First, the basic concept models for corresponding fractal of main wheat diseases are built with 3DMAX, and the data in models can be read in computer program. Secondly, combining with database (mainly weather data) and above basic models, 3D visualization model for wheat growth expresses the disease occurrence process in leaf, stem or ear. Finally the control over image display and 3D animation will be realized. Model development would adopt programming language Microsoft Visual C++ 6.0 and Open GL.

5 Conclusion

We described the implementation of 3D visualization system for wheat growth, development, yield formation, and occurrence rule and symptom for primary diseases, and the system could help the user understand the principle of wheat growth and diseases better. By the system, we can observe the process of growth, development and yield formation obviously and conveniently, at the same time, it will be a good tool to help farmer and technician to use crop growth simulation model better and provide decision-making support.

Due to the complexity of wheat growth process principle, this system might also has some shortages such as the morphology of root, leaf, stem, corncob weren't very realistic; the system couldn't simulate the population shade effect on the growth and so on. Additionally, disease occurrence rule depends on meteorological data which often are difficult to get and can't suits to the simulation location thoroughly. In the future, we will improve the function of the system by 3D digitalization of the plant and population, and more field experiment, which maybe solve above problems and make the system serve for the crop production better.

Acknowledgements

This research was supported in part by a grant from State High-tech Research and Development Project of China (Grant No. 2007AA10Z237, 2006AA10Z303, 2006AA10Z220), and Special Fund of Basic Scientific Research and Operation Foundation for Commonweal Scientific Research Institutes (2009J-04).

References

Zhao, C., Zhu, D., Li, H., et al.: Study on intelligent expert system of wheat cultivation management and its application. Scientia Agricultura Sinica 30(5), 42–49 (1997) (in Chinese)

Zhu, Y., Cao, W., Wang, Q.: A knowledge model- and growth model-based decision support system for wheat management. Scientia Agricultura Sinica 37(6), 814–820 (2004) (in Chinese)

Gao, L., Jin, Z., Huang, Y., et al.: Rice cultivational simulation-optimization-decision making system (RCSODS). China Agricultural Scientech Press, Beijing (1992) (in Chinese)

Hu, B., Zhao, X., Yan, H., et al.: Plant growth modeling and visualization-review and perspective. Acta Automatica Sinica 27(6), 816–834 (2001) (in Chinese)

Host, G., Isebrands, J., Theseira, G., et al.: Temporal and Spatial Scaling From Individual Trees to Plantations: A Modeling Strategy. Biomass and Bioenergy (11), 233–243 (1996)

Deng, X., Guo, X., Zhou, S., et al.: Study on the geometry modeling of corn leaf morphological formation. Journal of Image and Graphics 10(5), 637–641 (2005) (in Chinese)

Chen, G., Zhu, Y., Cao, W.: Modeling leaf growth dynamics in winter wheat. Acta Agronomica Sinica 31(11), 1524–1527 (2005) (in Chinese)

Ma, Y., Guo, Y., Zhan, Z., et al.: Evaluation of the Plant Growth Model GREENLAB-Maize. Acta Agronomica Sinica 7(32), 956–963 (2006) (in Chinese)

Guo, X., Zhao, C., Xiao, B., et al.: Design and implementation of three—dimensional geometric morphological modeling and visualization system for maize. Transactions of the CSAE 23(4), 144–148 (2007) (in Chinese)

Rossing, W.A.H., Daamen, R.A., Jansen, M.J.W.: Uncertainty analysis applied to supervised control of aphids and brown rust in winter wheat. Agricultural System 44, 449–460 (1994)

Brooks, D.: DESSAC and its winter wheat fungicide module. In: Proceedings of the sixth HGCA Research and Development Conf. 'Management through Understanding: Research into Practice', Cambridge, vol. 13, pp. 1–11 (1998)

Gonzalez-Andujar, J.L., Garcia-de Ceca, J.L., Fereres, A.: Cereal aphids expert system (CAES): Identification and decision making. Computers and Electronics in Agriculture 3, 293–300 (1993)

Guo, C.: Studies on the models for prediction of main diseases in wheat field in Luohe, Henan Agricultural University Master Degree Dissertations (2007) (in Chinese)

Li, S., Zhu, Y., Yan, D.: Study on digital maize management system based on model. In: Chunjiang, Z. (ed.) Progress of information technology in agriculture: proceeding on intelligent information technology in agriculture (ISSITA), pp. 240–243. China Agricultural Science and Technology Press, Beijing (2007) (in Chinese)

Ritchie, J.T., Otter, S.: Description and performance of CERES Wheat: A user-oriented wheat yield model. In: Willis, W.O. (ed.) ARS Wheat Yield Project. ARS-38, pp. 159–175 (1985)

Digital Modeling and Testing Research on Digging Mechanism of Deep Rootstalk Crops

ChuanHua Yang[1,2,*], Ma Xu[1], ZhouFei Wang[3], WenWu Yang[1], and XingLong Liao[1]

[1] Key Laboratory of Key Technology on Agricultural Machine and Equipment
(South China Agricultural University), Ministry of Education,
Guangzhou Guangdong 510642, China
[2] College of Mechanical Engineering, Jiamusi University,
Jiamusi Heilongjiang 154007, China,
Tel.: +86-454-8618685; +86-18903643399
chuanhua_yang@126.com
[3] Science College, South China Agricultural University

Abstract. The digital model of the laboratory bench parts of digging deep root-stalk crops were established through adopting the parametric model technology based on feature. The virtual assembly of the laboratory bench of digging deep rootstalk crops was done and the digital model of the laboratory bench parts of digging deep rootstalk crops was gained. The vibrospade, which is the key part of the laboratory bench of digging deep rootstalk crops was simulated and the movement parametric curves of spear on the vibrospade were obtained. The results show that the spear was accorded with design requirements. It is propitious to the deep rootstalk.

Keywords: Deep rootstalk crops, Parametric model, Virtual assembly, Simulation.

1 Introduction

Started from the 1940s, many overseas countries, such as the United States, West Germany, Japan, France, Italy and other developed countries have developed many kinds of rhizome crops harvest machinery. For example, they have already achieved the mechanized harvest of the long root crops (radish, beet and so on) and short root crops (potato , peanut , garlic , onion and so on), but have not reaped the crops which root depth is greater than 40cm. Our country, has introduced the excavation since the 1960s, which is mainly used to reap the short rhizome crops such as the peanut, potatoes, etc. in early days. As to the deep root crops, such as Radix Astragali, Radix Hedysari, skullcap, licorice, their harvest mainly on the artificially excavation. Research of the deep rhizome crops excavating machine started from 1980s, successively developed the model which is suitable for the rhizome that is smaller than 40cm, for example: Agricultural Experiment Station in Changzhi city in Shanxi Province

---

* Corresponding author: Ma Xu, Professor, Doctoral tutor, Key Laboratory of Key Technology on Agricultural Machine and Equipment (South China Agricultural University), Ministry of Education, Guangzhou ,Guangdong, 510642, China, maxu1959@scau.edu.cn

D. Li and C. Zhao (Eds.): CCTA 2009, IFIP AICT 317, pp. 382–388, 2010.
© IFIP International Federation for Information Processing 2010

developed 4SD-280-type vibrating root harvester, Heilongjiang provincial hydraulic research institute developed 4WZ-140-type, Weiyuan county of Gansu province introduced 4WG-120-type and so on; the model which is suitable for the rhizome that is bigger than 40cm, for example: agricultural research institute of Baicheng City, Jilin Province researched 4GKJ-11-type root tuber harvester, Agricultural Promotion Station in Weiyuan county, Gansu province developed 4YW-160-type rhizome crops excavator and so on. The design of these machines is primarily using the traditional design method, there are somewhat conservative dimension of product design, bigger product size, higher cost, longer design cycle, over-dependent on multiple rounds of physical prototypes trial, the level of the design constraints and other issues to raise. In addition, these machines have a single vibration, when carries on the deep earth excavation, needing overcome too big resistance, the power consumption is too large as well.

We used powerful CAD and CAE software – US PTC Corporation's Pro/E software for this machine to carry out parametric modeling, virtual assembly and other digitized designs, and has carried on the assembly interference testing, have realized the dynamic performance of pre-fabricated components, have predicted the product performance accurately , have reduced or avoided the physical prototype's trial manufacturing, enhanced a success rate of the design, so as to reduce the product's design development cycle, reduce the cost , improve the design quality.

2   The Digitized Model of the Deep Rhizome Crops Excavating Experimental Platform

Digital design technology is based on the product description of the digital platform, establishes the computer-based element of the digital product model and is used in the whole process of the product development, to reduce or avoid the use of physical model of a product. Among them digital modeling is the basis of digital design, and provides the technology infrastructure for the future design analysis, simulation and so on.

2.1   The Structure of the Deep Rhizome Crops Excavating Organization

Before establishing experimental model of the deep rhizome crops excavating experiment platform and virtual assembly, we must analysis the structure of the deep rhizome crops excavating organization and define the relation between each spare part. Take the model 1WZ deep rhizome crops excavating test-bed as study object, it mainly includes traction frame 1, coulter 2, the pivot axis of the coulter vibrating frame 3, digging share motor 4, coulter vibrating frame 5, coulter driving motor 6, coulter eccentric vibratory governing mechanism 7, the frame 8, depth wheel governing mechanism 9, digging share eccentric vibratory governing mechanism 10, depth wheel 11, digging share vibratory suspender 12, digging share 13, the branch arm of digging share 14, the junction panel of knife arm 15 and digging share vibratory support axis 16, as shown in Figure 1. The whole model of the deep rhizome crops excavating experiment table which completes finally through the virtual assembly, as shown in Figure 2.

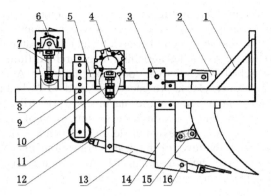

Fig. 1. The structure of the 1WZ- type deep rhizome crops excavating test-bed

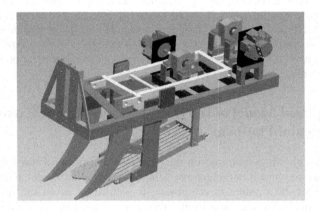

Fig. 2. The digital element model of the 1WZ-type deep rhizome crops excavating test-bed

There are two sets of lever system in this machine, coulter driving motor 6 and coulter vibrating frame 5 take the pivot axis of the coulter vibrating frame 3 as the pivot point driving the coulter 2 forms one set of the lever system, the active force arm is greater than the passive force arm, which makes the coulter 2 generates a greater vibration power in the approximate vertical direction, when the coulter driving motor 6 has smaller output power. The digging share motor 4 and digging share vibratory suspender 12 take the digging share vibratory support axis 16 as the pivot point driving digging share 13 forms the other set of the lever system, its active force arm is also bigger than the passive force arm, when the digging share motor 4 has smaller output power, which makes the point of the digging share 13 produces a greater vibration in the approximate vertical direction so as to achieve the saving power purpose.

2.2 Establishment of the Parts Entity Model

As the design and development platform, using the Pro/Engineer software which 3-D engineering software is based on the technology of parametric feature driven and characteristic modeling, directly designs the parts of the deep rhizome crops excavating organization, establishes the entity model of the parts. For most of the components, they were finished by using the feature modeling methods that Pro/Engineer provides, such as stretching, rotating, scanning and mixing. In the modeling process, we must first determine the order of the characteristic, because the order of the characteristic will have tremendous influence on the model results. Next, we must simplify the type of the characteristic and notice the association problems between the characteristics, namely establish appropriate fathers and sons relations characteristic. After establishing the entity model of the parts, if it is found that the parts cannot meet design requirements, we may amend the characteristic parameters of the parts to modify the design of the parts, which greatly improve the efficiency of the design.

2.3 Virtual Assembly of the Experimental Platform

Virtual Assembly (called VA for short) uses computer tools to realize the physical, without product or support process, to carry on or assist to process the related assembly project decision-making through the analysis, the pre-modeling, the visualization, the date expresses etc.

Total parts of the deep rhizome crops excavating experimental table, such as traction frame, coulter, driving motor, coulter vibrating frame, governing mechanism of the coulter vibrating frame, the adjusting device of the bottom shovel vibrating frame, the bottom shovel vibrating frame, the bottom frame, the plow arm, all the static connection parts to carry on the fixed assembly. Namely in Pro/E using automatic, matching, alignment, insert, coordinate, tangency, online dot and other constraint types in "the module laying unit" to place the components with the placed constrains, as shown in Figure 3; And the active connection part in which to carry on the connection assembly. Namely in Pro/E using pin joint, slide bar, cylinder, flat and other constraint types in the "connect" unit in the "module laying" to place the components with the link constraints, as shown in Figure 4.

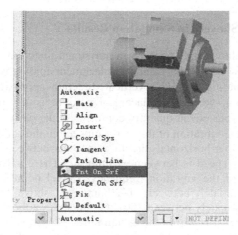

Fig. 3. Fixed constraint assembly of the experiment-table components

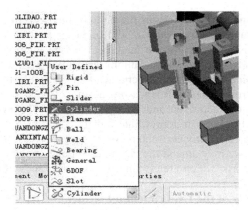

Fig. 4. Linking constraint assembly of the experiment-table components

3 Simulation Analysis of Shovel Key Components

3.1 Motion Analysis of the Shovel Point Simulation Model

After establishing the virtual prototype of the experimental table, we can measure the diversity parameters of different point in the bottom shovel, in the real situation to obtain the diversity displacement, velocity and acceleration change situation on the bottom shovel, through an analysis of the measure curve, to check whether the design is reasonable. In this case, the shovel point is the key force bearing point on the bottom shovel, so we must measure the displacement curve, the velocity curve and the acceleration curve for the analysis. The measure results of the horizontal movement direction of the bottom shovel tip are shown in the Figure 5, the measure results of the vertical movement direction of the bottom shovel tip is shown in Figure 6.

3.2 Analysis of the Shovel Tip Simulation Results

Known as Figure 8, shovel tip's position in horizontal direction approximately is between 185.45mm~-184.87mm, horizontal displacement amount is 0.58mm. Known as Figure 9, shovel point's position in vertical direction approximately is between -638.90mm~-634.42mm, vertical displacement amount is 4.48mm. Horizontal and vertical directions' displacement quantity conforms to the design requirements of the vibration amplitude, the vibration of this way helps to cut the soil into pieces.

Likewise, shovel tip's horizontal direction speed is between ±3.4mm/sec, shovel tip's vertical direction speed is between ±28mm/sec. These values meet the initially design requirements of the vibration frequency, shovel point's horizontal direction acceleration is approximately between 31~29mm/sec2 , shovel tip's vertical direction acceleration is approximately between 250~260mm/sec2 , although the acceleration in vertical direction is big, it is within the design range. Anyway, the kinematics simulation results of the bottom shovel tip are in accordance with the theory movement analysis, which meet the design requirements.

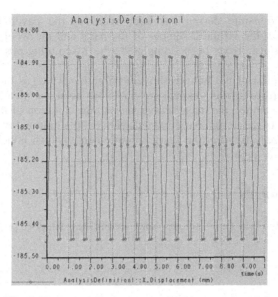

Fig. 5. Horizontal displacement of the bottom shovel tip

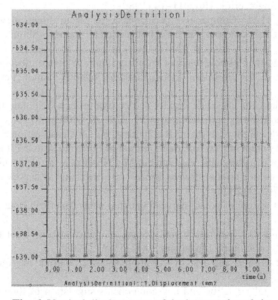

Fig. 6. Vertical displacement of the bottom shovel tip

4   Conclusion

(1)Under the support of the Pro/E platform, established the digital entity model of the deep rhizome crops excavating experiment table' parts, in the modeling process, the components dynamic design has been realized, the development efficiency has been enhanced, the virtual assembly of the parts and the whole machine of the experimental platform have been completed, the assembling-possibility has been improved.

(2) The simulation of the bottom shovel's vibration which is the key part of the deep rhizome crops excavating experiment table has been completed, the analysis of the shovel point's motion parameter curve shows that vibration of the bottom shovel tip is in the range of the design requirements, shovel tip's motion law is in accordance with the theory analysis, meet the design requirements, which is helpful to the deep rhizome crops excavation.

Acknowledgements

* State "Eleventh Five-Year" science and technology support projects (2006BAD 11A07).

References

Mcleod, C.D., Misener, G.C., Caissie, J.R.: A vertical lift digger for harvesting potatoes. Canadian agricultural engineering 31(1), 11–14 (1989)

Singh, R.D., Singh, H.M.: Comparative performance of potato digger elevator with conventional method of harvesting at farmer's fields. Potato Journal (Shimla) 31(3-4), 159–164 (2005)

Horvath, S., Kutassy, B.: Possible ways of reducing injury to potato tubers during mechanical harvest and storage. In: European Association for Potato Research [9th Symposium], pp. 174–175 (1984)

Baritelle, A.L., Hyde, G.M.: Specific gravity and cultivar effects on potato tuber impact sensitivity. Postharvest Biology and Technology 29, 279–286 (2003)

Baritelle, A.L., Hyde, G.M., Thornton, R., Bajema, R.: A classification system for impact-related defects in potato tubers. American Journal of Potato Research 77(3), 143–148 (2000)

Bentini, M., Caprara, C., Rondelli, V., Caliceti, M.: The use of an electronic beet to evaluate sugar beet damage at various forward speeds of a mechanical harvester. Transactions of ASAE 45(3), 547–552 (2002)

Brook, R.G.: Impact testing of potato harvesting equipment. American Potato Journal 70, 243–256 (1993)

Hu, Z., Peng, B.-l., et al.: Experiment and design of multifunctional root tuber joint harvest crops. Journal of Agricultural Machinery 39(8), 58–61 (2008)

Wen, X., Lian, Z.: Design and test of vibration excavat. Journal of Agricultural Machinery 37(10), 77–82 (2006)

Jie, L., Yan, C.: Digital modeling and key components simulation of joint harvest crops machine. Journal of Agricultural Machinery 37(9), 83–86 (2006)

Jia, J., et al.: Parametric modeling and virtual prototyping key components simulation of potato harvester. Journal of Agricultural Machinery 36(11), 64–67 (2005)

Zhenjiang Institute of Agricultural Machinery. Agricultural mechanics: Next List. China Agricultural Machinery Press, Beijing (1981)

In Internet-Based Visualization System Study about Breakthrough Applet Security Restrictions

Jie Chen* and Yan Huang

China Agricultural University College of Information and Electrical Engineering

Abstract. In the process of realization Internet-based visualization system of the protein molecules, system needs to allow users to use the system to observe the molecular structure of the local computer, that is, customers can generate the three-dimensional graphics from PDB file on the client computer. This requires Applet access to local file, related to the Applet security restrictions question. In this paper include two realization methods: 1.Use such as signature tools, key management tools and Policy Editor tools provided by the JDK to digital signature and authentication for Java Applet, breakthrough certain security restrictions in the browser. 2. Through the use of Servlet agent implement indirect access data methods, breakthrough the traditional Java Virtual Machine sandbox model restriction of Applet ability. The two ways can break through the Applet's security restrictions, but each has its own strengths.

Keywords: e-government, knowledge management, frameworks, e-governance.

1 Introduction

In network applications development, because of Java language have cross-platform, the program simple, suitable for network transmission, etc (Zhou Hang et at., 1997), applications prepared by Java is increasing. In many applications need to use Java Applet procedures to operate the client resources, but by default, even if the client to confirm the java procedure is "reliable", the browser will refuse the web Java program to operate client resources(William Stallings et at., 2001). We can through the following two methods to achieve this functionality: 1.Applet through Servlet agent indirect access to the database: in the Applet, we can use the url to establish a connection with the Servlet, and then through the Servlet access to the database client database and the results return to Applet, Indirectly through this process to achieve the Applet access to the client database. 2. Through the digital signature break the JVM sandbox, in java we can through increase the signature of Applet package to achieve Applet access to the client database.

2 Concrete Realization of Applet and Servlet Communication

The communication process include the following two aspects: achieve Applet Access to the Servlet and the transmission parameters and realization of transfer the data from Servlet to Applet.

---

* Corresponding author.

D. Li and C. Zhao (Eds.): CCTA 2009, IFIP AICT 317, pp. 389–392, 2010.

2.1 Achieve Applet Access to the Servlet and the Transmission Parameters

2.1.1 Create a URL Object
In JAVA Procedures, you can use the following method to create URL object:
URL servletURL=new URL (http://localhost:
8080/servlet/dbServlet.DbServlet).

2.1.2 Establish a Connection with the URL Address
After the successful creation of a URL object, you can type in the URL call openConnection () function to establish a connection. openConnection () function in the establishment of connections at the same time, to communicate the work of connection initialization:

URLConneCtion servletConnection=servletURL_()pen Connection();

2.1.3 Use URLConnection Object to Read and Write Operation

2.1.3.1 Using URLConneCtion Object Read the Information That Return from Servlet
After get URLConnection object, if the Servlet sent JAVA object to Applet, you can use URLConnection object openStream () methods to obtain input stream, and then generate a new ObjectInputStream object, use object ObjectInputStream readObject () method can get the JAVA object that Servlet returned.

If the Servlet sent Plain text to the, you can use URLConnection object getInputStream () method to obtain input stream, and then generate a new DataInputStream object, use the DataInputStream object readLine () method to obtain text that Servlet returned.

2.1.3.2 Using URLConnection Objects Delivery the Value to the Servlet
Applet sent the parameters to Servlet through the following two methods to achieve:

Through URL addresses add parameters GET method to achieve transmission parameters :

Another method is to obtain the output stream connection from the URLConnection, the output stream to be connected to the standard input stream which belong to the Common Gateway process (server-side), and then write relevant data into the output stream, when the transmission is over, close the output stream.

2.2 Realization of Transfer the Data from Servlet to Applet

Servlet request object's getParameter method can get the parameters that transmission from Applet:

String sql= request.getParameter(" sql");
the output stream Servlet that Servlet request object's getOutputStream () method get generate a new output stream, and then through the object's writeObject () method output JAVA type of object:

```
Class.forName( "sun:jdbc}:odbcdriver");
Connection  conn=DriverManager.getConnection(connectionString);
Statement st=conn.createStatement()
ResultSet rs=st.execute(sql)
```

```
dbStream= new ObjectOutputStream( response.getOut-putStream()) ;
／／Write the object...
dbStream.writeObject(rs);
```

Through the request's getWriter () method PrintWriter types of output, this object println () method can be output from the Servlet to Applet text:

```
PrintWriter out= response.getWriter0;
out.println(" <head> <title> DataCenter</title> </head>");
```

3 Applet's Digital Signature and Client Authentication

Java security guarantee by the following three aspects:

1, Language features (including border checks of array, type conversion, pointer-type variables).
2, Control Resource Access (including access the local file system, socket to connect and visit).
3, Code digital signature (digital signature to verify the source code and the code is complete).

This article discusses the combination of two technologies to achieve beyond the Applet security restrictions.

When a class file of the Applet is loading the JVM by the default class loader, JVM immediately loading a called for security manager (Security Manager) Class's sub-class Applet Security for it, by the manager to verify the operation. all the action of Code (such as file reading and writing) must be verified by the Security Manager, only be accepted the action can be completed, otherwise Security Exception will be thrown out Exception. Security Manager Class uses the Policy document to determine the code authority.

After JDK1.1, the JDK improved the division of permissions, the introduction of Permission Set concept. It designated the permissions detailed in every aspects, you can have a combination of selective permissions you need to meet special require-ments. Figure 1 shows such a division:

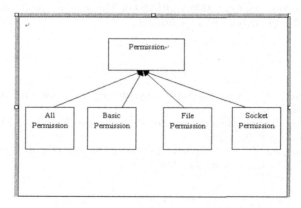

Fig. 1. The division of Java security permissions

Basic Permission in Figure 1 can be further broken down into more details of permissions, such as: AWT Permission, Runtime Permission and so on.

Java through a name suffix. Policy documents combination of these permissions. After The installation of JRE (Java Runtime Environment) there are two default permissions files, they are:

${java.home}/lib/security/java.policy
${user.home}/.java.policy

we have to create a policy document for the Applet, use the policy document the Applet will have the permission to read all the local documents .Only a policy document is not enough, the customer can not determine whether the executed code is your, there is no guarantee that the code in the course of transmission have not been malicious damaged. Therefore, digital signature technology needed to ensure these two aspects.

The realization of the system, we must first generate a key database, then we must have a certificate that be used when signature, finally, we use the generated certificate signature the Applet's jar files (Harvey M.Deitel et al.,2003).This completes digital signature the Applet. Digital signature technology to ensure the executed code is released by you and have not been malicious damaged.

4 Conclusion

Through the above analysis, the first method through the Servlet to achieve access client database, Servlet and Applet communication provides a language-level facilitate that the transmission of JAVA objects with each other, the second method use JDK tools digital signatures for Java Applet Program that do not have to modify existing Program, this is a practical, simple and reliable method.

Acknowledgements

This study was supported by the national 863 projects: Biological veterinary drug new product research and creation (2006AA10A208). The author would also like to thank Yan Huang, for her contribution to the project.

References

Hang, Z., WenSheng, C.: JAVA Language Programming. Publishing House of Electronics Industry, Beijing (1997)

Stallings, W.: Encryption and network security: Principles and Practice. Publishing House of Electronics Industry, Beijing (2001)

Deitel, H.M., Deitel, P.J., Santry, S.E.: Advanced Java 2 Platform How to Program. Publishing House of Electronics Industry, Beijing (2003)

Study of the Quality of Bee Product Tracking and Traceability System Based on Agent Technology

E. Yue*, Zhu Ye Ping, and Liu Sheng Ping

Research Institute of Agriculture Information Chinese Academy of Agricultural Science,
Beijing China 100081,
Tel.: +86-010-82109885; Fax: +86-010-82103120
eyue@mail.caas.net.cn

Abstract. This paper is the use of modern information technology, research the quality of bee product tracking and traceability system based on Agent. On the one hand, through establish the quality of bee product tracking and traceability system, to provide a technical support and general-purpose tools, it includes bee products from production to sales for the entire process digital, visual expression, design, control and management. On the other hand, as a key technology research and exploration, the use of Agent technology, proposed intelligence and cooperative control methods about the quality and safety of bee products, and for the control of the quality and safety of bee products to provide new methods and ideas.

Keywords: Bee Product, Agent, Cooperative, Tracking, Traceability.

1 Introduction

Food security is the most concerned about people, it is the most direct and practical interests for people. It is research in recent years one of the hot issues how to effective control and management food safety. China is the largest country about bee-keeping, production and export in the world's, but it is not apiculture power, the main reason is: beekeeping information relative occlusion, information asymmetry about bee production, processing, transportation, storage, marketing and so on, the quality of bee products difficult to control. On the other hand, the role of human factors greatly lead to loosely organized, information management methods stay in the database management and query stage. Therefore, it is necessary use of the modern methods of information management and technology in the bee produce, it can build a new information and intelligent management model about beekeeping in China, so that the level of quality and safety of bee products can assured and enhanced in China through effective tracking and traceability.

This paper is the use of modern information technology, research the quality of bee product tracking and traceability system based on Agent. As a key technology research and exploration, the use of Agent technology, proposed intelligence and

---

* Corresponding author.

D. Li and C. Zhao (Eds.): CCTA 2009, IFIP AICT 317, pp. 393–399, 2010.

cooperative control methods about the quality and safety of bee products, and for the control of the quality and safety of bee products to provide new methods and ideas.

From an engineering perspective, the bee product quality tracking and traceability system is a wide-area Internet-based technology and distributed artificial intelligence system. Agent technology can be used to solve the distributed system, it has many good features based on Agent and Multi-Agent technology application system, such as initiative, intelligence, interaction, collaboration and mobility and so on. In this paper is based on the Agent ideas and for the quality of bee product tracking and traceability system to explore.

2   Agent Summarize

Agent comes from distributed artificial intelligence (DAI) in 70s of the 20th century, and in other areas with a combination of various technologies, to expand from the DAI in its research field. It is generally believed that Agent should have the knowledge, goals and abilities. Agent knowledge is a described about its environment and the requirements of a solution of the problem by the user, it can obtain through the other Agent (in multi-Agent system) or their own learning. Agent goals is solution the problem of ways and means. Agent goals is can be defined as the form of multi-group: Agent = (M, K, A, I, L, S, G, F, C), in this: M= method, K=knowledge, A= property, I= reasoning mechanism, L=language, S=transmission of information, G=overall knowledge, F=inheritance mechanism, C=system services, this is a detailed description of Agent(Yao Li, Zhang Wei Ming,2002). With the increase distribution and complexity of system, the individual Agent often can not independently achieve a number of complex systems due to their own knowledge, computing resource constraints, Therefore, multi-Agent technology is the rapid development become a new focus about DAI research field. Multi-Agent System (MAS) consists of a number of Agent, the use of parallel distributed processing technology and modular design, the complex system is divided into relatively independent subsystems, through the Agent of cooperation and competition accomplish the complex issues to solution.

3   Bee Product Quality Tracking and Traceability System Architecture Based on Agent

3.1   System Architecture

Bee Products (such as: honey, Royal Jelly, etc.) from field to home, it have a number of links includes the acquisition of bee products, bee product processing, transport and sale, each link exists between each other the exchange of information. Due to geographical differences (the management are widely distributed In different regions), the time difference (there is a huge time difference about access to products, access to relevant information), making information management has become very complicated. On the other hand, for the bee products own characteristics, such as pests and diseases, medication, heavy metal pollution have a serious impact on the quality factors, these factors must be the focus of inspection about bee product.

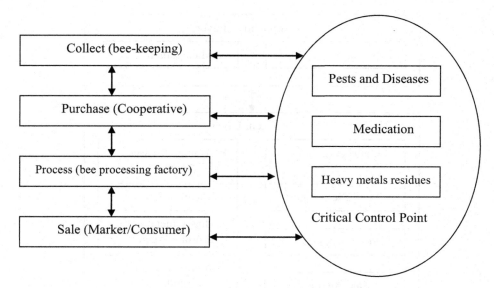

Fig. 1. Bee product quality tracking and traceability system architecture

Therefore, the quality of bee products tracking and traceability system can be decomposed as a dynamic system, no matter how large the scale, in theory, the system can be broken down into a number of interrelated sub-systems (such as acquisition system, processing systems, transport systems, marketing systems, etc.). There are also relatively independent subsystems in quality management in the actual work, the bee for the production, acquisition, processing, transportation and sales departments have the corresponding entities. On the other hand, As a result of its own characteristics, Critical Control Point(such as pests and diseases, medication, heavy metals) play a very important role in the quality of bee products tracking and traceability system, the advantages and disadvantages of these entities directly related to the quality of bee products. These entities can completed their own tasks and goals to achieve optimal, they have independent thinking and ability to fulfill its mandate, they can cognitive state that all other entities of the current, and can accept the request and orders of other entities to change their behavior mechanism.

Based on the above analysis, the author proposed a bee product quality tracking and traceability system modeling framework architecture. The entire model is built based on TCP/IP protocol on the conceptual level, it is divided into application-layer, Agent layer, network layer platforms.

(1) Application Layer: It responsible for user interaction. Its functions include two aspects: ① Accept the request of the user, ② Submit the query results in accordance with the requirements of users. Application layer will be the user's request in accordance with the Agent layer organize, and then put it to the Agent layer.

(2) Agent layer: it is the most complex part of the core in the whole system. For the Application Layer, in accordance with the requirements of users to generate user's

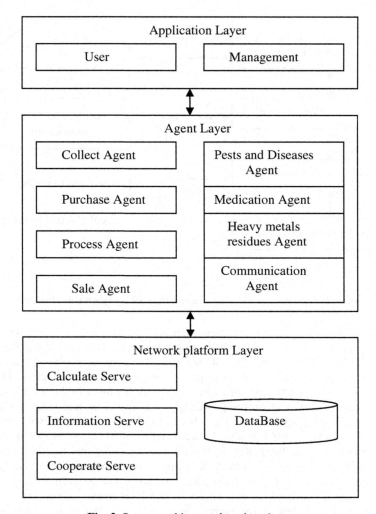

Fig. 2. System architecture based on Agent

Intelligent Agent object, At the same time, it also responsible for the management of all issued from the local Agent Object, and record their current location and status, and completed the collection of functions and returned to the application layer. For the network platforms layer, it transmission the Agent Object to the platform.

(3) Network layer platforms: It is the whole system of transmission pipelines, it provides the environment for the survival Agent, it is the basis for Agent Mobile, is a place of communication between the Agent. Platform network layer is based on the existing network communication protocols, it transmit data and receive data streams through consultations Port.

3.2 Multi-Agent Communication

Knowledge query and manipulation language (KQML) was widely used as a communication language of the Agent, it is becoming the standard for communication of Agent. KQML is a descriptive language about exchange knowledge and information, It defines the mode of impart information and Message handling agreement in Agent, By providing a standard of communication language to achieve the exchange of information and knowledge sharing between the Agent (George F Luger, 2004).

Agent communication model and multi-Agent collaboration is closely related to the organization. In essence, all the multi-Agent systems are distributed architecture, each of Agent are cooperation and collaboration, there is no relationship between the subordinate and control. Each of Agent have their own independent goals, aspirations and behavior, it can choose partners, and accept or reject the task. In this case, most of the problems about multi-Agent cooperative must be members of Agent own knowledge, reasoning, social, adaptation, learning ability to solve, which makes the design of Agent complicated entities (Wang Long, Zhang Yi, 2003).

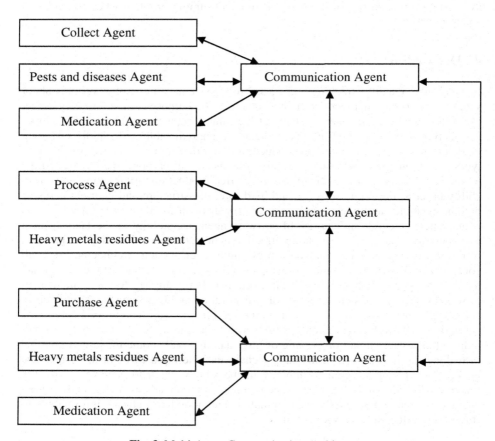

Fig. 3. Multi-Agent Communication Architecture

Bee Products Quality Tracking and traceability information system is a complex multi-Agent system, the organization of ideas from the Internet, as shown in Figure 4 using the communication mode.

First of all, the Agent classification, will be in the same geographic or Agent of the same functions of a domain Agent group, each Agent group establish a communications agent (Collaborative Communications Agent).

Second, Agent of the Agent group only communication with the communications agent (Collaborative Communications Agent) and find the object of communication.

Third, when there is a suitable Agent Object on the same Agent group, Agent by sending and receiving KQML messages to communicate with other Agent.

Fourth, When there is a suitable Agent object in other Agent groups, the communication process must join the other group of Agent communication agent(Collaborative Communications Agent). Communications agents do not know or explain the content of KQML messages, this improve the system's security and scalability.

The distribution and concentration of mixture communication construct a Union Agent group, which not only resolved the Agent to manage the complexity of communication, but also resolved the Agent time-consuming problem when the large of communication.

4 Discussion

Bee Products Quality tracking and traceability System is a complex systems about supply chain for all members of the management, the purpose is monitor every aspect of product quality in the supply chain, at the same time achieve maximum customer satisfaction. It faces the environment including unpredictable and rapidly changing market demand, product mix and production re-configurations, a variety of different types of computer hardware and software systems and different business rules and the production of technical specifications. In the Bee product quality tracking and traceability process, due to its software and hardware infrastructure and its users is distribution, dynamics and uncertainty, so it is very difficult to adapt to the needs of the management integration, The use of software agent technology in the tactical level and operational level to carry out an effective management will become the direction of the research. The main method is to use multi-agent system architecture, through consultations between the agent, to complete all the tasks together. At the same time, the mobile agent is a new research direction, it is being applied to solve the distributed and heterogeneous environment of collaboration and management. Mobile agent not only has the autonomy, communications and reflect, and it can move freely in the network, it can be brought about by their own information "push" out, but also will require the collection and processing of information " pull "come in, so it is possible to making the information in the supply chain into the rapid flow. In this paper, the application of mobile agent technology in the supply chain to achieve the quality tracking and traceability bee management, aimed at taking advantage of its own characteristics (autonomy, intelligence, initiative and mobility), allows information to flows more effectively in Agent group.

5   Conclusion

Based on the analysis of quality control of bee products, combined with bee management system characteristics, the introduction of agent technology, Put forward the theme of bee products quality management system model based on Agent, and through the introduction of agent technology to achieve the feasibility of the model. Described the system model of the structure and function modules in this paper. Through functions research of the Agent and analysis of communications patterns, it can be seen: bee product quality tracking and traceability information system based on Agent research and development have the prospect of important applications and practical value.

Acknowledgements

This research was supported by National Scientific and Technical Supporting Programs Funded by Ministry of Science and Technology of China (nyhyzx07-041).

References

Li, Y., Ming, Z.W.: Intelligent and Cooperative Information Systems. Publishing of Electronics Industry, Beijing (2002)

Luger, G.F.: Artificial intelligence. Publishing of Electronics Industry, Beijing (2004)

Long, W., Yi, Z.: Key Techniques to Realize Cooperation of Mobile Agent in MAS Environment. School of Computer Science and Engineering 32(2), 158–163 (2003)

Trappey, A.J.C., Trappey, C.V., Hou, J.: Mobile agent technology and application for online global logistic services. Industrial Management & Data System 104(1/2), 169–183 (2004)

Ming, Z.J., Peng, Z.X.: An agent-oriented requirement analysis and modeling method 4, 33–35 (November 2006)

Jing, Z.: Bee product quality and safety analysis of the key technology research and development. China apiculture 57(12), 30–32 (2006)

DeLoach, S.A.: Engineering Organization-based Multiagent Systems. In: Garcia, A., Choren, R., Lucena, C., Giorgini, P., Holvoet, T., Romanovsky, A. (eds.) SELMAS 2005. LNCS, vol. 3914, pp. 109–125. Springer, Heidelberg (2006)

Chang, M.-H., Harrington Jr., J.E.: Agent-based models of organizations. In: Handbook of Computational Economics II (2006)

Yi, S.C.: Computing based on Agent. Publishing of Tsinghua University, Beijing (2007)

Zhi, S.Z.: Advanced Artificial Intelligence. Publishing of Science (2006)

RFID Based Grain and Oil Products Traceability*
and Its Computer Implementation

Haiyan Hu[1,2,**] and Yunpeng Cui[1,2]

[1] Key Laboratory of Digital Agricultural Early-warning Technology, Ministry of Agriculture,
The People's Republic of China 100081
[2] Agricultural Infornmation Institute of Chinese Academy of Agriculture Science,
Beijing 100081, P.R.China,
Tel.: +86-10-82106263; Fax: +86-10-82106263
huhaiyan@mail.caas.net.cn

Abstract. Food safety is a widely concerned problem in current world. Traceability technology is an effective measure to solve the problem. This paper describes the study of the traceability of grain and oil products. Include the study contents, and a system we developed for traceability of grain and oil products, and the demonstration of the study. The system we developed was used in Luhua group and some supermarkets and get good feedbacks.

Keywords: RFID, traceability, grain and oil products.

1 Introduction

China is a type of plant based country. The major grain and oil products like wheat, rice, soybean and peanuts are widely consumed in China, and played a dominant role in people's diet. The quality of grain and oil concerned hundreds of millions of people's health, and economic development and social stability. So how to enhance grain and oil quality and safety management, rectify and standardize the market has now become the most important issue of grain and oil management.

The main grain and oil products have mainly three types of hazard sources, Chemical Hazards, Biological Hazards and Physical Hazards. "Traceability" can be divided into three parts: tracking, upward traceability and downward traceability according to the type of traceability behavior, traceability can also be divided into passive and active traceability. Traceability includes external and internal traceability, External traceability means the traceability among each node in the food chain, and internal traceability means the traceability inside an enterprise or an organization.

---

* The study was supported by the national 863 project named the study of global traceability technology of major grain and oil products quality, the number of the project is 2006AA 10Z268.
** Corresponding author.

D. Li and C. Zhao (Eds.): CCTA 2009, IFIP AICT 317, pp. 400–406, 2010.

In recent years, the United States, Japan, Australia and other developed countries, and the European Union have taken a lot of positive measures in food safety traceability and put them into practice.

Since China joined WTO, profound changes happened in food production and circulation area. The Ministry of Agriculture launched experimental work of urban agricultural products quality control system, and focus on the quality of agricultural products traceability system construction (Lin Ling, Zhou Deyi, 2005). In 2002, Beijing Vegetable Products quality traceability system experimental project, launch by the Ministry of Agriculture, used uniform packaging and product tag code; Shanghai Animal Husbandry Department put forward Shanghai animal immunity identification management method, and set up files for pigs, cattle, sheep and other livestock products (Cheng Hao, 2007), Nanjing, Tianjin and Shenzhen have also launched traceability system construction (Fan Hongping etc.,2007). The traceability system construction of food safety is at its initial stage in China (Chang Xiang, 2007). And So far, there is no traceability system about grain and oil products in China.

2 Implementation of the Traceability of Grain and Oil Products

2.1 The Goal of the Research

The goal of the research is to make a series of breakthrough on key technologies of grain and oil products traceability. Develop a number of hardware and software with independent intellectual property rights, and demonstrate these products in grain and oil production bases, warehouses, and markets, so find a Chinese characteristic way and establish the grain and oil traceability system in China.

2.2 The Research Route

Fig.1 shows the route of the research. The research includes 5 parts:
1. The general framework design of major grain and oil products traceability; 2. The coding identification and global files digitalization of grain and oil products; 3. Study of key technologies of global traceability; 4. The integration of key technologies and platform; 5. The demonstration. All the works can be divide into three stages: framework design, system design and system implementation.

2.3 The Key Problems We Solved in the Study

2.3.1 Study of the Framework of the Global Traceability of Grain and Oil Products
Based on HACCP (Hazard Analysis and Critical Control Point) and FMECA (Failure Mode Effect and Criticality Analysis), to confirm the hazard sources and Influencing factors of grain and oil products [1], and establish their key control points, find the key indicator of the traceability, and finally, construct the unit of the traceability.

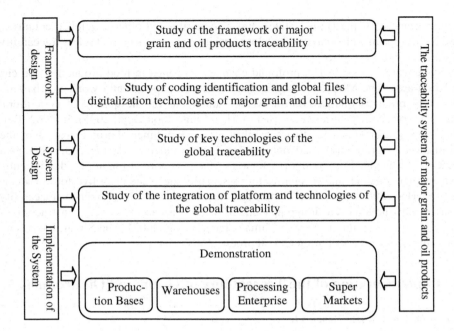

Fig. 1. The route of the research

2.3.2 Study of Coding Identification and Global Files Digitalization Technologies of Major Grain and Oil Products

We designed a major grain and oil products traceability classification and encoding standard, to implement the RFID (Radio Frequency Identification) based unique product identification. Based on the unique global code, we can construct the global digital files system to help the enterprises implement the digital management on production, processing, marketing and other aspects. In order to improve the data transfer rate, we use high frequency RFID to save the product information. Based on XML heterogeneous data conversion technology, we implement the data integration of external traceability and internal traceability, so the digital files of the grain and oil products chain can be exchanged transparently.

2.3.3 Study of Digital Products and Origin Authentication

Based on RFID digital certification, we implement the grain and oil products origin authentication. We implemented the digital origin information management through distributed database technology. We established a grain and oil products base database, to save the enterprises and origin related environment quality information, such as enterprise digital authentication certificate, the natural environment information and production environment information.

2.3.4 Study on RFID Middleware of Major Grain and Oil Products

We use web services architecture to implement the information delivery and filtration and RFID message aggregation between data collection layer and application layer, and RFID events management based event notification and subscribe.

We made a RFID data describe standard, a data accuracy requirement and a data format specification and RFID data layer interface standard, developed a series of RFID based middleware, established RFID database, so constructed a RFID data management system, and implemented XML based data exchange.

2.3.5 Study of Application Support Technologies of Traceability

According to the hazard sources, we use agent technology to monitor the KPI (key performance indicator) of the system, so the system can make rapid reaction and report for quality accident of grain and oil products. Through MEP (Message Exchange Pattern) based information exchange technology, the key traceability information and the complete report can be delivered to different user in time.

3 Implementation of the Grain and oil Products Traceability System

According to the study contents above, we developed a complete traceability system. the systems include 4 subsystems (Fig.2).

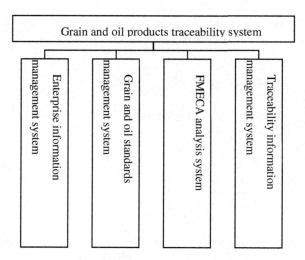

Fig. 2. The functional modules in the traceability system

Fig.3 is the portal of the traceability system. From this homepage, user can visit the related information, and navigate to the subsystem, subscribe related information through RSS, and retrieve the traceability information inside the application.

Fig. 3. The portal homepage of the system

Fig. 4. The standards management system in the traceability system

Fig.4 is the interface of the standards management system. In this system, user can retrieve, browse, and publish the related standards.

Fig.5 is the expert points function page of FEMCA analysis system. Experts can evaluate the basis of the traceability, the features of every phase, the control measure, and give every indicator a score.

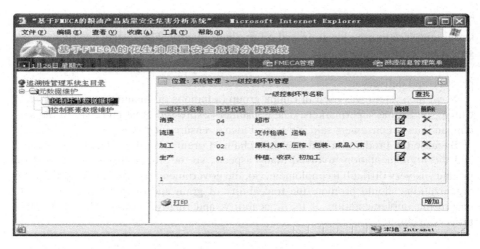

Fig. 5. The expert points function of FEMCA analysis system

Fig.6 is the traceability metadata management page of FEMCA system, which can define, modify, and delete the metadata used in traceability. These metadata will be used in every phase to describe the features of every product.

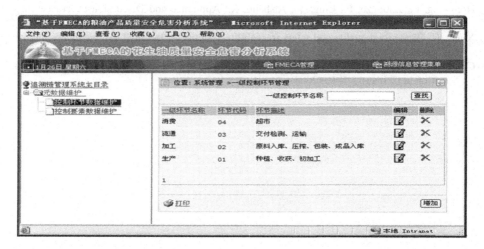

Fig. 6. Traceability metadata management sub-system

We also developed a terminal to support the mobile application of the system, user can visit the system and query related traceability information through mobile devices.

4 Conclusion and Outlook

The study of grain and oil products traceability technology and its implementation is a strong measure to monitor grain and oil products quality. The system can manage the global traceability information, locate the hazard source. For consumers, it provides a way to find the related products information in every phase of the product production, process, and delivery etc.

Now, the system was used in Luhua group (a famous oil manufacture enterprise in China), and some supermarkets. The consumers expressed great curiosity to the system, and many consumers said it's a good idea to ensure food safety.

Because the production and consume chain of grain and oil products is very long, and the implementation involved many aspects, so the traceability of grain and oil products is very difficult to implement. So, the government, the enterprises and related organizations should promote the traceability of grain and oil products together, to ensure the implementation of the traceability, and finally, promote the food safety process in china.

References

Ling, L., Deyi, Z.: On the Construction of Food Quality and Safety Traceability System. Commercial research (21), 41–44 (2005)

Hao, C.: Animal product safety control and traceability technologies. Modern Agriculture science and technology (13), 169–170 (2007)

Hongping, F., Zhongze, F., Ling, Y., Aisheng, R.: Appliance and Discussion of Traceability System in Food Chain. Ecological Economy 17(4), 30–33 (2007)

Xiang, C.: Three food quality and safety traceability systems for Beijing Olympic games will be put into use in August, http://www.chinanews.com.cn/ (July 09, 2007)

Chunhua, L., Shihong, L.: The status quo of FEMCA used in food security traceability. Food and Nutrition in China (6), 7–10 (2008)

Studies on the Pasting and Rheology of Rice Starch with Different Protein Residual

Qinlu Lin[1,*], ZhongHua Liu[2], Huaxi Xiao[1], Lihui Li[1], Fengxiang Yu[1], and Wei Tian[2]

[1] Faculty of Food Science and Engineering, Center South University of Forestry and Technology, Changsha, Hunan Province, P.R. China 410004
lql0403@yahoo.com.cn
[2] Faculty of Food Science and Technology, Hunan Agricultural University, Changsha, Hunan Province, P.R. China 410004

Abstract. Indica rice starch and japonica rice starch were used in the study. The protein contents of the two rice variety were respectively 0.43%, 0.62%, 0.84%, 1.08%, 1.25%. The pasting and rheological properties of samples were determined with Rapid Visco Analyzer and dynamic rheometer. The results indicated that, with the increase of protein content, the peak viscosity, breakdown viscosity and final viscosity of rice starch paste decreased, the setback viscosity increased and the pasting temperature did not change significantly. With the increase of protein content, the consistency coefficient of starch decreased, the corresponding yield stress also decreased, however, the flow behavior index increased with the decrease of consistency coefficient. At same temperature, the storage modulus G' was greater when the protein content was higher.

Keywords: starch, amylose, protein, content, pasting.

1 Introduction

Starch is generally applied in food processing. Pasting and rheological properties are important physical and chemical characteristics of starch, and consistency coefficient, flow-behavior index and yield stress are all important parameters of rheological properties. The rheological properties of liquid foods of starch will affect the quality of foods, such as hardness, stickiness, chewiness, etc. There is close relationship between the rheological properties of materials and its transportation, agitation, mixing and energy consumption. There have been many international and domestic studies on the factors affecting rheological properties of rice starch paste, rheological properties during the aging of rice starch paste, rheological properties of rice amylopectin, rheological properties of rice starch during the storage, rheological properties of modified rice starch, effects of food additives on rheological properties of rice starch, etc. It is considered that rice starch is composed of long-chain amylose and

---

* Corresponding author.

D. Li and C. Zhao (Eds.): CCTA 2009, IFIP AICT 317, pp. 407–419, 2010.

amylopectin, rice starch paste is a kind of pseudoplastic fluid with the property of shear thinning, which, generally, is beneficial for the application and processing. In certain range of shear rate, it can greatly decrease the viscosity of the materials and increase its fluidity so as to benefit the pumping transportation and processing, decrease energy consumption and increase productivity. However, the shear shinning should be small in some processing procedures to ensure the stable viscosity of the materials in order to prevent the rapid change of viscosity caused by the fluctuation of shear rate to guarantee the quality of half-finished product or finished product. However, the rheological properties of different rice varieties are different due to the differences in the contents of amylose and amylopectin as well as the characteristics and molecular conformation of starch molecules. The recent studies suggested that, besides the content of amylose, the protein content also greatly affected starch pasting and rheological properties. For example, Lim (Lim et al., 1999) reported that reducing the protein content in rice flour increases its peak viscosity. This was confirmed by Tan and Corke (Tan et al., 2002) who proposed that protein content is negatively correlated to peak viscosity and hot paste viscosity. Furthermore, Lyon (Lyon et al., 2000) found that protein content was negatively correlated to adhesiveness of cooked rice. However, in most of these studies, the prolamin proteins extracted from rice was added into rice starch to observe the effects on rheological properties of rich starch, which only showed the relationship between total protein content and physical characteristics of rice but not the effect the intrinsic protein content in rich starch on rich starch pasting. In this study, the indica rice starch and japonica rice starch were prepared by soaking rice flour in sodium hydroxide solution and five protein contents of indica rice starch and japonica rice starch obtained from the changes of soaking time, meanwhile, the main purpose of our study is to further investigate the effects of rice varieties and protein contents on rich starch pasting and rheological properties.

2 Materials and Methods

2.1 Starch Isolation

Rice starch was isolated from rice flour by soaking rice flour (10g) into sodium hydroxide solution (0.2% w/v; 200 ml) for 12h. The mixture was stirred at room temperature for 5 min, the slurry was centrifuged at 6000g for 5 min (Ibanez et al., 2007). The supernatant and any brown surface layer of the starch were removed and the lower white starch layer was washed with distilled water and centrifuged, the procedure was repeated for several time and the solid phase was washed using distilled water until the pH of the filtrate was 6.0–7.0. The solid phase was dried at 45 °C for 48h in a vacuum oven and moisture was allowed to equilibrate to a level of 13%. Pass through a 100-mesh sieve.

2.2 Defatting of Rice Starch

Starch samples were defatted before amylose determination in the experiments in order to minimize lipid contamination in the amylose determination. Defatting was carried out by adding 14 volumes of an ethyl ether/methanol (1:1; v/v) mixture to rice starch samples, and the resultant slurries were mixed thoroughly, allowed to stand for 1 h and centrifuged at 10000g for 10 min (Chrastil et al., 1994). The defatting procedure was repeated a total of three times to maximize removal of the lipid components.

2.3 Amylose Contents Determination

Amylose contents were determined by fully automated flow injection system (FIA-Star5000, FOSS, Ltd, Sweden.). This is a new type of continuous flow analysis technique. This technology is that a certain volume of sample solution is injected into a flowing and non-air spacing reagent solution (or water) containing the stream according to compared method and the work curve drawn by standard solution measured the concentration of a substance in the sample solution. After defatting, spread the rice starch in a thin layer on a dish or watch glass or place them in paper bags and leave for 2 days in the same room to allow evaporation of residual methanol and for moisture content equilibrium to be reached. Weigh in duplicate 100+/-0.5mg of the test sample into 100ml volumetric flasks. To the test portion, carefully add 1ml of ethanol 95% (V/V) using a pipette, washing down any of the test portion adhering to the side of the flask. Add 9.0ml of 1 mol/l sodium hydroxide solution using a pipette and mix. Then heat the mixture in a boiling water bath for 10 min to disperse the amylose. Allow to cool to room temperature without shaking for at least 2h (or over night). Make up to the metered volume with water and mix vigorously.

2.4 Protein Content

Protein content was calculated multiplying the nitrogen content of the starch sample by 6.25. The nitrogen content of starch was measured in duplicate with a Vario MARCO (Elementar Analysensysteme, Hanau, Germany).

2.5 Pasting Properties

The pasting properties of rice starch were determined with the Rapid Visco Analyzer (RVA super 4, Newport Scientific, Australia), and analyze with analysis software. Viscosities of starches were recorded with starch suspensions (Moisture Content 12.0%, sample 3.00g, water 25.00mL). Underwent a controlled heating and cooling cycle under constant shear where it was held at 50 ℃ for 1 min, heated from 50 to 95 ℃ at 5 ℃/min and held at 95 ℃ for 2.7 min, cooled from 95 ℃ to 50 ℃ at 5 ℃/min and held at 50 ℃ for 2 min. The initial speed of blender in 10s for 960 rpm, after that maintain 160 rpm. Viscosity value take Rapid Visco Units (RVU) for unit. Pasting parameters such as pasting temperature, peak viscosity, breakdown (peak viscosity-hot paste viscosity), final viscosity, setback (cold paste viscosity- peak viscosity) were recorded.

2.6 Measurement of Flow Behavior

Rice starch paste of 8% was put into the testing platform of dynamic rheometer (ARES, TA, Ltd, America). viscometry was performed using a controlled strain rheometer using parallel plate mould (40mm diameter and 1mm gap). After trimming off the over-loaded portion of samples around plates, the open side of samples was covered with a thin layer of silicon oil to prevent moisture loss. Apparent viscosity and shear stress with increasing shear rate (0–500 s-1) were obtained at 20 ℃ to characterize flow behavior. The Herschel–Bulkley equations (Rao, 1999) were fitted to obtain the parameters of flow curves. The multiple correlation coefficients, R2, show the degree of fitting of the Herschel–Bulkley model:

$$\tau = \tau_{0} + K \gamma_{n} \quad K, n \text{ is constant}$$

Where τ is the shear stress (Pa), $\tau 0$ is the yield stress (Pa), K is the consistency index (Pa.sn), γ is the shear rate (s-1), and n is the flow behavior index.

2.7 Measurement of Dynamic Viscoelasticity Behavior

The rheological properties of rice starch were measured by dynamic rheometer (ARES, TA, Ltd, America). Rice starch suspension of 20% was put into the testing platform of dynamic rheometer, the plane mould diameter is 40 mm and oscillation measurement procedure were selected, clearance was set up for 1.0mm, the strain was 2% and the angle frequency was 5 rad/s. After trimming off the over-loaded portion of samples around plates, the open side of samples was covered with a thin layer of silicon oil to prevent moisture loss. Experiment process : procedure increase temperature are from 20℃ to 100℃, which enables suspension system of rice starch to become paste, and then procedure decrease temperature from 100℃ to 20℃, and then test the changes of the storage modulus(G'), loss modulus(G") and loss tangent tanδ(tanδ=G"/G') of starch paste in the process of heating and cooling, the velocity of increasing and decreasing temperature respectively was 5℃/min.

3 Results and Discussion

3.1 Chemical Composition Analysis of the Rice Starches

Many factors may affect the pasting and rheological properties of rice starch, among which, rice varieties, fat content and protein content all play critical roles. In rice, the content of starch is 80% and fat content is about 1%~3%. However, the protein content is about 8% which is comparatively higher. The amylose content is different due to rice variety, which is one of the most important determinants of rice starch quality (Sudha et al., 2007). Therefore, the effects of amylose content and protein content on pasting and rheological properties of rice starch are particularly significant, and the effect of fat on starch characteristics is mainly due to the complexation between lipoids and amylose (Jean et al., 2008), which results in the decreasing in amylose content and affects the physical and chemical characteristics of starch at some degree. To perform defatting before experiment will minimize the effect of fat on starch characteristics. Indica rice starch and japonica rice starch were extracted by soaking rice flour into sodium hydroxide solution (0.2%, w/w), five protein contents (0.52%, 0.86 %, 0.95 %, 1.10 % and 1.20 %, respectively) were obtained through the changes of soaking time. The chemical compositions of indica rice starch and japonica rice starch are listed in Table1. As is shown in table1, there is not significant difference about the amylose content in the same rice starch with various protein contents.

Table 1. The chemical composition of indica rice starch and japonica rice starch

Sample	Moisture (%)	Fat content (%)	Protein content (%)	Amylose content (%)
indica rice starch	13.2	0.48	0.52	29.5
			0.86	29.5
			0.95	29.5
			1.10	29.6
			1.20	29.5
japonica rice starch	13.0	0.47	0.52	24.5
			0.86	24.5
			0.95	24.5
			1.10	24.5
			1.20	24.5

3.2 Effects of Rice Varieties and Protein Content on Pasting Properties of Rice Starch Paste

The amylose content is different from different rice varieties, and the amylose content directly affects the pasting properties of starch. As is shown in Table2., for indica rice starch and japonica rice starch, when the protein contents were both 0.52%, but amylose contents were 29.5% and 24.5%, respectively, the pasting temperatures of indica rice starch and japonica rice starch were 82℃ and 78℃, respectively, and the peak viscosity and breakdown viscosity of indica rice starch were lower than those of japonica rice starch, however, the final viscosity and setback viscosity of indica rice starch were higher than those of japonica rice starch, and the peak viscosity, breakdown viscosity, final viscosity and setback viscosity of indica rice starch were 5304, 3479, 4748 and -556RVU, repectively, and those of japonica rice starch were 5923, 3974, 4560 and -1363RVU, respectively. The bonding force between amylose molecules are strong due to the hydrogen bonds (John et al., 2008), and the pasting of starch with high amylose content was difficult than that of starch with low amylose content, therefore, the pasting temperature of indica rice starch was higher than that of japonica rice starch. The peak viscosity and breakdown viscosity of starch were negatively correlated with the amylose content, this is consistent with previous reports (El-Kha-yat et al., 2003; Varavinit et al., 2003), the starch granules swell with the increasing of temperature and the amylose in starch granules obstruct the swelling, thus, the peak viscosity and breakdown viscosity decrease with the increaseing of amylose content. The setback viscosity of indica rice starch was significantly higher than that of japonica rice starch, which suggested that indica rice starch is more liable to aging than japonica rice starch, in which, the aging is mainly due to amylose because the molecular movement of starch molecules of pasting slows down at low temperature and the amylose molecules trends to arrange parallelly and draw close to each other to combine through hydrogen bonds so that the starch with high amylose content is easy to retrogradation (Kyoko et al., 2007) and its final viscosity is also comparatively higher. As was shown by the RVA curves in Figure.1, the changing tendency of pasting properties of indica rice starch and japonica rice starch were substantially consistent when their protein content were both 0.52%, the viscosity increased with the increase of temperature during heating, and the viscosity began to decrease with further increase of temperature when the viscosity arrived the maximum value. During cooling, the viscosity of starch increased with the decreasing of temperature. In these processes, the starch molecules in starch granules combine to particle structures through hydrogen bonds, the strength of hydrogen bond decreases in hot water, the granules swell after absorbing water, the viscosity increase to the maximum value, Continued to heat, hydrogen bond and granules rupture, and a great amount of amylose leached out so that the viscosity decreases (Graeme et al., 2004). In the following cooling process, amylose begins to retrogradation with the decreasing of temperature and the hydrogen bonds between the chains of the molecules form again, which caused a significant increase in viscosity.

Table 2. Pasting properties of indica rice starch and japonica rice starch of different varieties and different protein content

Sample	protein content/%	Amylose content/%	PeakVisco (RVU)	Break-down (RVU)	FinalVisco (RVU)	Setback (RVU)	Pasting Temp/0C
indica rice starch	0.52	29.5	5304	3479	4748	-556	82
	0.86	29. 5	5247	3384	3960	-530	82
	0.95	29.5	5237	3255	3755	-485	82
	1.10	29.6	5119	3004	3520	-369	82
	1.20	29.5	4601	2663	3515	-286	82
japonica rice starch	0.52	24.5	5923	3974	4560	-1363	78
	0.86	24.5	5864	3744	3785	-1267	78
	0.95	24.5	5778	3551	3590	-1188	78
	1.10	24.5	5423	3148	3515	-908	78
	1.20	24.5	5349	2867	3455	-894	78

—japonica rice starch   ...indica rice starch

When the amylose contents were similar, rice starches with the different protein contents resulted significant difference in pasting properties, as was shown in Table2., protein content had little effect on pasting temperature, however, it significantly affected the peak viscosity, breakdown viscosity, final viscosity and setback viscosity of starch. Either for indica rice starch or for japonica rice starch, with the increasing of protein content in starch sample, the peak viscosity, breakdown viscosity and final viscosity of the starch all decreased significantly, however, the setback viscosity increased significantly. A decrease in the peak viscosity was also observed when prolamin were added to rice flour (Baxteer et al., 2004), likely due to the dilution effect of protein on the concentration of starch, since a negative correlation had been established between the protein content and the peak viscosity in rice flour (Lim et al., 1999; Tan et al., 2002). When the protein contents in indica rice starch and japonica rice starch increased from 0.52% to 1.20%, the breakdown viscosity and final viscosity of indica rice starch decreased from 3479 to 2663RVU and from 4748 to 3515 RVU, respectively, and the breakdown viscosity and final viscosity of japonica rice starch decreased from 3974 to 2867RVU and from 4560 to 3455 RVU, respectively. The breakdown is related to the ability of the starched to withstand heating at high temperature and shear stress, the decrease in breakdown viscosity might be due to the failure of complete pasting and swelling of starch granules induced by the reduction of water-absorption of starch granules caused by water-absorption of protein during pasting (Cristina et al., 2008). In the cooling process, the starch molecules without completely pasting and swelling could not rearrange regularly so that the final viscosity of starch paste decreased. In the studies on starch, the increase of setback viscosity usually related to the crystallization of amylose chains is caused by the liability of protein ease to retrogradation.

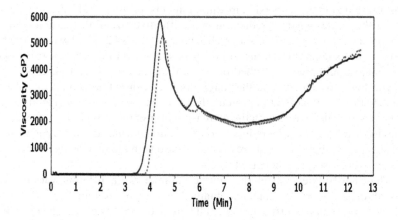

Fig. 1. RVA pasting curve of indica rice starch and japonica rice starch

3.3 Effect of Rice Varieties and Protein Content on Flowing Characteristics of Rice Starch Paste

The results from the previous studies indicated that high values of shear stress pointed to a high stability of the structural of the starch (Gibinski et al., 2006). According to that criterion, as was shown in Table3. and Figure.3, japonica rice starch had the more stable structure than indica rice starch. The flow behavior index, n, informed on the deviation from the Newtonian flow, for which n=1. That parameter for rice starch was below 1 pointing that the rice starches were pseudoplastic and shear thinning liquids

(Steffe, 1996), and its flow behavior index is below 1, which suggests that it is non-Newtonian fluid, the non-Newtonian behavior of starch solution was reported by many other researches (Evans et al., 1979; Nurul et al., 1999; Rao et al, 1997). As was shown by the flow curves in Figure 2, indica rice starch and japonica rice starch with different protein contents both presented shear shinning phenomenon which is the specific phenomenon of pseudoplastic fluid, the phenomenon suggested the decrease of apparent viscosity of the fluid with the increase of shear rate, the decrease with the increase of shear rate was quickly at the beginning, however, the decrease rate turned to stable gradually with the increase of shear rate when the apparent viscosity decreased to certain value, the apparent viscosity of japonica rice starch was greater than that of indica rice starch at the same shear rate, and the apparent viscosities of indica rice starch and japonica rice starch with protein content of 0.86% were highest at the same shear rate. As was shown by the flow curves in Figure.3, for the indica rice starch and japonica rice starch with different protein contents, the shear stress increased with the increase of shear rate, which was consistent with the characteristics of pseudoplastic fluid. It is accepted that the viscosity of a liquid is a function of the intermolecular forces that restrict molecular motion (Nurul et al., 1999). With the increasing of shear rate, disentanglement of long chain molecules occurs, as a result, the intermolecular resistance to flow (viscosity) is reduced, meanwhile, the highly solvated molecules or particles presented in the dispersion medium might be progressively sheared away with increasing shear rate causing a reduction in the effective size of the particles and hence a reduction apparent viscosity of starch (Bhandari et al., 2002; Holdsworth, 1971). The previous studies suggested that the viscosity of starch solution resulted from swelling starch granules and amylose leached out from the granules (Yue et al, 2008). For the starch with high protein content, the starch granules can not absorb adequate water to swell due to the dilution effect of protein, therefore, at the same shear rate, the apparent viscosity of starch decreased with the increasing of protein content. For the starch with high amylose content, the joint points among starch molecules usually trend to decrease and the shear stress between schlieres also decrease with the increasing of shear rate so that the apparent viscosity decreased.

The Herschel-Bulkley model flow curve parameters of different varieties of rice starch with different protein contents obtained with regression analysis were listed in Table3. The correlation coefficients were all among 0.985~0.999, which suggested that it was suitable to fit the flow curves of these rice starches with Herschel-Bulkley equation. For these rheological parameters, with the same protein content, the yield stress value of japonica rice starch was higher than that of indica rice starch. When the protein contents in indica rice starch and japonica rice starch increased from 0.52% to 1.20%, their yield stress values decreased from 85.84 to 51.24Pa and from 90.67 to 55.86Pa, respectively. Yield stress refers to the limited stress required for the initiation of flow. Consistency coefficient, K, from the Herschel–Bulkley model can also be taken as a viscosity criterion. With the same protein content, the consistency coefficient of japonica rice starch was higher than that fo indica rice starch. With the increase of protein content, the consistency coefficient of starch decreased, in which, the yield stress decreased with the decrease of interaction among starch molecules limiting the molecular movement. However, the change of flow behavior index was opposite, which increased with the decreasing of consistency coefficient. These results indicated that the indica rice starch with high amylose content and the starch with high protein content have good fluidity.

Table 3. Parameters of the Herschel–Bulkley models for the flow curves of rice starch with different varieties and diffetent protein contents

Sample	Protein content (%)	$\tau_{0(Pa)}$	K(Pa.sn)	n	R2
indica rice starch	0.52	85.84	88.85	0.252	0.998
	0.86	68.34	85.01	0.265	0.998
	0.95	56.594	75.33	0.299	0.997
	1.10	55.450	64.05	0.334	0.993
	1.20	51.24	59.68	0.344	0.994
japonica rice starch	0.52	90.67	171.2	0.236	0.993
	0.86	80.01	154.1	0.250	0.991
	0.95	69.93	132.95	0.267	0.985
	1.10	67.56	109.01	0.283	0.998
	1.20	55.86	101.1	0.287	0.988

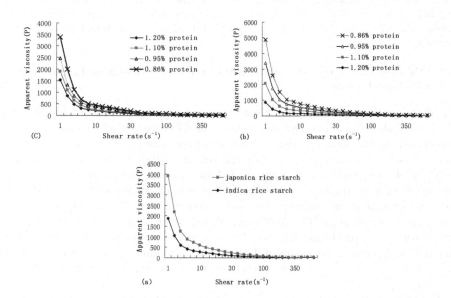

Fig. 2. Apparent viscosity as a function of shear rate for starches of different varieties and different protein contents: (a) indica rice starch and japonica rice starch with same protein content, (b) japonica rice starch with different protein contents, (c) indica rice starch with different protein contents.

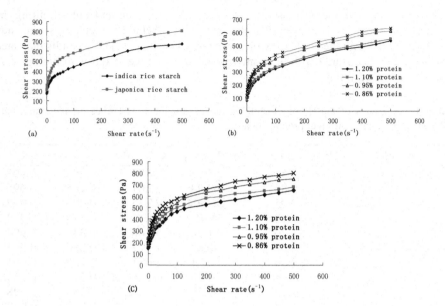

Fig. 3. Shear stress as a function of shear rate for starches of different varieties and different protein contents: (a) indica rice starch and japonica rice starch with same protein content, (b) japonica rice starch with different protein contents, (c) indica rice starch with different protein contents.

3.4   Effects of Rice Varieties and Protein Content on Viscoelastic Properties of Rice Starch Paste

The rheological properties of starch were measured by dynamic rheometer, then, the storage modulus G' and loss tangent tanδ (tanδ=G"/G') changing with temperature were determined by the dynamic oscillation measurement procedure of the rheometer, among which, the storage modulus (G') represented the elastic properties of rice starch paste and the loss modulus (G") represented the viscosity properties. In the dynamic measurement with rheometer, the changes of storage modulus G' can reflect the changes of hardness and strength of the gel, and the greater G' means higher hardness and strength (Rao et al., 1997). The changes of loss modulus G" can reflect the changes of viscosity properties of the gel, and The changes of loss tangent tanδ can reflect the changes of viscoelastic properties of the gel, lower tanδ value means that the elasticity of the gel is greater than the viscosity; conversely, the viscosity is greater than the elasticity.

The changes of storage modulus G' and loss tangent tanδ of indica rice starch and japonica rice starch with different protein contents in the processes of heating and cooling were listed in Table4. As was shown in Table4, during heating, the storage modulus G' of indica rice starch and japonica rice starch increased greatly at a certain temperature to a maximum and then dropped with continued heating. In the cooling process, G' increased with the decreasing of temperature. With the same concentration, protein content and changing rate of temperature, the storage modulus G' of indica rice starch was higher than that of japonica rice starch. The gels of all high amylose content rice starches have been credited with higher G' indicating a well-cross-linked network structures and increase in G' followed an exponential relationship with amylose content (Biliaderis et al., 1993). A significant increase in G'of rice starch on heating is caused by formation of a three-dimensional (3D) gel network developed by leached out amylose and reinforced by strong interaction among the swollen starch particles (Hsu et al., 2000; Vasanthan et al., 1996). The decreasing of G' might be attributed to the melting of the remaining crystallites, which resulted in swelling granules to become softer. In the cooling process, the amylose molecules cross linked through hydrogen bonds and formed the binding area, which enhanced the starch gel network and resulted the increase of G'. The response process of G" was similar as that of G' value.

Table 4. Effect of different varieties and different protein contents on the rice starch rheological parameters [G ' (Pa)]

Sample	Protein content (%)	Parameters	Temperature (^{0}C)										
			50	60	70	80	90	100	90	80	70	60	50
Indica rices tarch	0.86	G '(×103)	1.1	1.7	2.1	4.7	3.8	2.6	4.2	5.5	7.2	9.1	11
		tanδ	0.19	0.21	0.19	0.18	0.17	0.16	0.17	0.18	0.19	0.2	0.21
	0.95	G '(×103)	1.5	2.0	2.4	4.9	4.1	2.8	5.4	7.6	9.6	12	14
		tanδ	0.2	0.23	0.25	0.21	0.2	0.19	0.22	0.23	0.23	0.24	0.24
	1.10	G '(×103)	1.8	2.2	2.6	5.1	4.3	3.1	6.5	8.6	11	14	18
		tanδ	0.24	0.3	0.25	0.24	0.21	0.19	0.18	0.19	0.2	0.21	0.22
	1.20	G '(×103)	2.1	2.4	2.8	6.8	5.1	3.5	7.6	11	15	20	26
		tanδ	0.26	0.35	0.31	0.22	0.21	0.21	0.22	0.23	0.24	0.25	0.26
Japonica rice starch	0.86	G '(×103)	1.0	1.6	1.9	3.9	3.1	1.7	2.5	3.8	5.3	7.3	9
		tanδ	0.18	0.19	0.18	0.17	0.16	0.16	0.17	0.18	0.19	0.2	0.21
	0.95	G '(×103)	1.3	1.8	2.1	4.2	3.3	2.2	2.9	4.5	6.3	8.2	10
		tanδ	0.19	0.2	0.21	0.19	0.18	0.17	0.19	0.21	0.22	0.23	0.23
	1.10	G '(×103)	1.5	2.0	2.4	4.5	3.5	2.4	3.1	4.8	6.8	8.8	11
		tanδ	0.17	0.21	0.2	0.19	0.18	0.17	0.18	0.19	0.2	0.21	0.22
	1.20	G '(×103)	1.9	2.2	2.6	4.7	3.8	2.6	3.8	5.7	7.8	10	12
		tanδ	0.21	0.22	0.25	0.24	0.22	0.21	0.22	0.23	0.24	0.25	0.25

At same temperature, the G' value decreased with the decrease of protein content. The loss tangent tanδ were all less than 0.4, which suggested that G' values were all greater than G" values and the hardness and strength of starch gel were high. In the hearting process, the starch granules could not absorb adequate water to swell due to the water-absorption of protein so that the viscosity decreased. In the cooling process, the protein enhanced the formation of gel network which increased the hardness and strength of the starch paste.

4 Conclusion

The amylose content varies in the starches of different rice varieties. For the effects of rice varieties on rice starch pasting and rheological properties, besides the size and the swelling degree of the starch granules, amylose content is also a critical factor. For the starches with similar amylose contents but different protein contents, there are significant differences between their pasting and rheological properties. The pasting temperature of indica rice starch is 82℃ and the pasting temperature of japonica rice starch is 78℃. When the protein contents were same, the peak viscosity and breakdown viscosity of indica rice starch were lower than those of japonica rice starch, however, the final viscosity and setback viscosity of indica rice starch were higher than those of japonica rice starch. The peak viscosity, breakdown viscosity, final viscosity and setback viscosity of indica rice starch with protein content of 0.52% were 5304, 3479, 4748 and -556RVU, respectively, and the peak viscosity, breakdown viscosity, final viscosity and setback viscosity of japonica rice starch were 5923, 3974, 4560 and -1363RVU, respectively. With the increase of protein content in starch samples, the peak viscosity, breakdown viscosity and final viscosity of the starch all decreased significantly, but the setback viscosity of the starch increased significantly. With the same protein content, the consistency coefficient value of japonica rice starch was higher than that of indica rice starch, and the consistency coefficient of starch decreased with the increase of protein, however, the flowing-behavior index increased with the decrease of consistency coefficient. With the same protein content and changing rate of temperature, the storage modulus G' of indica rice starch was higher than that of japonica rice starch. The G' value decreased with the decrease of protein content. Loss factor tanδ were all less than 0.4, which suggested that G' values were all greater than G" values, the hardness and strength of starch gel were high, and the starch paste showed solid-like behaviors.

Acknowledgements

The authors gratefully acknowledge the financial support of the National Hi-Tech Research and Development Program of China (Grant No. 2006AA10Z341), and Science Foundation of Hunan Province (Grant No. 2007FJ1007).

References

Ibanez, A.M., Wood, D.F., Yokoyama, W.H., et al.: Viscoelastic properties of waxy and non-waxy rice flours, their fat and protein-free starch, and the microstructure of their cooked kernels. Journal of Agricultural and Food Chemistry 55, 6761–6771 (2007)

Graeme, B., Christopher, B., Zhao, J.: Effects of prolamin on the textural and pasting properties of rice flour and starch. Journal of Cereal Science 40, 205–211 (2004)

Lyon, B.G., Champagne, E.T., Vinyard, B.T., et al.: Sensory and instrumental relationships of texture of cooked rice from selected cultivars and postharvest handling practices. Cereal Chemistry 77, 64–69 (2000)

Biliaderis, C.G., Juliano, B.O.: Thermal and mechanical properties of concentrated rice starch gels of varying composition. Food Chemistry 48, 243–250 (1993)

Marco, C., Rosell, C.M.: Effect of different protein isolates and transglutaminase on rice flour properties. Journal of Food Engineering 84, 132–139 (2008)

Jean-Louis, D., Sylvie, D.A.: Rheological characterization of semi-solid dairy systems. Food Chemistry 108, 1169–1175 (2008)

Baxter, G., Blanchard, C., Zhao, J.: Effects of prolamin on the textural and pasting properties of rice flour and starch. Journal of Cereal Science 40, 205–211 (2004)

El-Kha-yat, G., Samaan, J., Brennan, C.S.: Evaluation of vitreous and starchy Syrian durum (Triticum durum) wheat grains: the effect of amylose content on starch characteristics and flour pasting properties. Starch/Stärke 55, 358–365 (2003)

Lim, H.S., Lee, J.-H., Shin, D.-H.: Comparison of protein extraction solutions for rice starch isolation and effects of residual protein content on starch pasting properties. Starch/Stärke 51, 120–125 (1999)

Evans, I.D., Haisman, D.R.: Rheology of gelatinised starch suspensions. Journal of Texture Studies 10, 347–370 (1979)

Nurul, I.M., Azemi, B.M.N.M., Manan, D.M.A.: Rheological behavior of sago (Metroxylon sagu) starch paste. Food Chemistry 64, 501–505 (1999)

Chrastil, J., Zarins, Z.: Changes in peptide subunit composition of albumins, globulins, prolamins, and oryzenin in maturing rice grains. Journal of Agricultural and Food Chemistry 42, 2152–2155 (1994)

Steffe, J.M.: Rheological methods in food process engineering, 2nd edn., pp. 328–331. Freeman, East Lansing (1996)

Ohishi, K., Kasai, M., Shimada, A., et al.: Effects of acetic acid on the rice gelatinization and pasting properties of rice starch during cooking. Food Research International 40, 224–231 (2007)

Yue, L., Charles, F.S., Ma, J.G., et al.: Structure-viscosity relationships for starches from different rice varieties during heating. Food Chemistry 106, 1105–1112 (2008)

Gibinski, M., Kowalski, S., Sady, M., et al.: Thickening of sweet and sour sauces with various polysaccharide combinations. Journal of Food Engineering 75(3), 407–414 (2006)

Sudha, M.L., Vetrimani, R., Leelavathi, K.: Influence of fibre from different cereals on the rheological characteristics of wheat flour dough and on biscuit quality. Food Chemistry 100, 1365–1370 (2007)

Rao, M.A.: Rheology of fluid and semisolid foods – Principles and applications, pp. 7–59. An Aspen Publication, Gaithesburg (1999)

Rao, M.A., Okechukwu, P.E., Da Silva, et al.: Rheological behavior of heated starch dispersions in excess water: Role of starch granule. Carbohydrate Polymers 33, 273–283 (1997)

Bhandari, P.N., Singhal, R.S., Kale, D.D.: Effect of succinylation on the rheological profile of starch pastes. Carbohydrate Polymers 47, 365–371 (2002)

Holdsworth, S.D.: Applicability of rheological models to the interpretation of flow and processing behavior of fluid food products. Journal of Texture Studies 2, 393–418 (1971)

Hsu, S., Lu, S., Huang, C.: Viscoelastic changes of rice starch suspensions during gelatinization. Journal of Food Science 65, 215–220 (2000)

John, S., Mounsey, E.D., O'Riordan: Characteristics of imitation cheese containing native or modified rice starches. Food Hydrocolloids 58, 184–193 (2008)

Lim, S.-T., Lee, J.-H., Shin, D.H., et al.: Comparison of protein extraction solutions for rice starch isolation and effects of residual protein content on starch pasting properties. Starch 51, 120–125 (1999)

Varavinit, S., Shobsngob, S., Varanyanond, W., et al.: Effect of amylose content on gelatinization, retrogradation and pasting properties of flours from different cultivars of Thai rice. Starch/Stärke 55, 410–415 (2003)

Vasanthan, T., Bhatty, R.S.: Physicochemical properties of small- and large-granule starches of waxy, regular and high-amylose barleys. Cereal Chemistry 73, 199–207 (1996)

Tan, Y., Corke, H.: Factor analysis of physiochemical properties of 63 rice varieties. Journal of the Science of Food and Agriculture 82, 745–752 (2002)

Research on Time and Spatial Variability of Soil pH in Sanmenxia Planted Tobacco Area

Hongbo Qiao[1], Gao Rui[1], Zhang Hui[1], Yanchun Chen[2], and Yongshi Su[2]

[1] College of information and management science, Henan agricultural university,
Zhenzhou 450002, Henan, China
[2] Sanmenxia Filiale Of Henan Tobacco Company, Sanmenxia, 472000, Henan, China

Abstract. Geostatistics combined with GIS spatial technology was applied to analyze the time and spatial variability of pH in topsoil(0-20cm) for planted tobacco region in Sanmenxia district. The results indicated that the pH value range form 6.5 to 8.8 and meet to the need of produce high quality tobacco, but the pH value of partial region is high. The pH value accord with logarithm normal distribution, variance coefficient is 15.2% and 4.5% of 2002 and 2007 year respectively. The semivariogram of pH was best described by the exponential model and spatial heterogeneity of pH were 55.77km and 92.39km. The Kriging interpolated method was applied to calculated the unobserved points and was used to generate the spatial and discrepancy map, analyzed the reason of the pH value increase and the method to improve soil. The research supply important method of the Sanmenxia high quality tobacco produce.

Keywords: soil pH; semivariance; spatial variability; GIS; Kriging interpolated method.

1 Introduction

Soil pH on soil fertility and a great influence on the effectiveness of nutrients, in the case at low pH value, P, Ca, Mg reduced the effectiveness of nutrients, on the contrary at high pH values, the micronutrient Fe, Mn, B for the crops of the state can not absorb. Tobacco on soil pH highly adaptive, in the pH 3.5 ~ 9.0 of the soil can grow normally, and the complete life cycle. However, the quality of flue-cured tobacco is in a certain range of pH values of the best, therefore, appropriate to the soil pH is the basis for the production of high quality flue-cured tobacco (Chen, et al.,1996).

Geostatistics at the beginning of the formation of the 1950's, based on the theoretical research work of France well-known statistician GMatheron in 1960's, some statistician formed a new branch of statistics. Statistics for some workers said: 'Geostatistics is based on regionalized variable theory, in order to function as the main tool for variation to study the spatial distribution of both random and structural, or spatial correlation and dependence of the natural the phenomenon of science' (Wang,1999), temporal and spatial variation of soil nutrients on the full study is accurate soil nutrient management and a reasonable basis for the high-fertilization. With the extensive application of 3S technology, the use of statistics and GIS technology to study the spatial variability of soil properties has become one of the hot soil science research (Liu, *et al.,*2004, Lian, *et al.,*2008, Wang *et al.,*2008).

D. Li and C. Zhao (Eds.): CCTA 2009, IFIP AICT 317, pp. 420–426, 2010.
© IFIP International Federation for Information Processing 2010

The use of Geostatistics and GIS (Geographical Information System, GIS) study of spatial variability of tobacco pH value of not more than is reported, only a few scholars were discussed. Qin studied in Guizhou Province and other tobacco PH value of soil spatial variability characteristics, results showed that the PH value of a strong spatial correlation, which is mainly affected by structural changes in the impact factor (Qin, *et al.,2007*). In this paper, the use of GIS and statistical methods to study the area of Sanmenxia tobacco pH value of temporal and spatial variation of soil characteristics, map soil pH value of the spatial distribution maps for the Sanmenxia and quality of tobacco production and provide a theoretical basis for soil improvement.

2   Material and Method

2.1   Study Area

Sanmenxia is the leading tobacco-producing areas of quality, is located in inland mid-latitude areas, a warm temperate continental monsoon climate. The annual average temperature 13.2 ℃, annual average 2354.3 hours of sunshine, frost-free period of 184 ~ 218 days, with an average annual rainfall 550 ~ 800mm, is suitable for the growth of flue-cured tobacco district.

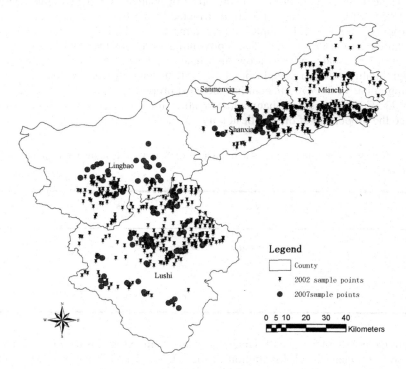

Fig. 1. Sample points in Sanmenxia City

2.2 Sample Collection and Analysis

According to the second national soil survey methods in collecting soil samples. Sanmenxia in the choice of representative tobacco soil, topsoil from 0 ~ 20cm soil samples, each sampling point of the area on behalf of 20 hm2, in samples collected at the same time positioning using GPS. 2.5:1 over the use of water and soil-pH extraction method of sample pH. In 2002 and 2007 collecting soil samples 697 and 299, respectively.

2.3 Data Processing and Analysis

In this paper, the use of SPSS (11.0) calculated the value of descriptive statistics, GS + (3.0) carried out statistical analysis, the use of ArcGIS (8.3) digital topographic maps and Kriging Interpolation.

3 Results and Analysis

3.1 Sanmenxia Tobacco Distribution of Soil pH Value

From Table 1, in 2002 tobacco-growing areas of change in soil pH values between 6.55 ~ 8.55, with an average of 7.76, distributed in the 6.6 ~ 7.5 and 7.6 ~ 8.5, the former level, a total of 251 samples, representing the total number of sampling points 35.9%, after a total of 449 samples, representing 64.1% of the sampling points. PH value of 7.5 is usually able to grow high quality tobacco. Accordingly, Sanmenxia has high-quality tobacco sample points is 251, accounting for the total number of sampling points of 35.9%, Mianchi has the most proportion of suitable points, mounting to 93.5%, followed by Shanxian(accounting for 24.8%), Lushi three (14.9%), Lingbao the proportion of the smallest, only 3.7%.

Table 1. The soil pH value grade and frequency of Sanmenxia Sanmenxia planted tobacco region(2002)

County	N	Range	X±S	5.6~6.5	6.6~7.5	7.6~8.5	8.6~9.0
Lingbao	110	7.45~8.3	7.88±0.16	0	3	107	0
Shanxian	157	6.55~8.45	7.74±0.47	0	39	118	0
Mianchi	184	7.1~7.7	7.4±0.13	0	172	12	0
Lushi	249	7.7~8.55	79.5±0.34	0	37	212	0
SUM	700	6.55~8.55		0	251	449	0

From table 2, the Ph value of Sanmenxia range from 6.6 to 8.8 in 2007, with an average of 7.92, centralized distribution in the 6.6 ~ 7.5 and 7.6 ~ 8.5 2, the former level, a total of 30 samples, accounting for sampling 10% of the total points, after a total of 239 samples, representing 79.8% of the sampling points.

Table 2. The soil pH value grade and frequency of Sanmenxia Sanmenxia planted tobacco region (2007)

County	N	Range	X±S	5.6~6.5	6.6~7.5	7.6~8.5	8.6~9.0
Lingbao	48	7.1~8.8	7.9±0.20	0	2	45	1
Shanxian	45	6.6~8.2	7.8±0.49	0	11	31	3
Mianchi	45	6.8~8.1	7.8±0.32	0	6	36	3
Lushi	99	7.1~8.8	8.2±0.37	0	11	65	23
SUM	299	6.6~8.8		0	30	239	30

3.2 Statistical Characteristics of the Soil pH Value

Data distribution is to use statistical methods to the spatial variability of soil characteristics of the premise of the analysis, only the data in line with the normal distribution only when the statistical analysis to meet assumptions. By the skewness and kurtosis tests indicate that in 2002 and 2007 the determination of soil pH values are in line with the requirements of normal distribution.

Table 3. The description statistics of soil pH in year 2002 and 2007

Year	model	C_0	C_0+C	C_0/C_0+C (%)	range(km)	R^2	RSS
2002	exponential	0.198	1.048	18.1	55.77	0.713	0.133
2007	exponential	0.190	1.156	16.4	92.39	0.560	0.423

3.3 Soil pH Value of the Spatial Structure Analysis

Semi-variance in the regionalized variable function is separated from the sample variance on the measure, the semi-variance function model and parameters to select the type of cross validation can be found (Hou, et al.,1998), the result as shown in table 4. 2002 and 2007 pH values of the semi-variance function model fit for the .0713 and .560 on the selected theoretical model to better reflect the spatial structure of soil elements.

The nugget value is far less than the sample spacing on the spatial scale differences in soil properties, which directly limit the size of spatial interpolation accuracy (Zhang, *et al.,* 2003). This study in 2002 and 2007 nugget value of 0.198 and 0.490, respectively, it showed that experiment error caused by the smaller variation of pH value in 2002. Sanmenxia soil pH value of spatial distribution is impacted by the topography, soil processes, parent material natural factors.

Nugget value/base value to determine the system variable degree of spatial correlation: If the ratio is less than 25%, indicating that the variable has a strong spatial correlation; if the ratio between 25% -75%, indicating moderate spatial correlation; if the ratio is greater than 75%, indicating that the space variables weaker spatial correlation (Cambardella, et al.,1994). Table 4 in 2002 and 2007 nugget value/base value of 18.1% and 16.4%, with a strong spatial correlation shows that the spatial variation of pH value caused mainly by parent material, topography and soil processes structural factors, this is accord with the nugget variance analysis results.

Table 4. Theoretical semivariogram model and corresponding parameters of soil pH value in year 2002 and 2007

Year	model	C_0	C_0+C	C_0/C_0+C (%)	range(km)	R^2	RSS
2002	exponential	0.198	1.048	18.1	55.77	0.713	0.133
2007	exponential	0.190	1.156	16.4	92.39	0.560	0.423

Range also known as the largest space-related distance, reflecting the range of spatial autocorrelation variable size, in the process of range with spatial correlation, on the contrary does not exist. Table 4 showed that the 2002 and 2007, the range is 55.77km and 92.39km, respectively, that pH value in the study area have a greater correlation between the extent to reflect the soil parent material, topography and structural factors such as soil type a greater impact. Fitting statistical models to the test by the F reach a significant level, that the sampling density and statistics to meet the needs of the interpolation.

3.4   pH Value of the Spatial Distribution Characteristics

According to the semi-variogram model, use of the optimal Kriging interpolation, plot the pH value spatial distribution map and the differential maps of Sanmenxia city in 2002 and 2007.

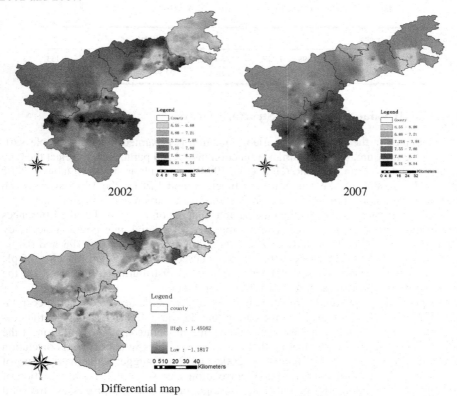

Fig. 2. Kriging estimates and difference for soil pH in year 2002 and 2007

In 2002 the soil pH values spatial distribution of the west, north higher than the east, south, there are some high pH value of the Lushi. compared to2002, there is the trend of pH value increased in 2007, that is, the degree of alkaline soil has increased. spatial distribution of pH as shown in Figure 2, the pH value of Lushi County Shahe, Panhe and Mianchi Potou has been increasing.

4 Conclusions

Suitable soil pH is conducive to root growth of tobacco, promote the growth of tobacco, enhanced resistance to disease, and improve yield and quality of tobacco (Tang, et al., 1999, Chen, et al., 1996). Therefore, the suitable pH value of tobacco production on the quality of great significance.

Statistical analysis results show that the pH value between 6.5 ~ 8.8 change in the concentration of 6.6 ~ 7.5 and 7.6 ~ 8.5 of sanmenxia, meet the needs of high-quality tobacco production, but to some local higher pH value need to be adjusted. The pH values obey the lognormal distribution, coefficients of variation were 15.2% and 4.5% in 2002 and 2007, respectively,with less variation. The best variance function is experimental model. Soil nugget value/base value were 18.1% and 16.4% in 2002 and 2007, with a strong spatial correlation, the spatial variability of the caused by parent material, topography and soil such factors.

The range of pH value were 55.77km and 92.39km in 2002 and 2007, respectively, that it have a greater correlation, the impact is caused by structural factors such as the soil parent material, topography and soil types.

Use of the optimal Kriging interpolation, plot the pH value spatial distribution map and the differential maps of Sanmenxia city in 2002 and 2007. As can be seen from the figure, pH values showed increasing trend, particularly in the eastern and southern part of the Lushi and Mianchi County. The reason may be to increase the amount of chemical fertilizers, natural precipitation decreased but ground water irrigation increased, the amount of organic fertilizer relative decreased.

Therefore, a higher pH value of the land, should pay attention to soil improvement and selected appropriate physiological acid nitrogen, potassium and superphosphate acid fertilizer. At the same time, humic acid fertilizer application to reduce pH value is an effective way. Studies have shown that humic acid on soil acidification, which is conducive to the soil of alkaline tobacco plant growth and development and the formation of good quality.

Acknowledgements

This research was funded by Henan Tobacco Company Projects Grant (200903-02).

References

Chen, J.J., et al.: Effects of pH in Rhizosphere on the absorption of inorganic nutritions in tobacco. Plant Physiology Communications 32(5), 341–344 (1996)
Wang, Z.Q.: Geostatistics and its application in ecology. Science Press, Beijing (1999)

Liu, X.M., et al.: Application of geostatistics and GIS technique to Characterize spatial variabilities of bioavailable micronutrients in paddys soils. Environmental geology 46, 189–194 (2004)

Lian, G., et al.: Spatial variability and prediction of soil nutrients in a small catchment of the loess plateau. Acta ecological snica 45(4), 577–584 (2008)

Wang, S.Y., et al.: Preliminary study on spttial variability and distribution of soil available microelements in Pinggu district of Beijing. Scientia agricultura sinica 28(10), 4957–4964 (2008)

Qin, Z.L., et al.: Spatial variability of pH and soil nutrient for planted tobacco leaf area in Guizhou Province. Chinese journal of soil science 38(6), 1046–1051 (2007)

Hou, J.R., et al.: Practical geological geostatistics, pp. 31–72. Geological Press, Beijing (1998)

Zhang, Q.L., et al.: Study on Spatial Distribution of Soil Quality and Quantitative Evaluation of Soil Fertility Quality under Middle Spatial Scale. Chinese Journal of Soil Science 34(6), 493–497 (2003)

Cambardella, C., et al.: Field-scale heterogeneity of soil properties in central Iowa soils. Soil Science Society of America Journal 58, 1501–1511 (1994)

Tang, L.N., et al.: Effect of soil acidity adjustment on growth and quality of tobacco. Journal of Fujian agricultural university 28(1), 71–76 (1999)

Chen, J.J., et al.: Studies on relationship of rhizospere pH value with yield and quality of Flue-Cured tobacco. Tropic subtropics soil science 5(2), 98–101 (1996)

Web N.0, the New Development Trend of Internet

Zhiguo Sun and Wensheng Wang

Agricultural Information Institute, The Chinese Academy of Agricultural Sciences,
Beijing, China, 100081
Key Laboratory of Digital Agricultural Early-warning Technology, Ministry of Agriculture,
The People's Republic of China

Abstract. This article analyzes the Internet basic theory, the network foundation environment and the user behavior change and so on, Which analyzes the
development tendency of existing partial Internet products in the future Internet
environment. The article also hot on the concept of cloud computing, Demonstrates the relation between Cloud Computing and Web 2.0   from the angle
of Cloud-based end-user applications, The possibly killing application in the
future was discussed.

Keywords: web2.0, SNS, P2P, Long Tail, Blue Ocean Strategy, Cloud Computing, e-zine, IM,HTML5.

Nobody can forecast whether the concept of web N.0 will be sensationalized. In fact,
it does not matter whether the concept of web N.0 will come into being. What is important is that the development of the internet will not slow down and we can not exactly forecast the development tend of the future internet. However, it is certain that
the future development of the internet will go beyond our imagination.

As we know, from web 1.0 to web 2.0, we have experienced the development from
passively receiving information to initiatively creating information. Web 2.0 is only a
symbol indicating the changing internet. Its most predominant characteristic is indi-
vidualization and decentralization. Besides, web2.0 emphasizes socialization, opening, share, participation and creation. What will happen in days of the foreseeable
web N.0? In this paper, the author has briefly explained the variance of some basic
environment of the internet (including some basic theories and network hardware),
analyzed the possible expression forms of some existing internet products and programs in the future internet environment and forecasted the new applications which
are likely to come into being.

1   The Variance of the Basic Environment of the Internet

1.1   The Introduction and Development of Basic Theories and Concepts

1.1.1   Six Degrees of Separation Has Been Introduced into the Internet
We According to the Six Degrees of Separation principle, any two individuals in the
world can find the other through at most six people, no matter whether they know

D. Li and C. Zhao (Eds.): CCTA 2009, IFIP AICT 317, pp. 427–434, 2010.
© IFIP International Federation for Information Processing 2010

each other. At present, the theory functions as a theoretical base of constructing community, telecommunication and cooperative products.

1.1.2 The Application of the Long Tail theory

The basic principle of the Long Tail theory: if the channel of storage and circulation is large enough, the market share possessed together by the unpopular products or products with less demands can compare with that of a few popular products. That is, innumerable small markets will pool the market energy which can be equal to the mainstream market. The internet provides a good ground for the application of the Long Tail theory, which has become a new type of economic mode and been applied in the network economy with success. The websites of Google including Adsense, Taobao and Amazon are successful and typical cases of Long Tail theory.

1.1.3 The Form of the New Thought of Blue Ocean Strategy

The Blue Ocean Strategy is 'innovation-centered', which emphasizes searching for, or innovating brand-new market space and business opportunities untainted by competition, that is, creating new changes through exploring new thoughts to discover particularities of the market and create new market space without competitors. In practice, the so-called new market is new to us, but it may have existed since a long time ago. But we simply did not explore it.

1.1.4 The Speed Theory and the Elevator Theory

The internet is an economy of speed. As for middle and small-sized enterprises and individuals, they may have no powerful techniques, but they always can sniff out opportunities earlier for their small size and less links of decision making. Speed is not enough to get them success. They also have to launch their products at a suitable time. As we know, in a full elevator, the later person getting into the elevator possibly arrives earlier when the elevator is opened. The theories of speed and elevator show that, if you want to get a success, you must be fast; however, it is possible that the person who moves later arrives early. It is crucial that you have to arrive there at your speed before the elevator is closed.

1.1.5 The Emergence of the Concept of Cloud Computing

Cloud computing is the development result of distributed processing, parallel processing and grid computing or it can be regarded as business realization of these computer science concepts. The basic principle of cloud computing is to distribute the computing resources and storage resources into a large number of distributed computers instead of local computers or remote servers. Thus, users can obtain the needed resources (hardware, platform, software) through the network in the mode which is demanded and can be easy expanded. Cloud computing is characterized by the following aspects: its super large scale, for example, the "cloud" some enterprises established is as many as millions of servers; virtualization, the users do not have to care about the specific running location of applications; high reliability, it is more reliable to use cloud computing than local computers; high expansibility, it can meet the demand of application and the increase in the scale of users at any time; low price, the common users can enjoy the top IT technology even with less cost. Both cloud computing and web2.0 is user-centered. Cloud computing emphasizes the concept of service, while web2.0 focus on share, user driver and value creation of users. Cloud computing will provide better platform for the users, besides, the web2.0 applications based on the cloud computing will be driven by more users, boosting the development of cloud computing.

1.2 The Improvement of Internet Infrastructure and the Reduction of the Technique and Capital Barriers

The broadband has been popularized. With the development of infrastructure, now in China the most popularized mode of ADSL broadband access has been upgraded to 8Mb from 512Kb. Besides, CNGI, the largest pure IPV6 network of the second generation, has been initially founded in China. If it is turned into civilian usages from scientific research, it can be imagined that the problem of broadband bottleneck in the existing network application will be settled. Based on the new broadband, the designers of internet products will not have to consider the broadband restriction when releasing new applications. What they only have to think about is market demand. Therefore, new killer-grade applications will come to the fore continuously.

The rapid development of infrastructure has reduced the charges of virtual hosting and surfing the internet. Besides, the fast increasing amount of registering domain names makes the register fee of single domain name go down in a continuous manner. Thus, netizens do not have to worry about high charges of surfing the internet. They can obtain their own web space and individualized domain names just as easily as eat foods and dress them. The most important is the reduction of technique barriers, for which common netizens can easily establish their own websites and blogs by the means of utilizing open source foundation programs.

1.3 The Changes of User's Behaviors

The behaviors of internet users change from curiosity and surfing to building their own websites and writing Blogs, from searching for information to sharing information. The Web1.0 with the representation of portal sites has created a large number of users. Among them, there exist new netizens with special demands and habits. Besides, the new type of internet emphasizing opening, share, participation and creation will attract and guide more senior internet users to join the team of constructing the internet and sharing information.

1.4 The Popularity of the Opening Spirit and the Emergence of Copyright Agreements like CC

Open source program helps many common internet users without technical background to easily set up their own websites and Blogs. the open content typical for MIT open course makes the user can easily access to resources, and numerous open API with the representation of Google Map API not only enable the web to have a better connection, but also brings convenience to the creation of users. What's more, the emergence of the copyright agreements such as GPL, CC and Copyleft etc. has ensured the possibility of Web2.0 and accelerated the innovation and the transmission of knowledge.

1.5 The Popularity of the Safe and Operational Online Payment

The characteristic of the internet is that users can complete much work without going out just through clicking the mouse. However, the benign development of the internet can not be separated from the support of funds, because the internet community is just the same as the realistic society, where both transaction and capital circulation occur. The popularity of safe and operational online payment represented by the online payment brands, such as Paypal, Alipay, 99Bill etc., has strongly supported the development of personal e-commerce, which really activates the internet.

2 Analysis and Forecast on the Future Development Trend

2.1 The Electronic Magazines Imitating Paper-Based Journal Format Will Disappear, and the Plane Medium Will Transfer to the Internet in a Comprehensive Manner

Actually, The existing products of electronic journal is immature and unscientific, because they only change the existing format of printing medium into an electronic one, applying sound and animation to enhance the sensory manifestation. In order to indulge the realistic browsing mode, the existing electronic journals intent to imitate the reading custom of the current plane magazines, which not only makes no practical sense but also restricts the development of electronic magazines. Besides, It ignores the diversified expression skills that computer is born with and the desire of readers for fresher and more exciting representing modes. As for common electronic journals, users often have to download exe files to the local, and then they can view them. However, as the users increasingly show less trust in the .exe files in the internet, they will be more and more cautious to download. It can be forecasted that in the near future, the existing electronic magazines imitating paper-based journal format will disappear. The internet, as a new type of media replacing paper, demands a new media form which adapts to its characteristics of opening and share.

The existing plane magazine and newspaper will transfer to the main war field of the internet in a comprehensive manner. A new type of media will come out, and the network magazine and newspaper will become more lifelike through the diversified representing skills of computers. With the popularization of online payment and the browsing techniques on controllable content being really mature, it can be forecasted that, in the near future, common users will possibly subscribe web newspaper and magazine. The possible problem is that, as more and more users themselves function as medium contributing their information and knowledge, users will not be willing to purchase the web newspaper and magazine with so much available internet information. Therefore, the plane medium used to mutual-way benefits (purchase fees of users and the advertising fees from enterprises) may possibly return to the old way of only charging ad fees.

2.2 Instant Message Software Will Integrate the Functions of Voice Telephone and E-Mail and Virtual Reality Technology Is Applied in the Instant Message Software. Meanwhile, the Status of E-mail and Common Fixed Telephone Will Be Weakened and Finally They Will Vanish

The function of video chat in Instant Message software will be developed in great progress. With the application of virtual reality, the chatting two parties even can feel touching the other. In addition, as more cities have paved WIFI hotspots, smart phones with WIFI functions can realize free voice communication through installing Instant Message software for phones. The Instant Message software which integrates better offline message and voice functions will drive E-mail and common fixed telephone to the margin. Users will be willing to use less software and account number to accomplish more communication tasks. At present, Google Wave which has not been formally released is doing available exploration in this aspect.

2.3 The Boundary between Browsing Software and Client Software Will Vanish

As a matter of fact, Microsoft is undertaking an exploration in this aspect at the operating system level. Although it is still distant from the formal application of XAML, under the existing technical conditions, now much browser application software has come into being, for example, the Thunder has released website download software. Besides, various manufactures of anti-virus software have released their own online anti-virus products for browsers. Google has integrated online Gtalk in Gmail and released Google Docs through purchasing writely. In fact, this paper is written in the online office. The realistic significance of the online office is that we can conduct online writing anywhere and anytime without taking with mobile storage equipment. What's more, the online office supports multi-people online cooperative writing, which is very convenient. What is more important is that HTML5 standard is more and more close to its release, which will provide faster response time of JAVAScript, more perfect semantic labels, stronger chart disposal and webpage interaction based on HMTL. Furthermore, it also can provide API for the development of the third party. With the foresaid strong webpage standards, the development of the application software based on browsers will become easier. We have reason to believe that more and more software will come into being in the form of browser.

2.4 The Existing News Portals will Disappear and the Search Engines Will Shoulder More Functions

In the future, nobody will be wiling to pick and read news from the ads on the distributing centers like Sina. More users will be distributed to pages of search engines where the engines select news automatically and news aggregation websites like Digg.com(Digg differs from the traditional news websites where editors select the most attracting news information on the day, while the news on Digg.com are completely dug from each corner of the internet). It can be forecasted that search engine possibly will release a mode of users digging news (it is learned from Digg) on engines selecting news automatically. Certainly, products like the existing Quick News of Google and the products of e-mail news subscription of Baidu will be popularized widely. And the subscription mode of RSS news will possess certain market.

During the early development stage of news portals, people of Sina cooperated with various traditional medium to search for news, but up to the later stage, because of its predominance of scale and the form of user monopolization, more large and small medium initially turn to Sina for cooperation on news. With the alteration of the user's behavior mode, the influential force of portals will decline; then, more and more traditional medium will gradually come to their sense and will no longer leach on to the channels of news supermarket like Sina. They turned to cooperating with search engines like Baidu. As a matter of fact, the news channel of search engines will serve as a transfer, which guides users in entering the pages of news providers while helping users obtain news. Besides, it helps traditional medium websites increase their flow, which is a win-win mode.

2.5 More Powerful Application of P2P Will Appear in the Next Generation of Internet Based on IPv6

The basic characteristic of P2P technology is distributed and decentralized. Each node of P2P functions as both client end and server. The technical characteristics of P2P technology happen to have the same view with the developing internet spirit, which is now an extremely powerful and arrogant technology on the internet. P2P flow which MP3 and video documents share and download has become the main body of the flowof broadband users. Instant Message software and internet phone based on P2P (like Skype) develop fast and P2Pflow medium (like pplive, QQ live, etc.) are springing up. P2P cooperative computing and grid are in the ascendant. Besides the mode of client/server, P2P will become another main mode of the internet. The emergence of Thunder, a kind of download software applied with P2P technology, has rapidly triumphed over various existing download software and become the most popular download software now. It is believed that, with the popularization of IPV6 in the future, the problem of honesty and security stopping the development of P2P will get solved. In the future, new top-grade internet applications will come into being in the field of P2P.

2.6 New Internet Products Will Be Applied in the Area of Enterprises, Because Cloud Computing Helps Each Enterprise Contact the Top IT Technology at a Very Low Cost

Many manufactures have observed the new development trend of the internet. Lots of enterprises have started to try to communicate better with users through establishing enterprise blog, which covers the release and popularization of products, clarifying the market information, exposing the process of research and development, and the publication of enterprise culture. Besides, more enterprises get closer to consumers through establishing enterprise blog, building up an new marketing image with an easy-going characteristic. The new enterprise network marketing mode will continuously get rid of the stale and bring forth the fresh. It can be forecasted that, with the persistent expansion of the development momentum of individual blog, enterprises will increasingly purchase and apply the soft ads in the content of individual Blogs. The devolution of public praise based on the internet will bring enterprises the most effective growth in payoff. In addition, the development of cloud computing is attracting some attention of enterprises. The novelty of cloud computing lies in that it almost can provide infinite cheap memory and computing ability. The technology of cloud computing will help middle and-small-sized enterprises greatly lower their cost. The newest and most powerful IT technology will no longer be only possessed by large-sized enterprises. Cloud computing helps each enterprise contact top IT technology at a very low cost.

2.7 SNS Websites Have Become the Internet Sites with the Largest Amount of Browse

At present, great progress has been made in the new type of SNS websites. At abroad, Facebook has been a widely known social website, while at home Kaixin0001.com has become the website which is most regularly used by the white collar through viral marketing. As more people register on SNS websites, the potential application lurking for years finally burst violently. According to the theory of elevator, when timing is mature, the later comers always surpass the formers. The theory is verified again in the field of SNS. Orkut and wealink of the early stage are fading out people's eyeshot, while Facebook and kaixin110.com are leading in the field. By adding more applications, the new type of SNS websites have integrated many functions of blog, instant message software, BtoC, CtoC, online office and game. Therefore, users will spend more time on these websites, which will become the internet sites with the largest amount of browse. It can be forecasted that a large merger among portal websites, instant message, SNS sites and CtoC about SNS application will put on anytime.

2.8 Semantic Search Engines Will Be Applied on a Large Scale

Semantic search engine is the future development trend of search engines. At present, semantic search system has attracted extensive attention. However, the development of semantic search engines is mainly limited by the development of semantic web and the technology dealing with natural languages. With the development of technology, semantic search engines will finally enter the stage of practical application, when computers will be equipped with artificial intelligence. The final design target of semantic search engines is to return to what the users want. At present, Microsoft has purchased Powerset, a semantic search engine, and released a brand-new search brand of Bing after integrating it.

Acknowledgements

The work is supported by the Academy of Science and Technology for Development fund project "intelligent search-based Tibet science & technology resource sharing technology", Special Project on of The National Department of Science and Technology "TD-SCDMA based application development and demonstration validation in agriculture informationization", Special fund project for Basic Science Research.

References

Andersen, C.: The Long Tail. Zhongxinchubanshe, Beijing (2006)
Chan Kim, W., Mauborgne, R.: Blue Ocean Strategy: How to Create Uncontested Market Space and Make Competition Irrelevant. Shangwuyinshuguan, Beijing (2005)

Miller, M.: Cloud computing. Jixiegongyechubanshe, Beijing (2009)

Zhiguo, S.: Web2.0, the coming of the new era of Internet based on individual. Agriculture Network Information (12), 97–100 (2005) (in Chinese)

Zhiguo, S.: Blog, knowledge sharing and personal knowledge management. Agriculture Network Information (10), 35–36 (2004) (in Chinese)

Baike, B.: Cloud computing [EB/OL] (in Chinese), http://baike.baidu.com/view/1316082.htm

Bo, H.: Playin' with IT (in Chinese), http://blog.donews.com/keso

Ren, L.: Liuren Blog (in Chinese), http://blog.donews.com/liuren

Dongmei, G.: From Copyright to Copyleft: GPL the birth, growth and future [EB/OL] (2006) (in Chinese), http://of.openfoundry.org

Wikipedia. Semantic search [EB/OL] (2009-5-30), http://en.wikipedia.org/wiki/Semantic_search

Wikipedia. Social network service [EB/OL] (2009-5-30), http://en.wikipedia.org/wiki/Social_network_service

Study and Improvement for Slice Smoothness in Slicing Machine of Lotus Root

Deyong Yang[1,*], Jianping Hu[1], Enzhu Wei[1], Hengqun Lei[2], and Xiangci Kong[2]

[1] Key Laboratory of Modern Agricultural Equipment and Technology,
Ministry of Education & Jiangsu Province, Jiangsu University, Zhenjiang,
Jiangsu Province, P.R. China 212013,
Tel.: +86-511-8; Fax: +86-511-8
yangdy@163.com
[2] Jinhu Agricultural Mechanization Technology Extension Station, Jinhu county,
Jiangsu Province, P.R. China 211600

Abstract. Concerning the problem of the low cutting quality and the bevel edge in the piece of lotus root, the reason was analyzed and the method of improvement was to reduce the force in the vertical direction of link to knife. 3D parts and assemblies of cutting mechanism in slicing machine of lotus root were created under Pro/E circumstance. Based on virtual prototype technology, the kinematics and dynamics analysis of cutting mechanism was simulated with ADAMS software, the best slice of time that is 0.2s~0.3s was obtained, and the curve of the force in the vertical direction of link to knife was obtained. The vertical force of knife was changed accordingly with the change of the offset distance of crank. Optimization results of the offset distance of crank showed the vertical force in slice time almost is zero when the offset distance of crank is -80mm. Tests show that relative error of thickness of slicing is less than 10% after improved design, which is able to fully meet the technical requirements.

Keywords: lotus root, cutting mechanism, smoothness, optimization.

1 Introduction

China is a country of producing lotus root, lotus root system of semi-finished products of domestic consumption and external demand for exports is relatively large. In order to improve efficiency, reduce labor intensity, the group work, drawing on the principle of the artificial slice based on the design and development of a new type of lotus root slicer (Bi Wei and Hu Jianping, 2006). This new type of slicer solved easily broken cutting, stick knives, hard to clean up and other issues, but the process appears less smooth cutting, and some have a problem of hypotenuse piece of root. In this paper, analyzing cutting through the course of slice knife, the reasons causing hypotenuse was found, and the corresponding improvement of methods was proposed and was verified by the experiments.

---

* Corresponding author.

D. Li and C. Zhao (Eds.): CCTA 2009, IFIP AICT 317, pp. 435–441, 2010.
© IFIP International Federation for Information Processing 2010

2 Structure of Cutting Mechanism of Slicing Machine

Cutting mechanism of slice lotus root is the core of the machine, the performance of its direct impact on the quality of slice. Virtual prototyping of cutting mechanism of slice lotus root (Fig.1)was built by using Pro/E, and mechanism diagram of the body is shown in Fig.2. Cutting principle of lotus slicer adopted in the cardiac type of slider-crank mechanism was to add materials inside , which can be stacked several lotus root, lotus root to rely on the upper part of the self and the lower part of the lotus press down so that it arrives in the material under the surface of the baffle. While slider-crank mechanism was driven by motor, the knife installed on the slider cut lotus root. In the slice-cutting process it was found that parallelism of the surface at both ends of part of piece lotus was not enough, which can not meet the technical requirements for processing.

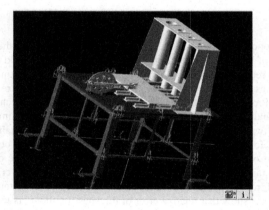

Fig. 1. Virtual prototyping of cutting mechanism

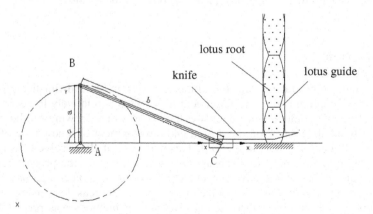

Fig. 2. Diagram of cutting mechanism

3   The Cause of the Bevel Edge

Uneven thickness and bevel edge of cutting were related with forces on the slice knife in the process of cutting. In accordance with cutting mechanism (Fig.2), without taking into account the friction and weight, the direction of force F of point C was along the link. Force F may be decomposed with a horizontal direction force component and a vertical direction force component. The horizontal force component pushed the knife moving for cutting, but the vertical force component caused the knife moving along the vertical direction. Because of the gap between the slider and the rail, the vertical force component made the blade deforming during the movement, and the knife could not move along the horizontal direction to cut lotus root, which caused the emergence of bevel edge. Thus, to reduce or eliminate the vertical force component in the cutting-chip was key to solve the problem of bevel edge and improve the quality of cutting.

When Crank speed was 60 ~ 90r/min, the horizontal and vertical direction of the force curve of point C connecting link and the blade hinge are shown in Fig.3 and Fig.4 respectively. As can be seen from the chart, with the crank speed improvement the horizontal and vertical direction of the force in point C also increased. The horizontal force changed relatively stable during 0s ~ 0.2s, which was conducive to cutting lotus, but the vertical force increased gradually. The more the vertical force was, the more detrimental to the quality cutting.

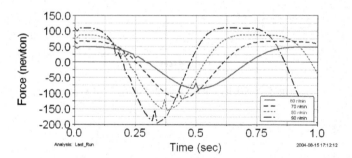

Fig. 3. Horizontal force of C

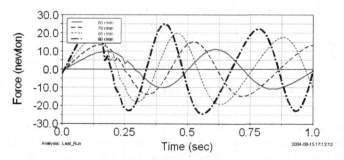

Fig. 4. Vertical force of C

4 Simulation and Optimization

If improving flatness of the slicer, the structure was optimized to reduce the vertical force component, so as far as possible the level of cutting blade.

When crank speed was 60 ~ 90r/min the velocity curve and acceleration curve of the knife center of mass are shown in Fig.5 and Fig.6 respectively. According to the speed curve, the speed of the knife center of mass was relatively large in a period of 0.2s ~ 0.3s.In accordance with the requirements that the knife should have a higher speed during cutting lotus, so this period time was more advantageous to cutting than other terms. According to acceleration curve, when calculates by one cycle, the acceleration value was relatively quite small in the period of time,0.15s~0.3s compared with other time section, which indicated that the change of velocity was relatively small, simultaneously the force of inertia was small, and the influence of vibration caused by the force was small to the slicer. Therefore, this period of time, 0.2s~0.3s, to cut lotus root piece was advantageous in enhances the cutting quality of lotus root piece.

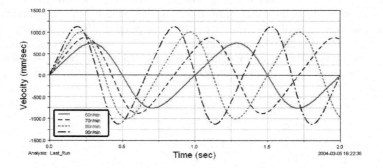

Fig. 5. Velocity curve of center of mass of knife

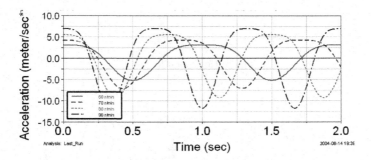

Fig. 6. Acceleration curve of center of mass of knife

Based on the above analysis, the vertical force component between link and the knife was the main reason for bevel edge. According to the characteristics of slider-crank mechanism, reducing the vertical force on the knife in the period of cutting time by altering crank offset was tried to enhance the quality of the cutting. When crank speed was 60r/min, the crank eccentricity was optimized. When the offset of the crank was 40 mm, 20 mm, 0 mm, -20 mm, -40 mm, -60 mm, -80 mm, -120 mm respectively, the mechanism was simulated and the vertical force curves under different crank eccentricity were obtained, as shown in Fig 7.

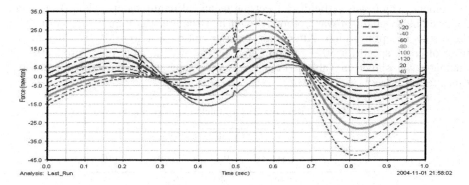

Fig. 7. Vertical force in different offset

Fig.7 indicates that: When the eccentricity was positive, the vertical force on point C increased gradually in 0.2s ~ 0.3s with the increase of crank offset; When the eccentricity was negative , the force decreased gradually first and then begun to increase along with the increase of crank offset. The force was almost zero when the eccentricity was -80 mm. So when the offset was -80 mm, the numerical of the force in 0.2s ~ 0.3s achieved the minimum and the quality of cutting was the best.

When the crank rotated in the other speed, there were the same optimization results. Fig.8 shows the curve of vertical force in the offset of 0 mm and -80 mm when the speed of crank was 80r/min. From the Fig 8 it is obvious that the vertical direction of the force of point C in 0.2 ~ 0.3s reduced a lot when the eccentricity is -80 mm. Therefore, the vertical force could be reduced by optimizing the slider-crank mechanism of eccentricity.

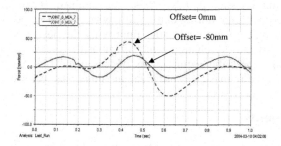

Fig. 8. Vertical force of C

5   Experimental Analysis

The relative error of thickness of lotus root piece reflects the quality of cutting, which is generally controlled of 10%. There always existed bevel edge phenomenon and the relative error of thickness was about 15% before structural optimization and improvement, which was difficult to meet the technical requirements. The offset in the slider-crank mechanism was optimized, and its structure was improved according to the results of optimization. After improvement cutting test were done in the conditions of crank speed for 80 ~ 110r/min and statistical data about the relative error of thickness was shown in Table.1. Four levels were separated in the experiment, three times for each level.

Table 1. Relative error of thickness of slicing

NO	Crank speed (r/min)			
	80	90	100	110
1	6.6%	6.4%	8.2%	9.5%
2	5.3%	6.1%	8.5%	9.2%
3	6.4%	7.9%	7.9%	9.4%
Average	6.1%	6.8%	8.2%	9.4%

It is derived from Table.1 that the relative error of the thickness of slices could meet the technical indicators when the crank speed was 80 ~ 110r/min, especially in the crank rotation speed 80r/min, 90r/min the relative error of thickness was less than 7%,and high quality was achieved.

6   Conclusion

The vertical force component acted on the knife in the process of cutting was the main reason for surface formation and bevel edge, so the key of improving the quality was to reduce the vertical force. Through slice knife velocity and acceleration simulation analysis the best time for slicing ,0.2s ~ 0.3s, was obtained. By optimizing the offset of the crank the vertical force during cutting time was greatly reduced when the offset was -80mm. Experiments were made after improving the design of lotus root slicer, which results showed that by changing the offset of the crank, the relative error of thickness could fully meet the requirements of less than 10%. So the problem was basically solved that the flatness was not ideal and there was the issue of bevel edge.

Acknowledgements

This work was financially supported by Agricultural Mechanization in three agriculture projects, Jiangsu Province (NJ2007-17) and Tackling Key Problems of Science and Technology of Taizhou City (BE2008385).

References

Wei, B., Jianping, H.: Study of Lotus Roots Slicing Techniques and Design of New Model. Journal of Agricultural Mechanization Research (12), 112–114 (2006) (in Chinese)

Enzhu, W.: The Simulation and Optimization on the new slicing machine of lotus root Based on Virtual Prototype Technology. Jiangsu University (2008) (in Chinese)

Ce, Z.: Mechanical Dynamics. Higher Education Press (1999)

Xiuning, C.: Optimal Design of Machinery. Zhejiang University Press (1999)

Liping, C., Yunqing, Z., Weiqun, R.: Dynamic Analysis of Mechanical Systems and Application Guide ADAMS. Tsinghua University Press, Beijing (2005)

The Research of Key Technologies of Streaming Media Digital Resources Transmission Based on CDN and P2P

Jichun Zhao[1,*], Feng Yu[1], Junfeng Zhang[1], Sufen Sun[1],
Jianxin Guo[1], Jing Gong[1], and Yousen Zhao[2]

[1] Institute of Information on Science and Technology of Agriculture,
Beijing Academy of Agriculture and Forestry Sciences, Beijing, China, 100097,
Tel.: 86-010-51503172
zhaojichun_0@163.com
[2] The Information Center of Beijing Municipal Bureau of Agriculture, Beijing, China

Abstract. Traditional system based on the current form of distance education network technology does not support the existence of dynamic multimedia data and real-time streaming data, as well as the existence of streaming media technology in the server bottleneck problem. The key technologies of streaming media digital resources transmission based on CDN (Content Delivery Network) and P2P (peer-to-peer) is discussed in the paper, and the P2P core algorithm is optimized, the CDN network deployment of the program and the edge server deployment strategies are discussed, which can address the situation of network congestion and improve user access to video and web site response time.

Keywords: P2P, CDN, Steaming media.

1 Introduction

Multimedia Network teaching now begins becoming a most important teaching way besides the traditional teaching way. This new teaching way aims at that anyone can learn anything at any time and from anywhere. Facing with the very large amount of multimedia data, many technologies should be involved in and variety of computer resources should be utilized if you want to transmit multimedia data efficiently on the internet.

Peer-to-peer is a communications model in which each party has the same capabilities and either party can initiate a communication session. Other models with which it might be contrasted include the client/server model and the master/slave model. In some cases, peer-to-peer communications is implemented by giving each communication node both server and client capabilities. In recent usage, peer-to-peer has come to describe applications in which users can use the Internet to exchange files with each other directly or through a mediating server.

On the Internet, peer-to-peer (referred to as P2P) is a type of transient Internet network that allows a group of computer users with the same networking program to

---

* Corresponding author.

D. Li and C. Zhao (Eds.): CCTA 2009, IFIP AICT 317, pp. 442–448, 2010.

connect with each other and directly access files from one another's hard drives. Major producers of content, including record companies, have shown their concern about what they consider illegal sharing of copyrighted content by suing some P2P users.

Meanwhile, corporations are looking at the advantages of using P2P as a way for employees to share files without the expense involved in maintaining a centralized server and as a way for businesses to exchange information with each other directly.

Short for content delivery network, a network of servers that delivers a Web page to a user based on the geographic locations of the user, the origin of the Web page and a content delivery server. A CDN copies the pages of a Web site to a network of servers that is dispersed at geographically different locations, caching the contents of the page. When a user requests a Web page that is part of a CDN, the CDN will redirect the request from the originating site's server to a server in the CDN that is closest to the user and deliver the cached content. The CDN will also communicate with the originating server to deliver any content that has not been previously cached.

This service is effective in speeding the delivery of content of Web sites with high traffic and Web sites that have global reach. The closer the CDN server is to the user geographically, the faster the content will be delivered to the user. CDN also provide protection from large surges in traffic. The process of bouncing through a CDN is nearly transparent to the user. The only way a user would know if a CDN has been accessed is if the delivered URL is different than the URL that has been requested.

A peer-to-peer (P2P) Contents Delivery Network (CDN) is a system in which the users get together to forward contents so that the load at a server is reduced. Lately, we have high-speed services for an access to the Internet such as the Asymmetric Digital Subscriber Line (ADSL). Some broadcasters may not have such services because they have only dial-up services and wireless services as PHS and a mobile phone to broadcast live. A problem with P2P CDN is its overhead to construct a distribution tree. It becomes a crucial problem when a broadcaster has only a low-speed access to the Internet, and we propose a P2P CDN system which reduces such an overhead. The key technologies of streaming media digital resources transmission based on CDN and P2P is discussed in the paper, and the P2P core algorithm is optimized, the CDN network deployment of the program and the edge server deployment strategies are discussed, which can address the situation of network congestion and improve user access to video and web site response time.

2 The Research Content

Web server based on the current form of distance education network technology does not support the existence of dynamic multimedia data and real-time streaming data, as well as the existence of streaming media technology in the server bottleneck problem. The key technologies of streaming media digital resources transmission based on CDN and P2P is discussed in the paper, and the P2P core algorithm is optimized, the CDN network deployment of the program and the edge server deployment strategies are discussed, which can address the situation of network congestion and improve user access to video and web site response time.

The key technology of P2P transfer research is that the core algorithm of P2P optimization to study the domain to support the autonomy of the P2P discovery and optimization of management agreements, so that the impact of interactive P2P networks to the whole net can be decrease minimum. P2P streaming media transmission node in the core agreement in accordance with P2P content (including files and streams) to deal with so slices, P2P users will be in accordance with these rules to complete the P2P sharing. The edge server pressure can be reduced greatly, and the file transfers and streaming media transmission can be improved efficiency.

CDN network deployment optimization program is to study the edge server deployment strategies, in order to solve the situation of network congestion problem and improve respond speed of user access to the site. The main technologies include distributed storage, load balancing, request redirection network and content management such as the core technology, and content management and global management of network traffic (Traffic Management) is the core of CDN. Through the user and the server load the nearest judge, CDN to ensure that the contents in a very efficient way for users to request services, which can be achieve fast and redundant for the purpose of accelerating the multiple sites.

3   The Research Objective

The text is research on P2P streaming media distribution system architecture, and establishment of P2P streaming media distribution model, in order to address the limited bandwidth which can not meet the increasing number of customer service demand. The key technology of P2P transfer research is that the core algorithm of P2P optimization to study the domain to support the autonomy of the P2P discovery and optimization of management agreements. CDN network deployment optimization program is to study the edge server deployment strategies, in order to solve the situation of network congestion problem and improve respond speed of user access to the site.

Research on key technologies of the P2P streaming media live, and the test system of CDN + P2P based on streaming media is set up, analysis and test the deployment of P2P algorithms and programs CDN.

4   The Key Technology

4.1   Research on P2P Streaming Media Distribution System Architecture

Research on P2P streaming media distribution system architecture provides a solution to the current distribution of streaming media in a limited bandwidth.

The paper is research on P2P streaming media distribution system architecture which uses the existing CDN network. The user side of the edge server is as a superseed provider, and at the same time user terminals provide the media distribution, content storage and effective management of resources, which can improve Unicast Quality of Service. P2P traffic will be severely restricted in the same region of the edge node through the distribution hybrid structure in order to avoid the flow hedge in the backbone of the Internet, and which increased manageability and high reliability of service.

4.2 P2P Live Streaming Media Key Technologies

Using member services and certificate realize building peer group and joining the inspection. Streaming media will not be affected and increased the capacity of the system through multi-source transmission and node monitoring service quality in the P2P network nodes dynamically changing environment. The P2P live streaming key is building a P2P network, data transmission, security and issues such as QoS guarantee.

4.3 P2P Cache Strategy and Replacement Algorithm

The technology advance in network has accelerated the development of multimedia applications over the wired and wireless communication. To alleviate network congestion and to reduce latency and workload on multimedia servers, the concept of multimedia proxy has been proposed to cache popular contents. Caching the data objects can relieve the bandwidth demand on the external network, and reduce the average time to load a remote data object to local side. Since the effectiveness of a proxy server depends largely on cache replacement policy, various approaches are proposed in recent years.

A proxy capable of transcoding is called transcoding proxy. Since the transcoding proxy lies between the server and the clients, this architecture can coordinate the mismatch between what the server provides and what the client prefers. The proxy-based approach has several advantages. First, an intermediate proxy has larger bandwidth to download the web objects and higher computing power to perform encoding/decoding, which enables proxy-based transcoding to effectively reduce the

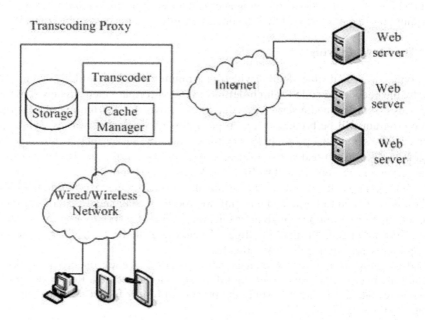

Fig. 1. The transcoding proxy structure

waiting time for clients to access the required version. Moreover, the proxy-based technologies can perform on-the-fly transcoding for dynamic content adaptation. Therefore, the proxy-based architecture combines the advantageous features of the client-based and server-based technologies. Figure.1 shows the architecture of the transcoding proxy. There are two primary components: transcoder and cache manager. After downloading the web objects from the remote servers, the transcoder should perform transcoding according to the clients' requirements. The transcoding results will be kept in the local storage (referred to as cache) of the proxy. Since the cache has limited capacity, the cache manager will perform cache replacement to evict the less valuable objects. Research topics in proxy-based transcoding include cache replacement schemes, system architectures, and proxy collaboration.

5   The Research Program -P2P + CDN Streaming Media Broadcast Model

5.1   System Function

Living channel list can be generated in the web page, a user clicks on different channels, the can watch different living channel programs.

According to the specific needs, the user can freely set the living parameters. The parameters are the size of the video window, video stream, and video encoding format. Real-time video surveillance can monitor the living content.

It can be set in the process of living access policy control, when a user exceeds the maximum on-line number of simultaneous users, other user access to be controlled.

5.2   The Flow of Living

The living flow: real-time satellite video programming or other video source in real time by the encoder to code for high-definition H.264 format video streams sent to the programs origins of media servers, media server shows the distribution of the origins of the program will be broadcast live to push down broadcast server, as well as programs can be sent to the program library for storage, the server may be broadcast live programming direct broadcast simultaneously live broadcasts can also be distributed CDN servers, CDN servers will be live broadcasts push terminals;

Living initiative launched by the terminal, terminal-demand programming flow: End users first need to log in and teaching web portal management system authentication, identity verification through after clicking on the appropriate program to receive the video on demand, terminal intelligent in choosing the best CDN server (recently and optimal server) to receive data streams.

End-user log in at teaching management system and click on programs will be teaching management at the server log information to do records, Report information to generate statistics in order to study the process of timely feedback. The structure of living is shown as Fig.2.

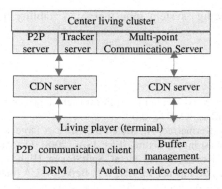

Fig. 2. The structure of living

6   The Test Effect

The video plays clearly, through its strong enabling technology to provide users with a more perfect enjoyment of the audio-visual situation and there is no buffer.

The drag response speed is fast, to enable users to select video was the moment, the system immediately begin to watch the speed of the user at least 3 times the speed of download videos and simultaneously broadcast buffer situation does not arise.

Living response is fast, the user at the request of streaming media services and switch channels hardly feel the delay, but sound and video always maintain smooth continuous, not a reproduction pause, such as jitter and discontinuous phenomenon.

Facilitate management: Modular design and fully technology application development, can achieve more than the business aspect as well as support and content providers (ICP) into programs such as outsourcing, co-operation of encryption programs, video streaming, data flow, cost flow interaction, the real implementation seamlessly between the various systems used in combination.

7   The Direction of Future Research

First of all, the need for further research question is how to generate effective coverage of P2P application-layer topology. Systems not only need to consider scalability and fault-tolerant, and also be taken into account the dynamic nature of nodes and heterogeneity, and in accordance with the status of physical network, and dynamically adjusted to optimize the topology and error handling. At the same time, minimize system maintenance costs in order to improve network utilization.

Security issues of P2P streaming media technology, one of research directions. How to node authentication, digital signature for the content providers and, in conjunction with digital rights management DRM technology to ensure the legitimacy of media content, whether for the protection of user privacy from being violated, or for P2P Streaming Media Technology Application and Promotion, has importance.

Combine Network coding P2P technology and streaming media technology to increase the calculation of the volume of network nodes for improving the network throughput and enhancing system robustness and stability has high research value and use value.

References

Hung, H.-P., Chen, M.-S.: Maximizing the profit for cache replacement in a transcoding proxy. In: Proceedings of IEEE ICME (2005)

Castro, M., Druschel, P., Kermarrec, A.: Metal. SCRIBE: A large-scale and decentralized application-level multicast infrastructure. IEEE JSAC 20(8) (October 2002)

Castro, M., Druschel, P., Kermarrec, A.: Metal. SplitStream: High Bandwidth content distribution in cooperative environments. In: IPTPS'0, Berkeley, CA (February 2003)

Tran, D., Hua, K., Do, T.: Zigzag: An efficient peer-to-peer scheme for media sreaming. In: Proc.of the IEEE IN FOCOM 2003, pp. 1283–1293. IEEE Computer and Communications Societies, NewYork (2003)

Hefeeda, M., Habib, A., Botev, B., Xu, D., Bhargava, D.B.: PROMISE:A peer-to-peer media streaming using collect cast. In: Proc.of the ACM Multimedia 2003, pp. 45–54. ACM Press, New York (2003)

Ford, B., Srisuresh, P.: Peer-to-Peer communication across iddleboxes, http://midcom-p2p.sourceforge.net/ draft-ford-midcom-p2p-01.txt

Syverson, P., Reed, M., Goldschlag, D.: Onion Routing Access Configurations. In: Proceedings of DARPA Information Survivability Conference and Exposition (DI: SCEX 2000), pp. 34–40 (2000)

Abrams, M., Standridge, C.R., Abdulla, G., Williams, S., Fox, E.A.: Caching proxies: limitations and potentials. In: Fourth International World-Wide-Web Conference (1995)

Han, R., Bhagwat, P., LaMaire, R., Mummert, V.P.T., Rubas, J.: Dynamic adaptation in an image transcoding proxy for mobile www browsing. IEEE Personal Communication (December 1998)

Padmanabhan, V., Wang, H., Chou, P.: Distilbuting Streaming Media Content using Cooperative Networking. In: Proc. of the imitational workshop on network and operation system support for digital audio and video, Miami, Florida (2002)

Influence of Sound Wave Stimulation on the Growth of Strawberry in Sunlight Greenhouse

Lirong Qi[1,*], Guanghui Teng[1], Tianzhen Hou[1], Baoying Zhu[2], and Xiaona Liu[1]

[1] College of Water Conservancy and Civil Engineering, China Agricultural University,
Beijing, P.R. China 100083,
Tel.: +86-10-62737583-1
525617qi@163.com
[2] Beijing Xiaotangshan National Agricultural Demonstration Zone, Beijing, P.R. China 102211

Abstract. In this paper, we adopt the QGWA-03 plant audio apparatus to investigate the sound effects on strawberry in the leaf area, the photosynthetic characteristics and other physiological indexes. It was found that when there were no significant differences between the circumstances of the two sunlight greenhouses, the strawberry after the sound wave stimulation grew stronger than in the control and its leaf were deeper green, and shifted to an earlier time about one week to blossom and bear fruit. It was also found that the resistance of strawberry against disease and insect pest were enhanced. The experiment results show that sound wave stimulation can certainly promote the growth of plants.

Keywords: environmental factors, sound wave stimulation, sunlight greenhouse, strawberry.

1 Introduction

Plants are stimulated inevitably by a variety of external environmental factors in the growth process and these stimulations have different extent influence to plants' growth, and then influence the crops' output and quality. As a flexible mechanical wave, the sound wave is a form of alternative stress and also a universal source of external stimulation to plants. Studies have shown that a certain frequency or sound intensity of the sound wave stimulation can promote the growth of plants. Scholars have done a lot of research on the role and mechanism of sound waves on plants. The approaches are used mainly in the form of music sound and pure tone (single frequency sine wave).

In music sound processing, music sound (natural sounds) had significantly improved the number of seeds sprouted compared to the untreated control, and there were no significant differences between harsh noise group and the untreated control (Creath et at., 2004). Under both light and dark conditions, sound up-regulated expression of the rbcS and ald by using classical music and single-frequency vibration signal (Jeong et at., 2008).

In pure tone processing, the hypocotyls' elongation and gene expression of Arabidopsis thaliana seeds were both improved by sound stimulus of about 50Hz and 90dB (Johnson et at., 1998). Chinese Academy of Sciences, Department of Applied

---

* Corresponding author.

D. Li and C. Zhao (Eds.): CCTA 2009, IFIP AICT 317, pp. 449–454, 2010.
© IFIP International Federation for Information Processing 2010

Chemistry of China Agriculture University and Department of Engineering Mechanics of Tsinghua University jointly find that a range of sound waves can stimulate tobacco's synchronization of cell division and promote DNA synthesis in the S-stage of cell division, and then improves plants growth and development (Li Tao et at., 2001). Sound wave stimulation can significantly enhance or inhibit the ATP content of Actinidia chinensis callus. Moderate sound stimulation can increase the activity of ATP synthase and is conducive to the level of energy metabolism of plants (Yang Xiaocheng et at., 2003; Yang Xiaocheng et at., 2007). By using QGWA-03 plant audio apparatus (frequency range: 100-2000Hz), tomato's yield increased by 13.2%, and its disease of grey mold decreased by 9.0% (Hou Tianzhen et at., 2009).

At present, the sound wave stimulation studies on the impact of plants are increasing, but the sound effect and mechanism are still controversial. To this end, QGWA-03 plant audio apparatus (PAA) was used to stimulate the strawberry growing in the sunlight greenhouse, and the sound effects to leaf area, photosynthetic characteristics rate and other physiological indexes were researched. This paper is our preliminary study of sound stimulation mechanism.

2   Materials and Methods

2.1   Test Materials and Design

The test was started in the sunlight greenhouse from November 2008 to January 2009 in Beijing Xiaotangshan National Agricultural Demonstration Zone. We selected 60 healthy strawberry seedlings (U.S. "Sweet Charlie") which grew in the same condition and transplanted them into white plastic flowerpots with medium loam. Then put the flowerpots into two sunlight greenhouses respectively (30 pots each) which were 80m apart. In the two sunlight greenhouses, conditions of structure, environment, irrigation control, and the relative position of flowerpots were basically the same between each other. We carried out the sound stimulation experiment (the PAA was put in the middle of the pots) in one building, and the other one used as control. Test arrangement is shown in Fig.1.

(a) the sunlight greenhouses   (b) strawberry seedlings in the flowerpots

Fig. 1. Layout of sound wave stimulation experiment on strawberry in sunlight greenhouse

The sound wave treatment was begun when the strawberry seedlings were transplanted into the greenhouses, and once every two days to play, and 9:00 to start dealing with each 3h. The frequency and volume were determined by the temperature and humidity of greenhouse. We measured the leaf area and photosynthetic indicators of strawberry at the beginning of growing season, squaring period, flowering period and fruiting stage. The production and disease resistance of strawberry were also determined in the fruiting stage.

2.2 Determinations

We used LI-3000 portable leaf area meter (LI-COR Inc. USA) and LI-6400 (LI-COR Inc. USA) portable photosynthesis meter to measure the leaf area and photosynthetic characteristics. Results are expressed as means ± SDs. Data were analyzed using Non-parametric test of two independent samples included in the SPSS version 13.0 software (SPSS Inc, Chicago, Ill). Statistical significance was set at $P<0.05$.

3 Results

3.1 Sound Effect to the Leaf Area of Strawberry

Fig.2 shows that in the beginning of growing season, squaring period and fruiting stage, the leaf area in the treatment are significantly greater than control ($P<0.01$); but there is no significant difference in the flowering period.

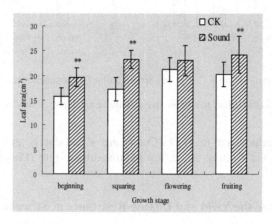

Fig. 2. Sound effect to the leaf area of strawberry

3.2 Sound Effect to the Photosynthetic Characteristics of Strawberry

Fig.3 shows that during the four stages, there is no significant difference of the intercellular CO_2 concentration (Ci) between the two groups, but the transpiration rate

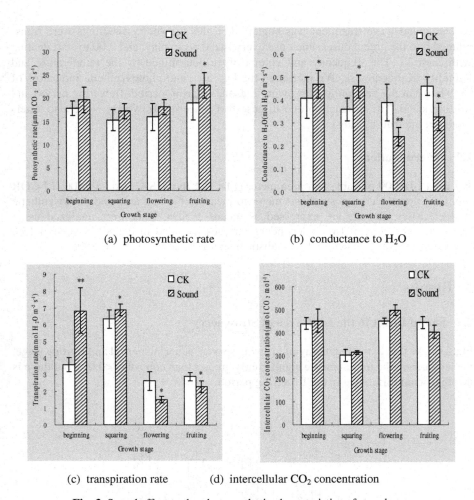

(a) photosynthetic rate

(b) conductance to H_2O

(c) transpiration rate   (d) intercellular CO_2 concentration

Fig. 3. Sound effect to the photosynthetic characteristics of strawberry

(Trmmol) and conductance to H_2O (Cond) are significantly greater than control (P<0.05). Only in the fruiting stage, the photosynthetic rate (Photo) has significant difference (P<0.05).

3.3 Sound Effect to the Yield and Disease Resistance of Strawberry

We surveyed the yield and disease situation in January 18, 2008. Table1. shows that the disease rate in the treatment is significantly less than control; but there is little effect on the yield.

Table 1. Sound effect to the yield and disease resistance of strawberry

Table 1. Sound effect to the yield and disease resistance of strawberry

	Yield (kg)	Yield per plant (kg)	Disease rate (%)
Sound stimulation	1.53	0.051	16.67
CK	1.63	0.054	50.00

3.4 Discussion

It is found that sound stimulation has great effect on the physiology of strawberry, and it has different impact in the different physiological stages.

Viewing in the entire growth period, sound waves do promote the leaf area of strawberry, but the effects on the Photo and Ci are not obvious. The trend of sound influence on Cond and Trmmol are both firstly increasing and then decreasing. It indicates that the Trmmol of strawberry leaf is mainly affected by stomata factors. Trmmol decreases in the latter growth stage. It shows that sound has little effect on the growth of strawberry after the flowering period and also explains the reason of production without significant changes. In addition, the experiment was not started in the breeding period, so it may be another reason for the un-improving production. During the initial growth period, the improvement of Trmmol promotes the transportation of water and mineral elements, so it increases the disease resistance (Wu Weihua, 2003).

4 Conclusion

In this paper, experimental results show that sound waves not only can promote the growth of strawberry, but can also increase the disease resistance. About the mechanism of sound waves improving the growth of plants, there are three possible reasons: environmental stress (including the sound waves stimulation) changes the fluidity and permeability of membrane; the signaling molecule of Ca2+ deliveries the stress signaling to other signaling molecules; the spread of stress signal causes related gene expression (Liu yiyao, et at., 2000). But we believe that there is phenomenon of spontaneous sound in plants. When the frequency between external vibration and plants spontaneous sound are consistent, the resonance will occur, thus promoting plants growth. We did the pre-test by using the He-Ne laser Doppler vibrometer to measure the sound frequency of Alocasia, and found that in normal growth conditions, plants' spontaneous sound frequency was in low-frequency range of 40-2000Hz (Luan Jiyuan, et at., 1995; Hou Tianzhen, et at., 1994). At the same time, we used low-frequency sound waves to stimulate more than 50 kinds of crops, and achieved remarkable effects (Hou Tianzhen, et at., 2009).

To sum up, we believe that the mechanism of sound effect to plants can be explained in two ways. From the biological point of view, sound may affect the characteristics and function of plant cell membrane, and gene expression. But from the physics point of view, the frequencies of sound vibration and plants spontaneous sound are in line, and then the resonance occurs. This experiment is only the initial discussion on the mechanism of sound stimulation to plants, and it is the foundation for exploring the mechanism from the perspective of plants' vibration characteristics.

Acknowledgements

This study was financially supported by the subject group of urban agriculture projects of Joint-Build Plan of Beijing Education Commission (item no. XK100190553). We thank the laboratory members (Guo Nan, Yu Ligen, et al.) for the help.

References

Sukumaran, C.R., Singh, B.P.N.: Compression of bed of rapeseeds: the oil-point. Journal of Agricultural Engineering Research 42, 77–84 (1989)

Davion, E., Meiering, A.G., Middendof, F.J.: A theoretical stress model of rapeseed. Canadian Agricultural Engineering 21(1), 45–46 (1979)

Tianzhen, H., Jiyuan, L., Jianyou, W., et al.: Experimental evidence of a plant meridian system: III. The sound characteristics of phylodendron (Alocasia) and effects of acupuncture on those properties. American Journal of Chinese Medicine 22, 205–214 (1994)

Tianzhen, H., Baoming, L., Guanghui, T., et al.: Application of acoustic frequency technology to protected vegetable production. Transactions of the Chinese Society of Agricultural Engineering 25(2), 156–159 (2009) (in Chinese)

Creath, K., Schwartz, G.E.: Measuring effects of music, noise, and healing energy using a seed germination bioassay. The Journal of Alternative and Complementary Medicine 10(1), 113–122 (2004)

Johnson, K.A., Sistrunk, M.L., Polisensky, D.H., et al.: Arabidopsis thaliana response to mechanical stimulation do not require ETR1 or EIN2. Plant Physiol. 116, 643–649 (1998)

Tao, L., Yuexia, H., Guoyou, C., et al.: Analysis of the effect of strong sound wave on plant cells cycles using flow cytometry. Acta Biophysica Sinica 17(1), 195–198 (2001) (in Chinese)

Yiyao, L., Bochu, W., Hucheng, Z., et al.: The biological effects of plant caused by environmental stress stimulation. Letters in Biotechnology 11(3), 219–222 (2000) (in Chinese)

Jiyuan, L., Tianzhen, H.: Principle and design of a laser Doppler Vibrometer for measuring acoustical characteristics of plants. Bulletin of Science and Technology 11(5), 266–267 (1995) (in Chinese)

Jeong, M.J., Shim, C.K., Lee, J.O., et al.: Plant genes responses to frequency-specific sound signals. Mol. Breeding 21, 217–226 (2008)

Weihua, W.: Plant physiology, vol. 4, pp. 64–65. Science and Technology Public, Beijing (2003) (in Chinese)

Xiaocheng, Y., Bochu, W., Chuanren, D., et al.: Effects of sound stimulation on ATP content of Actinidia chinensis callus. Progress in Biotechnology 23(5), 95–97 (2003) (in Chinese)

Xiaocheng, Y., Jianping, D., Bochu, W.: Effects of different sound frequency on roots development of Actinidia chinensis plantlet. Journal of Chongqing University (Natural Science Edition) 30(11), 72–74 (2007) (in Chinese)

Seeding Element Polarity Arrangement on Drum-Type Magnetic Precision Seeder

Qirui Wang[*], Jianping Hu, Qi Liu, and Junchao Yan

Key Laboratory of Modern Agricultural Equipment and Technology,
Ministry of Education & Jiangsu Province, Jiangsu University, Zhenjiang,212013, China,
Tel.: +86-15862938529
wangqirui2@126.com

Abstract. In order to improve Multi-polar magnetic structure's efficiency. Ansoft Maxwell of electromagnetic field finite element software is utilized to study the effect of different polarity arrangement of magnetic seeding element on the magnetic induction intensity of seeding element end and magnetic field distribution in the seeding air gap. As a result of the different electromagnetic coil current direction, magnetic polarity side is different, space structure of magnetic field distribution is also different. The suction metering device of the scope and efficiency is decided by the magnetic field distribution. According to the results of finite element analysis of magnetic field, research on magnetic drum seed-metering device with the design of the program components, and enhance the precision seed metering device performance. Studies have shown that the magnetic induction intensity of seeder in the same polarity arrangement such as NN or SS is lower than in the arrangement of cross with NS, and magnetic field is uniform along the circumference and non- uniform along the axial in the both the arrangement.

Keywords: Magnetic seeding element, Polarity arrangement, Magnetic induction analysis.

1 Introduction

Precision sowing that is a modern seed technology, is general one seed to one plug hole. At present, at home and abroad, advanced precision seeder mainly use airsuction and mechanical way. Such as Shuttle international companies of Holland's needle precision seeder. Features of air-suction precision sowing is that seed size not strict, higher planting accuracy, but the structural complexity, manufacturing costs are high. Mechanical structure is simple precision seeding, planting speed, but seed size is strict. For shortcomings of precision seeder, in this paper, the new principle of magnetic-type seed is raised.

Magnetic system is consisted of many magnetic seeding elements which are arranged by the form of 4 rows uniform along the circumference and the axial layout by way of distance of Plug hole. As shown in Figure 1, The principles of its absorbing

---

[*] Corresponding author.

D. Li and C. Zhao (Eds.): CCTA 2009, IFIP AICT 317, pp. 455–460, 2010.
© IFIP International Federation for Information Processing 2010

seeds is that the seeds coated magnetic material comes with the magnetic seeding element to do circular motion, when the magnetic components are passing the seed-box. For different particle size of the seeds, just change the solenoid current, precision seed can be realized. As a result of the direction of current is difference, the polarity of the seeding elements is different. Space composed of magnetic field distribution (K J Binns,B Sc and so on, 1997) is different and space distribution of magnetic field directly determines the metering device performance. In the following study the effects of the magnetic field of space with the different magnetic order (YANG and CHEN, 2009).

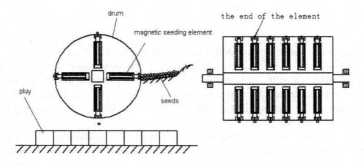

Fig. 1. Schematic diagram of Precision Seeder

2   The Study of Structure of the Magnetic Polarity

Effect of spatial distribution of magnetic field of Polarity is manly about magnetic circuit which is generated between magnetic components N, S pole. There are two main aspects, which the components of seed-polar components effect the space magnetic field. Two aspects are along the circumferential direction and the radial direction. Following we through the finite element analysis software (Pearson J, Squire P T, Maylin M G, et al. 2000),to study the magnetic field with different polarity and the spatial distribution.

2.1   Effect of Polarity Order of the Magnetic Seeding Elements with Circumferential Direction

There are two means that is NN magnetic order and NS magnetic order, about distribution of the direction of the polar circle (Dawson F P, Hideo OKA,1991). As shown in Figure 2, the following we analysis of different polarity on the case with the magnetic circuit, taking the condition of the 400 coil turns and current, the distribution of the magnetic line of force as shown in Figure 2.

   As figure 2 the polarity in different circumstances stations, magnetic circuit has changed dramatically, when we used NN magnetic order, most of the magnetic line of force through composed of the covers from the core and the magnetic circuit. A small number of magnetic lines pass through magnetic circuit between two ends of the core. When used NS magnetic order, most of the magnetic line of force through composed

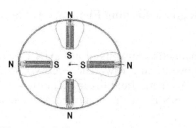
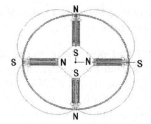

(a) NN same polarity arrangement   (b) NS different polarity arrangement

Fig. 2. Distributing of magnetic force line under two magnetic polar order

of the covers from the core and the magnetic circuit, The magnetic line of force losing in space along in the circumferential direction of seed magnetic circuit consisting of components. It can be seen the magnetic seeding elements along the circumferential direction have mutual effect. When used different magnetic orders magnetic seeding elements are interaction, and the space corresponding magnetic field changes.

Through the distribution of magnetic line of force with different polarity, magnetic interact between the components. The effect of specific size is the change in magnetic induction. Magnetic induction distribution along the circumference show in figure 3 and 4. Abscissa for the circle's circumference, unite is mm. Longitudinal coordinates for the magnetic induction intensity, unite is T. There are four seeding magnetic elements along circumferential direction .In figure 3, 4 raised at the two ends of a single convex, there are four convex. The convex position is position of seeding magnetic element's end face. As the figure 3 and 4, magnetic induction intensity is relatively uniform, when there four seeding magnetic elements along circumference. When used NN magnetic order, the face value of seeding magnetic elements is 0.058T. When used NS magnetic order, the face value of seeding magnetic elements is 0.060T. the face value of seeding magnetic elements change 2mT. Along the circumferential direction the polarity order of the magnetic seed components have a certain interaction. Therefore we used NS magnetic order along the circumferential direction.

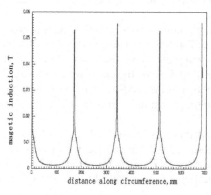
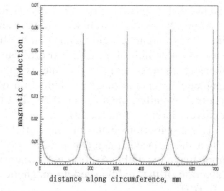

Fig. 3. The curve of magnetic induction intensity in the order of NN along the circumferential direction

Fig. 4. The curve of magnetic induction intensity in the order of NS along the circumferential direction

2.1   Effect of Polarity Order of the Magnetic Seeding Elements Along Axial Direction

The effect of polarity of the seeding magnetic elements along axial direction is the effect between the seeding magnetic elements in a row. There are six seeding magnetic elements in arrow. Now study the magnetic field diversification with different-polarity. There are also two means that is NN magnetic order and NS magnetic order. As shown in Figure 5,6.

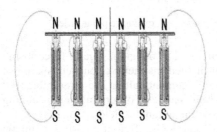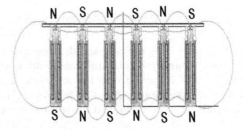

Fig. 5. NN order of space distribution of magnetic field lines of magnetic force

Fig. 6. NS order of space distribution of magnetic field lines of magnetic force

The following we analysis of different polarity on the case with the magnetic circuit, taking the condition of the 400 coil turns and current, the distribution of the magnetic line of force as shown in Figure 5 and 6. When we used NN magnetic order, as result of the air magnetic resistance, when adjacent seeding magnetic elements arrange with same polarity .Magnetic lines of force pass through the core and magnetic shield and form magnetic circuit. There is no magnetic circuit (POWELL J D , ZIEL INSKI A E, 1999) between the magnetic elements. When we used NS magnetic order, most of the space magnetic field lines are passing through the magnetic circuit between the core and the magnetic shield. As a result of adjacent magnetic polarity opposite, some losing magnetic force lines in the air through the magnetic circuit which is constituted by adjacent magnetic polarity. Thus to know: the effect between magnetic elements which in a row, is obvious.

Along the cylinder wall seeding magnetic elements of end-point draw a horizontal line, study magnetic field changes in the line. The length of roller is 240 millimeters. Abscissa is length of a roller and ordinate is magnetic induction. Magnetic induction intensity changes along a the line as figure 7 and 8.As figure 7 When we use NN magnetic order, the magnetic induction of the middle seeding magnetic elements is relatively smaller, the size is 0.02T . But the magnetic induction intensity on sides is relatively larger, about 0.029T.

The distribution of the magnetic field distribute as strong - weak - strong form. When the adjacent magnetic polarity is the same, they have effect of weaken each other magnetic field. As figure 8, when we used NS magnetic order, the magnetic induction of the middle seeding magnetic elements is relatively bigger, the most size is 0.070T.

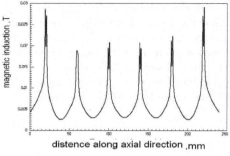

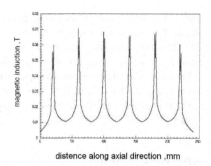

Fig. 7. The curve of magnetic induction intensity in the order of NN along the axial direction

Fig. 8. the curve of magnetic induction intensity in the order of NS along the axial direction

Table 1. Magnetic induction intensity of the seeding element end (T)

Polar Arrangement Type		The number of magnetic seeding element					
		1	2	3	4	5	6
NN	Measured(Bi)	2.89E-2	1.92E-2	2.21E-2	2.28E-2	2.42E-2	2.98E-2
	Bi-B min	0.97E-2	0	0.29E-2	0.36E-2	0.50E-2	1.06E-2
NS	Measured(Bi)	6.02E-2	7.37E-2	7.22E-2	6.98E-2	7.20E-2	6.10E-2
	Bi-B min	0	1.35E-2	1.20E-2	0.96E-2	1.18E-2	0.08E-2
	Bi(NS)- Bi(NN)	3.13E-2	5.45E-2	5.01E-2	4.70E-2	4.78E-2	3.12E-2

When the magnetic induction intensity on sides is relatively smaller, about 0.055T. The distribution of the magnetic field distribute as weak-strong - weak form. When the adjacent magnetic polarity is opposite, they have effect of weaken each other magnetic field. They have effect of enhance each other's magnetic field. As table 1 the magnetic induction intensity of six elements of difference polarity arrangement, we can know the most size of induction of seeding magnetic elements from 0.0298T to 0.0737T, Increase 147.32%. The least size of induction of seeding magnetic elements is from 0.0192T to 0.0602T, Increase 213.54%. The changing of magnetic induction of the end face of seeding magnetic elements is obvious.

3 Conclusions

(1) Adopting different polarity order along the circumferential direction, the magnetic induction intensity of the seeding elements end did not change significantly.
(2) Adopting the NN same magnetic order along the axial direction, the distribution of the magnetic field distribute as strong - weak - strong form; when we use NS magnetic order, the distribution of the magnetic field distribute as weak – strong- weak form.

(3) Adopting the same or different polarity order along the axial direction, the magnetic induction intensity of the seeding elements end had great difference. When the magnetic seeding elements were arranged across with the NS polarity order along the axial direction, its magnetic induction intensity at the end of the seeding elements was bigger than arrangement with the NN same polarity order.

(4) According to the research results, we suggest adopting a sequence of magnetic NS and increasing Coil number of magnetic seeding elements of two side of the drum.

Acknowledgements

This work was financially supported by the open fund of Jiangsu Provincial Key Laboratory of Modern Agricultural Equipment and Technology, Jiangsu University (NZ200604), by Eleventh Five-Year-Plan National Scientific & Technological Supporting Project (2006BAD11A10) and Nature Science Fund of Jiangsu province (BK2007088).

References

Binns, K.J., et al.: Compution treedimensional finiteelement solution of permanent magnet machines. IEEE Trans. on MAG17 (6), 2997–2999 (1997)

Yang, M.-z., Chen, H.: On Electromagnetic Fields Flux Density. Journal of Shanxi Institute of Education. IEEE. Trans. Magnetics 33(1), 3255–3258 (2009)

Pearson, J., Squire, P.T., Maylin, M.G., et al.: Biaxial stress effects on the magnetic properties of pure iron. IEEE. Trans. Magnetics 36(1), 3251–3253 (2000)

Dawson, F.P., Hideo, O.: Flux Control Model of A Ferrite Orthogonal Core. IEEE Tran. Magnetics 27(6), 5259–5261 (1991)

Powell, J.D., Ziel Inski, A.E.: Observation and si2 mulation of solid2armature railgun performance. IEEE Trans. on Magn. 35(1), 84289 (1999)

Powell, J.D., Ziel Inski, A.E.: Current and heat transport in the solid2armature railgun. IEEE Trans. on Magn. 31(1), 6452650 (1995)

Problems and Countermeasures on the Development of Precision Agriculture in Heilongjiang Province

Jinbo Zhang[1,2,*], Junfa Wang[1], and Caihua Li[1]

[1] College of Mechanical Engineering, Jiamusi University, Jiamusi,
Heilongjiang Province, P.R. China 154007,
Tel.: +86-454-8721580
zhangjinpo9872@sina.com
[2] The Key Laboratory of Terrain-Machine Bionics Engineering (Ministry of Education),
Jilin University, Changchun Jilin Province, P.R. China 130025

Abstract. The present development of precision agriculture is at the lower stage in Heilongjiang province, but it has taken effect in aspects of demonstration and extension. However there are many problems in the technology research and application, such as technology development lags, the key techniques supporting lags, the lack of "3S" technology service for agricultural application with independent intellectual property, the whole poor production conditions caused by the limited resource and the man-land conflict, the lacks of the agricultural funds and the higher level scientific researchers, the application difficulties between technique equipments of the high-tech content and the low cultural level of rural human resources, the inconvenient using of multi-function agricultural equipments for the dispersed land and complex landform and so on.

The above problems have limited the development of precision agriculture seriously in Heilongjiang province. Therefore, we should take the effective ways as soon as possible, depend on the existing experiences, techniques and equipments, and put forward the development strategy in accordance with regional condition to speed up the development of precision agriculture. There are many concrete countermeasures such as using demonstration experiences and production research results at home and abroad, introducing technique idea and partial equipments technique to promote independent innovation, proceeding demonstration experiments in the area with suitable conditions to spread of the idea of "precision agriculture", supporting the sustainable development of agriculture by new techniques and new equipments, enhancing communication and cooperation between department and subject, emphasizing the research of application basis, increasing th`e related course of precision agriculture such as GPS, GIS, RS and sensor technique and so on in colleges and universities, especially in agricultural universities, enhancing the development and application of computer and internet in the agricultural field, and so on.

Keywords: precision agriculture, development countermeasures, demonstration experiments.

---

* Corresponding author.

D. Li and C. Zhao (Eds.): CCTA 2009, IFIP AICT 317, pp. 461–465, 2010.
© IFIP International Federation for Information Processing 2010

1 Introduction

Precision agriculture is an agricultural development strategy. It is an idea on agriculture sustainable development to introduce high technology in the agriculture such as information, artificial intelligence and so on, which can utilize resources fully, decrease unnecessary capital input, reduce environmental pollution and obtain most economic benefit, social benefit and environment benefit. In the past few years, precision agriculture had got further research and practice, and formed some concrete production model, which is worth learning in our agriculture production. But because of imbalances in our agriculture production, implementation of precision agriculture under all kinds of natural conditions has great differences. Therefore according to the development level in our agriculture, it is necessary to take the suitable measure to develop precision agriculture based on the actual situation in every area.

At present implementation of precision agriculture is precision farming in fact, which is an intensive cultivation technology introduced the high technology of modern information and agronomy-agriculture engineering technology in the agriculture to get high yield, high efficiency, high quality and low consumption. That is to say, according to the previously determined index, the input of corps, operation and decision are controlled precisely, which can realize low input and high production. The basic characteristics of precision agriculture is a comprehensive system engineering that involves the accurate seed engineering, the accurate fertilization technology, the accurate seeding technology, the accurate irrigation technology, the dynamic crop monitoring technology and so on. Therefore to develop precision agriculture will promote the agricultural and rural sustainable development, even the whole human society.

2 Advantages of Developing Precision Agriculture in Heilongjiang Province

Heilongjiang province is an important commodity grain and soybean bases of china with its rich resources. Agricultural infrastructure construction has get the bigger achievement in Heilongjiang province, which has great potential market and wide development prospects. The advantages of developing precision agriculture in Heilongjiang province are, 1. regional superiority. Heilongjiang province has a flat and wide terrain with a large area of plain hinterland such as Sanjiang plain, Songnen plain and so on, which is convenient to form scale management and to work with the large modern farm implements. 2. talent superiority. Labor resources are abundant in Heilongjiang province. Research strength and the proportion of scientific and technical personnel have come out top in china. It is easy to get advanced talents. 3. technical superiority. Scientific and technical personnel have been begun to explore precision agriculture technology appropriately in Heilongjiang province based on introducing advanced technology and farm implements at home and abroad. Technology department of many university and local farms have researched deeply on precision drilling technology, variable fertilization technology, accurate irrigation technology and measuring-soil and formula technology, and also has get a lot of research results. 4. hardware superiority. Many farms such as Jiangsanjiang farm, Youyi farm and so on in Heilongjiang province, had been introduced advanced cultivation machinery and equipments, which had laid a hardware foundation and provided a supporting platform for extending and researching on precision agriculture. The above cases are the advantages of developing precision agriculture in Heilongjiang province.

3 Exiting Problems of Developing Precision Agriculture In Heilongjiang Province

The development of most agriculture still depends on natural conditions, the scale of famers is small and dispersed, and the area of per capita cultivated land is small in Heilongjiang province. Experiences on technology, production, and management are lack. The existing problems of developing precision agriculture are that, 1. Agricultural mechanization is at lower level. Agricultural infrastructure construction is relatively weak, and the degree of agricultural mechanization can't adapt to the development demand. Except for the large and middle farms, other farms have a low degree of agricultural mechanization, which can't meet the development of precision agriculture. 2. The development of agricultural informatization is slower. Infrastructure construction of information technology is relatively poor and need to strengthen. 3. Agricultural talents and famers' quality still need to improve. At present, the staff who is engaged in developing precision agriculture know only business not technology or know only technology not business. The comprehensive staff who knew technology and business are less. Farmer's education level is generally lower. Most of farmers don't know what precision agriculture is and how to develop it. They have not recognized the necessity and urgency of developing precision agriculture, and can't accept the previous high input. 4. Capital is lack. Precision agriculture is the capital intensive agriculture. In recent years, the main office of state farms and local governments at all levels increased the agricultural input, but the capital supporting for developing precision agriculture is far from enough. Specifically speaking, agricultural input is less, technology input is low, external capital input is small, the income per capita is low, and the capital for developing precision agriculture is lack seriously. 5. Technology and equipment of the independent development and independent innovation are less. Up to now, the new technology and new equipment of the independent development and independent innovation are less in Heilongjiang province, even the whole china. Some advanced tillage machinery and equipments are introduced from foreign countries, and the software for agricultural information and data processing is absent. The above questions are the "bottleneck" of developing precision agriculture in Heilongjiang province. It is a key to resolve the above questions for developing and extending precision agriculture.

4 Countermeasures of Developing Precision Agriculture in Heilongjiang Province

According to the advantages and existing problems of developing precision agriculture in Heilongjiang province and actual conditions, countermeasures of developing precision agriculture with Heilongjiang characteristics are as follow: 1. to improve emphatically mechanization level of the region suitable for developing precision agriculture during the follow years. The support dynamics of capital and technology should strength. To encourage farmers updates the existing small-sized or high-energy agricultural machinery and equipments. Country may provide favorable policy to the conditional area and increase the subsidy of buying advanced farm implements, which will lay a foundation for later developing and extending precision agriculture. 2. to strengthen the infrastructure construction of agricultural informatization. At present, most of farmers in Heilongjiang province learn of information by looking others, by television, by broadcast and by newspaper. Few people know the agricultural information by related website. From the point of view of modern agriculture, it is necessary to establish agricultural information network and ensure the information channel. There are many agricultural information websites like Heilongjiang Information Port, and they had given full play of the important role. Strengthening the infrastructure construction of agricultural informatization is to establish a system of omnidirectional agricultural resources and information network in a large range, to promote informatization of agriculture production, to pay attention to the agricultural basic information and agricultural expert's knowledge acquisition, to provide accurate timely effective information service for agricultural decision department, agricultural enterprises and farmer, to establish some demonstration area of agricultural information, then to promote the process of agricultural informatization and information technology industrialization, to track the development trend of agricultural informatization at home and abroad, to make research and experiment further the international and domestic advanced technique, to fend to line with international practice on technology research of agricultural informatization and application ability, and to set up a high-quality and multi-level talent team in a short time. 3. to emphasize the cultivation of talents and improve farmer's quality. Precision agriculture needs some agricultural production managers with higher technology culture quality, with the ability of skilled operating computer and intelligent mechanical equipment, and with abundant agricultural technology knowledge. On the basis of actual situation in Heilongjiang province, scientific and technical personnel should have the professional technique training. At the same time, it is necessary and essential to attract the studying abroad personnel and related technology talents, even to choose some good technology personnel to study the most advanced technology of precision agriculture at abroad. Farmer's agriculture knowledge should improve synthetically, especially should enhance the understanding of precision agriculture and make them recognize the necessity and urgency of developing precision agriculture. 4. to establish some demonstration area of precision agriculture. Extending precision agriculture massively needs plenty of capital supporting. Therefore, in view of the actual situation, demonstration area of precision agriculture should be established from good rural area or farm, and research and application of agricultural high technology should be enhanced. Government and related departments should support for capital. At the same time, demonstration area also may absorb the capital at home and abroad with applicable ways such as shareholding and cooperation. The scientific and technical personnel's treatment should increase. As a role of template and demonstration, demonstration area can drive other area so as to achieve the goal of extending and popularizing of advanced technology. 5. to research the technological equipments with self-owned intellectual property right. At the same time, the advanced technology equipments at home and abroad would be introduced.

In order to speed up the research process, the advanced foreign technology and equipments can be referenced, the independent research and innovation are carried out after mastering the key technology according to own practical situation. The achievement transformation must be accelerated and applied it to actual production as soon as possible. The operating mechanism of spreading and popularizing must be perfected, the scientific research achievements of precision agriculture should be extended propagandized with different forms, outstanding the important technology, mastering the key links, and carrying out the technological measures.

5 Conclusion

Apparatus The base and experience have provided a premise for speeding up the pace of advanced technology to transform traditional agriculture and achieve the development of precision agriculture in our province at this stage. At present, precision agriculture of our province still is at the exploration and research stage and does not have conditions of large scale and range extension. But leading development of precision agriculture will explore road and accumulate experience for development of precision agriculture and it can lay a foundation for the spreading and popularizing of precision agriculture of our country.

Acknowledgements

Author is grateful to professor Junfa,Wang associate professor Chuanhua Yang and lecturer Caihua Li for providing the support and help in this paper.

References

Yinsheng, Y., Xiufeng, A., Hongpeng, G.: The quantitative analysis of the feasibility for developing precision agriculture in jilin province. Journal of Agricultural Mechanization Research 5(2), 63–66 (2002)

Jianneng, C.: New intension of precision agriculture in the 21st century and its development strategy in China. Journal of Fujian Agriculture Universigy (Social Science Edition) 6(1), 8–11 (2003)

Zhigang, X., Shuguang, Z., Yongqiang, Y., Yan, L.: The present situation and tendency of precision agriculture in China. Journal of Agricultural University of HeBei 26(5), 256–259 (2003)

Quanbao, Y., Hua, L., Zhongyang, H., Hongcheng, Z., Qigen, D., Ke, X.: Precision agriculture and its development countermeasure in China. Guizhou Agricultural Sciences 30(4), 58–59 (2002)

Lijiao, Y.: Consideration of several issues on developing precision agriculture in China. System Sciences and Comprehensive Studies In Agriculture 18(4), 273–276 (2002)

Research on Regional Spatial Variability of Soil Moisture Based on GIS

Yongcun Fan[1,*], Changli Zhang[1], Junlong Fang[1], and Lei Tian[2]

[1] Department of Electrical Engineering, Northeast Agricultural University,
Harbin, Heilongjiang Province, P.R. China, 150030,
Tel.: +86-451-55190146; Fax: +86-451-55190238
ycfan@neau.edu.cn
[2] University of Illinois at Urbana Champaign
University of Illinois at Urbana Champaign, USA, 61801

Abstract. As one of soil dynamics properties, soil moisture content is an important factor of soil fertility which counts for much to crop growth situation and scientific irrigation management. A design plan of regional spatial variation of soil moisture measurement was introduced. Its main job includes the use of differential GPS technology for each sampling points in farmland, collecting data of high-precision geo-spatial information and soil moisture in farmland resorting on measure instruments of soil moisture, communicating the data between measuring instrument and portable data analysis devices or computer with cable or wireless network based on ZigBee technology, analyzing data of experimental farmland of the topography and terrain, processing and interpolating data of soil moisture content.

Keywords: regional spatial variability, soil moisture measurement, interpolation algorithm, GIS.

1 Introduction

The soil is not uniform continuum. (Hua Meng et al., 1992) Even in the region of same kind of soil, spatial character of soil moisture is evidently different at the same time. Soil moisture is not only one of the main factors that the crop depends on, but also the important premise that the fertilizer can be made use of effectively by the crop. The data collection of soil moisture is the most important content of research on regional spatial variability of soil moisture. With the help of GPS and measuring instrument of soil moisture, regional spatial soil moisture was measured. A new wireless network based on ZigBee technology is designed for transmitting the data that is collected by sensors in the research. Applying the IDW and Kriging in the spatial analyst of ArcGIS 9.0, we can get the distributing map of soil moisture.(Tang Anguo et al., 2006) It is very important and useful for adjusting precise fertilization and precise irrigation. It also offers the theoretical foundation of precise farming study for enhancing the yield.

---

* Corresponding author.

D. Li and C. Zhao (Eds.): CCTA 2009, IFIP AICT 317, pp. 466–470, 2010.

Here present the scheme of network based on ZigBee for transmitting the measure data. The test data can be transmit to the computer and processed by the Kriging interpolation. The distributing map of soil moisture is given by the Spatial Analyst of ArcGIS 9.0.

2   Theories and Methods of Spatial Variation of Soil Moisture

2.1   Regionalized Variables

If a variable is distributed in space, it is said to be "regionalized". Regionalized variables have two important characters: (1) The regionalized variable is a random function that is local and exceptional; (2) The regionalized variable is a general or structural. The soil moisture and the other variables of farming field information are regionalized variables. So they can be processed by regionalized theory to research their spatial variation.(Qiu Yang et al., 2001)

2.2   Semi-variance Function

The difference of sample space place was not involved in general statistics method which process information of farming field as absolute random variable. So the semi-variance function can be used in description of farming field information that is a method of Geo-statistics for the regionalized variables study.

Semi-variance function is one of the description functions for space variation, which can analyze the correlation of samples. Its expression is underlay as equation (1).

$$\gamma(h) = \frac{1}{2n}\sum_{i=1}^{n}(z(x_i) - z(x_i + h))^2 a \tag{1}$$

Where:

    h =distance of samples; as lag coefficient; s
    n=number of sample couple separated by h;
    z=attribute value.

When research the value of regional variation, a semivariogram should be given. For drawing semivariogram, theory model conform with semi-variance function is a precondition. On requirement of positive definiteness of Kriging equations, there are several common theory models: Spherical model, Exponential model, Gaussian model, Linear Still Value model etc.(Zhang X. F. et al., 1995)

2.3   Drawing Spatial Variability Map on Spatial Distribution

Spatial variability distribution map of soil moisture can be drawn with Kriging interpolation method after choosing the theory model of semi-variance function that can show the regularity of spatial variability. Kriging interpolation method is a kind of optimum unbiased estimate algorithm for unknown regional sample value utilizing the original values and structure of semi-variance function. Kriging interpolation method get the unknown sample value by setting weights to original sample values, which shows as equation (2).

$$Z(x_0) = \sum_{i=1}^{n} \lambda_i Z(x_i) \tag{2}$$

Where:

Z(x0)= unknown sample value;
Z(xi)= sample value around unknown sample value;
λi= weight of sample value i;
n= number of sample values.

In order to fulfill un-bias and optimization, the weight coefficient of Kiriging equations can be given by equation (3). (Li Jun et al., 2006)

$$\begin{cases} \sum_{j=1}^{n} \lambda_j \gamma(x_i, x_j) + \mu = \gamma(x_i, X) \\ \sum_{i=1}^{n} \lambda_i = 1 \end{cases} \tag{3}$$

Where:

$\gamma(x_i, x_j)$ = covariance functions of sample values;

$\gamma(x_i, X)$ = covariance functions between sample values and interpolating points;

μ = Lagrange multiplier.

3   Data Communication Network Design

When the data of soil moisture is collected by sensors, the most important course is data transmission from sensor or memory of measure device to computer. Here present a ZigBee network for data communication. There are wireless sensor node and central coordinator node (sink node) in wireless network. The wireless sensor node is integrated in measurement device of soil moisture which distribute in the farm field grid. The central coordinator node is integrated in PDA or monitor spot in the field. The data can be gathered by the central coordinator node when the soil moisture data is studied.

3.1   Wireless Sensor Node

Wireless sensor node provides two functions: collecting and processing data, communicating with other node. (Qu Lei et al., 2007) So RFD (Reduced-Function Device) is adopted for wireless sensor node whose structure is shown as figure 1.

The wireless sensor node is integrated in SOC with MG2455 chip. MG2455 chip produced by Radio Pluse Company of Korea is the core microprocessor of system, which embody 8051MCU, CC2430 chip of Chipcon Company and ZigBee protocol stack. There are two kind of address of every device in ZigBee network, one is 64 bit physical address of IEEE, the other is unique 16 bit net address of its PAN (Personal

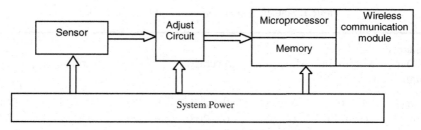

Fig. 1. Components in Wireless Sensor Node

Area Network). The latter address is distributed by parent device which own the certain address segment. The data query is main content of communication of wireless sensor node, whose frequency is low and quantity is small. It's fit for ZigBee application.

3.2 Central Coordinator Node

Central coordinator node (sink node) of ZigBee network is a FFD (Full Function Device), which can gather data from every wireless sensor node and transmit the data to data processing computer. Its structure diagram is shown as figure 2.

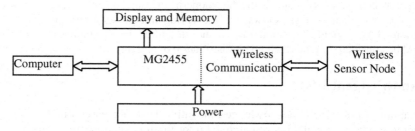

Fig. 2. Structure of Central Coordinator Node

4   Experiment Design and Data Analysis

The distribution of sample point in experimental farm field is regular square grid in order to gather accurate moisture information of soil. (Zhang Shujuan et al., 2004) The distance of sample point grid is 50m and there 20 sample points in the experimental farm field. When the data of sample point is gathered, spatial interpolation is used. This sample scheme is accurate and economical.

The experimental data is gathered once a month in June, July and August. Resorted to traditional statistics method, the eigenvalue of soil moisture data of experimental farm field is presented. The three series data are shown as table 1.

The all three times measure data show that every Coefficient of Variance is more than 100. It means that the variable extent of soil moisture in experimental farm field is strong. All data of soil moisture is checked by ArcGIS normal distribution test before the Kiring interpolation is applied. The result shows that most of data is accord

Table 1. Statistical Description of Soil Moisture Data

Sequence of Data Gather	Minimum	Maximum	Average	Standard Deviation	Variance	Coefficient of Variance (%)
June	0. 09	0. 357	0. 224	0. 267	0. 071	119.2
July	0. 086	0. 33	0. 208	0. 244	0. 06	117.3
August	0. 099	0. 45	0. 275	0. 351	0. 123	127.6

with normal distribution. Kiring interpolation's result shows the spatial distribution of soil moisture in the experimental farm field. The north region's soil moisture is less than the south region's. It is similar to topography of the experimental farm field.

5 Conclusions

The character of soil moisture accord with normal distribution and the spatial distribution of soil moisture accord with actual situation of topography. A ZigBee wireless sensor network is a quite good solution for communication of soil moisture measurement system, which can associate with the modern measure method and interpolation algorithm. The data analysis shows that the Kiring interpolation can provide accurate situation of farm field. The randomicity and structure of soil character's distribution should be considered in the research on distribution of soil moisture.

Acknowledgements

The authors would like to acknowledge Innovative Team Research Fund of Northeast Agriculture University and my research team for funding and research support.

References

Meng, H., Jian, W.: Soil Physics. China Agricultural Press, Beijing (1992)

Anguo, T., Xi, Y.: ArcGIS geographical system spatial analyze experiment tutorial. Science Press, Beijing (2006)

Yang, Q., Bojie, F., Jun, W., et al.: Spatial variability of soil moisture content and its relation to environmental indices in a semi-arid gully catchment of the Loess Plateau, China. Journal of Arid Environments 49(8), 723–750 (2001)

Zhang, X.F., Van, J.C., Eijkeren, H., et al.: On the weighted least-squares method for fitting a semivariogram model. Computers and Geosciences 21(4), 605–608 (1995)

Jun, L., Songcai, Y., Jingfeng, H.: Spatial interpolation method and spatial distribution characteristics of monthly mean temperature in China during 1961-2000. Ecology and Environment 15(1), 109–114 (2006)

Lei, Q., Shengde, L., Xianbin, H.: ZigBee technology and application. BeiHang University Press, Beijing (2007)

Shujuan, Z., Hui, F., Yong, H.: Sampling strategies of field information on precision agriculture. Transactions of The Chinese Society of Agricultural Machine 35(4), 88–92 (2004)

A Study on Using AD5933 to Realize Soft-Sensing Measurement of Conductance and Capacitance

Chuanjin Cui, Haiyun Wu, and Yueming Zuo*

College of Engineering and Technology, Shanxi Agricultural University,
Taigu, Shanxi Province, P.R. China 030801,
Tel.: +86-0354-6288400-8305
Zyueming88@Yahoo.com.cn

Abstract. A conductance and capacitance test system was designed in this paper. AD5933 a new chip for impedance testing was used in it. By designing the test system, and testing the impedance of a model, which is a resistor in parallel with a capacitor, under different frequency (range from1000Hz to 100kHz). The information of impedance, real part, and imaginary part of the model under the frequency points were get. Based on this, the internal relations between conductance, capacitance, impedance, real part and imaginary part were carefully analyzed, using the soft-sensing method established the capacitance support vector machine (SVM) regression model on MATLAB. The regression model is the relation between resistance, impedance and capacitance. The test results showed that the test relative error is small in the capacitance range (0.1 pf~33 pf) and the conductance range (9.05*10-3s~3.8*10-5s), the biggest, smallest and average test error of capacitance are -2.9%, 0.2%, 1.27% respectively. The study showed that using AD5933 and the soft-sensing method to realize the multi-parameter test is feasible.

Keywords: AD5933, conductance, capacitance, soft-sensing measurement, support vector machine.

1 Introduction

Conductance and capacitance are very important electrical parameters, the wide use of them in the agricultural materials nature test will promote the development of agricultural modernization and automation, so the research of their test methods has an important practical significance. However, in the actual measurement, resistance and capacitance are usually coupled together, Expressed as a parallel system of resistance and capacitance, because capacitance measurement is always affected by temperature, parasitic capacitance, dielectric medium leakage resistance, so it is difficult to measure it accurately. There are four traditional methods of capacitance measurement just as follows:

Resonance method. The shortcomings of this approach are the distributed capacitance of wire and coil has a big effect on the measurement results. In addition,

---

* Corresponding author.

D. Li and C. Zhao (Eds.): CCTA 2009, IFIP AICT 317, pp. 471–478, 2010.
© IFIP International Federation for Information Processing 2010

the method can not realize automatic measurement, and can not apply to online measurement. (Xiangjun Z et al., 2001).

Oscillometry method. This method was divided into two types, RC Oscillation and LC Oscillation. The former ability in anti-parasitic capacitance is poor, the stability of oscillation frequency is poor too, the sensitivity of small capacitance changes is low. While the latter test range is from a few hundred kHz to several hundred MHz, suitable for high-loss materials capacitance measurement. (Zeljko I et al., 2005).

Alternating current bridge method. This method will have a larger non-linear when test in a point far away from the equilibrium position, the output impedance is high, the output voltage is very small, so it is difficult to measure. In addition, this method needs manual adjustment, the operation is too complex. (Wanguo Liu et al., 2005; Inglis A D et al., 2003).

Charge-discharge method. In this method of measurement the parasitic capacitance has a serious effect on the results. (Sell B et al., 2002).

These commonly used capacitance measurement methods have some week points as mentioned above. Therefore, it is necessary to study new ways to achieve accurate capacitance measurements in the case of RC coupling.

This paper aims to solve the capacitance measurement in the case of RC coupling, we used AD5933, a new chip for impedance testing, produced by the American company Analog Devices in the year 2005. By designing test circuit, using anti-interference measures, such as shielding and grounding to improve the accuracy and stability of the test system.

At the same time, soft-sensing technique was also applied in the capacitance measurement. Soft-sensing technique, also known as soft instrumentation technology (MCAVOY TJ et al., 1992). That is to measure the variables which are easy to measure and relevant to the measured which are hard to measure directly, and based on the mathematical relationships between easily measured variables and the measured to complete the inference and estimation of the measured. In this paper soft-sensing technique was used, the support vector machine (SVM) regression mathematical model of resistance, impedance and capacitance was established on MATLAB platform, the capacitance measurement was realized. (Huichun Liu et al., 2002).

2   The Design of Hardware Circuit

2.1   The Measurement Principle

AD5933 has a voltage output pins Vout Fig.1. It issued a certain frequency sinusoidal scanning signals to external impedance z (ω) for incentives. After going through the sample the signal was amplified, filtered, sampled by the ADC, then using discrete Fourier transform to process the data, and calculate the impedance. The sinusoidal scanning signals were generated by the DDS (direct digital synthesizer) a internal part of chip AD5933, and has a resolution of less than 1Hz. Clock frequency provided for the DDS can be generated by external passive oscillator or inner active oscillator, Fig.1. Signals synthesized by DDS were processed by digital-to-analog conversion and amplification, then the signals can be used as the needed scanning and incentives signals.

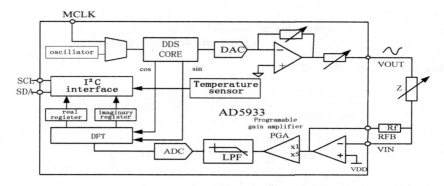

Fig. 1. Function structure diagram

After going through the test material, then the scanning incentive signals were amplified, filtered, A/D converted to transform to digital signals. The output digital signals from A/D converter were processed by Discrete Fourier Transform (DFT), the formula is as follows:

$$X(f) = \sum_{n=0}^{1023}(x(n)(\cos(n) - j\sin(n)))\qquad(1)$$

Where f is the frequency of scanning points, x(n) is the output of A/D converter, cos(n) and sin(n) were generated by DDS at the frequency f. The calculation results of X(f) is generally a complex number, with a real part R and imaginary part I. These are the easy measured variables. The measured impedance module Magnitude and the phase angle Phase can be calculated by the following formulas:

$$Magnitude = \sqrt{R^2 + I^2}\qquad(2)$$

$$phase = Tan^{-1}\left(\frac{I}{R}\right)\qquad(3)$$

So we can get the following test methods. First measure a known impedance, get its R and I, and calculate the Magnitude, then to obtain a gain factor using the following formula. This is the so-called calibration process.

$$gain \cdot factor = \frac{\left(\dfrac{1}{impedance}\right)}{Magnitude}\qquad(4)$$

After getting the gain factor, it becomes bridge of solving the impedance test. It is clear that the mathematical formula should be (5). By measuring the unknown impedance we get the R and I, then get them into formula (2) and (5), we can calculate the unknown impedance.

$$impedance = \frac{1}{(gain \cdot factor \times Magnitude)} \tag{5}$$

Then according to （6） （7）

$$\frac{1}{impedance} = admittance = G + jBc = \frac{1}{R} + j\omega c \tag{6}$$

$$\tan^{-1}\left(\frac{Bc}{G}\right) = -phase = -Tan^{-1}\left(\frac{I}{R}\right) \tag{7}$$

So the conductance G can be calculated out.

2.2 Impedance Measurement Circuit

Fig. 2 is the circuit diagram of impedance test using AD5933. In the figure the parallel circuit composed of Rx and Cp is the model need to be measured, the test system was controlled by PC. ADP3303 provide a stable supply voltage of 3.3v to AD5933. There are two ways to provide a clock frequency for the AD5933. one is as shown in Fig.2 the external passive oscillator clock will provide a reliable external clock frequency 16MHZ for the system, the other is as shown in Fig.1 the inner active oscillator provide a reliable frequency 16.776MHZ for the system. By choosing the item in the control interface Fig.4, we select the internal clock frequency. The measured data of impedance Z, R and I were downloaded to the PC for preservation through the USB interface.

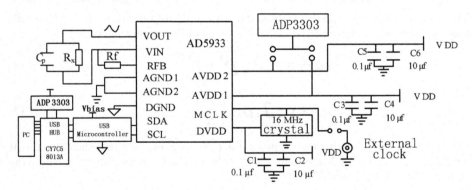

Fig. 2. Test circuit diagram

3 Measurement of Conductance and Capacitance

3.1 Conductance Measurement

In the test, we use paprallel structure of different resistors and capacitors as test objects. Tested 210 different combinations composed by 7 different resistors and 30 different capacitors.

In order to get a greater change in impedance by making a small change in the capacitance, that is to use the easily measured variables to accurately reflect the measured capacitance value, so in the design of parallel coupling model, first we did the theoretical calculations on the MATLAB software and found that when Rx is around 500KΩ and Cp around 30pf will meet the requirements. Given laboratory conditions, we designed a parallel model as shown in Fig.3 (Rx in parallel with Cp), where Rb is 512.2K, C1 is 11.1pf, C2 is 20.6pf, while the variable resistor r and the variable capacitor c is the value we want to measure, all the metal film resistors have an accuracy of 1%, and all the ceramic capacitors have an accuracy of 10%, their values were   measured by a resistance and capacitance meter UNI-T ® UT601 (accuracy: R ± (0.8% +1), capacitance ± (1% +5))produced by You li de company for comparison.

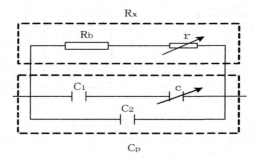

Fig. 3. Parallel coupling model of resistor and capacitor

Access the test system to a computer through USB, and open the control interface, estimating the value of the impedance to be tested, program frequency sweep parameters into relevant registers, and choose appropriate impedance Z and the feedback resistor Rf Fig.2 to calibrate the system, and then got the gain factor. make sure that the ratio of Rf/Z maintained between 0.2 and 0.066, this is because the highest voltage of the internal A/D converter is 2V calculated from the following Formula Fig.1. In order to let the A/D converter work in linear area, it should be no more than 2V, so the $Vout \times \dfrac{Rf}{Z} \times PGA$ ratio of Rf/Z should be maintained between 0.2 and 0.066. In this test we adopted one impedance test range(150 kΩ~200 kΩ), the calibration value r is 2.403 kΩ and c is 7.9 pf Fig.3, Rf is 18.37 kΩ.The rest parameters setting by control interface Fig.4 are as follows: start frequency is 30000Hz, delta frequency is 5Hz, number of increments is 300, number of setting time cycle is 15. PGA value is 5, scanning voltage is 2vp-p, and 16776000Hz internal clock frequency.

We tested 210 different combinations which composed by 7 different resistors r and 30 different capacitors c. After getting the data of impedance, real and imaginary parts under the frequency point 30kHz, the conductance can be calculated by formula (6) and (7), then made a comparison with the value measured by UT601 in Table 1.

Table 1. Conductance value

UT601 measured value (mS)	9.05	1.623	0.664	0.417	0.194	0.100
measured value (mS)	8.834	1.647	0.653	0.431	0.193	0.101
Relative error (%)	2.39%	1.66%	1.66%	-3.36%	0. 52%	-1%

In addition, this experiment was carried out in the environmental temperature 20 ℃, so there is no need to make temperature compensation.

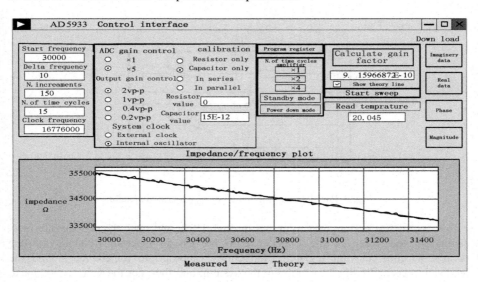

Fig. 4. Control interface

3.2 Capacitance Measurement

A svm loop program was established on the matlab plantform, that is the two elements regression model, Using 140 sets of data (resistance r, impedance Z and capacitance c) from the 210 sets of data to training the two elements support vector

Table 2. Capacitance value

UT601 measured value(pf)	0.7	1.1	1.5	3.3	4.9	5.1	6.5	10.8	12.5	15.4	18.1	21.8	25.7	33	41.4
Predicted Values(pf)	0.71	1.09	1.54	3.35	4.94	5.07	6.59	11.03	12.55	15.56	18.48	21.71	25.65	33.96	41.69
Relative error(%)	-1.4	0.9	-2.7	-1.5	-0.8	0.6	-1.4	-2.1	-0.4	-1.0	-2.1	0.4	0.2	-2.9	-0.7

machine (SVM) capacitance regression model, by adjusting the relaxation factor, the optimization model was selected and use the model to predict capacitance in the 70 remaining data sets, the result was good, parts of the data were shown in Table 2.

3.3 Results and Discussion

The measured conductance value was the value calculated from the measured impedance, real part and imaginary part. Seen from Table 2, the biggest relative error of measurement is -3.36%, the smallest is 0.52% and the average relative error is 1.86%, and meets the testing requirements.

For the capacitance measurement, after calculating the conductance, a two elements support vector machine (SVM) capacitance regression model was established by using 140 sets of data (resistance r , impedance Z and capacitance c), and trained the model several times, then use it to predict the capacitance, choose 15 predict capacitance results at random. As shown in Table 2, and can be seen from the table, the biggest relative error is -2.9%, the smallest is 0.2% and the average relative error is 1.27% compared with the conductance measurement the error is obviously low, in theory, the capacitance is calculated on the base of conductance, the error should be increased, but the actually it is reduced, that is because of the support vector machine (SVM) model and its ability in data processing. SVM have a solid theoretical foundation, it can establish a good model by a small sample set, after training the model, it will have a good prediction capability, it's strong robust can reduce random error and the effect caused by structural error, further improved the accuracy and had the advantages which Neural network does not have.

4 Conclusion

As AD5933 integrated the hardware (circuit) and the signal processing, use it to compose the test system can save a lot external circuits as well as software cost, and the test operation is easy, and AD5933 also has a merit of strong function, low prices and able to test online, has a broad application prospects. The measurement system composed with AD5933 can not only the complete the test of impedance, phase angle, but also can complete the measurement of conductance and capacitance. A single chip achieved a multi-parameter measurement. In addition, SVM plays a very important part in improving the measurement accuracy. The combination of AD5933 test systems and SVM regression model successfully applied in conductance and capacitance measurement. This is the foundation for the manufacture of related portable test instruments, at the same time, the test method can be applied to test the agricultural materials conductance and capacitance.

Acknowledgements

This work is supported by a research grant from the National Science Foundation, NSF 30871445.

478   C. Cui, H. Wu, and Y. Zuo

References

Xiangjun, Z., Xianggen, Y., Deshu, C.: A novel technique for measuring rounding capacitance and grounding fault resistance in ineffectively grounded systems. Power Engineering Review 21(3), 65–67 (2001)

Zeljkol, I., Mark, F.: An interface circuit for measuring capacitance changes based upon capacitance-to-dutycycle (CDC) Converter. IEEE Senors Journal 5(3), 403–410 (2005)

Liu, W.-g., Yang, G.-l., Xiao, Q.: Application of Small Capacitance Measuring Techniques in Gravimeter. Journal of Chinese Inertial Technology 13(1), 68–71 (2005) (in Chinese)

Inglis, A.D., Wood, B.M., Cote, M., et al.: Direct determination of capacitances tandards using a quadrature bridge and a pair of quantized hall resistors. IEEE Trans. on Instrumentation and Measurement 52(2), 559–562 (2003)

Sell, B., Avellan, A., Krautschneider, W.H.: Charge-based capacitance measurements (CBCM)on MOS devices. IEEE Trans. on Device and Materials Reliability 2(1), 9–12 (2002)

Mcavoy, T.J.: Contemplative Stance for Chemical Process Control. Automatica 28(2), 441–442 (1992)

Liu, H., Ma, S.: Present situation of SVM. Journal of Image and Graphics 7(6), 618–623 (2002) (in Chinese)

Wireless Sensor Networks Applied on Environmental Monitoring in Fowl Farm

Fangwu Dong[1,*] and Naiqing Zhang[2]

[1] Department of Mechanical and electrical, Zhejiang Textile & Fashion College,
NingBo, Zhejiang Province, P.R. China 315211,
Tel.: +86-0574-86329628
Dongfw01@163.com
[2] Zhejiang Textile & Fashion College, NingBo, Zhejiang Province, P.R. China 315211

Abstract. Aiming at the real time monitoring requirement of poultry farms on the environment, a online monitoring system is proposed for poultry farms on the environment based on ZigBee, its application of ZigBee wireless networks and sensor technology. supply a network structure of monitoring system, monitoring system node controller of data acquisition, data transmission and control node, which is TI's CC2430 based on ZigBee technology. CO_2 sensors use TGS4161, temperature and humidity sensors use SHT75 to detect environmental parameters. designed circuit diagram of parameter testing node and system master control node, CC2430 as a data processing chip. through the analysis of data transmission of system, simplifying the ZigBee protocol stack, designed data transmission protocols and communication formats of the system. given program flow chart of sensors nodes and main node. practical application shows that the performance ratio cable monitoring system is better, Especially in real-time systems and anti-jamming, it so superior on the current forms of environmental monitoring SCM cable system which cost lower than the SCM cable control system about 30%.Successfully achieved the Monitoring of fowlery's CO_2 concentration, temperature, humidity and other environmental parameters for large-scale poultry farming, and to provide a new monitoring environment technologie.

Keywords: ZigBee, CC2430, fowlery environment, Network protocol, Monitoring system.

1 Introduction

With large-scale intensive poultry model increasingly promotion and popularization of farming, in order to improve the yield and quality, prevent poultry diseases, we need an monitoring system for the poultry house temperature, humidity, CO_2 concentration and illumination parameters [1]. At present, the domestic fowlery environmental parameters automatically monitor system, generally use the single-chip or industrial computer systems, monitoring of signal transmission using Cable transmission [2], but this way exists many defects. Such ashighercosts, complicated

---

* Corresponding author.

D. Li and C. Zhao (Eds.): CCTA 2009, IFIP AICT 317, pp. 479–486, 2010.

system, poor anti-interference, limit the universal promotion in the poultry production. ZigBee [3] wireless sensor technology applied an IEEE802.15.4 standard [4] of a short-range wireless sensor network technology, is a combination of sensor and wireless network technology, is a new intelligent control technology, with low-cost, ad hoc network, small size, Strong real-time, low power consumption, anti-interference, good features such as embedded[5], can be widely used in industrial and agricultural production.

2   Monitoring and Control System Design

2.1   Design of Poultry House Parameters Monitoring and Control System

The fowlery parameters of the principle of wireless monitoring system as shown in Fig.1, the system is composed of the sensor node, environmental parameter control node, data processing and the main nodes.pology Network using star-shaped structure, the main node is connected to control PC,set it as FFD (Full Functional Device) nodes or NC (network coordinator);and sensor or control of the implementing agencies are integrated from the node, the node is set as RFD(Reduced Functional Device). Node control chip use TI, (Chipcon) Corporation' chip which is based ZigBee technology with SOC (system-on-chip) features a small footprint wireless system-on-chip CC2430 [6], it can according to consumer needs to be flexible and set to FFD, NC or RFD [7]. Sensor node regularly detect the poultry house environment parameters,when receive the master node's directions,will sent the monitored environmental parameters through wireless transmission to the master node; Host transceiver node after receiving data from the controller(or host) for the corresponding treatment, compared with setting datas and output corresponding action commands, transmitted to the various environmental parameters control nodes, such as temperature control, ventilation control, humidity control, etc. the corresponding control equipment will make the stability of poultry house environment parameters in the set range. It can also input the environmental parameters detected into the computer storage system.

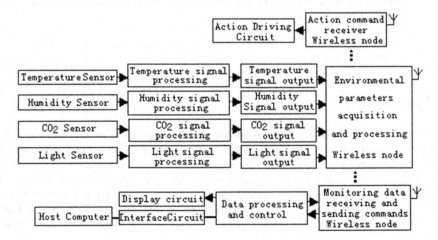

Fig. 1. Block diagram of the fowlery environment monitoring system

2.2 Sensor and Data Acquisition Node Design

Parameters of the fowlery environment monitoring system sensor node circuit as shown in Fig.2. the detection of CO2 concentration is useing sensors TGS4161. The sensor has a small size, long life, selectivity and good stability characteristic, and also with low temperature and humidity-resistant properties, can be widely used in automatic ventilation system for CO2 gas or long-term monitoring applications [8]. CO2 Sensor output weak voltage thourgh the amplification of amplifier U2, and then output to the U5 amplifier P0_3 to A/D conversion. PR1 adjust amplifier gain so that the concentration of the output signal voltage change between 0 ~ 3 V. U5 conducted by A/D conversion, stored in designated memorizer cell CC2430. In order to enable the sensor maintain the temperature in the most sensitive State, the general need to provide heat heater heating voltage.

Temperature and humidity testing apply digital temperature/humidity sensor SHT75 [9]. The sensor has a small size, simple and reliable, low cost, digital output, the avoidance of debugging, and calibration-free interchange ability of strong features, integrated A/D converter and memorizer, in the measurement process can be automated calibration of relative humidity. U4's DATA and SCK pin, respectively connected with the U5's P0_0, P0_1 pin, U5's P0_1 pin controlled the SCK pin of U4, from U4 memorizer to read the temperature or humidity data. And then make temperature/humidity parameters stored the memorizer cell designated in CC2430. Testing the use of light for illumination photodiode detection, Photosensitive components D1's signal Amplified through the U3, the input to U5's P0_2 to A/D conversion.

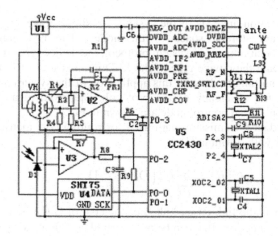

Fig. 2. Basic circuit of the fowlery environment sensor node

2.3 System Control and Transceiver Node Design

System control and wireless transceiver node of the basic circuit shown in fig.3. Mainly are formed of the power supply, controller, RS232 interfacer, display circuit and the wireless transceiver circuit. through the connection of RS232 serial port chip

MAX232 with other devices, U1 provide 3V power for u3, LCD display for TBG128064F type with Chinese font display module. When Receiveing data, the sensor node received send data stored in the transceiver circuit in the FIFO, and then entered into the data storage area designated unit of CC2430 with DMA mode. CPU received data of various sensor node, and then according to a certain algorithm to derive the average, compare with the setting datas, If it exceeds setting range, will send control commands to the appropriate control node. Data transmission with DMA mode to the RF of the FIFO, then sent through a wireless circuit sent.

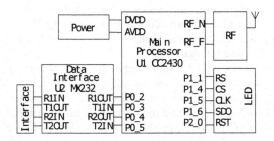

Fig. 3. Basic circuit of system control and transceiver node

3   The Design of System Network Protocol and Software

3.1   The Design of Network Protocol and Data Frames

Through the application analysis of the system, we simplify the ZigBee protocol for saving the program storage space of the nodes. The beaconing and security mechanisms have been omitted in system protocol, the device type of the FFD node is set to FFDNBNS and the device type of the RFD node is set to RFDNBNS[10]. The primitive that has nothing to do with this application of the system node is omitted in order to improve the efficiency of the protocol. The protocol and the realization primitive of the sensor and control node are shown in Fig.4.

In the system, data transmission path used for routing cost metric of comparison, the path cost show with formula (1):

$$C\{P\} = \sum_{i=1}^{L-1} C\{[D_i, D_{i+1}]\} \tag{1}$$

The $C\{[D_i, D_{i+1}]\}$ is the link costs, link L of the cost of C (L) is the value in [0...7] within the scope of functions that can be used formula (2). Formula p^l is transmission probability in the link packet . Through the MAC layer of each frame the LQI average to estimate p value.

$$C\{L\} = \left\{ \min\left(7, round\left(\frac{1}{p_l^4}\right)\right) \right. \tag{2}$$

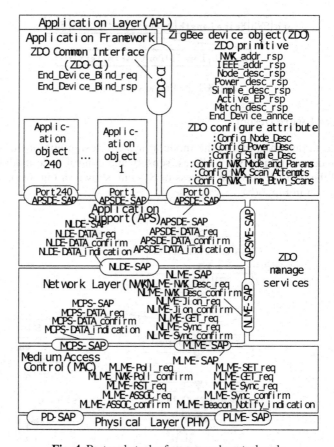

Fig. 4. Protocol stack of sensor and control node

In the data transmission process, routing algorithm used in classification of data frames transmitted along the tree. if a router's address is A, depth is d, when the target device is currently receiving equipment or its sub-equipment, next skip address N is available to the formula(3):

$$N = A + 1 + \left[\frac{D - (A+1)}{Cskip(d)} \right] \times Cskip(d) \tag{3}$$

If the NWK layer of routing node to receive data frame of the MAC layer non-broadcast frame, NWK layer to determine data frame the purpose of this address is for the current device the logic address, If it is, data frame will processing be transmitted to the upper layer of node, otherwise, data frame will be transmitted to other node.

MAC layer of ZigBee data packet format is used in data frame format of system data transmission[11]. The structure is shown in Fig.5. The definition of frame load of data packet is node ID which is binding with sensor nodes port ID plus port parameters.

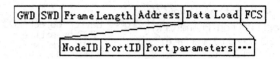

Fig. 5. Data Frame structure

3.2 System Software Design

System Program design applying the modular system design approach, is composed of the host node and transceiver modules, sensors and control node module. the main program and the transceiver node process of System, shown in Fig. 6(a). Sensor node flow chart shown in Fig. 6(b).

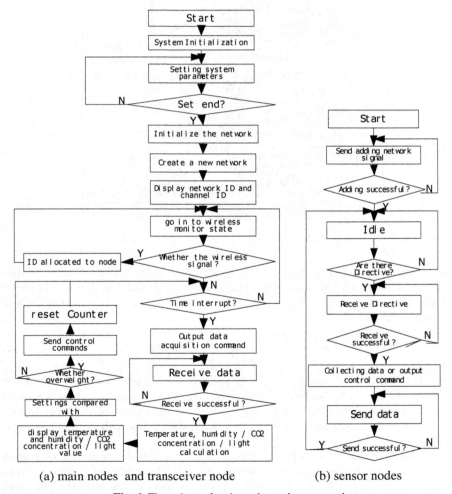

(a) main nodes and transceiver node   (b) sensor nodes

Fig. 6. Flow chart of main nodes and sensor nodes

4   System Performance Test Results and Discussion

After the completion of the design of the system, the application of test. choose two 500m2 fowlery area, placed 10 sensor node, the distance between each other for about 20m, routing node location relatively fixed, the successful implementation of the environmental monitoring birdhouse. Table 1 for the system with a SCM (single-chip microcomputer) cable control system using the comparison test parameters and test results show that this system in single node in data transmission distance less than SCM cable monitoring system, and other parameters are higher than using SCM cable monitoring system.

Table 1. System test parameters and comparison of SCM monitoring system

Test items of the detection	CO_2 concentration measurement	CO_2 concentration measurement accuracy	Temperature measurement accuracy	Temperature detection range	Humidity measurement accuracy	Humidity detection range	Light detection range	Data transmission distance	Response time	Distortion rate(%)	Node power consumption	Node cost
Cable mode	0~5000PPM	±5%	±0.3℃	-10~+40℃	±2%HR	0~100%HR	0.01~20000LX	50~100m	>1S	2.5%~5%	>5W	1
wireles s mode	0~5000PPM	±5%	±0.1℃	-15~+45℃	±1%HR	0~100%HR	0.01~30000LX	20~40m	>0.5S	<1.5%	>3W	0.7

Data transmission distance for the CC2430 the supply voltage of 3V, the data distortion rate <1.5% distance CO_2 concentration measurement accuracy is the CO_2 conventition of 1000ppm accuracy

About data transmission distance, it can increase distance by set up network routing method node transmission. Since the wireless transmission of the frequency of 2.4GHz, at all nodes need to take the appropriate anti-jamming measures. In order to improve the detection sensitivity, Heating voltage must be stable in the range of 5.0±0.2VDC.

5   Conclusion

In this paper, the design based on the ZigBee wireless sensor parameter of the fowlery monitoring systems, with low-cost, reliable operation, widely area of application characteristics, it can improve the production and quality of Poultry farms, improve the degree of automation, lower production costs, reduce labor intensity. and has a certain significance, The technology in poultry production has a good promotional value.

Acknowledgements

Funding for this research was provided by Ningbo advanced textile technology and apparel Key Laboratory of CAD project (2007ZDSYS-A-002) (P. R. China). the project work also with the help of colleagues ,at the same time, The authors thanks their for theirs kind guidance and encouragement.

References

Zhu, Q.: China's poultry industry and the difficulties currently facing towards the future. Guide To Chinese Poultry 22(13), 3 (2005)

Ying, Y.: Large-scale Laying hens farm automation layer monitoring system set up and application. Heilongjiang Animal Science and veterinary Medicine 9, 33 (2005)

ZigBee/IEEE 802.15.4 Summary.pdf [EB/OL] (2003),
http://www.sinemergen.com/zigbee.pdf [2009-4-24]
IEEE Standards Association IEEE.std.802.15.4, 802.15.4-2003.pdf [EB/OL] (2003),
http://standards.ieee.org/getieee802/download/
802.15.4-2003.pdf [2009-5-6]
Yu, H.-b., Zeng, P., Liang, W.: Intelligent Wireless Sensor Network, vol. 1, pp. 6–12. Science Press, Beijing (2006)
Chipcon(Ti), cc2430.pdf [EB/OL] 2001.11,
http://focus.ti.com.cn/cn/lit/ds/symlink/cc2430.pdf [2009-5-15]
Dong, F.-w., et al.: Application of Wireless sensor networks in dissolved oxygen concentration of freshwater aquaculture automatic monitoring system. Journal of Anhui Agricultural Sciences 28(33), 14345 (2008)
Wang, Y.-j., Zhao, H.-q.: CO2 sensor TGS4160 Principle and Application. International Electronic Elements 2, 64 (2004)
Cheng, W., Xin, Z., Yunhe, Z., et al.: The Development and Application of Portable Instrument for Measuring Temperature, Humidity & Dew Point. Modern Scientific Instruments 5, 62 (2007)
Shen, Z., Li, Q.: An Approach to Wireless Sensor Networks Protocol based on Zigbee Technology. Control & Automation 24(12), 165 (2008)
IEEE Standards Association IEEE.std.802.15.4, 802.15.4-2003.pdf [EB/OL] 2003,
http://standards.ieee.org/getieee802/download/
802.15.4-2003.pdf [2009-5-28]

Research on Non-destructive Comprehensive Detection and Grading of Poultry Eggs Based on Intelligent Robot

Shucai Wang[1,*], Jinxiu Cheng[2], and Youxian Wen[1]

[1] Department of Engineering and Technology, Huazhong Agricultural University,
Wuhan, Hubei Province, P.R. China 430070,
Tel.: +86-13387580932; Fax: +86-27-87285346
wsc01@mail.hzau.edu.cn
[2] Department of Foreign Languages, Huazhong Agricultural University,
Wuhan, Hubei Province, P.R. China 430070

Abstract. This study presents a light-duty automation system for the automatization of detecting and grading poultry eggs, namely SIRDGE, short for System of Intelligent Robot Detecting and Grading Eggs. This system combines crack detection, inner quality detection and eggs conveying in detecting and grading, which are all performed by a joint-robot to fully automate egg detection and gradation. The hardware components and software structure of the system are presented. Necessary explanation of the main hardware and the software is made, which includes video image capturing and camera calibration, image processing and feature extraction, motion control and route planning of the robot, MCU control for the vacuum sucker, knocking sound processing and crack detection, color image processing and inner quality detection, etc. Property testing of SIRDGE has been done in the research.

Keywords: poultry eggs, detecting, grading, intelligent robot.

1 Introduction

Selling and deep processing of each egg after automatic cleaning, detecting, grading and coding enable consumers to learn about such information as the production date, brand and quality grade of each fresh egg so as to value the eggs according to quality. Moreover, it can prevent cross infection in production process so as to fulfill the traceability of the products. In other words, it not only increases the producers' profit but also ensures the consumers' interests.

In the countries with highly automatized technology such as the US, Japan and the Netherlands, processing facilities of fresh eggs include: pneumatic-sucking collecting and conveying device, cleaning and disinfection machine, coating drier, classifying and packing machine, and strikes code machine. These facilities process eggs respectively without human contact and ensure full automatization, high precision and non-destruction in processing, grading and packing. For example, FPS MOBA full

---

* Corresponding author.

D. Li and C. Zhao (Eds.): CCTA 2009, IFIP AICT 317, pp. 487–498, 2010.

automatic classifying and packing machine by Holland Hot-Cheers Ind. Inc. can be used along with dirty eggs detection system, crack detection system, inner-blood spot and eggshell color detection system, and ultraviolet radiation sterilization system. In the event of the production demand, a system with cleaning, drying, oil coating and reassuring function can be outfitted. The products can be outfitted with full automatic material feeder as well as central collecting and conveying system connected with the henhouse.

The general methods for poultry eggs automatic detection are machine vision and acoustic features of eggs. In general, machine vision is for cracks or inner-quality detection while acoustic features of eggs for crack detection.

Whether for the export of fresh eggs or for deep egg processing, it is necessary to clean, disinfect, detect and grade eggs. At present, most enterprises in China are still employing manual work in fulfilling these tasks; as a result, there is a remarkable gap between China and developed countries both in processing quality and in efficiency.

The whole set of oversea mechatronic device for poultry eggs detection is too expensive for the small and medium-scaled enterprises in the area of eggs production and processing to afford. Under such circumstances the author uses modern computer technology achievements (machine vision, intelligent robot, etc.) to excogitate a light-duty system of intelligent robot detecting and grading eggs (SIRDGE for short), which is cheap, agile, flexible and hence fit for the small and medium-scaled enterprises. Under the guidance of machine vision, the robot can judge the position and directions of each egg, snatch and knock at eggs automatically, detect cracks by knocking sound signals, identify and grade the freshness and sizes of eggs by the image features of eggs under transmission light (backlight), and finally put the eggs in corresponding grades according to the graded signal.

2   The Structure and Principles of SIRDGE

SIRDGE is composed of location system (including scene video image capturing, processing and feature extraction of eggs location), conveying system (including robot motion control, manipulator and vacuum sucker system), crack recognition system (including sound producing, data acquisition and processing), and inner-quality detection system (including illumination, color image capturing and image processing).

The workflow of the system is illustrated in Fig 1. A scene camera captures the population image of eggs. Through image processing, the central coordinates and maximum axis direction of each egg are obtained. Thus joint angular displacement of the robot is solved by coordinate transformation and robot kinematic of the data. With a vacuum sucker the manipulator moves to the center of an egg, sends signals to MCU controller for the vacuum sucker, starts the vacuum suction's gas path, and then sucks the egg. According to the planned motion track, the robot carries the egg to the top of the knocking device. A microphone acquires and sends the knocking sound on the eggshell to DSP for processing, i.e. diagnosing the egg's cracks by a corresponding recognition model, the result of which is sent to a robot motion controller. If the egg is cracked, the egg will be put in the eggcrate for cracked eggs. Otherwise, the manipulator will move to the color camera and obtain the color image of the egg in the backlight. Through image processing the freshness of the egg is identified

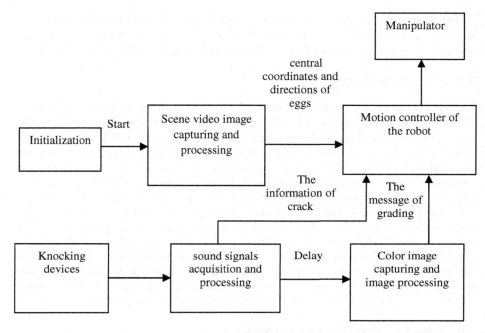

Fig. 1. The working flow of SIRDG

and graded by a corresponding recognition model. Finally, when the grading information is sent to the robot motion controller, the manipulator puts the egg in the corresponding eggcrate.

3   The Hardware Components of SIRDGE

3.1   Video Image Capturing Equipment

The video image capturing equipment consists of CCD camera, image acquisition card, image display card, controlling computer and so on. Black and white CCD camera image processor is LG 1/3″CCD, the system signal PAL standard pattern, the sensitization area 4.9 mm×3.7 mm, the scanning frequency 15.625 kHz (horizontal) and 50 Hz (vertical), and the pixel 500×582. The image acquisition card, DK-2000 Video Capturing Card by Tomoos Inc., is based on PCI bus, also compatible with windows PNP (Plug and Play), of which maximum display resolution is 640 × 480, 24 bit color,   sampling frequency 14.7 MHz.

3.2   The Robot Reality—The Manipulator

The study adopts a Japan-made precise universal robot called Move Master-EX which has a light joint opened-chain linkage with 5-DOF (5-degree-of-freedom).

The main technical parameters of Move Master-EX are as follows: the maximum speed is 1.23 m/s, the maximum load on wrist joint 1.00 kg, repeated positioning accuracy 0.05 mm, moving scope (θ1, θ2, θ3, θ4, θ5) = (300°, 130°, 110°, 180°, 360°).

3.3 Robot Motion Controller

The robot motion controller of Move Master-EX consists of motion controller, IBM-PC or its compatible computer, servo motor with incremental encoder on the robot reality, driver, driver power supply of the driver, +12 V ~+24 V DC supply (for interface card power supply), original point switch, positive/negative limit switch and so on. Move Master-EX uses AC servo motor and has multiple ways of motion control. The system adopts Open-architecture PMAC Motion Controller by Delta Inc., in which each clip can synchro-control 8~32 motion axes and parallel them to achieve multiple axes coordinate motion. The core of PMAC is composed of Motorola DSP56K Digital Signal Processor and FPGA to achieve control calculation of high performance. The system adopts PMAC Motion Controller of PCI Switchboard, which provides RS232 serial communication port, language C math library and Windows DLL(dynamic-link library). Such application modules as data processing required by control math and system, interface display, and users interface are integrated to construct the control system that meets the requirement of poultry eggs detection and grading.

3.4 Robot End Effector—Vacuum Sucker

The original end effector of Move Master-EX Robot is two semi-circular chucks. Because the conveying subjects of the study are extremely friable poultry eggs, the original end effector of Move Master-EX is replaced with vacuum sucker made from silicone rubbers. Experiment shows that the reliable, soft and controllable sucker is good at sucking and discharging poultry eggs.

The gas circuit control of the vacuum sucker adopts AT89S52 MCU, which corresponds with the host computer via MAX232, uses interruption program, sends processing result from pin 2 and 3 of AT89S52 MCU (P1.1and P1.2) to two 4N25 optocouplers. If P1.1 or P1.2 sends low level, the connected optocoupler flows, the coil of DS2Y SSR (Solid State Relay) energizes, the main contact of SSR switches on, the controlled electromagnetic valve energizes, the gas circuit flows and the eggs can be sucked up or put down.

3.5 Crack Detection Device

When the eggs are sucked by the sucker and moved along with the manipulator to crack detection point, they shadow the lights of the optoelectronic sensor. Then the optoelectronic sensor produces a voltage signal. The signal is sent through current processing circuit to the interrupt interface of MCU (AT89C51). The MCU accepts this interrupt application, sends pulse signal, drives stepper motor via driving circuit, which further drives the knocking stick to knock on the eggshell. This device can fulfill multipoint knocking and knocking force control. The knocked eggshell produces sound signal, which is converted into electric signal via sound sensor (microphone). The sound signal enters the interrupt interface of DSP (TMS320VC33) after being amplified, filtered and getting through threshold trigger circuit. The DSP accepts this interrupt signal and starts A/D converters, which convert the filtered signal into digital signal by sampling, perpetuating, quantization and coding.

3.6 Inner Quality Detection Equipment

Inner quality detection hardware consists of illumination, color camera, image acquisition card and computer. The system uses Beijing Daheng DH-VRT-CG200 Image Acquisition Card (Parameters: video source illumination is 20, contrast 20, gray 4, hue 18, saturation 50, image sharpness 3, resolution 640×480, pixel depth RGB24) to convert the video signal from CCD into digital signal that can be recognized by computer, sending the signal via computer PCI bus to computer inner storage of video signal by DMA controller so as to complete the external image acquisition by the computer.

4   Software Algorithm of SIRDGE

The main program of SIRDGE software is made up of 5 molds. The module of scene image capturing and processing, the module of motion controller of the robot, the module connected to AT89S52, the module communicated with DSP and the mode of color image capturing and processing. The control and crack recognition of the vacuum sucker are accomplished respectively by AT89S52 MCU and DSP hardware.

4.1 Video Image Capturing and Camera Calibration

The flow of the image acquisition card in image capturing is as follows: beginning (initialization), parameter setting, acquired images to the screen or memory, and closing (resource releasing). First the disc installation program along with the image acquisition card is run before programming and the needed library DSStream.DLL is obtained. In the program the relevant include file (.h) is used, and Static Linking for Library (.lib) is added to project file to be used by Compiler Monolithic in linking. The function library file offered by the image acquisition card is added to VC program, and before the image acquisition system works properly, the appointed image acquisition card must be opened first. The last parameter setting before closing is read out from initialization file for initializing, the needed system setting is obtained, and quotation handle of the card is constructed to be quoted by various functions.

This study employs the method of linear camera calibration, which enables the central pixel coordinates of the eggs in the images to correspond with the absolute coordinates of the eggs' center in reality under the robot coordinates through the conversion from image pixel coordinates to image plane coordinates, from image plane coordinates to camera coordinates and from camera coordinates to robot coordinates (absolute coordinates).

4.2 Image Processing and Feature Extraction

Gray images acquired by the acquisition card is greatly influenced by lights. When several eggs are close to each other, the division between eggs is difficult to be distinguished in the images. Furthermore, image noises also have an effect on sequent positioning accuracy. As a result, the acquired gray images must be processed, including gray threshold transform, image denoising, image dilation and erosion, image connected components label as well as feature extraction of eggs' central coordinates and maximum axis direction.

From the gray histogram of several images in the experiment it can be seen that double peaks appear in the histogram, in which the host peak (255) is the egg while the subordinate peak (about 50) is the background. The bottom of the valleys is between 120 and 250. Image threshold transform is convenient for sequent processing. Adopting the transform function as follows

$$f(x) = \begin{cases} 0 & x < 250 \\ 255 & x \geq 250 \end{cases} \qquad (1)$$

After image threshold transform, the image becomes black and white image.

From the optimal gray threshold transform, area of the egg and background areas can be clearly distinguished. However, tiny glistening spots (dirt, paper scraps, water stain, etc.) appear to be white like the egg area. Moreover, image noises make small white dots appear in the background areas and small black dots in the egg area. In order to eliminate noises and obtain ideal images, this study applies Gauss Function of 2D mean value discretization to image smoothing.

Owing to the existence of connecting on the edge between eggs in the images, erosion is applied to the images in order to detach all close eggs. The function of erosion in mathematical morphological calculation is to eliminate boundary points of an object. With white dots of the structure element [3]×[3], an egg image processed above is eroded and a pixel will be lost on the egg edge. When there is fine connecting between two egg images, two eggs can be detached as long as the structure element is big enough.

While programming, image erosion calculation is fulfilled by a function named ErosionDIB () . ErosionDIB () has several parameters: LPSTR lpDIBBits (the pointer pointing to original image), LONG lWidth(the width of the original image), LONG lHeight(the height of the original image), int nMode (the mode of erosion) and int structure [3][3](structure element).

After the image is eroded, the size information of the egg area is lost. For egg size reverting, the image can be dilated again with the same structure element. It has been found in the experiment that the result of direct dilation is image reverting and egg reunion. The reason is that closed angles have been left in the original joint in the eroded image, which are also dilated in dilation and result in egg reunion. Cross use of smoothing and threshold transform is a feasible method to eliminate closed angles. Therefore, the study first smoothes and threshold transforms the eroded image, and then dilates the image with the structure element the same as erosion.

In order to extract the location coordinates and angle of each egg for the robot to snatch it precisely, each egg must be labeled. Before labeling the egg areas have been distinguished by all the above mentioned means. Therefore, labeling each egg is to label the connected components in the image.

Locating connected components in an image is one of the commonest calculations in machine vision. Dot composition in connecting area indicates optional area.

Connecting label algorithm can locate all the connected components in the image and distribute the same label to all the dots in one connected component. The study employs four connected components sequential algorithm to label each egg areas.

After the eggs are labeled, the features of the location coordinates and azimuth angles of each egg in the image can be extracted. The ratio of first-order matrix and total pixel area serves as the coordinates of each egg's center under the image coordinates. For a binary image with n lines and m columns, pixel point in Line i and Column j is B(i,j), and the labeled area is

$$A = \sum_{i=0}^{n-1} \sum_{j=0}^{m-1} B(i, j) \tag{2}$$

And the central coordinates of eggs in the image coordinates are:

$$\overline{x} = \frac{\sum_{i=0}^{n-1} \sum_{j=0}^{m-1} jB(i, j)}{A}$$

$$\overline{y} = -\frac{\sum_{i=0}^{n-1} \sum_{j=0}^{m-1} iB(i, j)}{A} \tag{3}$$

Because an egg is approximately oval, its major axis is defined as the direction of the egg and the angle between normal line of the major axis and Axis X is defined as the direction angle of the egg. Minimal second-order moment axis sharing the same direction with axes of minimal moment of inertia on 2D plane is defined as major axis for the long objects in the image. The second-order moment axis of an image object is such a line: the distance sum of square from all its points to the line is minimal. The distance square r_{ij}^2 from pixel points to the line is calculated for the binary image B(i,j) of each egg's labeled area. In order to avoid numerical ill-conditioned problem when the line is approximately vertical, the line is expressed as polar coordinates $\rho = x\cos\theta + y\sin\theta$. In this formula θ is the angle between the normal line of the major axis and Axis X, ρ is the distance from the line to the origin, as shown in Fig 2. The distance sum of square from all the egg's points to the line is minimal.

The function of the distance sum of square from all the egg's points to the line is

$$x^2 = \sum_{i=0}^{n-1} \sum_{j=0}^{m-1} r_{ij}^2 B(i, j) \tag{4}$$

$$r_{ij}^2 = (i\cos\theta + j\sin\theta - \rho)^2$$

After seeking extremum, θ and ρ can be worked out so as to determine the direction of the egg.

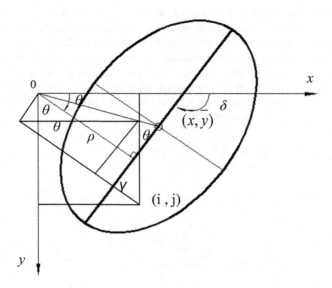

Fig. 2. The definition of the direction of the egg

4.3 Robot Motion Control and Route Planning

The robot manipulator motion in picking up eggs can be decided by pose of manipulator nodes sequencing. Each node is described by homogeneous transformation of sucker center coordinates (tool coordinate system) relative to workpiece coordinate system. The corresponding joint variable can be calculated by Kinematics Converse Solution program.

According to the origin of workpieces and major axis direction of the eggs in workpiece coordinate system worked out by the host computer in image processing system. The transforms coordinates is $_T^B T = _S^B T \, _G^S T \, _T^G T$. In this formula $_T^B T$ is the transform from tool coordinate system to spacecraft-referenced coordinates, $_S^B T$ is the transform from reference coordinates to spacecraft-referenced coordinates, $_G^S T$ is the transform from workpiece coordinate system to reference coordinates, and $_T^G T$ is the transform from tool coordinate system to workpiece coordinate system. Hence the host computer obtains coordinate transformation matrix of the robot end effector (sucker) from the current point to the next one. Then the host computer obtains rotation angles of all the robot's joints$\theta 1 \sim \theta 5$ by means of robot kinematics converse solution, and gives corresponding control orders to motion control card according to the given acceleration and top speed. According to the control orders the motion control card accelerates the electromotor of every joint of the robot at the given acceleration to top speed or starts to decelerate until it stops at the set place. The sensor sends the position signal to the host computer, which gives orders to MCU of the sucker pneumatic system after receiving the signal. The MCU controls electromagnetic valve to switch on the gas path and suck up or put down the eggs. In order to fully accomplish detection and grading in a process of sucking up and putting down an egg, the planning is carried out to the robot motion path. The complex course in which jumping-off point and terminal are both variational in detection and grading is simplified into three easily fulfilled steps: the jumping-off point is variational but the terminal immovable (from sucking the egg to crack detection position); the jumping-off point and terminal are both immovable (from crack detection position to quality defect detection position); the jumping-off point is immovable but the terminal variational (from quality defect detection position to the corresponding eggcrate the egg is carried into). Such treatment enables the robot motion control to be realized easily.

In order to make use of the dynamic link libraries, Pmacu.h, m_hPmac.lib and PmacPc.dll (in the path of Windows\Dll of the disc attached to the product) ,of motion control card , #include "pmacu.h" is first added to the program, then project-setting-link is chosen in VC menu, then m_hPmac.lib is entered in object/library modules, and finally the function in dynamic link libraries of motion control card can be used in the program.

5   Property Testing of SIRDGE

SIRDGE is composed of the above software. The system gives the scene image gray threshold transform, smoothing using Gauss template, image dilation and erosion, image connected components label, the result of which is shown in Fig 3.

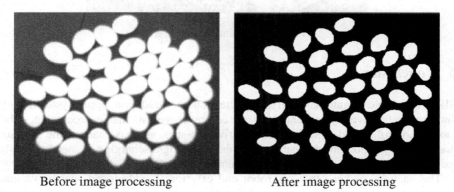

Before image processing   After image processing

Fig. 3. Comparison of pre-and post-image processing

In order to extract the absolute coordinates of the eggs' center, nine calibration points are used to calibrate the scene camera. Then nonlinear regression models are used to obtain the conversion matrix parameter from the central coordinates of the eggs in the images to the absolute coordinates of the eggs' center in reality space.

The rotation matrix is

$$R=\begin{bmatrix} 0.02 & -1.08 & 0.14 \\ 1.25 & 0.05 & 0.08 \\ 0.12 & 1.03 & 1.16 \end{bmatrix} \tag{5}$$

The shift matrix is

$$T = \begin{bmatrix} -352 \\ 447 \\ 598 \end{bmatrix} \qquad (6)$$

The robot motion experiment has been conducted in Intelligent Robot Institute, Central China University of Science and Technology. Fig 4 is the experiment photo.

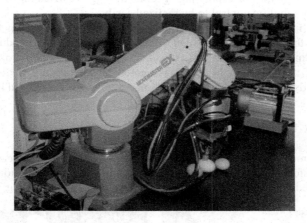

Fig. 4. The picture in experiment

6   Conclusion

SIRDGE is able to automate fully egg detection and gradation, apply systematically and comprehensively the technology on computer vision, image processing, robot motion control, sound detection and negative pressure, and has changed the situation that huge operation mechanism or manual charging/discharging is needed in the existing egg detection and gradation machinery.

The vacuum sucker serves as robot end effector. The designed sucker, vacuum suction gas path and MCU have good effects in conveying, sucking and discharging poultry eggs.

The ratio of first-order matrix of digital image and total pixel area serves as the coordinates of each egg's center under the image coordinates. The method of linear camera calibration is employed. Minimal second-order moment axis sharing the same direction with axes of minimal moment of inertia on 2D plane is defined as maximum axis for the egg. This study applies optimal threshold transform and Gauss Function of 2D mean value discretization to image smoothing, uses several times of image dilation and erosion, makes image segmentation, and employs four connected components sequential algorithm to label the same egg area in the image. The corresponding algorithm obtained can meet the working demand for leading the robot to locate each egg precisely.

By means of robot kinematics converse solution, the absolute coordinates of the eggs' center and maximum axis direction are transformed into rotation angles of all the robot's joints. In order to fully accomplish detection and grading in a process of conveying, the planning is carried out to the robot motion path. The complex course in which jumping-off point and terminal are both variational in detection and grading is simplified into three steps: the jumping-off point is variational but the terminal immovable; the jumping-off point and terminal are both immovable; the jumping-off point is immovable but the terminal variational. Such treatment enables the robot motion control to be realized easily.

Acknowledgements

Funding for this research was provided by Chinese National Programs for High Technology Research and Development (project number: 2007AA10Z214). The first author is grateful to professor Xinhan Huang, the Intelligent Robot Institute and the Central China University of Science and Technology for providing him with a Japan-made precise universal robot called Move Master-EX. to do experiments.

References

[1] Hot Cheers Ind. Inc. Full automatic egg-graders [EB/OL] (2005-05-16), http://hotcheers.diytrade.com [2006-11-25]
[2] Elster, R.T., Goodurm, J.W.: Detecting of cracks in eggs using machine vision. Transactions of the ASAE 34(1), 307–312 (1991)
[3] Goodurm, J.W., Elster, R.T.: Machine vision for cracks detection in rotation eggs. Transactions of the ASAE 35(4), 1323–1328 (1992)
[4] Cho, H.K., Won, Y.K.: Crack detection in eggs by machine vision. In: 6th Int. Conf. on Computers in Agriculture, May 15-16, pp. 777–784. St. Joseph. Mich., ASAE (1996)
[5] Nakano, K., Motonaga, Y., Mizutani, J.: Development of non-destructive detector for abnormal eggs. In: Workshop on Control Applications in Post-Harvest and Processing Technology, vol. (1), pp. 71–76 (2001)
[6] Kuchida, K., Fukaya, M.: Nondestructive prediction method for yolk: albumen ratio in chicken eggs by computer image analysis. Poultry Science 78(6), 909–913 (1999)
[7] Das, K., Evans, M.D.: Detecting fertility of hatching eggs using machine vision. Transactions of the ASAE 35(6), 2035–2041 (1992)
[8] Sinha, D.N., Johnston, W.K., Grace, C.L.: Acoustic resonance in chicken eggs. Biotechnology Prog. 8(3), 240–243 (1993)
[9] Ketelaere, E.B.D., Coucke, P., Baerdemaeker, J.D.: Eggshell crack detection based on acoustic resonance frequency analysis. Agricultural Engineering Res. 76(2), 157–163 (2000)
[10] Shucai, W., Yilin, R., Hong, C., Lirong, X., Youxian, W.: Detection of cracked-shell eggs using acoustic signal and fuzzy recognition. Transactions of The Chinese Society of Agricultural Engineering 21(4), 130–133 (2004)
[11] Jianying, L., Jiayan, C., Youchun, D., Yilin, R., Shucai, W., Lirong, X., Dongjiao, C., Youxian, W.: Model for automatic detection of eggshell crack. Transactions of The Chinese Society of Agricultural Engineering 21(9), 189–192 (2005)

[12] Lirong, X., Youchun, D., Hong, C., Youxian, W.: Study On the Automatic Detection System of Duck Egg Classification. Journal of Huazhong Agricultural 19(4), 133–137 (2000)

[13] Qiaohua, W., Yilin, R., Youxian, W.: Study on Non-destructive Detection Method for Fresh Degree of Eggs Based on BP Neural Network. Transactions of the Chinese Society for Agricultural Machinery 21(1), 158–162 (2006)

[14] Qiaohua, W., Yousheng, Y., Wangyuan, Z., Xiaowei, C., Jianying, L., Youxian, W.: Research on the Grading Model of Duck Egg Bulk. Transactions of the Chinese Society for Agricultural Machinery 17(6), 202–205 (2001)

[15] Lirong, X., Shucai, W., Yilin, R., Youchun, D.: The Revision of Grade Model on Egg's Weight. Journal of Agricultural Mechanization Research (2), 35–37 (2006)

[16] Shucai, W.: A Study of Autonomous Robot Detecting and Grading Eggs, pp. 75–89. Huazhong Agricultural University, Wuhan (2006)

[17] Minghong, R.: Realizing Image Acquisition with AVICAP.DLL in VC++. Computer Knowledge and Technology (Academic Exchange) (24), 1133–1134 (2007)

[18] Zhenyou, Z., Aijie, X., Tao, L., Shanben, C.: High-speed calibration method for the relationship of the eye-in-hand of robot vision. Optical Technique 30(2), 150–154 (2004)

[19] Yuting, J., Zhongke, Y.: Optimal Template for Image Smooth Denoising. Journal of Transportation Engineering and Information (3), 68–71 (2004)

[20] Yunde, J.: Machine Vision, pp. 196–204. Publishing House of Science and Technology (2004)

[21] Hong, X., Fangwen, R.: Numerical stability and morbidity problem. China Science and Technology Information (11), 322–323 (2006)

[22] Ganmin, T., Zhongke, S.: Measuring and Analyzing Industrial Noise Based on DSP. Microprocessors (2), 28–30 (2006)

[23] Meihu, M., Jiansheng, X., Farong, G., Yongchang, Z.: Research of Egg-Vacuum -Sterilization Preservation-Technology (EVSPT). China Poultry (8), 24–26 (2001)

A Pvdf Sensor for Monitoring Grain
Loss in Combine Harvester

Jiaojiao Xu[*] and Yaoming Li

Key Laboratory of Modern Agricultural Equipment and Technology,
Ministry of Education & Jiangsu Province, Jiangsu University, P.R. China, 212013,
Tel.: +86-13952882634
xujj525@163.com

Abstract. Grain loss sensor is an important monitoring module to realize the intelligent automatic control in combine harvester. Polyvinylidene fluoride piezoelectricity film has been one of the most widely used macromolecule materials at present. Based on its piezoelectric effect, in this paper, a grain loss monitoring system for combine harvester by using the PVDF sensor was designed. The system is composed of the PVDF sensor, intelligent display, SCM and so on. The working principle was analyzed and the measuring accuracy was dynamic calibrated in laboratory. The polyvinylidene fluoride piezoelectricity film was selected as the sensing material and the signal processing circuit was mainly composed by charge amplifier, band-pass filter. In order to verify the accuracy of the sensor, the dynamic calibration experiments were carried out. By adjustment of the feeder's vibration parameters, the different feeding velocity can be gained. In the dynamic calibration experiments, the sensor was fixed at the height of 19.5mm and the fixing angle of $33°~35°$, the feeding velocity of 60~80 grain per second, 1000samples of full grain were randomly selected in the experiments, and the results indicated the detecting error was less than 5.7%. Using the different mixture ratio of the full grain, the imperfect grain, short stalk to do multi-group experiments, the result shows that the sensor can distinguish the grain from the mixture stalks. In the rig-tests, the monitoring sensor was fixed on the caudal region of the cylinder concave grid of the combine harvester in order to monitor the threshing compounds. The precision error of the monitoring sensor in combine is less than 8% through field experiments. So this advanced sensor can realize the aim of monitoring the entrainment loss in combine harvester.

Keywords: PVDF, Grain Loss Sensor, Accuracy Calibration, Combine Harvester.

1 Introduction

Polyvinylidene fluoride (PVDF) is a new type of piezoelectric polymer material which has many advantages such as high piezoelectric voltage constant, light weight, low density, better toughness, wide frequency response, low acoustic impedance,

---

[*] Corresponding author.

D. Li and C. Zhao (Eds.): CCTA 2009, IFIP AICT 317, pp. 499–505, 2010.

better sensitivity and stability, easy processing and installation(e.g., Xu Hongxing et al.,1999; Ju dianshu et al., 2004). The piezoelectric constant d of PVDF is higher than quartz as the value of more than 10 times and the value g is 20 times higher than PZT, while the low acoustic impedance is about 3.5×10^{-6} which is only 1/10 of the PZT piezoelectric ceramic. The frequency response is wide at room temperature in the range of 10-5~109Hz and the response is flat. It means that from the quasi-static, low frequency, high frequency, up to ultra-high frequency ultrasound can convert mechanical and electrical effects. So it can be processed into thin films and be used as sensors, because the vibration response of the system has little effect.

Based on its piezoelectric effect, in this paper a set of monitoring system which can monitor the gain loss in combine harvester by using the PVDF sensor was designed. The monitoring system is composed by the PVDF sensor, intelligent display and so on. The working principle and measuring accuracy was analyzed and dynamic calibrated in laboratory.

2   Dynamic Calibration Experiment

2.1   The Measuring Principle of PVDF Sensor

Polyvinylidene fluoride is a kind of organic macromolecule polymers. Through heating, cold stretching and processing in the Curie point of temperature in order to demonstrate the polarization piezoelectric and thermoelectric properties. The pressure principle of PVDF piezoelectric films as shown in Figure 1 (e.g., Cheng Qihua and Li yongxin, 2007), the direction of uniaxial tension is the x-axis, the z-axis is perpendicular to the polarization surface and parallel to the polarization direction, the y-axis is perpendicular to the x-axis and z-axis. When the PVDF membrane is acted by external force, the corresponding charge will appear on both sides of the film and the charge will led by the metal conductive layer. By measuring the amount of charge, it can achieve the measurement of pressure. In practice application (refer as Eq. (1)) only on the axial stress the relationship is between charge and plane (e.g., Tomasz Janiczek, 2001).

$$q(t) = q(t_0) + \iint_{\Omega} [\int_0^t [(\sum_{j=1}^3 d_{3j} \frac{\partial \sigma_{jj}(x,y,t)}{\partial_t}) d_t] dxdy - \int_0^t \frac{q(t)}{RC} dt \qquad (1)$$

Where: $q(t_0)$ =initial voltage, V. d_{3j} =piezoelectric constant, pc/N

$\sigma_{jj}(x,y,t)$ = stress

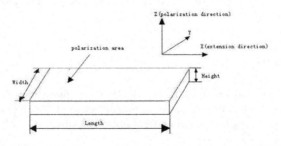

Fig. 1. Principle diagram of pressure measurement on PVDF

2.2 Design of Dynamic Calibration Experiment

In the dynamic calibration experiments, the piezoelectric sensor is used by 5 sheets of PVDF film that each module's thickness is 50mm and the size is 20mm*100mm. Fig.2 is the circuit schematic diagram of single channel.

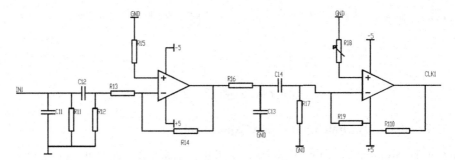

Fig. 2. Circuit schematic diagram of single channel

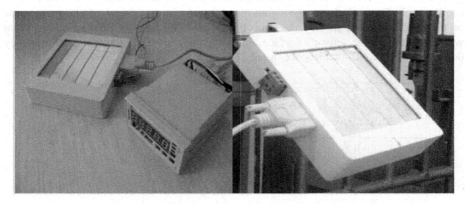

Fig. 3. Design of the monitoring sensor

The circuit schematic diagram includes filter circuit, frequency discriminating circuit, amplification system and shaping circuit. Fig.3 is the design of the monitoring sensor. Because the whereabouts of threshing mixture has different height, the different contact forces for the sensor can cause different frequency response. So the sensor is installed on the bracket which the fixing height can be adjustable in order to regulate the whereabouts of a suitable calibration height though the contact forces on the sensor. The installing angle of sensor at the same time can also be adjusted and the different seeds, stem and others film has different impact forces on the PVDF piezoelectric (e.g., Li Junfeng, 2006). Therefore different angles must be adjusted to determine the optimum installation angle. The impact situation of the threshing compound in combine harvester was simulated using a magnetic vibration-actuated feeder. When the threshing grain and other compound drop on the sensor's film, the impact signals can be detected through the data wire and the amount of grain can be shown in the computer. Using the different mixture ratio of the full grain, the imperfect grain, short stalk to do multi-group experiments, the result shows that the sensor can distinguish the grain from the mixture stalks. Fig.4 is the Circuit structure of monitoring sensor.

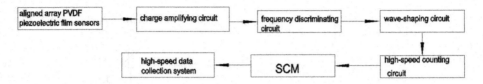

Fig. 4. Circuit structure of sensor

2.3   Results of Dynamic Calibration Experiment

When the different threshing mixture including grains, stalks and others drop on the sensor's surface, as the grain and the stalk has different impact force on the sensor's surface, so the different frequency and amplitude of the signal will be reflected in the oscillograph. Figure 5 shows the signal maps that the impact forces of grain and straw on the PVDF sensor's surface. As a result that miscellaneous residual are so light that when they fall on the sensor's surface, they don't has any trigger signals and thus they don't have any significant impact signal.

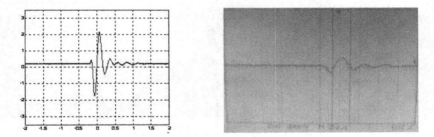

Fig. 5. The impact signal of grains and straws

The abscissa axis is time and the vertical axis is voltage. So in the proceedings, the cutoff frequency of band-pass filter and the amplitude threshold voltage of the frequency discriminator must be reasonable set. And the chaotic signals can be extracted from the impact of grain, then the signals will be counted by program and they can be accurately recorded the amount of grain losses.

3   Analysis of Experiments Results

3.1   Analysis of Dynamic Calibration Experiment

In dynamic calibration experiments, different moisture content of rice grains and different length of straws are used as the calibration materials. The moisture content are respectively 14.72%,18.32%,23.19%. The same moisture content of 39.33% has different length of straws. The lengths are 10mm,20mm,30mm. The thousand-grain weight of the full grains is 28g ~ 32g. The sensor was fixed on a bracket which can adjust its height and the sensor's fixing angle was also can be adjusted.

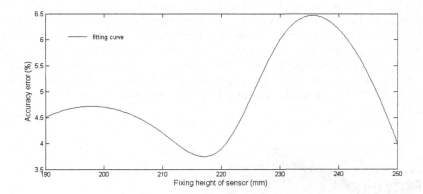

Fig. 6. Calibration curve of variable height

The impact situation of the threshing compound in combine harvester was simulated using a magnetic vibration-actuated feeder. When the threshing grain and other compound drop on the sensor's film, the impact signals can be detected through the data wire and the amount of grain can be shown in the display. As shown in Figure 6, the calibration curve is obtained. When the sensor's fixing height between 190mm and 225mm, the sensor can meet the testing requirements (e.g., Wan Jianguo et al.,1998).

Similarly, as shown in Figure 7, from the calibration curve, when the reasonable fixing angle between 32^0 and 43^0 the sensor can meet the testing requirements and the accuracy error is steady.

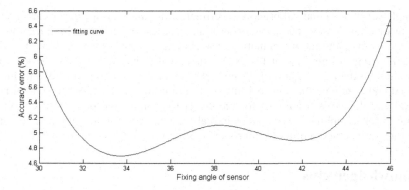

Fig. 7. Calibration curve of variable angle

3.2   The Fixing Position of Sensor in Rig-Test

The monitoring sensor can be fixed under the combine threshing cylinder to collect the grain signals from threshing mixture. Figure 8 shows the actual view of sensor's installing place. The sensor can be installed under the separation device at three positions .Figure 9 shows the left view of sensor, the sensor installs on the bracket and the location is at the end of concave. The threshing probability curve and mathematical model are established through the threshing and separation process along the axial points of the combine cylinder. In order to find out the mathematical relationship between the amount grains monitored from the sensor and the loss grains from the rowing straw port, the high-speed data acquisition system collect the amount grains and using the mathematical model in order to get the entrainment losses in the harvest process.

 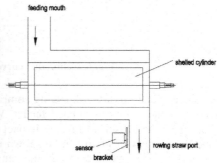

Fig. 8. Sensor's installation position   **Fig. 9.** Left view of the installation position

4   Conclusion

Through bench tests and calibration experiments, the monitoring sensor can detect the amount of grain losses. The monitoring system bases on sensor techniques and computer technology and the mathematical model of science as a guide in order to achieve the monitoring of entrainment losses. Fundamentally this method and device change the traditional manual method of testing entrainment losses and it can save manpower, material and eliminate the non-real-time test of entrainment loss. At the same time this method and device is propitious for the driver to monitor the process of entrainment loss during the combines' work status in real-time and achieve the automatic control of operating parameters in order to improve the efficiency of the combine harvester.

Acknowledgements

This work was supported by the National Key Science & Technology Pillar Program during the Eleventh Five-Year Plan Period (2006BAD11A03).

References

Qihua, C., Yongxin, L.: Application of PVDF sensor to the collision stress detection on the projectile base of artillery. Journal of Test and Measurement Technology 21(1), 90–94 (2007)

Dianshu, J., Zhi, Z., Jinping, O.: Study on strain-sensing of PVDF films. Journal of Functional Materials 4(35), 450–456 (2004)

Junfeng, L.: Improvement design of the structure of combine harvester grain loss sensor and laboratory calibration. Agricultural equipment & vehicle engineering 184(11), 10–13 (2006)

Janiczek, T.: Analysis of PVDF transducer signals stimulated by mechanical tension. Journal of Electrostatics 51(52), 167–172 (2001)

Jianguo, W., Jijun, Z., Lihua, S.: Preliminary search on application of PVDF piezoelectric film for monitoring structures. Chinese Journal of Sensor and Actuators 12(1), 18–24 (1998)

Hongxing, X., Zuting, L.: Developments of PVDF Piezoelectric Films. Journal of Jiangsu University of Science and Technology 20(5), 88–91 (1999)

Quantification Model for Estimating Temperature Field Distributions of Apple Fruit

Min Zhang[1,*], Le Yang[1], Huizhong Zhao[2], Leijie Zhang[1],
Zhiyou Zhong[1], Yanling Liu[1], and Jianhua Chen[1]

[1] College of Food Science and Technology, Shanghai Ocean University,
Shanghai, P.R. China 201306,
Tel.: +86-21-61900392; Fax: +86-21-61900392
zhangm@shou.edu.cn
[2] College of Urban Construction and Environment Engineering,
University of Shanghai for Science and Technology, Shanghai, P.R. China 200093

Abstract. A quantification model of transient heat conduction was provided to simulate apple fruit temperature distribution in the cooling process. The model was based on the energy variation of apple fruit of different points. It took into account, heat exchange of representative elemental volume, metabolism heat and external heat. The following conclusions could be obtained: first, the quantification model can satisfactorily describe the tendency of apple fruit temperature distribution in the cooling process. Then there was obvious difference between apple fruit temperature and environment temperature. Compared to the change of environment temperature, a long hysteresis phenomenon happened to the temperature of apple fruit body. That is to say, there was a significant temperature change of apple fruit body in a period of time after environment temperature dropping. And then the change of temerature of apple fruit body in the cooling process became slower and slower. This can explain the time delay phenomenon of biology. After that, the temperature differences of every layer increased from centre to surface of apple fruit gradually. That is to say, the minimum temperature differences closed to centre of apple fruit body and the maximum temperature differences closed to the surface of apple fruit body. Finally, the temperature of every part of apple fruit body will tend to consistent and be near to the environment temperature in the cooling process. It was related to the metabolism heat of plant body at any time.

Keywords: apple fruit, thermocouple, temperature, quantification model.

1 Introduction

Precision agriculture is one of the tendency of modern agriculture. Temperature is an important factor of affecting plant's growth. So the plant body's temperature should be described as accurately as possible when the environmental around plant was regulated and controlled. For example, it is obviously more accurate using the body

---

* Corresponding author.

D. Li and C. Zhao (Eds.): CCTA 2009, IFIP AICT 317, pp. 506–512, 2010.

temperature of plant than the air temperature to judge whether the plant was suffering chilling injury in the chilling process. Ansari and Haghighi studied the problem of heat transfer of the fruits and vegetables under cold storage environmental conditions (Ansari et al., 1999). Lisowa and Shyam studied the impact parameters and measurement methods of the thermal properties of the fruits (Lisowa et al., 2002; Shyam et al., 2003). Becker studied the models of thermophysical property of Food (Becker, 1999). Miklos studied simultaneous heat and mass transfer within the maize kernels during drying (Miklos et al., 2000). Li and Yang studied heat and mass transfer characteristics of the vegetables and seeds in the drying process (Li Yebo et al., 1996; Yang Junhong et al., 2001). However, most of these studies neglected specific lives characteristics of the fruits and vegetables. In the modern scientific research of plant life, using physics model, many complex phenomena of lives successfully could be explained. Moreover, the present phenomenological science of plant life could be quantified as it as possible. The operating mechanism of the environment in living creature and the processes and principles of biological adaptation could be further understood. Also it could provide new models and ideas for revealing and resolving other complex issues of plant life science.

The object of this work was to develop a more accurate mathematical model to describe the dynamics temperature distribution of fruits and vegetables under different conditions. It could prove up the mobile forms of the internal heat of fruits by combination of thermal physics methods. It would be verified through the experimental and theoretical research of apples' internal temperature distribution of each moment in the cooling process. The results would find out the general rules of characteristics of heat transfer and lay a theoretical foundation to further explore plant stress physiology, including plant burn and heat resistance, plant frost damage and frost resistance, plant chilling injury and chilling resistance and even low-temperature preservation of vegetables and fruits.

2 Theory and Methods

2.1 Theoretical Analysis on Temperature Field

The heat transfer thoery of engineering thermo-physics was used to study the energy change within apple fruit. It took into account heat exchange of representative elemental volume, metabolism heat and external heat. The fruit organization was considered as a heat exchange control volume. A quantification model of transient heat conduction was bulit through the analysis of micro-control volume heat exchange by the heat conduction theory and the energy conservation law. The basic model could be expressed in Eqution (1)

$$\rho_m c_m \frac{\partial T_m}{\partial t} = k_m \nabla^2 T_m + q_m + q_a \tag{1}$$

Where: $\rho_m c_m \frac{\partial T_m}{\partial t}$ is the energy change from the change of organization temperature, $k_m \nabla^2 T_m$ is the heat exchange between micro-control volume and the environment by thermal conductivity, q_m is the metabolic heat of organization, q_a is the absorption external heat by absorption.

For the sphere fruit with radius R, which is under the third boundary condition, with uniform inner heat source and with variable properties, a complete mathematical description of non-steady-state thermal conductivity of fruit was given as follows

$$\frac{\partial T}{\partial t} = a\left(\frac{\partial^2 T}{\partial r^2} + \frac{2}{r}\frac{\partial T}{\partial r}\right) + \frac{q_v}{\rho c} \qquad (t>0, 0\leq r\leq R) \qquad (2)$$

Initial condition,

$$T(r,0)=T0 \qquad (t=0, 0\leq r\leq R) \qquad (3)$$

Boundary conditions,

$$-\lambda\frac{\partial T(r,t)}{\partial r}\Big|_{r=R} + \alpha(T_\infty - T_0) + \frac{q_v}{2} = \rho c \cdot \frac{\Delta r}{2} \cdot \frac{\partial T}{\partial t} \qquad (t=0, r=0) \qquad (4)$$

$$\frac{\partial T(r,\tau)}{\partial r}\Big|_{r=0} = 0 \qquad (t=0,r=0) \qquad (5)$$

Where: T is sample temperature, τ is heating time, λ is thermal conductivity of the sample, a is temperature coefficient of the sample, ρc is sample specific heat capacity of sample, qv is the respiratory heat of unit volume, α is convection heat transfer coefficient of boundary, R is sample radius.

In order to simplify the numerical calculation, apple organization was regarded as the same quality uniformly. The equivalent thermal parameters of thermal property was adopted to calculate. Properties of fruit were only affected by temperature (Zhang Min et al., 2004). However, in a range of temperature difference, they could be regarded as homogeneous. In the cooling process, major components were not changed from biochemical reactions. Some parameters of thermal property of apple and environment were shown in Table 1.

Table 1. Parameter of thermal properties of apple and environment

parameters	reference values
diameter (m)	0.09
water content (%)	87.6
initial temperature (°C)	20
frozen storage temperature (°C)	3
thermal conductivity (W/mK)	0.0032T- 0.5703
specific heat capacity(KJ/m3K)	10.581T+799.85
respiratory heat (W/m3)	$e^{0.0958T-23.884}$
convective heat transfer coefficient (W/m2K)	30

2.2 Measurement System

The measurement system was shown in Fig.1. It mainly included constant temperature box (VELP Scientifica, Italy) with ±0.5℃ precision, thermal probe measurement system and data acquisition system. The thermal probe measurement system mainly included micro-thermocouple probe, thermos full of ice water with 0℃ , transformer-oil vessel for heat transfer enhancement, compensation terminal in the mixture of ice and water, and so on. The data acquisition system mainly included 2700 data acquisition instrument (Keithley Instruments, USA), computer interface, and corresponding computer memory system.

At first, the micro-thermocouple probe should be calibrated to determine the test error. Low temperature thermostat bath DC2006 (produced by Shanghai Instrument Factory) was used to calibrate measurement terminal of each thermocouple at the temperature of 0℃ ∼ 30℃. And then the actual measured temperatures were amended. Computer P, I, D was used to control for DC2006. The Pt100 was used as temperature sensor. The circulating pump in the DC2006 flew in 10L/min.

Secondly, the calibrated micro-thermocouple probe was penetrated the apple sample vertically at 20℃ environment. Point 1-4 were put into sample, Point 5 was on the epidermal position of the sample, Point 6 was exposed in constant temperature box to obtain its environmental temperature. Then the test system and computer systemwas turned on and stabilized for 60 min, which was adjusted at the temperature of 3℃±0.5℃ and relative humidity of 85 ∼ 90%.

Thirdly, the apple sample penetrated by the calibrated micro-thermocouple probe was put into the constant temperature box with temperature of 3℃±0.5℃ and relative humidity of 85 ∼ 90%.

Finally, the temperature data at each moment and each section of apple fruit were collected by the Keithley acquisition instrument and stored on the computer to deal with at the same time.

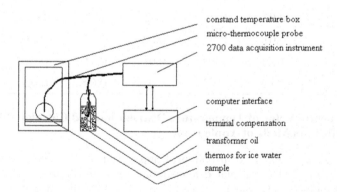

Fig. 1. Experimental equipment of the refrigeration apple's inner temperature distribution

3 Measured Results and Discussion

3.1 Probe Calibration

The calibration results of multipoint temperature measurement probe in the low temperature thermostat bath DC2006 were listed in Table 2. It shown that the average temperature deviation of every point was less than 1.5℃ at the range of 0℃ and 30℃. The maximum fluctuation temperature deviation after amended could be controlled less than 0.03℃(see Table 3).

Table 2. Calibration deviation of multipoint temperature measurement probe

Temperature	Point 1	Point 2	Point 3	Point 4	Point 5	Point 6
0℃	-0.80	-0.89	-0.98	-1.05	-1.07	-0.87
5℃	-0.93	-1.02	-1.21	-1.03	-1.25	-1.06
10℃	-0.87	-1.15	-1.36	-1.22	-1.15	-1.02
20℃	-0.90	-1.16	-1.44	-1.21	-1.36	-1.23
30℃	-1.03	-1.17	-1.38	-1.23	-1.45	-1.27

Table 3. Calibration deviation of multipoint temperature measurement probe after amended

Temperature	Point 1	Point 2	Point 3	Point 4	Point 5	Point 6
0℃	0.02	0.02	0.02	0.02	0.02	0.02
5℃	0.02	0.02	0.02	0.02	0.01	0.02
10℃	0.01	0.01	-0.02	-0.02	-0.01	-0.01
20℃	-0.01	-0.01	-0.01	-0.01	-0.02	-0.02
30℃	-0.03	-0.02	-0.02	-0.02	-0.02	-0.02

3.2 Experimental Value of Temperature Distribution and Calculated Value of Quantification Mode of Apple Fruit

The determined temperature distribution of apple fruit in constant temperature box was shown in Fig.2. The calculated temperature value by the quantification model including Formula(2) to Formula(5) was shown in Fig.3. So the following conclusions could be obtained:

First, the quantification model can satisfactorily describe the tendency of apple fruit temperature distribution at the cooling process.

Then, there was obvious difference between apple fruit temperature and environment temperature. Compared to the change of environment temperature, a long hysteresis phenomenon happened to the temperature of apple fruit body. That is to say, there was a significant temperature change of apple fruit body in a period of time after environment temperature dropping. And then the change of temerature of apple fruit body in the cooling process became slower and slower. This can explain the time delay phenomenon of biology.

After that, the layer temperature differences increased from centre to surface of apple fruit gradually. That is to say, the minimum temperature differences closed to centre of apple fruit body and the maximum temperature differences closed to the surface of apple fruit body. Finally, the temperature of every part of apple fruit body will tend to consistent and be near to the environment temperature in the cooling process. The temperature distribution of apple fruit body was related to the metabolism heat of plant body.

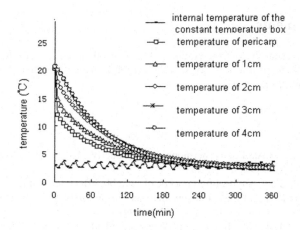

Fig. 2. The experimental temperature variation of apple fruit

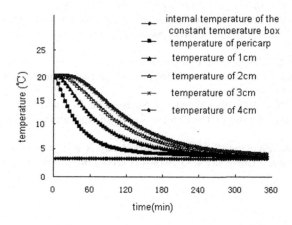

Fig. 3. The simulation temperature variation of apple fruit

4 Conclusion

In this paper, the heat transfer theory of thermal physics is applied. It takes into the characteristics of life-activity apple. A computer quantification model of transient heat conduction was provided to simulate apple fruit temperature distribution in its temperature dropping process. Contrasted between the results of numerical analysis and the temperature field distribution of actual test, the conclusion could be got by experiment: the quantification model can satisfactorily reflect the tendency of apple fruit temperature distribution at the dropping temperature process. And we can further explore plant stress physiology, including plant burns and heat resistance, plant frost damage, plant chilling injury and frost resistance and vegetables, fruits and other low-temperature preservation, and so on.

Acknowledgements

Funding for this research was provided by the Natural Science Foundation of China, Grant No 30771245 and the key construction course project of Shanghai Educational Committee, Grant No 6700308.

References

Becker, B.R., Fricke, B.A.: Food Thermophysical Property Models. Int. Comm. Heat Mass Transfer 26, 27–35 (1999)
Ansari, F.A., Khan, S.Y.: Application concept of variable effective surface film conductance for simultaneous heat and mass transfer analysis during air blast cooling of food. Energy Conversion and Management 40(5), 567–574 (1999)
Lisowa, H., Wujec, M., Lis, T.: Influence of temperature and variety on the thermal properties. Int. Agrophysics 16, 43–52 (2002)
Yang, J.H., Chu, Z.D., Meng, X.L.: Internal moisture content diffusion diffusion mechanism of seeds drying. Journal of Tianjin University 34(2), 142–145 (2001)
Miklos, N., István, C., Attila, K.: Investigation of simultaneous heat and mass transfer within the maize kernels during drying. Computers and Electronics in Agriculture 26(2), 123–135 (2000)
Shyam, S., Sablani, M., Shafiur, R.: Using neural networks to predict thermal conductivity of food as a function of moisture content, temperature and apparent porosity. Food Research International 36, 617–623 (2003)
Li, Y.B., Yu, Q.L., Zhao, L.H.: Experiment and Study on the Heat and Mass Transfer in Potato During Drying. Transactions of The Chinese Society of Agricultural Engineering 12(4), 62–65 (1996)
Min, Z., Bailing, Z., Zhiqiang, S.: Heat Transfer Analysis in Postharvest Fruit. Journal of Northeast Agricultural University 4(35), 418–421 (2004) (in Chinese)

A Wireless Real-Time Monitoring Node of the Physiological Signals for Unrestrained Dairy Cattle Using Wireless Sensor Network

Xihai Zhang, Changli Zhang[*], Junlong Fang, and Yongcun Fan

Engineering college, Northeast Agriculture University, Harbin,
Heilongjiang Province, P.R. China 150030,
Tel.: +86-451-55191146
xhzhang@neau.edu.cn

Abstract. A newly developed smart sensor node that can monitor physiological signals for unrestrained dairy cattle is designed through modular design and its advantages are compact structure and small volume. This sensor node is based on a MSP430F133 micro-controller; the digital sensor includes temperature sensor (DS18B20-America) and vibration-displacement sensor (DN series China); transmission of the digital data uses the nRF903. The results show that this node can collect physiological signals for unrestrained dairy cattle and then send it to upper network node. This research can provide better hardware platform for further researching the communication protocols of wireless sensor networks.

Keywords: dairy cattle, physiological signals, wireless sensor network, ZigBee.

1 Introduction

The physiological signals always reflect the presence status of dairy cattle, such as increased activity amount that maybe indicate estrus or abnormal body temperature that maybe imply malfunction or disease. However, at present, the monitoring of physiological signals for dairy cattle almost depends on dairyman's observation and judgment by experience, which is not quite accurate and timely. Therefore, the accurate and timely monitoring system of physiological signals should be established for allowing us to better understand of cattle status.

Some techniques have been used successfully for recording and storing physiological signals in unrestrained animals, often based on the mounting of some type of tape recorder or data logger package on the animal in Ref. (Hahn et al., 1990; Feddes et al., 1993; Eigenberg et al., 2002). In a few cases, it is difficult to realize wireless real-time monitoring signal by these techniques (Lowea et al., 2007).

A wireless sensor network is very feasible for on-line data collection tasks and animal monitoring, which can greatly extend our ability to monitor the physiological signals of unrestrained livestock from remote locations. However, there are a few

---

[*] Corresponding author.

D. Li and C. Zhao (Eds.): CCTA 2009, IFIP AICT 317, pp. 513–518, 2010.
© IFIP International Federation for Information Processing 2010

obstacles that must be overcome before a general purpose sensor network can be developed. These obstacles arise from the limited energy, computational power, and communication recourses available to the sensors in the network (Silva et al., 2005).

Based on these considerations, this paper presents a wireless sensor network node for real-time monitoring the estrus and body temperature of unrestrained dairy cattle. This system consists of cattle module, which is responsible for the amplification and transmission of body data. The system adopts the ZigBee wireless technology, which enables long battery lifetime and offers the opportunity to build up complex wireless networks of sensors.

2 Hardware Design

The system consists of two main modules: the first module is responsible for the amplification and transmission of acquired cattle physiological signals, and the second module receives data and connects the system to a host. The system architecture is shown in Fig. 1.

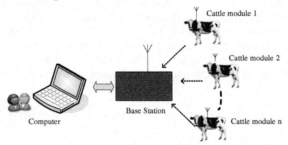

Fig. 1. System architecture

The cattle module is placed on the cattle, which measures physiological signals through sensors, digitalizes data, and transmits it to the base module. This module consists of three main parts. (1) digital sensor includes temperature and vibration-displacement sensor; (2) digital amplification; (3) microprocessor; (4) transmission of the digital data. Those modules use an I2C bus that can be connected to the different digital sensor instruments. The hardware of node is shown in Fig.2.

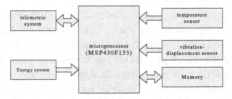

Fig. 2. The composing frame of data collector

2.1 Digital Sensor

2.1.1 Temperature Sensor

The DS18B20 communicates over a 1-Wire bus that by definition requires only one data line for communication with a central microprocessor. It has an operating temperature range of -55°C to +125°C and is accurate to ±0.5°C over the range of -10°C to +85°C. In addition, the DS18B20 can derive power directly from the data line, eliminating the need for an external power supply (Wang et al., 2007).

The 1-Wire bus system uses a single bus master to control one or more slave devices. The DS18B20 is always a slave. When there is only one slave on the bus, the system is referred to as a "single-drop" system; the system is "multidrop" if there are multiple slaves on the bus.

The communication circuit between 18B20 and MSP430F133 is shown in Fig.3. The VDD is connected to external supply. The 1-Wire bus requires an external pull up resistor of approximately 5kΩ; thus, the idle state for the 1-Wire bus is high. If for any reason, a transaction needs to be suspended.

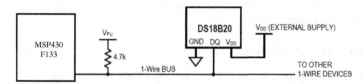

Fig. 3. The interface circuit between 18B20 and MCU

2.1.2 Vibration-Displacement Sensor

According to monitor requirement, we use vibration-displacement sensor (DN series-China). It has high detection sensitivity and good anti-interference ability. DN series sensor is not mechanical fatigue, false trigger, and so on for working at solid detection mode. The main parameter is as follows:

(1) Sensitivities: 0.19;
(2) Volume: 0.8×2×2.2 cm, it is shown in Fig.4.
(3) Test direction: omnidirectional;
(4) Working temperature:- 30℃~+65℃

ND-2
Fig. 4. ND-2

2.2 Microprocessor

Microprocessor uses MSP430F133. The MSP430F133 captures signals of temperature and vibration-displacement, convert them to digital values and process and transmit the data to a base module. The timers make the configurations ideal for industrial control applications such as ripple counters, digital motor control, EE-meters, hand-held meters, etc. The hardware multiplier enhances the performance and offers a broad code and hardware-compatible family solution.

2.3 Transmission Module

Transmission Module uses nRF903. The nRF903 is a true single chip multi-channel UHF transceiver designed to operate in the unlicensed 433MHz, 868MHz and 915MHz ISM-/LPRD-bands. Multi-channel operation, excellent receiver selectivity and sensitivity, high bandwidth efficiency and blocking performance make the nRF903 suitable for wireless links where high reliability is a key requirement.

The device features GFSK modulation and demodulation capability at an effective bit rate of 76.8kbit/s in 153.6kHz channel bandwidths. Transmit power can be adjusted to a maximum of 10dBm which is available for all frequency bands and channels. Antenna interface is differential and suited for low cost PCB-antennas. All necessary configuration data is programmed by a 14-bit configuration word via a SPI.

3   Design of Node Software

Node software realizes its functions of data collecting, data processing under the guide idea of modularization. A characteristic of this software is its good utility, expansibility and operating.

3.1 Temperature Collection

The transaction sequence for accessing the DS18B20 is as follows:

Step 1. Initialization;
Step 2. ROM Command (followed by any required data exchange);
Step 3. DS18B20 function command.

It is very important to follow above sequence every time the DS18B20 is accessed, as the DS18B20 will not respond if any steps in the sequence are missing or out of order.

3.2 Programing of nRF903

Step 1. Initialization : Setting registers in chip, working frequency and transmit power.
Step 2. Entering normal working state: working transceiver conversion control, sending data and state transition.

3.3 ZigBee-Stack

The ZigBee protocol is designed to communicate with experimental modules without information loss, using minimum microprocessor program memory space. It is a distributed medium access control protocol that provides robustness against single node failure and supports for flexible topologies, in which nodes are partially connected and not all nodes need to have a direct communication with a host. This process repeats until a maximal interval of data receiving was reached (Sinem et al., 2006).The system structure of ZigBee-stack is shown in Fig.5.

ZigBee-Stack operates in an OSAL operation system. The task in OSAL adds to this system by API, which could realize the multitask mechanism. The OSAL task

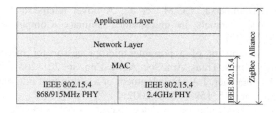

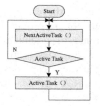

Fig. 5. ZigBee system structure diagram

Fig. 6. OSAL task scheduling mechanism

scheduling is shown in Fig.6. NextActiveTask( ) is task event query function what returns the ActiveTask. When we design software, we can decide if executing to corresponding ActiveTask( ) by the value of ActiveTask.

4   Conclusion

In this paper, a novel WSN node for monitoring physiological signals in unrestrained cattle was designed, which based on MSP430F133, DS18B20 and nRF903. The sensor node is designed through modular design and its advantage is compact structure and small volume. The node can collect physiological signals for unrestrained dairy cattle and send it to upper network node. This research can provide better hardware platform for further researching the communication protocols of wireless sensor networks. The next work will complete the whole system, including many cattle modules, station module and host (or PDA).

Acknowledgements

Funding for this research was provided by Heilongjiang youth science and technology special funds project (No. QC2009C18); Innovation Talent Research Fund of Harbin Science and Technology Bureau (2008RFXXN003); Innovation Team Fund of Northeast Agriculture University (CXZ010-1).

References

Silva, A.C.S., Arce, A.I.C., Souto, S., et al.: A wireless floating base sensor network for physiological responses of livestock. Computers and electronics in agriculture 49, 246–254 (2005)

Sinem, C.E.: Power efficient and delay aware medium access protocol for sensor networks. IEEE Transactions on Mobile Computing 5(7), 920–930 (2006)

Hahn, G.L., Eigenberg, R.A., Nienaber, J.A., et al.: Measuring physiological-responses of animals to environmental stressors using a microcomputer-based portable data logger. J. Anim. Sci. 68, 2658–2665 (1990)

Lowea, J.C., Abeyesinghea, S.M., Demmersa, T.G.M., et al.: A novel telemetric logging system for recording physiological signals in unrestrained animals. Computers and electronics in agriculture 57, 74–79 (2007)

Feddes, J.J.R., Deshazer, J.A.: Development of a portable microprocessor for measuring selected stress responses of growing pigs. Trans. Am. Soc. Agric. Eng. 36, 201–204 (1993)

Eigenberg, R.A., Brown-Brandl, T., Nienaber, J.A.: Development of a respiration rate monitor for swine. Trans. Am. Soc. Agric. Eng. 45, 1599–1603 (2002)

Wang, W., Zebin, F.: Design of wireless temperature detect based on CC2430. Electronic Engineer 33(8), 78–80 (2007) (in Chinese)

The Application of Wireless Sensor Networks in Management of Orchard

Guizhi Zhu

Institute of Sci-Tech Information, Guangdong Agricultural Academy of Sciences, 510640, Guangzhou, China

Abstract. A monitoring system based on wireless sensor network is established, aiming at the difficulty of information acquisition in the orchard on the hill at present. The temperature and humidity sensors are deployed around fruit trees to gather the real-time environmental parameters, and the wireless communication modules with self-organized form, which transmit the data to a remote central server, can realize the function of monitoring. By setting the parameters of data intelligent analysis judgment, the information on remote diagnosis and decision support can be timely and effectively feed back to users.

Keywords: wireless sensor networks, orchard management, real time, database management system.

1   Introduction

Wireless Sensor Networks (WSN) is self-organized networks whose nodes are capable of sensing, gathering, processing and communicating data. Compared to the cable network can be several kilometers long, WSN don't depend on any preexisting infrastructure. So it can be deployed where it is unsafe or unwise for field studies. More recently, the availability of inexpensive sensors that are able to measure physical phenomena like temperature, pressure, light, humidity, or location of objects has enabled the development of WSN. WSN has drawn the attention of the research community since it arised in 1970, driven by a wealth of theoretical and practical challenges. Panchard.J(2008) compares two types of WSN architecture and apply them to india agriculture for guiding irrigating. An WSN application to real-world habitat monitoring was developed by Mainwaring et al(2002). Wang et al (2006) presents an overview on recent development of WSN and gives some examples of WSN applied in agriculture. Zhang W.Y et al(2007) proposes a coverage strategy to solve power management problem. Prabhat Ranjan proposes the possibility of monitoring parameters which impact the quality of life in rural india. Ian F. Akyildiz et al(2006) explores wireless multimedia sensor networks (WMSNs) architectures, and survey algorithms, protocols, hardware for WMSNs.

This paper develops an architecture of WSN and describes the components in detail. WSN is deployed in the orchard to fulfill these requirements: collecting the data

D. Li and C. Zhao (Eds.): CCTA 2009, IFIP AICT 317, pp. 519–522, 2010.

information, such as temperature, light, soil moisture, humidity. Transmitting the data to a remote central server beyond the orchard. Developing a database management system running in the remote central server for data receiving and management. From the gathered information, the decision for orchard management can be made rightly.

2 System Design

We now describe the system architecture. It contains several components: sensor network, GPRS/INTERNET, remote central server, base station. Sensor nodes communicate with each other by self-organized format to make up sensor network, and transmit their data to the base station. The base station is responsible for transmitting data to remote central server through GPRS. A platform for orchard management system is established and runs in the remote central server. Finally, The data is displayed to users through a web-based user interface. Mobile devices, such as PDA and laptop, implement command to WSN through GPRS. The system architecture is as shown in Figure 1.

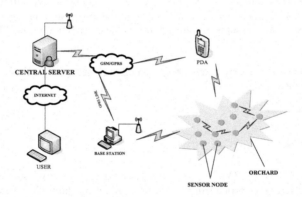

Fig. 1. System architecture

The operating system running in nodes is TinyOS, which is widely used. Power consumption is one of the main considered issues in the design of WSN. There is an efficient protocol used in WSN. We concentrate on low cost, low energy consumption sensors that will be densely deployed and provide detailed parameter information in the surrounding environment of orchard.

The Web-based Data Management plays an important role and is responsible for storing all information in WSN, such as sensor location, data received from a particular node. The remote central server stores and displays data to users, also, one can acquire and send commands to the WSN, thus enhance interaction with the environment.

Scientific users hope to enable widespread environmental monitoring and collection of experimental data.

3 Deployment and Orchard Management

We proceeded to the deployment of WSN in agricultural base in 2008, which is a high-tech agricultural base of the Institute of Sci-Tech Information, Guangdong Agricultural Academy of Sciences. The agricultural base is an area of 700 acres of hilly areas. Fixed telephone and power supply is available. There is a radio relay station in the north-west of the base, so communication signals are very well.

We deployed 20 different sensors among orchard to collecting the data information, such as temperature, light, soil moisture, humidity. Zigbee technology forms connections between the nodes. The platform used is MICA2, which was produced by Crossbow. The nodes are generally stationary after deployment except for a few mobile nodes. The sensors collect data and transmit data every 5 minutes. Our main goals are prolonging the life of the network and preventing connectivity degradation through aggressive energy management as the batteries cannot usually be replaced because of operations in hostile or remote environments. Dynamic environmental conditions requiring the system is adaptive to changing connectivity and node failure.

A database management system is developed running in the remote central server for data receiving and management. From the gathered information, the decision for orchard management can be made rightly by users. A map of the deployment in orchard is depicted in Figure 2.

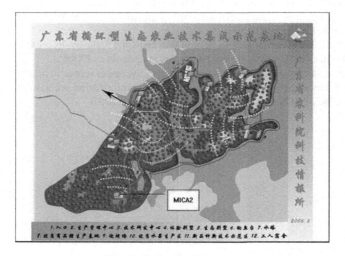

Fig. 2. Orchard deployment

4 Discussion

Applications of WSN in agriculture are still at its early development stage. This paper has made a description of WSN, the system has been used in orchard, running in an automated fashion. Deployment issues are covered in detail. At present, the system hasn't be used widely. It's important to improve the energy efficiency and reduce the cost of sensors.

In our deployment, sensors and central server is equipped with a GPRS unit, the system achieves remote monitoring by WSN and the GPRS/INTERNET, which overcomes the limitations of traditional on-site closed monitoring. The future development direction of the system is develop the decision-support system.

Acknowledgements

This work was supported by a grant from Guangdong Natural Science Foundation (8151064001000009), Guangdong Provincial Public Laboratory on Wild Animal Conservation, Management, Key Technologies R & D Program of Guangdong Province of China (2008A020100026, 2007A020300004-7).

References

Mainwaring, A., Polastre, J., Szewczyk, R., Culler, D., Anderson, J.: Wireless Sensor Networks for Habitat Monitoring
Akyildiz, I.F., et al.: A survey on wireless multimedia sensor networks. Computer networks, 1–40 (2006)
Panchard, J.: Wireless Sensor Networks for Marginal Farming in India, PhD Thesis (2008)
Ranjan, P.: Sensor networks to monitor quality of life in rural area,
    http://intranet.daiict.ac.in/~ranjan/research/papers/
    nsc_93rd_talk.pdf
Wang, N., et al.: Wireless sensors in agriculture and food industry-Recent development and future perspective. Computers and electronics in agriculture (2006)
Zhang, W., et al.: Energy-efficient coverage stategy in wireless sensor networks based on clustering by k-coverage. In: Proceedings of the 4th information technology in agriculture (ISIITA), Beijing,China, October 26-29, pp. 546–550 (2007)

Research of Rice-Quality Based on Computer Vision and Near Infrared Spectroscopy

RuoKui Chang[1], WeiYu Zhang[1], Jing Cui[2], YuanHong Wang[3,*], Yong Wei[1],
and Yuan Liu[1]

[1] Department of Electromechanical Engineering, Tianjin Agricultural University,
Tianjin, P.R. China 300384
[2] Department of Agronomy, Tianjin Agricultural University, Tianjin, P.R. China 300384
[3] Department of Horticulture, Tianjin Agricultural University, Tianjin, P.R. China 300384,
Tel.:13652126792; Fax:86-22-23781291
changrk@163.com

Abstract. A rapid and nondestructive way to measure protein and amylose content of rice was put forward based on near infrared(NIR) spectral technology. The NIR spectra were acquired from 13 varieties of rice with the wavelength from700 to 1100nm. The objectives of the present study were to establish forecasting model to find out the relationship between the absorbance of the spectrum and the main components of rice. By using the machine vision-based method, the rice appearance quality can be studied. On the basis of the evaluation criteria, 13 different kinds of rice were classified. And according to the usage of neural network, the detection model was established, so it can lay the foundation for the prediction grade of the unknown kinds of rice in the future.

Keywords: near-infrared spectroscopy; appearance quality; machine vision; artificial neural network.

1 Introduction

Rice is the major food crops of China. Rice-quality is the basic characteristic of rice as a commodity in the circulating process. It mainly includes internal quality and appearance quality. The amylose content (AC) and the protein content (PC) are the two central indicators of the internal quality of rice. The amylose content is one of the most important elements affecting the rice cooking and processing properties. The amount of it will directly affect the cooking and eating quality of rice. So it is often used as the evaluation index of grain quality characteristics of the parboiled rice. Protein is the major nutrient of rice, but the rice with high protein content will become light yellow, and can be degenerative during the process of storage. It makes the quality of cooking and appearance of rice lower (Mo huidong et al., 1993;Liu jianxue et al.,2001).So the protein content of rice is often regarded as one of the most indexes for the evaluation of grain quality. As a main commodity trait of rice, the appearance quality determines

---

* Corresponding author.

D. Li and C. Zhao (Eds.): CCTA 2009, IFIP AICT 317, pp. 523–531, 2010.
© Springer-Verlag Berlin Heidelberg 2010

524   R. Chang et al.

the market price of rice to a large extent. At present, the evaluation method of the rice quality in China is still at the level of naked eye observation. The test results not only lack objectivity and repeatability, but also time consuming and laborious for operation. It has the lower cognition rate of the trait parameters which reflect the appearance quality of rice, such as head milled rice rate, chalkiness degrees. It greatly reduces the competitiveness of China's rice in the international market.

For these reasons, near infrared spectroscopy is used for the building of the forecasting model of amylose and protein content of rice, the non-destructive test of the main components of rice is realized, the testing time is shortened, the internal quality of rice is quickly and easily determined, it is convenient for high-quality breeding; the machine vision technology can be used to identify the form of rice quickly and effectively, and on the basis of the evaluation criteria, different kinds of rice are classified.

2   Materials and Methods

2.1   Preparing Test Materal

The test samples of rice are provided by China-Japan Joint Center on palatability and quality of rice. 13 kinds of samples are selected for the inspection. 20g of each kind of rice is selected; the rice samples can be gained through browning and milling for 2 minutes. Chemical calibration value of amylose and protein content are gained by iodine colorimetry and Kjeldahl method respectively, these values are provided by China-Japan joint center. It is shown in Fig.1.

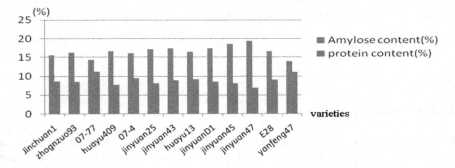

Fig. 1. Amylose content and protein content of different varieties of rice

2.2   Non-Destructive Test of Amylose and Protein Content of Rice

2.2.1   The Composition of NIRS and the Testing Process

The detection system of amylose and the protein content is made of two parts, computer and near infrared spectroscopy instrument. The USB 2000 miniature fiber optic Spectrometers, which is produced by ocean optics, inc. is used. The wavelength coverage is 200-1100nm. It is mainly composed of the S2000 miniature fiber optic spectrometer, an A/D converter, our operating software, a light or excitation source, and sampling optics.

The testing process is as follows: The light or excitation source sends light through an optical fiber to the single grain. The light interacts with the sample. Then the light is collected and transmitted through another optical fiber to the spectrometer. The spectrometer measures the amount of light and the A/D converter transforms the analog data collected by the spectrometer into digital information that is passed to the software, providing us with application-specific information.Fig.2 shows us the system connection diagram.

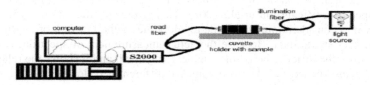

Fig. 2. Connection diagram of NITS detection system

2.2.2 Spectral Acquisition

The near-infrared spectroscopy of substance is a reflection of its molecular structure. Fig.3 shows the molecular absorption peaks correspond to the form in the vibration of the rotation group. The absorption peak with XHn (O-H, N-H, C-H) functional groups are in the region domination of the near-infrared spectroscopy(Zhang jun et al., 2003) So three points are selected, which are the frequency characteristics of protein contains, respectively, 785nm (third harmonic generation), 910nm (C-H stretching vibration), 1020nm (N-H stretching vibration, the second octave). The characteristic frequency of amylase is at 990nm (O-H stretching vibration, the second octave) (Li minzan et at., 2006) The peak absorption of Near Infrared Reflectance Spectroscopy of protein and amylase content are collected one after another from 13 kinds of single-grain samples.

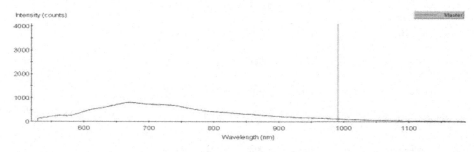

Fig. 3. The relationship between absorption peak and wavelength

2.2.3 Preprocessing Spectral Data

Spectroscopy will inevitably produce errors in the determination procedure; the mathematical pre-processing method can be used to deal with the specific samples. The abnormal samples can be removed, spectral noise can be eliminated, data variables can be selected, spectrum of information is purified, and all non-target effects of factors on the spectrum are reduced. It will lay the foundation for establishing spectral correction model and forecasting the concentration of the components of unknown samples.

First of all, a number of points around the smoothing point are selected to get the best estimation of result. Its purpose is to eliminate the random error of high-frequency. Secondly, as a result of the influence of derivative spectra on the elimination of the baseline drifting or the moderating of the background interference, it can also offer the transformation of spectral outline which is higher and clearer than the original spectrum (Zhang jun et al., 2003) .Taking the specific circumstances of the laboratory into consideration, we decide to adopt the method of first-order differential to preprocess spectral data. Because the actual measurement of spectrum is discrete spectrum, after the differentiation of it, table 1 is gained, and it shows the relationship between the absorbance and chemical values at different wavelengths.

Table 1. The relationship between the absorbance and chemical values at different wavelengths

various	Protein					Amylase		
	Content(y)	lgy	$lg\chi_{785}$	$lg\chi_{910}$	$lg\chi_{1020}$	content(y1)	Lgy1	$Lg\chi_{990}$
jinyuan25	8.1	0.908	1.273	0.520	0.318	17.14	1.234	0.420
jinyuanD1	8.6	0.934	0.981	0.417	0.246	17.34	1.239	0.311
07--4	9.5	0.977	1.192	0.502	0.297	16.08	1.206	0.365
jinyuan43	9	0.954	1.216	0.509	0.307	17.42	1.241	0.381
huayu13	9.2	0.963	6.863	0.473	0.463	16.46	1.216	2.363
jinyuan45	8.2	0.913	6.513	0.381	0.080	18.53	1.267	3.813
E28	9.1	0.959	6.584	0.447	0.982	16.59	1.219	2.784
jinyuan47	6.9	0.838	1.060	0.442	0.260	19.42	1.288	0.338
jinchuan1	8.7	0.939	4.392	0.468	2.192	15.45	1.188	2.092
huayu409	7.8	0.892	9.002	0.402	0.430	16.55	1.218	1.969
07-77	11.2	1.049	3.245	0.530	0.645	14.25	1.153	2.012
zhongzuo93	8.5	0.929	1.179	0.475	0.284	16.22	1.210	0.385
yanfeng47	11.1	1.045	8.322	0.404	0.610	13.91	1.143	4.223

2.2.4   The Establishment of Forecasting Models

Near infrared spectroscopy is composed of the absorbance of many wavelengths, the absorbance of each wavelength is the volume, which changes with different specimen. Therefore the spectrum can be regarded as the composition of many variables of multi-variable data. Protein and amylose content of rice are selected as objective functions respectively in this experiment, the absorbance value of absorption data of absorbance band is chosen as independent variables, the partial least squares (PLS) linear regression model of rice is established.

Assuming the prediction model between the rice protein content (y) and absorbance (x) is as follows:

$$y = -0.0144x + 0.9126 \tag{1}$$

Assuming the prediction model between the amylose content (y) and absorbance (x) is as follows:

$$y = -0.0027x + 1.3251 \tag{2}$$

Table 2 shows the results of calibration test of quantitative analysis model of protein and amylase content of rice.

Table 2. The results of calibration test of quantitative analysis model of protein and amylase content of rice

content	Calibration set				Test set			
	Ingredients range	R	SEC	Relative Standard error(%)	Ingredients range	R	SEP	Relative Standard error (%)
amylose	16.08-19.42	0.69	0.03	2.10	13.91-16.55	0.69	0.02	1.93
protein	6.9-9.5	0.61	0.16	7.12	7.8-11.2	0.64	0.04	4.43

Note: SEC is the standard error of estimate from the calibration; SEP is the standard error of predication.

2.3   The Establishment of Machine Vision Systems

2.3.1   Image Sampling of Rice

By using color image analysis system, the image sampling of rice of single grain is completed. The color image analysis system is composed of computer system, camera system and color image analysis software package. The objects can be analyzed quantificationally by using computer multi-media. 100 images of single-grain of 13 varieties of rice are selected respectively in this test. Using electron microscopy analyzer, we can detect morphology parameters of rice and analyze the morphological characteristics of rice. These features include: the largest and smallest diameter, ratio of grain length to width, the outline of the complexity, the average gray level, the average optical density, the average transmission rate and so on. After image processing, parameters of these patterns are quantitative.

2.3.2 Appearance Indicators Analysis and Data Processing

The appearance quality refers to the indicators which are related to the characteristics of rice, such as the chalkiness, translucency, and grain shape. Chalkiness refers to the opaque parts of grains(Li tianzhen et.al.2005), it is a kind of optical feature caused by loose filling and inflating of starch and albumen grains in the endosperm of rice. The chalkiness of rice refers to the percentage of the shares of chalky areas in the projected areas of the whole rice when the chalky rice is placed flatly. It is a defect of the grain structure. It can influence the appearance of the rice, and also, it can make the grains broken in milling process. Thus, it is an undesirable characteristic of rice, and it has become one of the objects which scientists need to research emphatically and overcome. Chalkiness degree is used to describe the degree of chalkiness. It refers to the parts of chalky areas and the percentage of total areas(Sun Ming et.al.2002). The formula is as follows:

$$\text{Chalkiness ratio (\%)} = (\text{part chalkiness area} / \text{total area of grain}) \times 100\% \qquad (3)$$

$$\text{Chalky rice ratio (\%)} = (\text{containing rice grains} / \text{total grains}) \times 100\% \qquad (4)$$

The translucency of rice is an indicator to describe the transmission characteristics of grains. Variety is the main factor affecting the translucency of rice. Cultivation and processing conditions also have an impact on it. The grain shape is described with the length, width and the ratio of it, as shown in formula 3. In the past the grain shape was judged as quality indicators, it is more reasonable to judge it as classified parameters currently.

$$\text{length/ width ratio of grain} = \text{average length of grain (mm)} / \text{average width of grain (mm)} \qquad (5)$$

In accordance with the national rice quality standard (GB / T 17891-1999), the grade of external characteristics can be divided into three levels, according to head milled rice rate, chalky rice percentage, chalkiness degree, translucency, grain shape etc.

If two or more indicators are not up to standard, but can up to the next rank, they will be reduced to a lower rank; if any of the indicators can't meet the three requirements, it can't be used as high-quality rice. The grade judgment of appearance of rice is shown in table 3. By using electron microscopy analyzer, we can collect and analyze the data of single grain to obtain outside characteristic parameters, as is shown in table 4.

Table 3. Indicators of appearance quality

Item	Translucency and gloss	Chalkiness size (%)	Chalky rice ratio (%)	Grain shape
First Class	Translucent glossy >0.7	<5	<5	>1.7
Second-class	Translucent 0.61-0.7	<10	<10	>1.6
Third class	Poor transparency 0.46-0.6	<20	<20	>1.5

Table 4. Analysis of different data characteristics of rice

various	grain length to widthratio(LWR)	Average translucency	Chalky rice percentage(%)	Appearance degree
Jinchuan1	1.530121	0.801545	3	II
zhongzuo93	1.520617	0.710783	8	III
07-77	1.706824	0.785607	21	O
huayu409	1.479549	0.777721	6	III
07-4	2.07547	0.709472	3	I
jinyuan25	1.638583	0.761274	6	III
jinyuan43	2.011645	0.780593	3	I
huayu13	1.639941	0.792892	12	III
jinyuanD1	1.603584	0.714055	9	III
jinyuan45	1.577334	0.772294	14	III
E28	1.929455	0.704889	1	I
yanfeng47	1.665421	0.753477	16	III
jinyuan47	1.738327	0.787858	15	III

Note: O stand for out of high-quality rice.

Table 4 shows the appearance quality of 07-4, jinyuan43, E28 are better than others.

2.3.3 The Building of Forecast Models

As a new method of information processing, artificial neural network has highly parallel characteristics and the features of input signals. The mapping relationship between input and output can be learned from it, and associative memory can be integrated. When a particular input signal is given, the appropriate output signal can be renewed. And when a new input signal is given, an appropriate output signal can be generated. As it is difficult to establish a linear relationship between the appearance characteristics of rice and grade, in order to rapidly and accurately determine the grade of rice, we use three-tier BP neural network. We use the main components, like translucency, chalkiness size, grain shape, which are extracted from the images of single grain rice as the input layer of BP network. The output layer corresponds to the level of grading standards, so the mapping relationship can be established from P3 to y1.

The 300 single grain of rice were selected randomly, in which 250 samples for training, 50 as testing samples, Fig.4 shows the network structure.

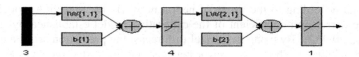

Fig. 4. The structure of ANN

The correlation coefficient of the calibration model established by appearance indicators is 0.9621. The finished models are used to forecast the prediction set, the correlation coefficient of the prediction is 0.9134. The calibration set or the prediction set, it can be seen that the correlation between the forecast value and the true value is significant. The using of artificial neural network has a good effect on establishing models.

Table 5. Predicted values and true values of appearance quality degree

subject	Predicted value(degree)	True value (degree)	relative error	subject	Predicted value(degree)	True value (degree)	relative error
1	2.1774	2	0.0887	5	2.1544	2	0.0772
2	2.1544	2	0.0772	6	1.9000	2	-0.05
3	1.1342	1	0.1342	7	1.2198	1	0.2198
4	2.8976	3	0.03413	8	1.7894	2	0.1053

Note: the data listed above are the results of 8samples among 50 unknown samples

Table 5 shows the forecast value, the true value and the relative error of 8 samples, which are selected from prediction set.

From table 5, we can see that the relative error between the forecast value and the true value is too small. The average relative error is 5%.

3 Results and Discussion

In the experiment, the near-infrared spectral analysis technology is used for the selection of 13 kinds of intact rice samples, the spectrogram of protein and amylase of rice is gathered, By extracting the absorbency from the functional group with the characteristic value of 785nm、910nm、990nm and 1020nm, the forecasting model of quantitative determination of protein and amylose content can be established. In the calibration set, the determination coefficient R of amylose and protein is 0.69 and 0.61 respectively, and the relative standard error is 2.1% and 7.1% respectively. The correlation coefficient r of the validation set is 0.69 and 0.64 respectively. This NIR model is used to predict the contents of amylase and protein in 200 samples, When using the standard error of predication (SEP) to assess the deviation of analysis results is 0.02 and 0.04 respectively, the relative standard error is 1.95% and 4.43% respectively. These results would provide convenience for the breeders to select varieties of high-quality rice to facilitate breeding.

The machine vision technology was used to obtain three appearance parameters of rice. On the basis of the evaluation criteria, 13 kinds of rice are divided into three grades. And according to the usage of neural network, the detection model is established, the correlation coefficient of the calibration model established by appearance indicators is 0.9621. By using the model to forecast the prediction set, the correlation coefficient is 0.9134. The relative error between the forecast value and the true value is relatively small. The average relative error is 5%, so it can lay the foundation for the prediction grade of the unknown kinds of rice in the future.

Acknowledgements

Funding for this research was provided by Tianjin Agricultural University (P. R. China).

References

Wenchuan, G., Xinhua, Z.: Research Progress of Machine Vision Technology in Grains Identification and classification. Cereal & Feed Industry 6, 50–51 (2002)

Minzan, L.: Spectroscopy Analytical Technique and Application. Science Press, Beijing (2006)

Tianzhen, L., Boqing, Z.: Research on rice- quality inspection basing on computer vision technology. Cereal and food industry 12(4), 50–53 (2005)

Jianxue, L., Shouyi, W., Ruming, F.: Rapid Measurement of Rice Protein Content by Near Infrared Spectroscopy. Transactions of The Chinese Society of Agricultural Machinery 32(3), 68–70 (2001)

Ming, S., Yun, L., Yiming, W.: Computer Vision Based Rice Chalkiness Detection Using MATLAB 18(4), 146–149 (2002)

Jinchun, X.: The Study of Nondestructive Detection with NIR Spectroscopy in Single Rice Qualities. Master's thesis, China Agricultural University 6, 2–4 (2003)

Guoqing, X.: The rice quality and evolution indicators (1). Hunan Agricultural 2, 15 (1995)

Guoqing, X.: The rice quality and evolution indicators(1). Hunan Agricultural 3, 16 (1995)

Jun, Z., Yongmei, Z., Fangrong, W.: Discussion on some regular methods for cereal near infrared spectra analysis. Journal of Changchun Post and Telecommunication Institute 21(1), 4–8 (2003)

Soil Suitability Evaluation for Tobacco
Based on Grey Cluster Analysis

Gao Rui[1], Qiao Hong-bo[1], Zhang Hui[1], Su Yong-shi[2], and Chen-Yanchun[2]

[1] College of Information and Management, Henan Agricultural University,
Zhengzhou 450002, China
[2] Sanmenxia Branch of Henan Tobacco Company, Sanmenxia, 472000, China

Abstract. Suitability evaluation of soil for tobacco is the base of spatial analysis and optimization disposition. It provides scientific basis for reasonable development of soil for tobacco. Taking soil in San Menxia city of Henan province as a study of object, five factors which had relation of tobacco growth are adopted, grey cluster method are carried out to appraise suitability evaluation. The result shows that some soil area of Lingbao, Mianchixian and Shanxian are higher suitability areas that contain higher organic matter and K contents ,while the soil area of Lushixian has lower organic matter and N contents are less suitability area. Compared with traditional methods, grey cluster method determined the weight based on factors. The result is objective reasonably.

Keywords: soil nutrients; suitability evaluation; grey cluster method.

1 Introduction

Soil planting tobacco suitability evaluation aims to optimally allocate crop planting through measuring the coupling of designating crop to given tobacco land, and considers simultaneously soil physical features and current/future land use patterns (Qiu, et al., 2005) .Soil condition is the base of top quality tobacco engineering and the main environmental factor that affects the quality of tobacco leaves(Zhang, et al.,2003).Soil suitability evaluation for tobacco has important practical significance that not only fully and rationally use natural resource and further develop land potential, but also promote social economic sustainable development. Qiu researched dynamic soil suitability evaluation by using artificial neural network method and discussed how to realize soil suitability evaluation in time and space (Qiu, et al., 2002); Liu studied the implementation method of expert system for soil suitability evaluation (Liu, et al., 2001); Chen studied the soil fertility suitability evaluation for tobacco in Henan province by using the methods of AHP and fuzzy mathematics (Chen, et al.,2007); Wang analyzed spatial variability characteristics of tobacco-planting soils and made suitability evaluation (Wang, et al.,2008).Soil suitability evaluation used the methods of parametric, geo-statistics(Chen, et al.,2003), fuzzy mathematics(Chen, et al.,2007) had been carried out in china. But the evaluation with the method of grey cluster is rarely reported.

D. Li and C. Zhao (Eds.): CCTA 2009, IFIP AICT 317, pp. 532–538, 2010.

The tobacco-growing distribution area is from 300 to 1500m above sea level, topography complex and soil nutrient is of great spatial variability. This study take Sanmenxia soil as study object use grey cluster method to evaluate soil nutrient suitability to detect soil nutrient. The result will provide reference for rational distribution of tobacco, balanced fertilization and tobacco sustainable development.

2 Material and Method

2.1 Study Area

Sanmenxia is the leading tobacco-producing areas of quality, is located in inland mid-latitude areas, a warm temperate continental monsoon climate. The annual average temperature 13.2 ℃, annual average 2354.3 hours of sunshine, frost-free period of 184 ~ 218 days, with an average annual rainfall 550 ~ 800mm, is suitable for the growth of flue-cured tobacco district.

2.2 Sample Collection

The method in collecting soil samples was used according to the second national soil survey. Sanmenxia in the choice of representative tobacco soil, topsoil from 0 ~ 20cm soil samples, each sampling point of the area on behalf of 20 hm2. In 2007 collecting soil samples 299 were made.

2.3 Study Method

Grey fixed weight cluster method is the method that according to whitenization weight function of index gathering a number of observed objects into several defined types, it empowers the cluster index prior. Grey fixed weight cluster can be carried out based on the following steps (Liu, *et al.*, 2004):

Firstly, according to available indicators of observation, we define k subclass whitenization weight function of j index $f_j^k(\bullet)$ (j=1, 2, $\wedge$, m; k=1,2,$\wedge$,s).

Secondly, we determine cluster weight of every index η_j (j=1,2, $\wedge$,m).

Thirdly, according to the whitenization weight function $f_j^k(\bullet)$ (j=1,2, $\wedge$,m; k=1,2,$\wedge$,s), cluster weight η_j (j=1,2, $\wedge$,m)derived from the previous two steps ,as well as observation x_{ij} (i=1,2, $\wedge$,n ; j=1,2, $\wedge$,m) of object i on j index, we can calculate the coefficient of grey fixed weight $\sigma_j^k = \sum_{j=1}^{m} f_j^k(x_{ij}) \bullet \eta_j$, i=1,2, $\wedge$,n ; k=1,2, $\wedge$, s.

Finally, if $\max_{1 \le k \le s} \{\sigma_i^k\} = \sigma_i^{k^*}$ you can conclude that object i belong to grey category k^*.

3 Soil Suitability Evaluation

3.1 Selection of Evaluation Index

According to checking related literature and listening to experience and views of experts , based on the principles of influence that made tobacco quality and growth, variance degree, stability and operability, we finally select 5 indexes: PH, organic, available N, available P and available K.

3.2 Delineation of Suitability Standard

According to the evaluation of soil nutrient standards (Ren, *et al.*, 2007, Li, *et al.*,2008) and the actual situation of tobacco-growing areas in Sanmenxia, each index will be divided into three levels: most appropriate, suitable and unsuitable corresponding to 1~3(Table 1).

Because soil PH value changes between 6.6~8.8 in tobacco-growing areas of Sanmenxia, the data should be standardized to positively correlate with evaluation standard

Table 1. Soil nutrient suitability evaluation standard

index	unsuitable	suitable	most appropriate
organic(g·kg^{-1})	<5	5-10	>10
available N(mg·kg^{-1})	<25	25-40	>40
available P(mg·kg-1)	<10	10-20	>20
available K(mg·kg^{-1})	<80	80-150	>150
PH	<4.5	4.5-5.5	5.5-8.0
	>8.5	8.0-8.5	

3.3 Identification of Whitenization Weight Function

We will divide five evaluation index into three gray class: most appropriate, suitable, unsuitable in accordance with their respective evaluation criteria, then j(j=1,2,3,4,5)index k(k=1,2,3)gray class indicate the j index the k gray class. Then, we set whitenization weight function of j index k gray class:

$$f_j^k(\bullet)\,(j=1,\ 2,\ 3,\ 4,\ 5\ ;\ k=1,\ 2,\ 3)$$

Commonly used whitenization weight functions are typical, upper limit measure, moderate measure, lower limit measure and triangle whitenization weight function in grey system theory. Here, we select upper limit measure whitenization weight function measure the most grey class, moderate measure whitenization weight function measure the suitable grey class, lower limit measure whitenization weight function measure the unsuitable grey class. Based on the soil conditions, we determine the whitenization weight function as follows:

$$f_1^1[0,20,-,-],\ f_1^2[0,13,-,20],\ f_1^3[-,-,11,13]$$
$$f_2^1[0,65,-,-],\ f_2^2[0,50,-,65],\ f_2^3[-,-,45,50]$$
$$f_3^1[0,20,-,-],\ f_3^2[0,14,-,20],\ f_3^3[-,-,10,14]$$
$$f_4^1[0,250,-,-],\ f_4^2[0,180,-,250],\ f_4^3[-,-,120,180]$$
$$f_5^1[0,2.2,-,-],\ f_5^2[0,2.0,-,2.2],\ f_5^3[-,-,1.6,2.0]$$

Based on the above whitenization weight function we can write specific function expression. Here only given whitenization weight function expression of soil organic index three grey classes and other expression can be written similarly.

$$f_1^1(x) = \begin{cases} 0, & x < 0 \\ \dfrac{x}{20}, & x \in [0,20] \\ 1, & x > 20 \end{cases}$$

$$f_1^2(x) = \begin{cases} 0, & x \notin [0,20] \\ \dfrac{x}{13}, & x \in [0,13] \\ \dfrac{20-x}{20-13}, & x \in [13,20] \end{cases}$$

$$f_1^3(x) = \begin{cases} 0, & x \notin [0,13] \\ 1, & x \in [0,11] \\ \dfrac{13-x}{2}, & x \in [11,13] \end{cases}$$

3.4 Cluster Results and Soil Suitability Evaluation

Based on the survey data of tobacco fields , combined with observations and recommendations of relevant experts, the weight of organic, available N, available P, available K, PH value are: $\eta_1 = 0.45, \eta_2 = 0.15, \eta_3 = 0.08, \eta_4 = 0.150, \eta_5 = 0.17$.

Whitenization weight function $f_j^k(\bullet)$ $(j=1,2,\ \wedge,\ m;\ k=1,2,\wedge,s)$, cluster weight $\eta_i(j=1,2,\ \wedge,m)$ and observation x_{ij} $(i=1,2,\ \wedge,n\ ;\ j=1,2,\ \wedge,m)$ of object i for j index calculate the grey fixed cluster coefficient. $\sigma_i^k = \sum\limits_{j=1}^{m} f_j^k(x_{ij}) \bullet \eta_j, i = 1,2,\wedge,n; k = 1,2,\wedge,s$·

The cluster coefficient of three types can be made according to sample data and the above steps as shown in table 2.As the sample points located in every township, the township as a unit were to be analyzed. From the result, the areas that have the most

Table 2. Soil suitability evaluation cluster coefficient in Sanmenxia area

serial number	township	most suitable cluster coefficient	suitable cluster coefficient	unsuitable cluster coefficient
1	guandaokouzhen	0.69	0.91	0.10
2	duguanzhen	0.81	0.58	0.17
3	hengjianxiang	0.69	0.78	0.65
4	fanlizhen	0.69	0.78	0.27
5	wenyuxiang	0.67	0.78	0.70
6	dongmingzhen	0.75	0.72	0.26
7	shahexiang	0.65	0.78	0.32
8	panhexiang	0.69	0.95	0.24
9	mutongxiang	0.87	0.37	0.06
10	mokouxiang	0.72	0.84	0.17
11	xujiawan	0.86	0.54	0.00
12	shuanghuaishu	0.75	0.64	0.15
13	shizipingxiang	0.85	0.24	0.23
14	guanpozhen	0.92	0.29	0.00
15	wayaogouxiang	0.68	0.73	0.23
16	tanghexiang	0.57	0.79	0.87
17	wulichuanzhen	0.83	0.53	0.17
18	zhuyangguanzhen	0.84	0.57	0.07
19	caiyuanxiang	0.72	0.92	0.01
20	dayanwaxiang	0.64	0.87	0.30
21	gongqianxiang	0.70	0.69	0.11
22	xilicun	0.63	0.76	0.62
23	chencunxiang	0.69	0.84	0.35
24	guoyuanxiang	0.77	0.67	0.13
25	hongyangzhen	0.81	0.45	0.14
26	potouxiang	0.68	0.83	0.35
27	rencunxiang	0.81	0.46	0.19
28	tianchizhen	0.61	0.74	0.21
29	xiyangxiang	0.66	0.86	0.42
30	yangshaoxiang	0.76	0.87	0.24
31	yinghaozhen	0.66	0.71	0.31
32	chuankouxiang	0.76	0.69	0.08
33	suncunxiang	0.74	0.79	0.28
34	wumiaoxiang	0.86	0.45	0.00
35	zhuyangzhen	0.87	0.37	0.00

suitability are: Duguanzhen, Dongmingzhen, Mutongxiang, Xujiawan, Shuanghuai-shu, Shizipingxiang, Guanpozhen, Wulichuanzhen, Zhuyangguanzhen of Lushixian; Gongqianxiang of Shanxian; Guoyuanxiang, Hongyangzhen, Rencunxiang of Mian-chixian; Wumiaoxiang, Zhuyangzhen of Lingbao. The Soil of these 15 townships organic and soil K content is high, soil pH and N content is medium. The number of sample point account for 29.4%in total.

Suitable soil areas are: Guandaokouzhen, Hengjianxiang, Fanlizhen, Wenyuxiang, Shahexiang, Mokouxiang, Wayaogouxiang of Lushixian; Caiyuanxiang, Dayan-waxiang, Xilicunxiang of Shanxian; Chencunxiang, Potouxiang, Tianchizhen, Xiyangxiang, Yangshaoxiang, Yinghaozhen of Mianchixian; Chuankouxiang, Su-cunxiang of Lingbao. Although soil K content is high in these areas, soil organic content is lower in some parts, soil improvement should be pay attention to in the future. The point number is 210(70.4%).

Tanghexiang locate in Lushixian has poor soil suitability. The soil organic content is lower, pH value is higher , N and P content is lower.

Generally, soil planting tobacco suitability is good in Sanmenxia that suitable for the growth of high-quality tobacco leaf. In recent years, soil in this area pH value, N content has an upward trend, pay attention to balanced fertilization to ensure harmonious proportion of N, P, K content.

4 Conclusions

Gray clustering method for soil suitability evaluation is a new attempt, by choosing its evaluation index, determining whitenization weight function to carry out soil nutrient gray clustering. In this paper, using above method, the study of Sanmenxia tobacco soil suitability assessment carried out. The results show that almost 30% area of this region has better soil suitability. Less than 1% has poor foil suitability, rest area has medium suitability. Taking into account the soil itself is a complex multi-phase organic-inorganic complex, involving the interaction of various factors, the evaluation results must be combining with the actual production, index weight and the determination of whitenization weight function is still value the experience of using the traditional method. The evaluation results basically tallies with actual production. Soil pH value and available N content are higher in Sanmenxia, Attention should be given to improve soil, reduce soil pH value, CL content, control N content, complement P content and stabilize K content.

References

[1] Qiu, B.W., et al.: Fruit tree suitability assessment using GIS and multi-criteria evaluation. Transaction of the CSAE 21(6), 667–690 (2005)
[2] Zhang, Y., et al.: Oriental tobacco from different region and the intrinsic chemical composition analysis of aroma substances. Chinese Tobacco Science (4), 12–16 (2003)
[3] Qiu, B.W., et al.: Dynamic assessment of regional land resource suitability based on geographical information system. Journal of Soil 39(3), 301–307 (2002)
[4] Liu, Y.Z., et al.: Establishment of GIS-supported expert system for soil suitability evaluation. Chinese Journal of Soil Science 32(5), 193–196 (2001)

[5]   Chen, H.S., et al.: Comprehensive evaluation of soil fertility suitability for tobacco based on GIS in Nanyang city, He'nan province. System Science and Comprehensive Studies in Agriculture 23(4), 498–502 (2007)

[6]   Chen, H.S., et al.: Comprehensive evaluation system of soil fertility suitability in Xuchang tobacco planting regions based on GIS. Journal of Henan University (Natural Science) 39(1), 51–56 (2009)

[7]   Wang, Z.F., et al.: Spatial variability of soil nutrient and its suitability evaluation in tobacco planted area. Journal of Southwest University (Natural Science Edition) 30(1), 98–102 (2008)

[8]   Chen, W.H., et al.: Spatial variability analysis for environmental information of fields. Eco-Journal 24(2), 347–351 (2004)

[9]   Chen, H.S., et al.: Comprehensive fertility evaluation of soil for tobacco plantation in Henan province based on GIS. Chinese Journal of Soil Science 38(6), 1081–1085 (2007)

[10] Liu, S.F., et al.: Gray system theory and its application. Science Press, Beijing (2004)

[11] Ren, Z., et al.: Practical method of appraising soil suitability of planting tobacco in Guizhou. Journal of Anhui Agriculture Science 35(31), 9955–9956 (2007)

[12] Li, X.H., et al.: Feasibility of climate and soil in Enshi tobacco growing areas. Chinese Tobacco Science 29(5), 18–21 (2008)

Animal Disease Diagnoses Expert System Based on SVM

Long Wan[*] and Wenxing Bao

Department of Computer Science and Engineering, The North University for Ethnics,
Yinchuan Ningxia, P.R. China 750021
oneloong@gmail.com

Abstract. Livestock breeding farms usually distribute in remote areas, with relatively poor condition of disease diagnosis. Generally, it is hard to carry out disease diagnosis rapidly and accurately. But the farms can diagnose animal disease quickly and accurately by the animal disease diagnoses expert system. It could ensure a sound development of the stockbreeding industry. This paper proved the practicality of support vector machine (SVM) which is used in the animal disease diagnoses expert system in theory by studying the disease diagnosis expert system based on SVM. And the experiments proved that SVM can make the disease diagnose accurately.

Keywords: SVM, disease diagnosis, expert system.

1 Introduction

Diagnosing animal disease quickly and accurately has the economic effectiveness. Livestock breeding farms usually is relatively poor condition of disease diagnosis. Generally, it is hard to carry out rapidly and accurately disease diagnosis. It requires veterinary experts a lot of experience and theoretical knowledge to diagnose the disease. The expert system for animal diseases diagnosing can meet the farms for the urgent needs of veterinary experts, since there are very few experts at the farms. SVM (Vapnik et al., 1998) is a new machine learning, which developed on the basis of VC dimension of statistical theory and structural risk minimization (SRM). SVM has many advantages in the problems in dealing with the pattern recognition of limited samples, nonlinear and high dimensional. To get the best generalization ability, SVM can find the best compromise between the model complexity and the model study ability by limited samples information. In nature, animal disease diagnosis is a problem of pattern classification and recognition. It can judge the disease based on symptoms of animal suffered. In this paper, we designed the model of animal disease diagnoses expert system which was based on SVM and was used to diagnose the cow diseases. It shows that the method is practical and effective. And this practice provides a new approach for animal disease diagnosis.

---

[*] Corresponding author.

D. Li and C. Zhao (Eds.): CCTA 2009, IFIP AICT 317, pp. 539–545, 2010.

2 About Expert System

Expert System is an intelligent computer program. The intelligence mainly expressed that can mimic human experts' thinking to solve complex problems in the field (Li et al., 1996). Diagnosis expert system for animal diseases belongs to diagnosis expert system in modern medicine. Its mission is to quickly and accurately judge the type of disease from observed symptoms of the disease.

A typical disease diagnosis expert system is composed of knowledge base, reasoning engine, explanation unit, control strategies and human-machine interface, as shown in Fig 1.

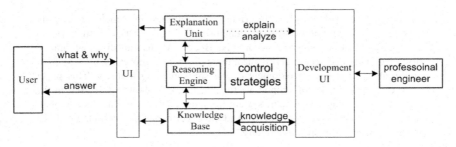

Fig. 1. System structure of animal disease diagnosis

Knowledge base: it is the core of the system, which stores professional knowledge of disease diagnosis, the principle knowledge and the data of disease cases, etc. Control strategy: it decides that system should be taken what way to find the reasoning rules. Reasoning engine: it obtains the inputting symptoms of the disease and analyzing, reasoning them and getting the result of disease diagnosis. Explanation unit: in order to make users to understand the reasoning process, it is responsible for answering various questions use raised. It gives an understandable and adequately explanation based on reasoning rules and reasoning conclusions. Human-machine interface: it is an interface of exchange information between the system and users.

Traditional rule-based expert system has its unique advantages: the description of rules is correspond with the process of knowing about problems and is more easily understood. However, over-reliance on rules to reason, the system has inevitable shortcomings: the rules of reasoning must be exact matches. It does not have the ability of self-learning and predictive, can not be self-improvement in the process of reasoning, and even lack effective treatment measures when there are some unavailable information in known information (Li et al., 1996). Some intelligence algorithm, such as neural networks and genetic algorithms, originate from the theory which is based on samples tended to be infinite, but the size of samples can not have a lot in practice. Hence the expert systems, taking these algorithms, can not get good performance in practice.

3 Fundamental Principle of SVM

At present, machine learning is one of the basic tasks in artificial intelligence field and statistics theory is its main analysis method. Its basic idea was to find the hidden rules through training samples, and using these rules to predict unknown data. V.Vapnik and others proposed a kind of learning machine support vector machine, which is based on limited samples' statistics learning theory. According to SRM theory, SVM can avoid local smallest point and solve over fitting problem effectively, so it has good extending capability and accurate classification. The basic idea of SVM can be illuminated by two kinds of linearly separable cases (training sets). As showed in Fig 2, filled points and hollow points stand for two kinds of samples. If they are linearly separable, the result of machine learning is a hyperplane (a straight line at two dimensions) or called discriminant function, which can divide training samples into categories: positive and negative. Obviously, according to the requirements of experience risk minimize (ERM), there are infinite hyperplanes like that. For training samples, though some hyperplanes have good classification capability, have bad predicting capability, such as hyperplane P_1 in Fig 2. According to the requirements of SRM, the result of learning should be the optimized hyperplane P_0 that it can not only classify two kinds of training samples accurately but also make the biggest classification margin. Making the biggest classification margin is controlling extending capability, which is one of core ideas of SVM.

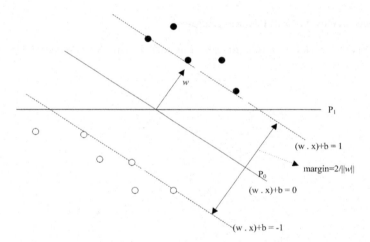

Fig. 2. Classification hyperplane on linearly separable

4   Animal Disease Diagnosis Expert System Model Based on SVM

4.1   Model Principle

The diagnosis methods based on disease symptoms still belong pattern classification problem virtually. It mainly searches the mapping relations between disease symptom and disease category. Animal disease diagnosis expert system based on SVM via training disease samples to obtain the mapping relations, and take disease symptom knowledge to implicitly represent on an optimize classification hyperplane in high dimensional features space. Moreover, the hidden rules among samples can be obtained and we can use them to diagnose diseases.

SVM was proposed for two classification problems primly, but it can be extended to multi-classification problems by adopting the combination method of binary classification. In this model, we choose the one-versus-one (OVO) (Blom et al., 1984, 1988) method, which has good classification capability. In processing of knowledge acquisition, we take the set, S: (x_1, y_1), (x_2, y_2),..., (x_l, y_l), of disease samples provided by experts and users as the training samples of SVM, and $x_i \in R^n, y_i \in \{\text{set of disease pattern (m)}\}$. The method constructed decision functions among disease categories, which can construct N = m(m-1)/2 decision hyperplanes. Each decision function is trained by two categories of corresponding samples. Max-wins-voting strategy is adopted when a disease sample x is classified. In the process if classification decision functions judge x belong to i category disease and the votes of i category disease add 1, otherwise, the votes of j category disease add 1. After N decision functions make the judgments, the disease category which has the most votes, is the diagnosis result of the unknown sample. For this model, we set a separate SVM for each disease according to the category of disease, and then all classified SVM are integrated, combined to a SVM classification group. The SVM group can reduce operation complexity of single SVM and improve reasoning efficiency and exactness of disease diagnosis.

4.2 Procedure of Animal Disease Diagnosis

The Fig 3 is the model of animal disease diagnosis based on SVM (MADDS)

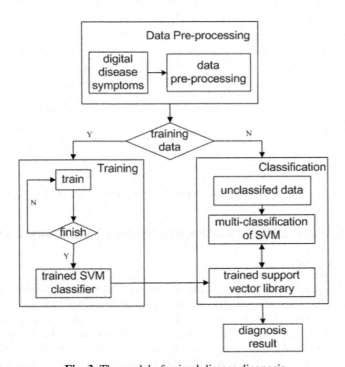

Fig. 3. The model of animal disease diagnosis

The model is composed of data pre-processing module, training module and classification of multi-value module. First, the model obtains symptoms of the disease which users have observed. Then data pre-processing module carries out data pre-processing, and works out the input vector form which correspond to SVM. If the vectors are the training data, the model will train them. And the results of training, a number of supporting vectors, will be saved in the knowledge vector library. The function of the three modules is as follows:

1.   Data pre-processing module of MADDS:

In the procedure of disease diagnosis, characteristic extraction is the key step. It usually converts the description of the symptoms of disease into the digital information, which the computer can handle, using standardized criteria. In response to the features of disease diagnosis, we created a relation tree of disease-symptoms about disease and disease symptoms information of disease, as Fig 4 shows. All animal diseases information can be classified, arranged, and formed a unified knowledge statement when using the tree structure as the standard.

In the course of disease diagnosis, the symptoms of the user inputting will be converted into a unified data format by the module. In this procedure, after the model normalizes the data using some of the basic concepts of the data reliability analysis, it can reduce the differences between the sample data and improve the convergence performance of the classifier.

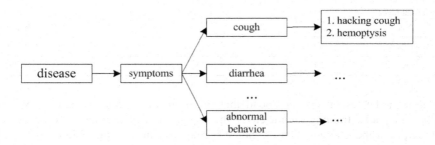

Fig. 4. Relation trees of disease-symptoms

2.   Training module of MADDS:

Since the diagnosis of animal diseases is a multi-classification problem, we constructed the multi-classification of SVM by the OVO method. The module chooses radial basis function as the kernel function of the classifier and used cross validation to train the data.

3.   Multi-valued classification module of MADDS

The module will classify the unclassified data by using the trained support vector library.

5   Experimental Study and Results

In this experiment, we used some data of cow disease to test. According to cow disease symptoms categories, the symptoms were classified as 26 categories, corresponding to 26 components of the support vector machine. Each disease sample data of training and testing should be filled in the corresponding information according to the format of Table 1.

Table 1. Vectors of disease symptom

	1.psychiatric condition	2.body surface	3.physical features	...	26.behaviors
Sample (preprocessed data)	0.5	0.015625	0.00076293	...	0.015625

In the test, we collected five kinds of cow disease. Each disease had 35 cases. We used 20 cases of them to be train, and other 15 cases to be tested by Lib-SVM (Chang et al., 2005). The test's results are shown in Table 2.

Table 2. Result of test

Disease category	Number of training sample	Number of test sample	Number of accuracy	accuracy
cold	20	15	11	73.33%
malignant catarrhal fever	20	15	15	100%
rabies	20	15	13	86.67%
poxviruses	20	15	15	100%
ketosis in cattle	20	15	15	100%

Through the test, it can be found that the more detailed of symptoms was described, then after training sample cases, the more accurate the disease was diagnosed. This required the sample of cases should include as much as possible all the symptoms information of such disease. By the experimental verification, the model can make the diagnosis accurately and rapidly.

6 Conclusion

In this paper, we put forward a model of animal diseases diagnosis expert system based on SVM from the perspective of animal disease diagnosis. Because SVM is established on the principle of structural risk minimization, it has a strong ability of generalization. We attempt to make use of its strong generalization ability to resolve the difficulty of the rapid diagnosis processed because of the complexity and diversity of the symptoms of animal diseases. Experimental results showed that the model can be carried out animal diseases diagnosis more accurately, rapidly on the condition of small samples. It shows that animal disease diagnosis expert system based on support vector machines is a good application in the field of diagnosis of animal diseases.

Acknowledgements

This work is supported by the National Key Technology R&D Program of China under Grant No. 2007BAD33B03.

References

Vapnik, V.: Statistical Learning Theory. Wiley, New York (1998)

Blom, H.A.P.: A Sophisticated Tracking Algorithm for ATC Surveillance Data. In: Proc. International Radar Conf., Paris, France (1984)

Blom, H.A.P., Bar-Shalom, Y.: The Interacting Multiple Model Algorithm for Systems with Markovian Switching Coefficients. IEEE Trans. Automatic Control AC-33(8), 780–783 (1988)

Li, X.R.: Hybrid Estimation Techniques. In: Leondes, C.T. (ed.) Control and Dynamic Systems: Advances in Theory and Applications. Academic Press, New York (1996)

Chung, C.C., Jen, L.C.: LIBSVM: a library for support vector machines (2005)

The Research and Application of Virtual Reality (VR) Technology in Agriculture Science

Feng Yu[1], Jun-feng Zhang[1], Yousen Zhao[2], Ji-chun Zhao[1,*], Cuiping Tan[1],
and Ru-peng Luan[1]

[1] Institute of Information on Science and Technology of Agriculture,
Beijing Academy of Agriculture and Forestry Sciences, Beijing, China, 100097,
Tel.:86- 010-51503135
yuf@ari.ac.cn
[2] The Information Center of Beijing Municipal Bureau of Agriculture, Beijing, China

Abstract. The construction of multi-media information resources of crops and livestock is discussed for interactive three-dimensional animation education in the paper. The animation technology, three-dimensional visualization technology and digital entertainment pattern are combined with modern agricultural science technology knowledge and the production process. Based on the application of virtual reality technology, animation and digital entertainment form, including crops, such as corn, cucumber, etc., livestock such as pig farms, chicken farms, etc., farmers and young people are more willing to acceptable, easy to understand and easy to master knowledge of agriculture, and the VR technology is researched in the text.

Keywords: Virtual Reality, three-dimensional visualization, animation technology.

1 Introduction

Virtual reality, virtual reality (VR) as early as 1965 by the Ivan Sutherland and others raised, development in the 1980s and 1990s, is considered a new technology. VR is a comprehensive integrated technology, computer graphics, human - computer interaction, sensor technology, artificial intelligence, it is used in areas such as computer – stereo vision, hearing, olfaction, etc. enable the user through the appropriate device, naturally on the virtual world experience and interaction, enables users to produce a kind of right. At present, its application has been involved in scientific research, education and training, engineering design, commercial, military, medical, film and television, and so on many areas, is recognized as experts in the 21st century society one of the development of technology.

2 The Meaning of VR

Virtual reality system in theory should have the following four aspects of meaning: ① Immersion. Participants in the integration to create a virtual computer environment,

---

* Corresponding author.

D. Li and C. Zhao (Eds.): CCTA 2009, IFIP AICT 317, pp. 546–550, 2010.

and do not feel the external environment itself. ② Multi-Sensory. Virtual reality technology ideal person should have all the perceptual features, including visual, auditory, force, tactile perception, including taste, smell, such as perception. Due to restrictions on sensor technology, the current virtual reality technology can offer is limited to visual perception, auditory, force, tactile and so on. ③ Interactive. Participants in a virtual environment with a variety of objects are participating in Circumstances. For example, participants can hand-catching virtual objects, when the hands are holding something of the feeling and can feel the hardness and the weight of objects, the object was captured with the hand movements of the mobile. ④ autonomy. Autonomy refers to a virtual environment to follow the objective law of the material performance. System in their own object models and rules are in accordance with the requirements of computer users in their own campaigns.

3   The Structure of VR Technology

Virtual Reality refers to the use of a computer-generated simulation of the environment, and through a variety of special equipment allows the user inputs to the environment, the user directly with the natural environment technology interaction. VR technology allows users to use people's natural skills of virtual objects in the world to inspect or operate, while providing visual, hearing, touch and other natural and intuitive perception of the real-time. Visibility, virtual reality gratitude is made up of people, machinery, and environment three parts:

The environment is generated by a computer can provide a visual, hearing, touch, smell and taste sensory world, which can be a true realization of the real world, and can also be a virtual concept world.

Machine is the computer systems and three-dimensional interactive equipment, which used a helmet with a three-dimensional, data gloves, three-dimensional mouse users, such as wearing the device. In reality there are settings in the environment-sensing devices, such as cameras, various sensors.

People are simulated environment participant which is dealt by the computer processing the action data that can bring immersive feel.

A typical virtual reality system is mainly composed by the 6 parts: computer-generated virtual reality and processes systems, software application systems, and virtual reality related to the theory and technology, input and output man-machine interface devices, users and databases. VR system architecture as shown in Figure 1:

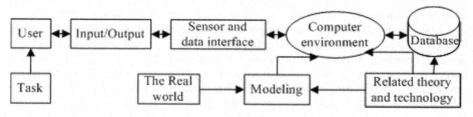

Fig. 1. The structure of VR system

In the virtual reality system, the computer is responsible for generating the virtual world and the realization of human-computer interaction. As the virtual world is a high degree of complexity, making deep into the virtual world a large amount of computing, therefore, virtual reality has higher requirements for the computer system configuration. This shows that the computer is the heart of VR systems. For the realization of the virtual world and the natural interaction, it is necessary to adopt special input and output devices to identify various forms of user input and generate real-time feedback to achieve a dialogue between man and computer. Theories and technologies related to the establishment of a virtual environment to provide the necessary theory, technology and so on. The function include: the virtual world of geometric objects model, physical model, behavior model, three-dimensional virtual stereo generation, model management techniques and real-time display technology, virtual world database, such as the establishment and management of several parts. Mainly used in the virtual world database storing the virtual world of information in all aspects of objects. Eventually the virtual reality systems are provided to user applications. The key technologies of Virtual Reality involved are: the scene of large-scale three-dimensional data modeling techniques; dynamic real-time three-dimensional vision, hearing, such as generating technologies; three-dimensional positioning; three-dimensional interactive software and system integration technology.

4   VR Applications in Agriculture

Virtual agriculture is a virtual reality technology applied to agricultural, which refers to the computer in the implementation of substance in virtual soil adsorption, emissions, the migration process, animal and plant growth process, the result of expression, assimilation, alienation, and so on. Researchers explore various stress conditions, human intervention conditions on the role of these procedures. Virtual technology applied to agriculture in the field of scientific research, teaching, agricultural resources, planning, production, circulation of goods and agricultural machinery design and manufacturing, etc.

4.1   Virtual Plant

Virtual plant is, following the development of information technology in the past 20 years, rapidly developed areas of research, widely used in the agronomy, forestry, ecology, remote sensing multi-fields, etc. virtual plant is to use VR technology in 3D simulation of plants and plant morphology of the growth process. Using virtual plant technology, on the computer screen design, simulation of the whole plant crops or even the whole group of plants throughout the life cycle, do not have to take a long time crops, it saves time, manpower and money. Such as virtual fertilization and fruit pruning, breeding of plant type analysis, etc. You can implement   the   implementation of the crop of dynamic growth process on your computer; obtain the various parameters of dynamic data, instead of traditional agriculture situation which is difficult to quantitative study, which can provides reference data for the intelligence agriculture and delicate operation agriculture. You can implement a plant of three-dimensional structure of visualization, provides methods and means to structure and research on crop Physiology and ecology of the relationship between crop plant type design and genotypes, etc. so far, virtual plant research including plant parts and root portion.

4.2 Virtual Reality Technology Application in Agricultural Machinery Design and Manufacture

Due to the characteristics of agricultural machinery design, in the development process, use the computer to draw a lot of new products and parts must be in the manufacturing process before you can learn about the product performance. When performance is not required to meet the requirements, product design will be modified, resulting in product design cycle length, wasted property, it is difficult to adapt to changing market demands. Virtual reality three-dimensional modeling technology is to improve the level of agricultural machinery product design an effective way. Three-dimensional technology can display the image of intuitive product appearance and difficult to quantify the expression of two-dimensional works view and the key parts of complex surface, allowing designers make it easier to exam design products, but also makes it easier to understand the product characteristics. Engineer in product processing before, comprehensive evaluation virtual products that could make him feel the future of the product - related performance and feasibility of providing decision making and optimizing ahead of the implementation plan.

4.3 Virtual Reality Technology Application in Teaching and Agriculture Popularization

The application virtual reality technology is the multimedia and interactive computer technology, combining the results of education in creative and effectiveness have been breakthroughs. students see crop growth conditions dynamic process from any angle, as they move the mouse and key, they can see, hear and control over their observations on the screen angle,   who roaming in which obtained better results than the traditional way. We set up virtual farm with virtual plant model technology, which enables farmers on the computer management virtual crops and planting virtual farmland, by changing environmental conditions and crop cultivation practices, they visually observation on the status changes in crop growth. virtual technology applied to the achievement of agricultural science and technology promotion, farmers will make it easier to understand and grasp advanced agricultural land management techniques, so the process of teaching have become more and more intuitive, vivid image, which greatly enhances teaching effectiveness and increases their learning interest.

5 Conclusion

Above all, virtual reality technology improve the using efficiency of   agricultural production, which can improve the efficiency of agricultural resources comprehensive utilization, which can simulation agricultural product market transactions and agricultural production management, which can realize the agricultural and technical education, training, research etc. Virtual reality technology can permeate various fields of agricultural production .Therefore, It has great significance that virtual reality technology application was researched in agriculture, It is extremely complex and very long cycle life science research on space-time quantitative analysis of coordinate system, which can greatly reduce the research cycle and also can get direct experimental results. However, virtual reality technology application have a scientific method, as do the guidance of an effective supplementary means of scientific research in order to better promote the development of agriculture.

References

Baomin, Y.: The application of Distributed VR environment in composes campaign body exercise simulating. Computer Simulation 14(4), 17–201 (1997)

Pemmaraju, S.V., Pirwani, I.A.: Good Quality Virtual Realization of Unit Ball Graphs. In: Arge, L., Hoffmann, M., Welzl, E. (eds.) ESA 2007. LNCS, vol. 4698, pp. 311–322. Springer, Heidelberg (2007)

Shen, W.j., Zhao, C.j., Shen, Z.r., Guo, X.y.: Virtual Reality Technology and Its Application in Agriculture. Research of Agricultural Modernization, 9 (2002)

http://www.sgi.com/products/software/opengl/tech_info.html

http://en.wikipedia.org/wiki/OpenGL#Specification

http://vr.isdale.com/WhatIsVR/frames/WhatIsVR4.1.html

Shuqiu, Z., Lei, L., Xinyu, G.: VRML Realization of Virtual Plants Growth. Journal of Capital Normal University (Natural Science Edition), 12 (2003)

Amditis, A., Karaseitanidis, I., Mantzouranis, I.: Virtual Reality Research in Europe: Towards Structuring the European Research Area, Product Engineering (2008)

Springer, S.L., Gadh, R.: State-of-the-art virtual reality hardware for computer-aided design. Journal of Intelligent Manufacturing 7(6), 12 (1996)

Research on Vegetation Dynamic Change Simulation Based on Spatial Data Mining of ANN-CA Model Using Time Series of Remote Sensing Images

Zhenyu Cai and Xiaohua Wang

School of Economics and management, Hebei University of Engineering, Handan, Hebei, 056038, China

Abstract. Dynamic change of vegetation has become a very sensitive problem in China due to climate variability and human's disturbances in the Yellow river basin. Dynamic simulation and forecast of vegetation are regarded as an effective measure to decision support for local government. This paper presents a new method to support the local government's effort in ecological protection. In integrates cellular automata (CA) -artificial neural network (ANN) model with Geographical information system (GIS) and remote sensing. The proposed method includes three major steps: (1) to extract control factors; (2) to integrate CA and ANN models; (3) to simulate the selected area using CA-ANN model. The results indicted that the integrated approach can rapidly find condition of future vegetation cover that satisfy requirement of local relative department. It has demonstrated that the proposed method can provide valuable decision support for local government. the result indicts that NDVI of the vegetation has an increasing trend and characteristics of distribution concentration trend, but the change rate is become lower from the year 2007 to 2014 compared with the changes from the year 2000 to 2007.

Keywords: CA-ANN model; control factors; Yellow river basin; vegetation changes.

1 Introduction

Vegetation is the most significant natural resources, however, Yellow river basin, especially in arid and semi-arid regions, is facing various changes caused by human's activities and climate changes. These changes could have a significant impact on the carbon cycle, regional economy and climate (Le Houerou, 1996; Angell and Mc Claran, 2001). The results of the eco-environmental vulnerability and its changes in the Yellow River Basin showed that the ecology become worse in some regions in the year from 1990 to 2000. There were many factors affecting the eco-environmental vulnerability, such as structural geology, meteorological conditions, sediment discharge, human activity, and forest degradation. The driving forces of the eco-environmental vulnerability changes were the rapid population growth, governmental policy, and vegetation degradation. Grassland degeneration, sandy desertation, and soil erosion had seriously affected the sustainable development of eco-environment. Study on

D. Li and C. Zhao (Eds.): CCTA 2009, IFIP AICT 317, pp. 551–557, 2010.

changes simulation of future vegetation is undoubtedly helpful for resource management and environmental conservation.

At present, study on future vegetation changes is often based on statistical analysis data of several points, investigation and experiences of the planner (Xu et al., 2000). It is virtually impossible to effectively manage and monitor the resources using field investigation methods. In the past two decades, satellite remotely sensed data have been widely used for vegetation investigation because of their advantages of frequent revisit, global coverage and lower cost (Langley et al., 2001).

A cellular automata (CA) model is a dynamics model using local interactions to simulate the evolution of a large system, where time are considered as discrete units and space is represented as a regular lattice of two or more dimensions. A CA model is capable of representing those non-linear, spatial and stochastic processes (Wolfram, 1984). More recently, there are growing literatures on the applications of CA models in urban growth, land use change and vegetation dynamics (Balzter et al., 1998; Batty et al., 1999; White and Engelen, 2000). Li and Yeh (2000) demonstrated the potential of integrating a CA model and remote sensing techniques for agricultural protection area zoning and land use planning. About neural network, its an important characteristic is its capability to learn from the data being processed. The network weights are adjusted in the training process, which can be executed through a number of learning algorithms based on backpropagation learning (Ripley, 1996; Haykin, 1999; Zhou, 1999; Lee et al., 2004; Gomez and Kavzoglu, 2005; Yesilnacar and Topal, 2005). Because of complexity of underlying surface of yellow river basin, tempro-spatial change simulation of vegetation using CA-ANN model according to spatial advantage of CA and learning ability of neural network.

2 Methods

2.1 The Study Area

The study areas selected for this research are located individually in the up course, middle course and low course of Yellow river basin, and which mainly include vegetation of various growth condition and sorts, such as farm field, grassland, forest. The approximate latitudinal and longitudinal ranges of the study areas of up, middle and low courses are "N(35.48-37.32), E(102.5-104.2)","N(37.71-39.20), E(107.40-109.33)" and "N(35.31-37.02), E(102.54-104.43)" individually. Annual mean precipitation and an annual average temperature of the three study regions have very different. More than 90% of the land in the up course region is covered by vegetation, the land in the middle course region is covered by sparse vegetation of loess plateau and the land in the low course region is covered by farm field and plantation.

2.2 Data Collection

Time series of MODIS image (the 16-day composite data) acquired from the year 2000 to 2007, In addition, essential ancillary data including topographic maps, land use, transportation and soil maps were also obtained from relative department of local government. All the data layers were registered to the same coordinate system and resampled to the same pixel resolution of 250 m.

2.3 The Proposed Approach

The proposed approach, to utilize the advantages of both ANN and CA models, is to simulate vegetation conditions using remote sensing techniques. Specifically, this approach includes three major steps (Fig. 2). The first step is to extract control factors from multi-source data. The second step is to produce candidate CA-ANN model based on advantage of ANN and CA model. The third step is future vegetation simulation using CA-ANN model.

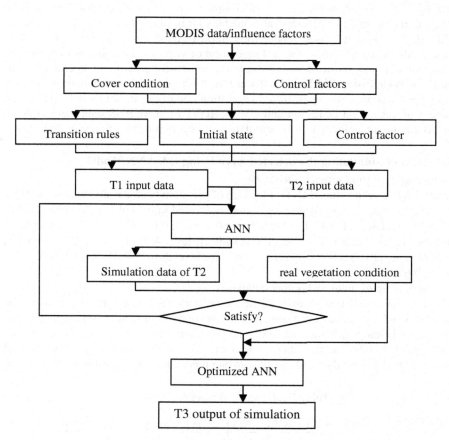

Fig. 1. The flowchart of the proposed approach

2.3.1 Extracting Control Factors from Multi-source Data

Vegetation indices are widely used in remote sensing of grassland since healthy vegetation has a high spectral response in near-infrared bands of remote sensing images

(Langley et al., 2001). Vegetation indices generally involve the direct or modified forms of the ratio between the infrared band and the red band. Among vegetation indices, Normalized Differing Vegetation Index (NDVI) is regarded as one of the most useful indices to measure vegetation vigor. Coefficient of variation (CV) of time series of NDVI can reflect fluctuation of vegetation from 2000 to 2007 and which include also influence of some control factors such climate, slope of underlying surface. So CV is selected as control factors.

2.3.2 Establishing of CA-ANN Model

Cellular automata, a complex discrete and local grid dynamic system, is a representative model for simulating complex system. It has the characters such as the research frame of "from bottom to up", powerful compute function, intrinsic parallel calculate faculty, dynamic spatial and temporal concept, so it is suitable for complex system, to build cellular automata based model for complex Geo-Spatial system simulation is focused. About spatial simulation and prediction of vegetation condition, integrate the method of CA and artificial neural network, constructing artificial neural network prediction model using MATLAB software as development plot, simulating and predicting the dynamic changes of future vegetation condition. In the model, spatial cell database of vegetation condition and CV of NDVI from MODIS data from 2000 to 2007 as input data and result of the simulation of vegetation condition as output data.

2.3.3 Data of Simulating the Selected Area Using CA-ANN Model

(1) Processing of time series MODIS data

NDVI data (mean value of DOY 177-273) of every study area is divide into 5 classes according to vegetation condition: up course is divided into level 1: 0.00-0.06, level 2: 0.06-0.24, level 3: 0.24-0.42, level 4: 0.42-0.60, level 5: 0.60-0.85; middle course: level 1: 0.00-0.06, level 2:0.06-0.15, level 3: 0.15-0.21, level 4: 0.21-0.3, level 5: 0.3-0.55; low course: level 1: 0.00-0.06, level 2: 0.06-0.25, level 3: 0.25-0.45, level 4: 0.45-0.65, level 5: 0.65-0.9.

(2) Setting up of input and output databases

Training and prediction of artificial neural network need input and output data. The input data need CV data and cell data, and which compose of a piece of record. The cell data in record include self attribute (S0), 8 neighbor attributes (S1, S2, S3, S4…..S8) and spatial attribute (ID). The records of output data include self attribute (S0) and spatial attribute (ID). All these data and records can be implemented by the software MATLAB.

Table 2-1. Structure of input data tables

ID	N0S	N1S	N2S	N3S	N4S	N5S	N6S	N7S	N8S	FACTOR

Table 2-2. Structure of output data tables

ID	NS

(3) Process of simulation and prediction

1) Training, verification and predicting using ANN

After ANN studies using data tables of above input and aim data, input data of the verification database and aim database need to provide ANN, then we compare results of prediction with verification data tables, at last evaluate precision of simulation result, if the result can satisfy precision requirement, the ANN-CA model begins to execute the prediction of future vegetation condition, or the result can't satisfy precision requirement, we adjust control factors and relative parameters and retrain.

2) Structure of ANN and parameters

① Design of ANN structure

In input layer, regard attribute of original cell, its 8 neighbors and control factor CV as input data, the number of nerve cells of competitive layer is set N0.5 (N is the number of input samples). Nerve cells of output layer present 5 kinds of condition of future vegetation simulation.

② Set of ANN parameters

Step of ANN training is 10, and minimum mean square deviation is 0.01. When times of train reach 1000 or mean square deviation is less than 0.01, the training ends and the initial study efficiency is 0.5.

③ input vector and aim vector of ANN

The input data tables of vegetation condition correspond to input vectors of ANN, and aim data tables correspond to aim vectors of ANN.

3 Analysis of the Results

According to above analysis and disign, we select time series MODIS data (mean value of NDVI (DOY from 177 to 273)) from 2000 to 2007 as study objects, and the result (figure 3,4,5 and table 3-1, 3-2, 3-3) indicts that NDVI of the vegetation has an increasing trend and characteristics of distribution concentration trend, but the change rate is become lower from the year 2007 to 2014 compared with the changes from the year 2000 to 2007.

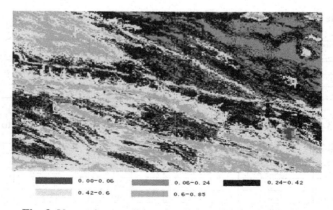

0. 00–0. 06   0. 06–0. 24   0. 24–0. 42
0. 42–0. 6   0. 6–0. 85

Fig. 2. Vegetation condition simulation of up course in 2014

Fig. 3. Vegetation condition simulation of middle course in 2014

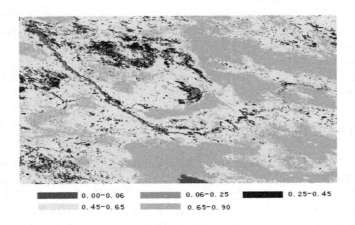

Fig. 4. Vegetation condition simulation of middle course in 2014

Table 3-1. Vegetation condition comparison of up course from 2000 to 2014

year \ NDVI(level)	0-0.06(0)	0.06-0.24(1)	0.24-0.42(2)	0.42-0.60(3)	0.60-0.85(4)
2000	0.41%	40.53%	28.05%	22.73%	8.38%
2007	0.30%	22.23%	34.55%	29.64%	13.29%
2014	0.28%	12.13%	38.47%	33.06%	16.06%

Table 3-2. Vegetation condition comparison of middle course from 2000 to 2014

year \ NDVI(level)	0-0.06(0)	0.06-0.15(1)	0.15-021(2)	0.21-0.30(3)	0.30-0.55(4)
2000	0.27%	22.15%	49.97%	23.52%	4.22%
2007	0.30%	5.60%	25.01%	45.40%	23.70%
2014	0.25%	2.67%	14.46%	48.71%	33.81%

Table 3-3 Vegetation condition comparison of low course from 2000 to 2014

year \ NDVI(level)	0-0.06 (0)	0.06-0.25(1)	0.25-0.45(2)	0.45-0.65(3)	0.65-0.9(4)
2000	0.433%	1.41%	29.19%	46.89%	22.08%
2007	0.437%	0.92%	11.03%	55.74%	31.87%
2014	0.46%	0.53%	3.81%	58.79%	36.41%

Conclusion and Expectation

The CA-ANN model is a kind of new method to research changes of future vegetation. It is proved effective to integrate advantage of CA and ANN for simulation of condition of future vegetation, but simulation of the changes of vegetation condition is very complex, and many collected factors need to consider and analyze.

References

Le Houe rou, H.N.: Climate change, drought and desertification. Journal of Arid Environment 34, 133–185 (1996)

Angell, D.L., McClaran, M.P.: Long-term influences of livestock management and a non-native grass on grass dynamics in the desert grassland. Journal of Arid Environments 49, 507–520 (2001)

Xu, Z.X., Zhao, M.L., Han, G.D.: Eco-environmental deterioration and strategies for preventing it in Inner Mongolia. Grassland of China 5, 59–63 (2000)

Langley, S.K., Cheshire, H.M., Humes, K.S.: A comparison of single date and ultitemporal satellite image classification in a semi-arid grassland. Journal of Arid Environment 49, 401–411 (2001)

Wolfram, S.: Universality and complexity in cellular automata. Physica D 10, 1–35 (1984)

Balzter, H., Braun, P.W., Hler, W.K.: Cellular automata models for vegetation dynamics. Ecological Modeling 107, 113–125 (1998)

Batty, M., Xie, Y., Sun, Z.: Modeling urban dynamics through GIS-based cellular automata. Computer, Environment and Urban Systems 23, 1–29 (1999)

White, R., Engelen, G.: High-resolution integrated modeling of the spatial dynamics of urban and regional system. Computer, Environment and Urban System 24, 383–400 (2000)

Li, X., Yeh, G.O.: Study on zoning of agricultural land protection using cellular automata model. China Environmental Science 20, 318–322 (2000)

Ripley, B.: Pattern Recognition and Neural Networks. Cambridge Univ. Press, Cambridge (1996)

Haykin, S.: Neural Networks: A Comprehensive Foundation, 2nd edn. Prentice Hall, New Jersey (1999)

Zhou, W.: Verification of the nonparametric characteristics of backpropagation neural networks for image classification. IEEE Trans. Geosci. Remote Sens. 37, 771–779 (1999)

Lee, S., Ryu, J., Won, J., Park, H.: Determination and application of the weights for landslide susceptibility mapping using an artificial neural network. Eng. Geol. 71, 289–302 (2004)

Gomez, H., Kavzoglu, T.: Assessment of shallow landslide susceptibility using artificial neural networks in Jabonosa River Basin. Eng. Geol. 78(1-2), 11–27 (2005)

Yesilnacar, E., Topal, T.: Landslide susceptibility mapping: a comparison of logistic regression and neural networks methods in a medium scale study. Hendek region (Turkey). Eng. Geol. 79, 251–266 (2005)

Author Index